AA003124

Environmental Division

Presentations at the 2008 AIChE Annual Meeting

Philadelphia, Pennsylvania
16 - 21 November 2008

ISBN: 978-1-61567-224-0

Printed from e-media with permission by:

Curran Associates, Inc.
57 Morehouse Lane
Red Hook, NY 12571
www.proceedings.com

Some format issues inherent in the e-media version may also appear in this print version.

Printed by Curran Associates, Inc. (2009)

AIChE
3 Park Avenue
New York, NY 10016-5991

Phone: (203) 702-7660
Fax: (203) 775-5177

www.aiche.org

TABLE OF CONTENTS

35a Revealing Microbial Strain Dynamics Case Study In An Oil-Absorber Bioscrubber System

Ines R. Baptista, Michalis Koutinas, Athanasios Mantalaris, and Andrew G. Livingston. Chemical Engineering and Chemical Technology, Imperial College London, Prince Consort Road, SW7 2AZ, London, United Kingdom

The stability of microbial communities is an important aspect of biodegradation strategies, as it can often decide the success or failure of biological treatment technologies (BTT). Recent studies on the dynamics of microbial communities within bioreactors have demonstrated that functional stability is not necessarily correlated to community stability [1]. However, in order to associate functional perturbations to community changes, the community has to be stable under constant operating conditions. Otherwise, it would not be reasonable to assume a cause-effect relation, as the community could change independently of functional perturbations. We have identified bacterial strains which remain stable during the biodegradation of halogenated compounds, which suggests that xenobiotic compounds can induce a strong selective pressure in bacterial communities and enhance strain stability [2].

We have then used these strains to further study the relationships between microbial strain dynamics and bioreactor system performance, using an Oil-Absorber-Bioscrubber as a model system. This bioreactor configuration was developed recently within our group to control fluctuating loads of dichloroethane DCE [3]. This configuration consisted of a bioscrubber combined with a packed oil-absorber column containing vegetal oil placed upstream, where the waste gas was bubbled through. This set-up prevented the contact between the liquid medium, containing the biomass, and the organic phase. Industrial (BTT) are often exposed to random variations in pollutant concentrations profiles and alternating pollutant compositions. These dynamic conditions can inhibit microbial activity and induce long re-acclimation periods, which can in turn compromise the treatment performance. In order to prevent bacterial inhibition and improve BTT robustness, it is interesting to combine the study of microbial dynamics with the bioreactor engineering development.

The aims of our study were therefore to: I) assess the stability of bacterial strains with specific metabolic pathways for the degradation of halogenated organic compounds, through the application of biomolecular techniques; ii) apply these stable strains to the treatment of waste gas under a sequentially alternating pollutant (SAP) regime, using the oil-absorber bioscrubber configuration.

The long-term stability of a microbial strain able to degrade chlorobenzene (CB) was a monitored over a full year, under constant operating conditions and a non-sterile environment. Fluorescence in situ hybridization (FISH) and denaturing gradient gel electrophoresis (DGGE) were applied to monitor the specific degrader and the overall community changes. It was observed that the non-sterile environment led to the natural out competition of this initial strain for a more competitive species within the initial weeks of operation. The succession was attributed to the competitive kinetic behaviour of the novel strain belonging to the Pandoraea genus, which exhibited faster growth and higher substrate affinity than the initial strain [4]. This strain was further applied to the treatment of SAP in an oil-absorber bioscrubber system.

The performance of a combined absorber-bioscrubber system, degrading a gas stream contaminated with fluorobenzene (FB) and chlorobenzene (CB), was studied under different cyclic feeding regimens, simulating industrial conditions. The absorber was applied as a strategy to buffer pollutant inlet concentrations into the bioscrubber, and as well, to provide a maintenance feed of compounds being desorbed from the oil to the bacterial culture [3]. In order to analyze the performance of the oil-absorber-bioscrubber system (OAB), it was compared to a bioscrubber-only (BO) configuration, operated under the same functional conditions. In both configurations, the bacterial communities were closely monitored by FISH and DGGE. The bioscrubber performance results have shown that the total organic discharged (TOD) was significantly lower in the OAB system when compared to the BO configuration. In the case of FB, the OAB configuration prevented the discharge of 1165 g m-3 to the environment, while in the CB case it prevented the release of 300 g m-3. This result has shown that the oil-absorber had a positive effect on the removal of FB and CB, and that it could be used as a strategy to deal with sequentially alternating polluting and fluctuating loads. The FISH and DGGE results contributed to the understanding of the OAB strategy success, revealing that this configuration sustained a more active community, which responded quickly to pollutant changes.

Overall, this study demonstrated the remarkable long-term stability of a novel CB degrading strain and highlighted the worth of the application of biomolecular techniques to follow microbial dynamics in bioreactors. Additionally, this study has shown that the combined oil-absorber bioscrubber system is a robust technology, offering an effective solution to the biological treatment of waste-gas streams during SAP treatment conditions.

References:

1. Fernandez A, Huang S, Seston S, Xing J, Hickey R, Criddle C, Tiedje JM. 1999. How stable is stable? Function versus community composition. Appl Environ Microbiol 65:3697-3704.

2. Baptista IIR, Peeva L, Zhou N-Y, Leak DJ, Mantalaris A, Livingston AG. 2006. Stability and performance of Xanthobacter autotrophicus GJ10 during 1,2-dichloroethane biodegradation. Appl Environ Microbiol 72:4411-4418.

3. Koutinas M, Martin J, Peeva LG, Mantalaris A, Livingston AG. 2006. An oil-absorber-bioscrubber system to stabilize biotreatment of pollutants present in waste gas. Fluctuating loads of 1,2-dichloroethane. Environ Sci Technol 40:595-602.

4. Baptista IIR, Zhou N-Y, Emanuelsson EAC, Peeva LG, Leak DJ, Mantalaris A, Livingston AG. 2008. Evidence of species succession during chlorobenzene biodegradation. Biotech Bioeng 99:68-74.

35b Molecular Analysis of Bioaugmentation and Biostimulation of Biological Acid Mine Drainage Remediation Systems

Sage R. Hiibel[1], Luciana P Pereyra[2], Elizabeth Perrault[3], Amy Pruden[2], and Kenneth F Reardon[1]. (1) Chemical and Biological Engineering, Colorado State University, 100 Glover Building, 1370 Campus Delivery, Fort Collins, CO 80523-1370, (2) Civil and Environmental Engineering, Colorado State University, 100 Glover Building, 1370 Campus Delivery, Fort Collins, CO 80523-1370, (3) Cell and Molecular Biology, Colorado State University, 100 Glover Building, 1370 Campus Delivery, Fort Collins, CO 80523-1370

The biological treatment of acid mine drainage using sulfate-reducing bioreactors is an an attractive passive, *in situ* remediation approach. Acid mine drainage is characterized by low pH and elevated levels of sulfates and trace metals. The bioreactors consist of organic substrate, typically wood chips or compost, which supports a complex microbial community. This community includes cellulose degrading and fermentative bacteria that breakdown the cellulosic materials into smaller organic molecules. Sulfate-reducing bacteria utilize the small organics to produce bicarbonate and hydrogen sulfide, which then precipitate the metals in the drainage.

The goal of this work is to investigate the effects of bioaugmentation and biostimulation on sulfate-reducing bioreactors treating acid mine drainage. Anaerobic columns were inoculated with manure from a local dairy farm and fed simulated acid mine drainage. Each column was packed with a an organic substrate mix (beech wood (22 wt%), pine chips (11 wt%), and alfalfa (2 wt%)), limestone (5 wt%), and silica sand (45 wt%). Sulfate and metal removals were determined by ion chromatography and inductively coupled plasma absorbance emission spectroscopy, respectively.

The microbial communities of each column were evaluated at the beginning of the experiment and after the columns had reached steady-state (as determined by sulfate reduction) using a suite of molecular methods. The active members of microbial community were identified using Active Community Profiling, a technique developed in our lab that uses capillary electrophoresis single-strand conformation polymorphism (CE-SSCP) to determine the 16s rRNA to 16s rDNA ratio. Key functional groups of the microbial community were also quantified using quantitative real-time PCR (Q-PCR). Q-PCR targets included family 5 and family 48 glycosyl hydrolases for cellulose degraders , iron only hydrogenases for fermenters, the alpha subunit of the dissimilatory sulfite reductase (*dsrA*) gene for sulfate reducers, and the alpha subunit of the methyl coenzyme M reductase gene for methanogens.

To explore the effect of bioaugmentation, in addition to the dairy manure one pair of columns were inoculated with cellulose-degrading bacteria and another pair with sulfate-reducing bacteria. Biostimulation effects were investigated by feeding ethanol to one pair of columns and carboxymethyl cellulose to another pair to stimulate sulfate-reducing and cellulose-degrading bacteria, respectively, in addition to the simulated acid mine drainage.

Bioaugmentation of columns with cellulose-degrading bacteria resulted in decreased startup times and increased short- and long-term remediation performance compared to those columns only inoculated with dairy manure. Multiple species identified as cellulose degraders had increased 16s rRNA:rDNA ratios, and the total proportion of cellulose degraders was significantly greater. The columns bioaugmented with sulfate-reducing bacteria also had a decreased startup time and increased short-term remediation performance, but their long-term performance was comparable to the non-augmented columns. The drop in performance was correlated to a decrease in *dsrA* genes.

The columns supplemented with a readily available cellulose source had elevated proportions of cellulose degraders, fermenters, and sulfate reducers as determined by the glycosyl hydrolases, iron hydrogenases, and *dsrA* genes, respectively. These columns had a decreased phylogenetic diversity of cellulose degraders and increased levels of short- and long-term remediation performance. Columns supplemented with ethanol followed a similar trend, having elevated proportions but decreased phylogenetic diversity of sulfate reducers, based on *dsrA* Q-PCR results, and increased short- and long-term remediation performance; the relative proportions of cellulose degraders and fermenters did not change significantly in these columns.

Based on the results of this work, bioaugmentation with cellulose-degrading or sulfate-reducing bacteria does not appear to offer a viable option for long-term remediation performance of bioreactors treating acid mine drainage. Despite decreases in the phylogenetic diversity of the respective functional groups, the biostimulation of cellulose degraders and sulfate reducers gave promising long-term remediation performance results. Additional studies exploring the robustness of the bioaugmented and biostimulated columns with respect to an environmental perturbation are underway.

35c Formulation of Microbial Consortium for Btex Biodegradation

Karthiga Nagarajan and Kai-Chee Loh. Department of Chemical and Biomolecular Engineering, National University of Singapore, 4 Engineering Drive 4, Singapore, 117576, Singapore

The mono-aromatic group of hydrocarbons, namely, Benzene, Toluene, Ethyl Benzene and o-Xylene (BTEX) are the most commonly cited environmental pollutants. Though natural ecosystems have self purification capacities, it is essential to adopt proper treatment methodologies to control their discharge into the natural environment. The conventional physico-chemical treatment methods for BTEX are neither cost effective nor easy to operate and also cause secondary pollution problems. Biodegradation, on the other hand, emanates as a promising approach for the removal of BTEX from industrial effluents.

In this research, we aimed at developing a concoction of two bacterial cultures, Pseudomonas putida F1 and Pseudomonas stutzeri OX1, for the effective biodegradation of BTEX compounds. It is anticipated that mixed cultures will perform better than single cultures. Pure substrate and mixed substrate studies conducted to elucidate the biodegradation capacities of each species showed that the biodegradation capability of the species complemented each other synergistically.

Based on the biodegradation profiles of the pure cultures, a number of parameters were chosen as heuristics to mix the bacteria depending on the effluent BTEX composition. The heuristics were used to formulate mixed microbial consortium for three different concentration levels of BTEX compounds. The mixed community of the two species demonstrated better biodegradation capability than the individual pure culture. It was found that the initial inoculum size and proportion of both the species in the consortium played a significant role in improving the efficiency of BTEX removal.

A novel genome based methodology was developed to differentiate and characterize the two species in the constructed consortium. This methodology helped in optimizing the species proportions based on the population stability during biodegradation. The methodology also threw light on understanding the population dynamics of the mixed culture, microbial interactions, and biodegradation capacity of each species in the mixed substrate system.

Degradation and Biosorption of Envirinmental Endocrine Discruptor Di-(2-ethylhexyl) phthalate (DEHP) by *Gordonia* sp. YK1

Cheol-Hee Kang, Byoung-In Sang, Youngsoon Um,

Center for Environmental Technology Research, Korea Institute of Science and Technology, Seoul, Korea

Abstract

In this study, the isolation of DEHP-degrading microorganism was successfully achieved by establishing highly enriched aerobic cultures capable of degrading DEHP. DEHP in the culture with the isolated microorganism was rapidly degraded with the growth of cells and disappeared in 48 hr (more than 90% degradation). Based on 16S rDNA analysis, the isolated DEHP-degrading microorganism belongs to the *Gordonia*. Because the DEHP-degrading microorganism, *Gordonia* sp. YK1, was attached to DEHP droplets and formed aggregation in the liquid medium during the degradation of DEHP, biosorption of DEHP by *Gordonia* sp. YK1 was investigated with 50 ppm DEHP by detecting DEHP concentration in supernatant and biomass phases. With 11 mg biomass and 50 ppm DEHP, more than 60% of DEHP was removed by biosorption in 15 min, and adsorbed DEHP was degraded rapidly (73% degradation of DEHP in 30 min). Interestingly, cell attachment to DEHP oil drops was observed even with the autoclaved biomass of *Gordonia* sp. YK1. Since it is known that there are long aliphatic chains of the mycolic acids in cell wall for the *Gordonia* genera, strong hydrophobicity of the cell surface of *Gordonia* sp. YK1 due to mycolic acids is likely to be involved in DEHP biosorption and, consequently, effective DEHP biodegradation by facilitating DEHP uptake.

Keywords: Di-(2-ethylhexyl) phthalate (DEHP), biodegradation, *Gordonia* sp. YK1

Introduction

The phthalic acid esters (PAEs) are a group of chemical compounds that are mainly used as plasticizers to make a hard plastic into a flexible plastic. The common PAEs used in plastic industry are di-(2-ethylhexyl) phthalate (DEHP), di-phentyl phthalate (DPP), di-butyl phthalate (DBP), and di-ethyl phthalate (DEP). DEHP has been classified as a priority pollutant with relatively low acute toxicity. Exposure to high concentration of DEHP has been found to cause a wide range of adverse effects in experimental animals, including cancer, liver damage and reproductive system.

Microbial degradation is believed to be the main degradation process for DEHP in aquatic and terrestrial environments, such as sewage, soils, sediments, and surface waters [1]. In recent years, a few researches have reported on DEHP degradation by *Microbacterium* [2], *Rhodococcus* [3], and *Gordoina* [4]. In this study, we report the isolation of a DEHP-degrading bacterium from environmental samples. Simultaneous biosorption and degradation of DEHP by *Gordonia* sp. YK1 was also investigated

Materials and methods

Reagents and chemicals

Di-(2-ethylhexyl) phthalate (DEHP) was purchased from Sigma-Aldrich (Dioctyl phthalate, 99.9%, Aldrich). Di-butyl phthalate (DBP) and di-ethyl phthalate (DEP) were bought from Junsei chemical (99.9%, Japan). The *n*-hexane and ethyl acetate were bought from J. T. Baker. All solvents used were HPLC-grade and all other chemicals were reagent-grade.

Media and culture conditions

Sediment sample was collected from Han River in Seoul, Korea. Enrichment cultures were established in M9 mineral medium containing (per liter): 6 g Na_2HPO_4, 3 g KH_2PO_4, 0.5 g NaCl, 1 g NH_4Cl, 0.015 g $CaCl_2$ ($2H_20$), and 0.5 g $MgSO_4$ ($7H_20$) [5]. The final pH of the medium was adjusted to pH 7.0. After the medium was sterilized, cultures were incubated at 30℃ in a shaking incubator at 200 rpm.

Enrichment cultures for DEHP degradation

Ten grams (milliliters) of sediment was added into a 250 ml flask with 30 ml M9 mineral medium and 500 mg/L DEHP. The cultivation was prepared in several replicates and incubated at 30℃ in a shaking incubator at 200 rpm. The enrichment cultures were transferred several times, and the fourth transferred one was spread on nutrient agar plates. Different types of colonies were tested for their capacity to grow on DEHP as the only carbon and energy source.

Biodegradation of DEHP by the isolated microorganism

The isolated DEHP-degrading microorganism, *Gordonia* sp. YK1, was inoculated into a 250 ml flask with 100 ml M9 mineral medium and 0.1 % yeast extract. After 24 hr of cultivation, the bacterial suspension was washed two times with M9 mineral medium (pH 7.0) and then added into a flask containing M9 mineral medium supplemented with DEHP (50 mg/L, 500 mg/L and 1000 mg/L). The residual concentration of DEHP was measured by the GC-FID method as described in analytical methods.

To investigate the effects of pH on DEHP degradation, pH was adjusted to 3.0 to 9.0 with 5N HCl or 5N NaOH. The effect of temperature on DEHP degradation was investigated by incubating cultures at 20, 25, 30, 35, and 40℃ and the residual concentration of DEHP was measured periodically.

Biosorption and Biodegradation of DEHP by *Gordonia* sp. YK1

The cultivated *Gordonia* sp. YK1 washed two times with M9 mineral medium (pH 7.0). Ten milliliters of DEHP-degrading cultures (OD_{600}=1.0) were incubated with 50 mg/L of DEHP in a shaking incubator (200 rpm) at 30 ℃. The cultures were centrifuged periodically to analyze the concentration of DEHP in supernatant and biomass phase.

Scanning electron microscopy (SEM)

The isolated strain was placed on PTFE membrane filter (Whatman, Maidstone), washed three times with 0.2 M phosphate buffer (pH 7.3) and fixed with 100 mM HEPES buffer and 2% (v/v) glutaraldehyde for 2h at room temperature. After gradual dehydration with ethyl alcohol (30%, 50%, 70%, 80%, 90%, 95%, and 100%), dehydration was done again with iso-amyl acetate (70%, 80%, 90, 95%, and 100%). The sample was dried for overnight at the room temperature [6]. The dried sample was coated on gold (Hitachi E1030 auto fime coater). For observation, a FEI XL-30 FEG Scanning electron microscopy (FEI) was used.

16S rDNA analysis

The isolated strain was identified by the amplification and sequencing of bacterial 16S rDNA. After extracting whole genomic DNA, bacterial 16S rDNA was amplified using the universal primer 27F (*E.coli* numbering 8-27: 5'-AGA TGA TGA TCC TGG CTC AG-3') and 1492R *(E. coli* numbering 1492-1513: 5'-GGT TAC CTT GTT ACG ACT T-3') [7]. The 16S rDNA gene was sequenced and aligned to the Ribosomal Database Project (RDP Database) [8] and NCBI blast search (National Center for Biotechnology Information) to identify the isolated microorganism.

Analytical methods

Twenty milliliters volume of DEHP-degrading cultures were acidified with 6N HCl to attain pH 2.0~4.0. The samples were then extracted with ethyl acetate using two times volume of cultures (i.e, 40 ml). After shaking it for 10 mi, the ethyl acetate layer was separated from the water phase and filtered with anhydrous sodium sulfate (Na_2SO_4, Sigma). N_2 gas was used to concentrate ethyl acetate layer until the volume reached to 20 ml. The ethyl acetate layer was collected in 2 ml GC vials and DEHP concentration was analyzed by GC-FID (ACME 6000 gas chromatograph with a flame ionization detector, Yonglin instruments, Korea). The column used was the HP-1 capillary column (length, 30 m; inner diameter, 0. 32 mm; film thickness, 0.25 ㎛). GC/TOF-MS conditions were as follows: the initial temperature was set at 40 ℃ for 3 min, increased by 10 ℃/min to 300 ℃, and it was maintained 2 min at 300 ℃. Injector and detector temperatures were 250 ℃ and 300 ℃, respectively. Helium was used as carrier gas at 3 ml/min. The inlet was set to split mode (1:10) and 2 μl of extracted sample was injected.

To analyze intermediates, the rest of ethyl acetate layer was dried with N_2 gas completely and the residue was dissolved 1 mL ethyl acetate to inject to GC/TOF-MS (HP-6890N, Agilent Technology; Leco Pegasus Ⅲ TOF-MS (Time-of-Flight-mass spectrometer), Leco corporation) with DB-5 capillary column (length, 17 m; inner diameter, 0. 25mm; film thickness, 0.25 ㎛) and the same temperature conditions as described above. The DEHP intermediates were identified with the score matching to the standard library and retention time.

Results and Discussion

Isolation and identification of a DEHP-degrading microorganism

Isolation of DEHP-degrading microorganism was successfully achieved by establishing highly enriched aerobic cultures capable of degrading DEHP. DEHP in the culture with the isolated microorganism was rapidly degraded with the growth of cells and disappeared in 48 hr (more than 90% degradation). The cell was an aerobic, gram-positive bacterium with a shape of rods (1.0-2.0 um) (Fig. 1).

The 16S rDNA sequence of the isolated DEHP-degrading microorganism, designed as YK1, was determined and compared with the sequences of previously reported 16s rDNA genes. The isolated microorganism (YK1) revealed a 99.6% (1483 bp out of 1499 bp) similarity with 16S rDNA sequences of *Gordonia sihwensis* (DSM44576) [9]. As shown in the phylogenetic tree in Fig.2, the isolated bacterium and all of the representatives of the genus *Gordonia* were clearly grouped in a single cluster. Based on the 16s rDNA gene sequence, the isolated bacteirum belong to the genus *Gordonia*.

Fig. 1. SEM photograph of *Gordonia* sp. YK1

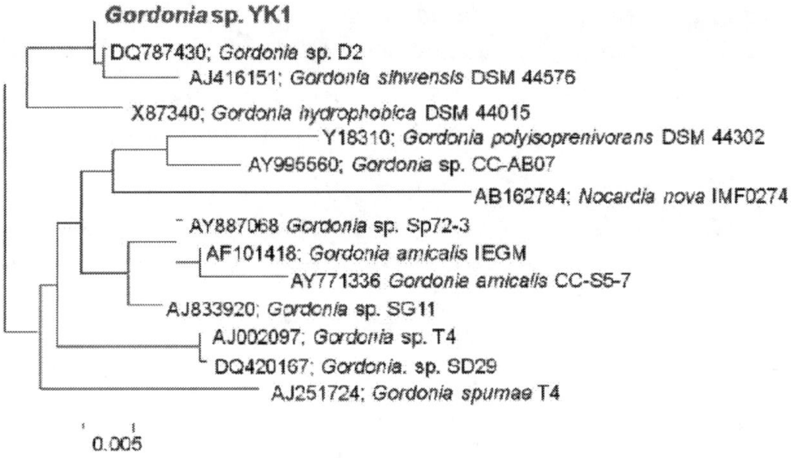

Fig. 2. Phylogenetic tree based on Neighbor-joining method with 16S rDNA sequences of *Gordonia* sp.YK1

Biodegradation of DEHP by the *Gordonia* sp. YK1

As shown in Fig. 3, the isolated microorganism, *Gordonia* sp. YK1 was capable of degrading DEHP as a sole carbon source in M9 medium with the growth of cells under aerobic condition. *Gordonia* sp. YK1 degraded more than 90% of the 50 mg/L, 500 mg/L and 1000 mg/L DEHP after 36 and 48 hours, respectively. The half life of DEHP by *Gordonia* sp. YK1 was 1 hr with 50 mg/L, 15 hr with 500 mg/L and 24 hr with 1000 mg/L. Other phthalate diesters, such as DBP and DEP, were tested for degradability as a sole carbon source. In M9 mineral medium supplemented with each of 500 mg/L DBP and DEP, *Gordonia* sp.YK1 exhibited rapid degradation activity and it was able to degrade more than 90% of the added DBP and DEP in three days under the same conditions that were used for DEHP (data not shown).

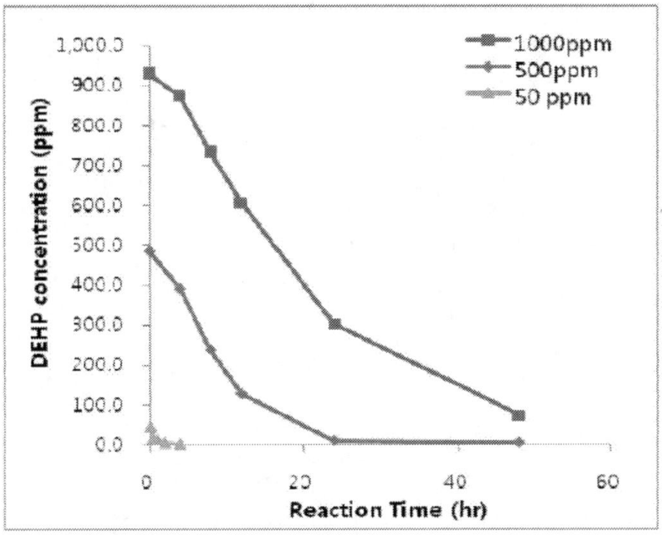

Fig. 3. Biodegradation of DEHP by *Gordonia* sp. YK1

Effect of pH and temperature on DEHP degradation by *Gordonia* sp. YK1

Gordonia sp. YK1 degraded more than 95% of the 500 mg/L at pH ranging from 5.0 to 9.0 after 48 hours (Fig. 4). As shown in Fig. 5, the *Grodoina* sp. YK1 degraded well at temperatures ranging from 20 to 40℃. Based on the results, *Gordonia* sp. YK1 can degrade DEHP effectively in a wide range of pH and temperature.

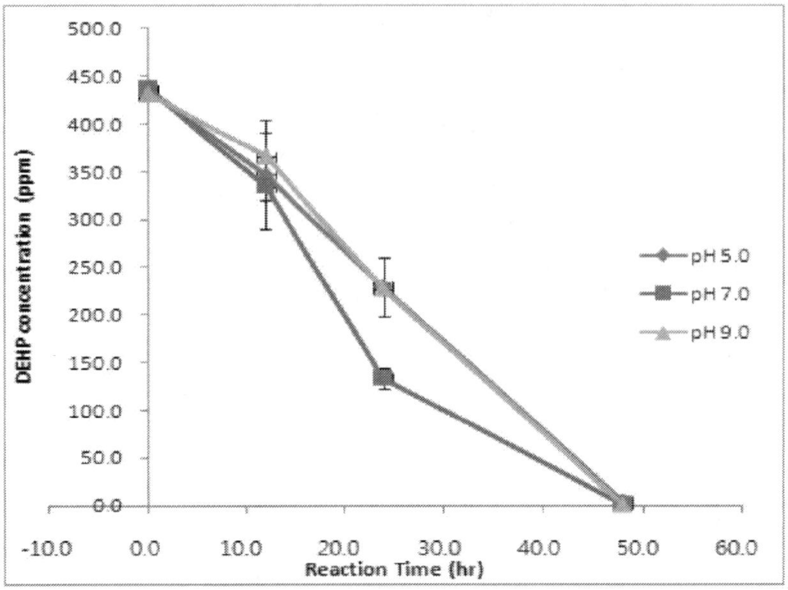

Fig. 4. Effect of pH on DEHP biodegradation by *Gordonia* sp. YK1

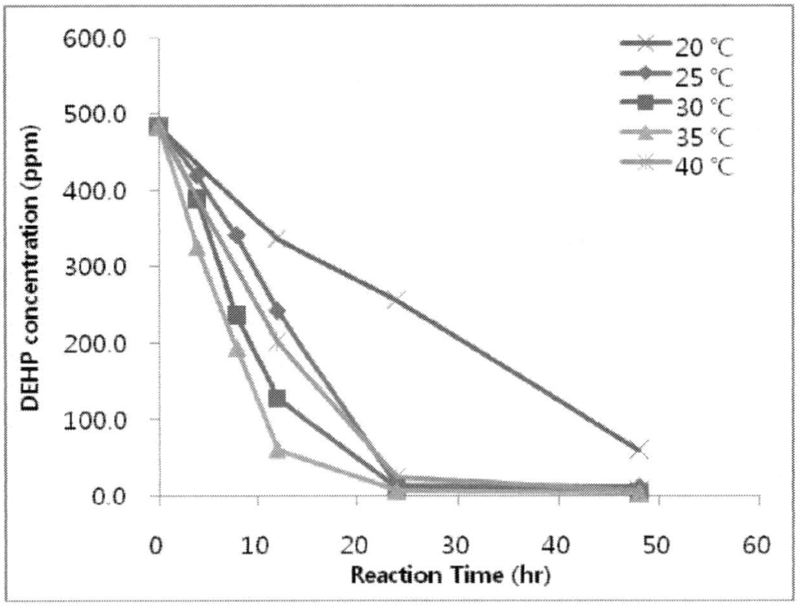

Fig. 5. Effect of temperature on DEHP biodegradation by *Gordonia* sp. YK1

Biosorption and Biodegradation of DEHP by *Gordonia* sp. YK1

Simultaneous biosorption and degradation of DEHP by *Gordonia* sp. YK1 was investigated with 50 ppm DEHP by detecting DEHP concentration in supernatant and biomass phases (Fig. 6). With 11 mg biomass and 50 ppm DEHP, more than 60% of DEHP was removed by biosorption in 15 min, and adsorbed DEHP was degraded rapidly (73% degradation of DEHP in 30 min).

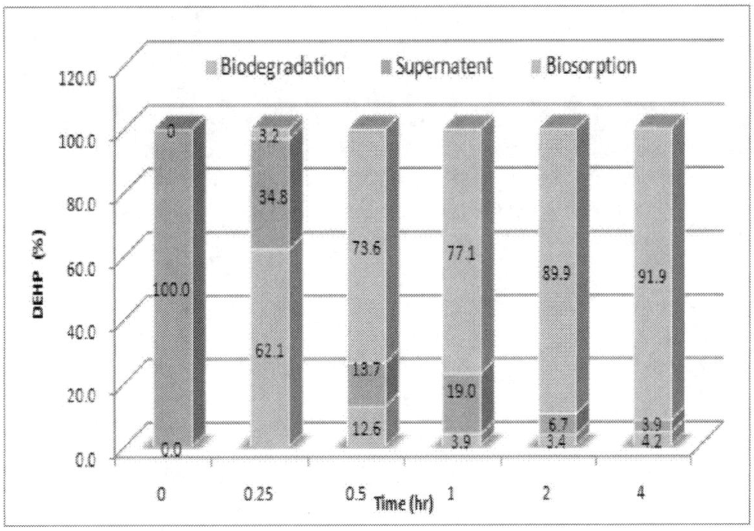

Fig. 6. Biodegradation and Biosorption of DEHP by *Gordonia* sp. YK1

Interestingly, cell attachment to DEHP oil drops was observed even with the autoclaved biomass of *Gordonia* sp. YK1. As shown in Fig. 7, with 11 mg biomass and 50 ppm DEHP, more than 50% of DEHP was removed by the autoclaved cells in 30 min. Since it is known that there are long aliphatic chains of the mycolic acids in cell wall for the *Gordonia* genera, strong hydrophobicity of the cell surface of *Gordonia* sp. YK1 due to mycolic acids is likely to be involved in DEHP biosorption and, consequently, effective DEHP biodegradation by facilitating DEHP uptake.

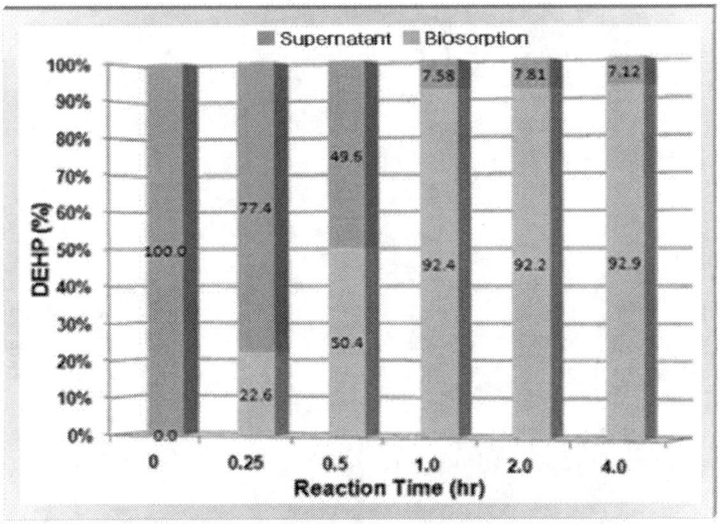

Fig. 7. Biosorption of DEHP by autoclaved *Gordonia* sp. YK1

Characterization of aerobic DEHP degradation intermediates

Intermediates were extracted from cultures and characterized by GC/TOF-MS analysis. The intermediates were identified to be MEHP and phthalic acid. Fig.7 shows the

GC/TOF-MS spectra with the extracted intermediates from 1day cultures.

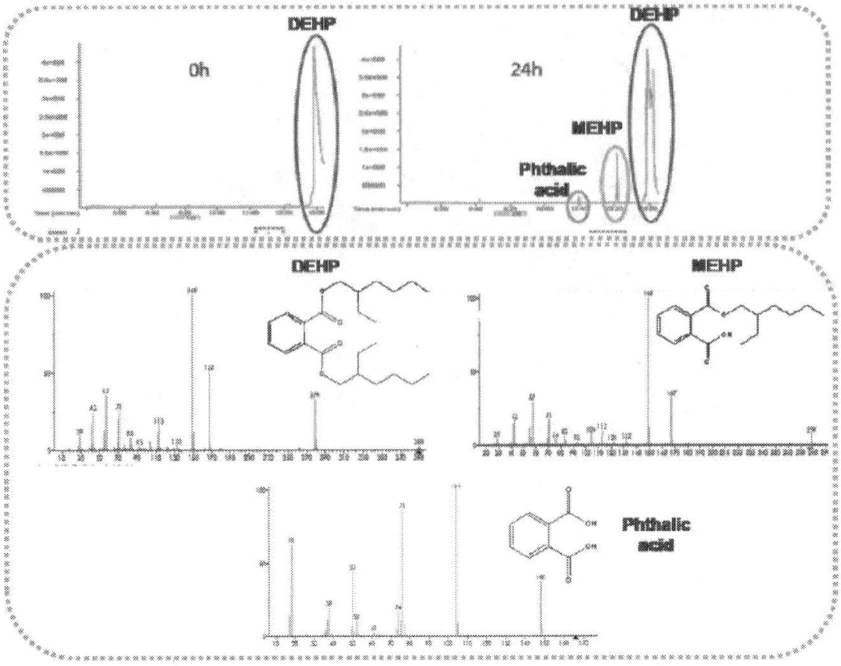

Fig. 7. GC/TOF-MS spectra of metabolic intermediates

On the basis of GC/TOF-MS results, *Gordonia* sp. YK1 is likely to degrade DEHP to produce MEHP and then PA as shown in Fig. 8.

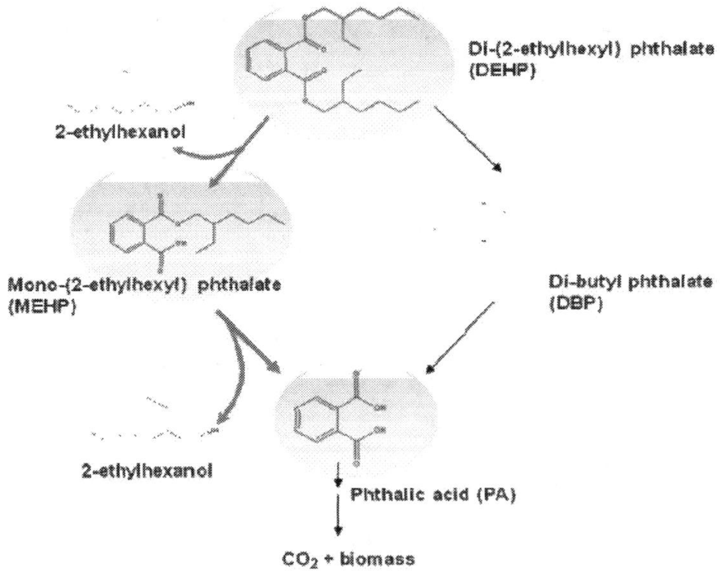

Fig. 8. Possible DEHP degradation pathway by *Gordonia* sp. YK1

Acknowledgements

This work was supported by the research program of The Korea Research Council of Fundamental Science & Technology.

Reference

1. Staples, C. A., D. R. Peterson, T. F. Parkerton, and W. J. Adams (1997) The environmental fate of phthalate esters: a literature review. *Chemosphere*, 35: 667–749.
2. Chen, J. A., X. Li, J. Li, J. Cha, Z. Qiu, Q. Zhao, C. Xu, and W. Shu (2006) Degradation of environmental endocrine disruptor di-2-ethylheyl phthalate by a newly discovered bacterium, *Microbacterium* sp. strain CQ0110Y. *Appl. Environ. Microbiol*, 72:2394-2399.
3. Nalli, S., D. G. Cooper, and J. A. Nicell (2006) Metabolites from the biodegradation of di-ester plasticizers by *Rhodococcus rhodochrous*. *Sci. Tota. Environ.*, 366: 286-294.
4. Chao, W. L., and C. Y. Cheng (2007) Effect of introduced phthalate-degrading bacteria on the diversity of indigenous bacterial communities during di-(2-ethytlhexyl) phthalate (DEHP) degradation in a soil microcosm. *Chemosphere*, 67: 482-488.
5. Ausubel, F. M., R. Brent, R. E. Kingston, D. D. Moore, J. G. Seidman, J. A. Smith, and K. Struhl (1994) *Current Protocols in Molecular Biology*, John Wiley & Sons, NY, USA.
6. Kunichika, N., S. Hashimoto, H. Ito, J. S. Edmonds, A. Yasuhara, and N. Morita (2005) Microbial treatment of Bis (2-Ethylhexyl) phthalate in polyvinyl chloride with isolated bacteria. *J. Biosci. Bioeng.*, 99: 115-119.
7. Lane, D. J. (1991) 16S/23S rRNA sequencing. pp. 115-175. In Nucleic Acid Techniques in Bacterial Systematics.
8. Maildak, B. L., G. J. Olsen, N. Larsen, R. Overbeek, M. Mc-Caughey, And C. R. Woese (1996) The ribosomal database project (RDP). *Nucletic Acid Res*. 24: 84-85.
9. Kim, K. K., C. S. Lee, R. M. Kroppenstedt, E. Stachebrandt, and S. T. Lee (2003) *Gordonia sihwensis* sp. nov., a novel nitrate reducing bacterium isolated from a wastewater treatment bioreactor. *Inter. J. System. Evol. Microbiol.*, 53: 1427-1433.

35e Bioleaching of Metals from Electronic Waste

Mercedes A. Rivero-Hudec[1], Manbir Sodhi[2], and Diana Goglia-Arora[1]. (1) Chemical Engineering, University of Rhode Island, Kingston, RI 02881, (2) Industrial and Systems Engineering, University of Rhode Island, Kingston, RI 02881

Estimates suggest that electronic waste is being generated two to three times faster than other waste streams. Estimates also suggest that between 2000 and 2007 about 500 million computers became obsolete and were discarded. Printed circuit boards are the essential components of most electronic waste, and metals in printed circuit boards are potentially toxic; when not disposed of properly, these metals can leach into soil and water and seep into watersheds. Chemical leaching has been the most commonly used method to extract these metals from waste.

Bioleaching is the extraction (mobilization) of metals from materials by microorganisms. In this study we have investigated the bioleaching of metals from computer circuit boards using Acidiphilium acidophilum. Batch cultures of A. acidophilum were exposed to different concentrations of shredded printed circuit boards; the amounts of copper, zinc, aluminum and nickel released into the culture medium were measured by inductively coupled plasma mass spectrometry. Our preliminary results indicate that A. acidophilum bioleach copper and aluminum from printed circuit boards, thus offering an alternative approach to chemical leaching; the results for zinc and nickel are inconclusive at this point.

35f Control of Wood Decay Fungi by Brominated Furanones

Kristen Manchester[1], Yongbin Han[2], Susan Anagnost[3], Yan-Yeung Luk[2], and **Dacheng Ren**[1]. (1) Department of Biomedical and Chemical Engineering, Syracuse University, 121 Link Hall, Syracuse, NY 13244, (2) Department of Chemistry, Syracuse University, 1-014 Center Sci & Tech, Syracuse, NY 13244, (3) Construction Management and Wood Products Engineering, SUNY College of Environmental Science and Forestry, 204 Baker Laboratory, 1 Forestry Drive, Syracuse, NY 13210

Growth of fungi on lumber can result in costly structural damage, loss of building materials due to rot, as well as detrimental impacts on the indoor air quality. Recent discoveries have shown that brominated furanones from marine red alga have remarkable activities against the colonization of both prokaryotes and eukaryotes; thus, they are promising reagents for fungal control. In this study, the effects of natural and synthetic furanones on the growth of three fungal species that commonly cause white and brown rot of North American timber were examined.

The growth of six fungal strains, *Gloeophyllum trabeum* (GT598 and GT599), *Chaetomium globosum* (wildtype and CG572), and *Trametes versicolor* (TV7B2-59 and TV57MAD277), in the presence and absence of brominated furanones, was evaluated. Furanones were found to inhibit the growth of all the six strains studied. The results indicate that brominated furanones are effective at preventing the fungal growth at concentrations as low as 1 μg/mL.

IMPORTANCE AND METHODS FOR REGULATION OF BUILDING RESOURCES IN AN ENVIRONMENTALLY CONSCIOUS WORLD

Dana Lackey and Dr. Robert W. Peters, University of Alabama at Birmingham, Birmingham, AL

Abstract

As energy costs nationwide are spiraling upward virtually out of control, more and more of the per capita salary goes into the gas tank, and as a result, less state revenue is generated from consumer spending. Therefore, universities and other state-funded establishments have been forced to make massive cutbacks, and almost all are seeking methods for conserving their own energy, thereby reducing expenditures. The University of Alabama at Birmingham (UAB) is no exception, as a $55 million budget cut in state funding has been proposed for the 2009 fiscal year. Therefore, enormous pressure has been placed on the UAB Facilities Management Department to identify ways to reduce energy-related expenditures, and one of the most important ways to do this is to regulate the resources that people within buildings use, including lighting, heating/air conditioning (AC), and water. This regulation also has favorable consequences for the environment, in terms of less contribution to global warming due to a reduction in burning of fossil fuels.

Building resources can be regulated a number of ways, from the extremely simple, such as lowering or raising thermostats depending on season or turning off light switches, to the more complex, such as implementation of motion sensors and lighting ballast controls. Other, long term solutions, such as implementation of green roofs or solar panels, typically require years of research to be effective. But, with environmental consciousness on the rise, and the skies ahead looking gloomy financially, the question that has been posed by building academic deans to the Facilities Management Department is, "what can we do about this now?"

With this in mind, the Civil, Construction, and Environmental Engineering Department of UAB has conducted collaborative research with the Facilities Management Department to perform occupancy surveys in a large number of buildings across campus on after-hours usage on weekdays, and on weekends. The aim of these surveys was to find out how efficiently building resources have been used, by counting the number of people occupying each room, and by noting whether lights have been left on if the room is unoccupied. Other "low hanging fruit" such as incandescent lighting, vending machines, freely-running water, and other non-efficient energy usages were also noted.

This data was used to provide overall recommendations to the Facilities Management Department and academic deans of measures they can take in order to reduce energy costs. Academic buildings that do not have large amounts of literature, for example, can withstand a 7-hour setback of air conditioning and dehumidification systems from 10:00 p.m. to 5:00 a.m. on weeknights, and possibly longer especially on weekends (48 hours from 6:00 p.m. on Friday to 6:00 p.m. on Sunday, for example). Research efforts could still be supported by the AC/humidity system; only large gatherings would find such a change uncomfortable. As for lighting, usage of lights only when the room is occupied or when natural lighting is unavailable is a favorable change at the building dean's discretion, but more long-term measures include motion sensors and layered lighting systems, as well as restricting access to certain floors at certain times of day. These measures have achieved favorable results in buildings where they have been implemented.

Introduction

The costs of operating each building at a large college campus such as UAB include a number of key utilities, including electricity, water, and natural gas. Each follows a distinct pattern, both in its monthly patterns each year and in its overall changes from year to year.

Electricity

The cost of electricity is usually cyclical in nature, but on the whole is rising year after year. For example, the Humanities Building had total electricity expenditures of $116,421 from March to December 2006, but $135,896 from March to December 2007. This rise also came despite the fact that there was a decrease from 2,003,951 kilowatt-hours from March to December 2006 to 1,879,222 kilowatt-hours during March-December 2007. Thus, the cost per megawatt-hour increased as seen in the graph below in Figure 1.

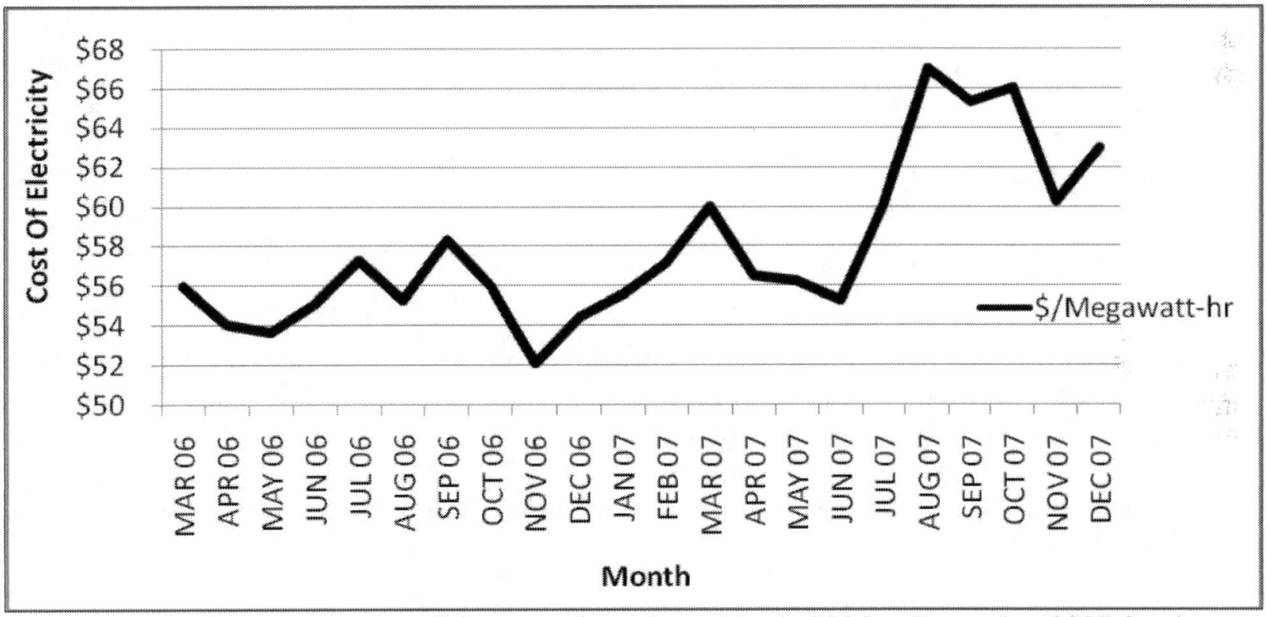

Figure 1. Electricity cost per Megawatt hour from March 2006 to December 2007 for the Humanities Building.

This indicates that there was a significant increase in the price of electricity independent of usage, especially during the time period from late summer to early fall.

Water

Costs of water supplied to buildings as a utility appear to be somewhat more random than electricity, but the same upward trend from one year to the next can be seen. In the Humanities building, total water costs for September-December 2006 were $4,736.93, and for the same period in 2007, they were $4,425.64, a 7% decrease. So, how can they be increasing? When you consider the unit of consumption, 100 cubic feet (CCF), there is a decrease from 526 CCF to 453 CCF, a 14% decrease. When the amount consumed decreases and the cost only decreases relatively half that amount, it indicates that total cost per unit increased, as can be shown in the graph in Figure 2.

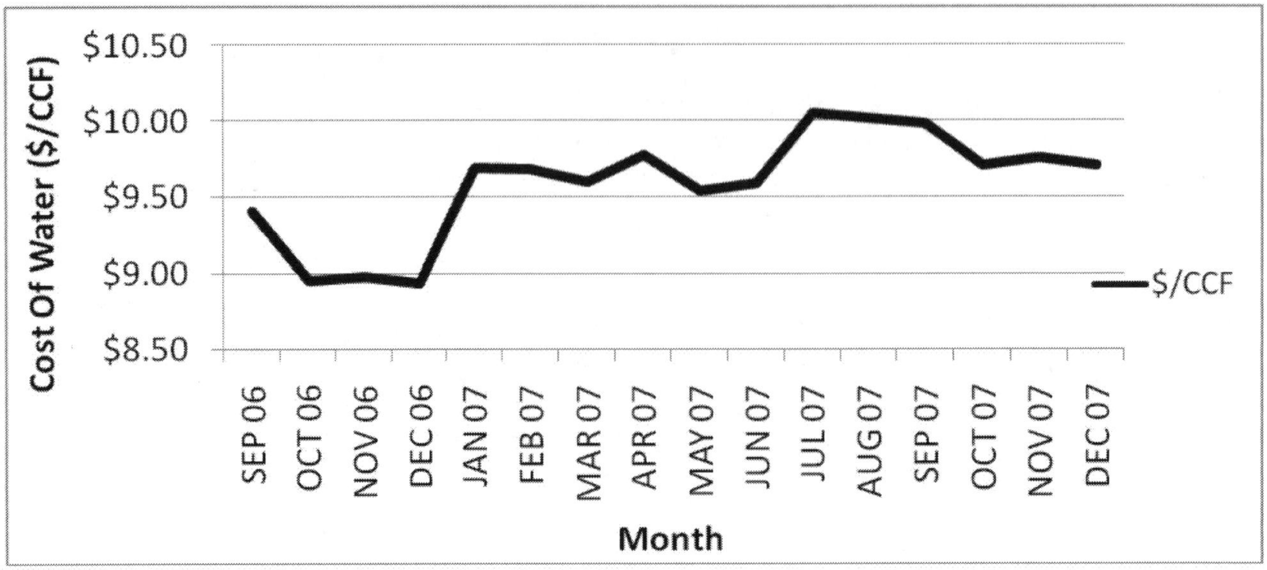

Figure 2. Water cost per CCF from September 2006 to December 2007 for the Humanities Building.

Thus, water costs are also rising steadily from one year to the next, although there is an encouraging sign in that actual water use dropped in 2007. This may be due to the fact that the drought was still considerably intense at this time.

Natural Gas

Natural Gas may be the only exception to the rule, as its usage was 27% higher from March-June 2007 over the same period in 2006, but its cost was only 22% higher for the 2007 period. As the graph in Figure 3 reveals, natural gas costs per unit are more or less stable, though there can be some fluctuation from month to month.

For July–October '07, the result was actually a division by zero, since 0 CCF were reported for those months.

Environmental Concerns

Energy concerns today are higher perhaps than ever before, and in Alabama, with its record high temperatures and drought levels in 2007, the effects of global warming are ever more of a concern [National Weather Service, 2008]. Carbon dioxide is the chief greenhouse gas in terms of emissions per year, and energy-related sources emit nearly 6 billion tons of carbon dioxide into the atmosphere [Energy Information Administration, 2008]. Since coal combustion generated 57% of the electricity in Alabama in 2005 [Birmingham Newschart, 2007], and burning 23 kg of coal produces 44 kg of carbon dioxide [Carnegie Mellon University, 2003], control of electricity usage is proven to reduce coal combustion, and thus decrease production of carbon dioxide which contributes greatly to global warming.

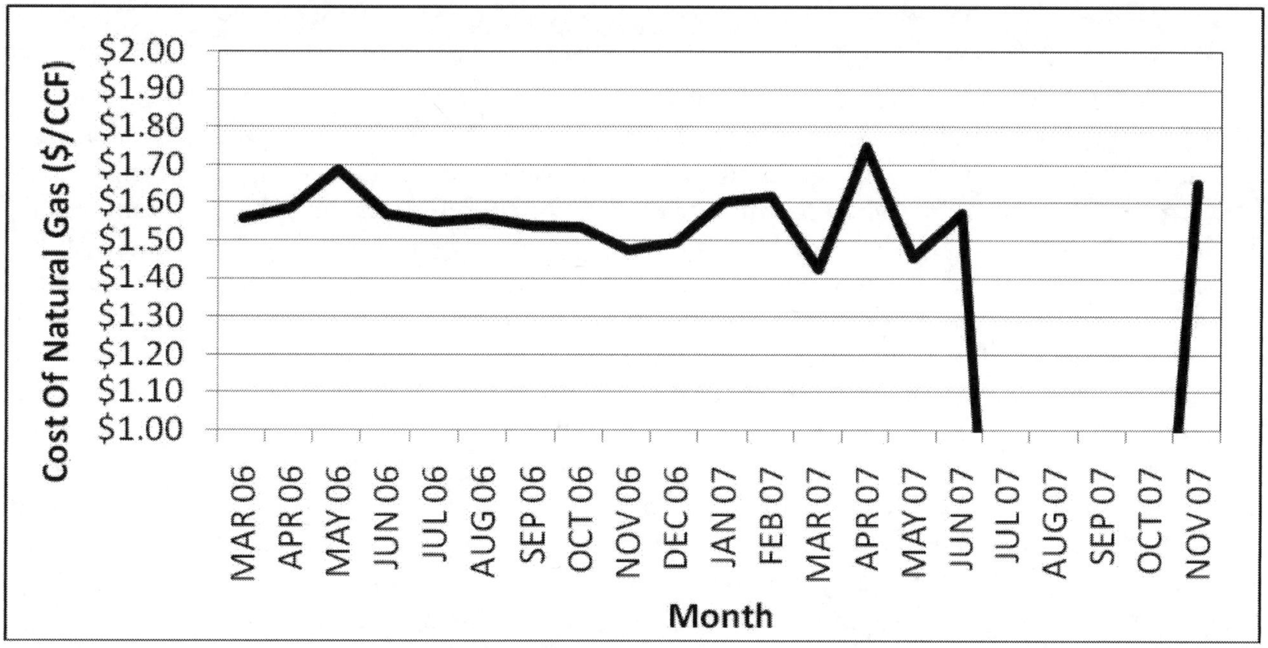

Figure 3. Natural gas cost per CCF from March 2006 to November 2007 for the Humanities Building.

For these reasons, the Facilities Management Department at the University of Alabama at Birmingham commissioned a study in which graduate students of the Civil, Construction and Environmental Engineering (CCEE) Department were tasked with researching ways that the building deans and administrators could reduce electricity use even further, and thus keep the utility bill down and greenhouse gas emissions to a minimum.

Brief Background

A similar study was performed in 1981 by George A. Jackins and Michael E. Scruggs. This study encompassed 11 buildings on the UAB campus, three of which were also surveyed in the more recent effort [Jackins, 1983a-e]. The objectives and goals of each of the two studies were also remarkably similar, though their methods differed. The 1981 studies involved an in-depth walkthrough of each building, including visual inspections of Heating, Ventilation, and Air Conditioning (HVAC) systems, lighting systems, and a complete cost analysis of the building as it was, and as it would be if their operation, maintenance, and energy conservation measures were adopted. The 2007–2008 studies involved 20–30 different walkthroughs of each building; however, due mainly to a lack of access to complex building systems by CCEE graduate students, only two major types of data were collected from each building:

1. Number of occupants in each room and corridor, and
2. If the room was unoccupied, noting whether or not lights were on in the room.

Additionally, notes were made of any "low-hanging fruit" (incandescent lights, vending machines, etc.), but these were not reported for each room in the building like the two main criteria were.

Project Goals/Objectives

The goal of the study was to determine how efficient each building was in terms of energy use by its occupants. A total of 14 buildings were covered in this study; in 2007, the Education Building, Hill University Center (HUC), Business and Engineering Complex (BEC), Sterne Library, Hoehn Engineering Building, Humanities Building, Scrushy Building, Ryals Building, University Boulevard Office Building (UBOB), and Campbell Hall were surveyed, while in 2008, Lister Hill Library, the Peters Building, the Worrell Building, and the Center for Biological Sciences and Engineering (CBSE) were surveyed. Each entire building that was surveyed in 2007 was surveyed roughly every two hours on weeknights from 6:15 p.m. to 12:15 p.m. and on weekends from 7:00 a.m. to 11:00 p.m. A little more freedom was granted with the buildings surveyed in 2008, so they were surveyed at varying times, roughly every two hours during the period of 5 p.m. to 9 or 11 p.m. on weeknights and every 2–3 hours during the period of 9 a.m. to 9 p.m. on weekends. As each room was surveyed, a symbol was recorded for that room for that nominal time: if the room had occupants, the number of occupants was recorded, and if not, the percentage of lights on in the room was estimated and recorded as follows: "P" for 1–50%, and a zero with a slash through it (Ø) for 51–100%. Using this method, the only building resource that could not be implied was water, due to the fact that it is not inherently used by occupants of rooms like electricity for lighting and natural gas for heating are.

The time period covered by each study was roughly 6 hours on weeknights and 16 hours on weekends (a total of 62 hours) for 2007 studies, and 4–6 hours on weeknights and 12 hours on weekends (a total of 44–54 hours) for the 2008 studies, or that portion thereof in which the building was open (HUC, Sterne, Lister Hill), By taking the cost of the resources that were expended during the month of the study, prorating it for the actual after-hours time period covered by each study, and dividing it by the average number of occupants witnessed over all studies, an expenditure per occupant figure can be calculated. A comparison of this will determine which buildings are being used less efficiently during this period. In addition, a qualitative judgment can be made about what time each weeknight HVAC systems can be set back, which varies from building to building, based on at which point occupancy tapers off significantly. Finally, a comparison of the number of fully lit rooms against the number of total rooms should give an idea of what buildings have the worst problem with lights being left on when rooms are not in use.

Results and Discussion

The results for the entire study cannot be displayed within a limited number of pages, but they can be summarized and analyzed in four key ways that will each point out a feature of the building's efficiency:

1. After-hours utility cost per occupant
2. Time of drop-off in occupancy
3. Percent of building fully lit in after-hours
4. Utility cost per gross square foot

After-Hours Utility Cost per Occupant

Table 1 shows a calculation of total expenditure on utilities for the after-hours period of each building for each person in average after-hours occupancy (AAHO). AAHO is defined as the mean of all results for total occupancy of the building regardless of nominal time. The amount of each utility bill is also prorated by the following equation:

$$\$Prorated = \frac{\$Bill * Survey\ Hours}{Total\ Hours\ in\ Month}$$

For surveys that took place in two different months, a simple average of number of days (usually 30.5) and bills was used. By dividing this value by the AAHO, a cost-per occupant (CPO), can be determined. Finally, a rank is given to each building, with lowest per-person occupancy amounts being the most efficient in terms of its CPO.

Table 1. Determination of Utility Cost per Occupant and Rank.

Building	AAHO	Study Month(s)	MHrs	PHrs	Prorated, ($)	CPO, ($)	Rank
Education	60.26	2/07	672	62	2,249.71	37.33	2
Ryals	6.53	11/07	720	62	3,568.36	546.77	10
BEC	97.97	10/07, 11/07	732	62	3,087.33	31.51	1
HUC	26.69	10/07, 11/07	732	62	3,175.22	118.96	5
Humanities	21.95	11/07	720	62	1,476.11	67.26	4
Hoehn	3.55	10/07, 11/07	732	62	764.98	215.33	7
UBOB	9.89	10/07, 11/07	732	62	1,397.08	141.19	6
Scrushy	1.82	11/07	720	62	1,413.80	778.61	12
Lister Hill	45.81	4/08	720	77	2,220.22	48.47	3
CBSE	3.70	4/08, 5/08	732	44	2,247.28	607.37	11
Peters	7.90	5/08	744	54	1,805.65	228.56	8
Worrell	2.67	5/08	744	44	1,282.25	480.84	9

Campbell Hall and Sterne Library could not be analyzed due to the fact that data was missing. Classroom buildings such as BEC, Humanities, and Education buildings and the Lister Hill Library, tend to be on the low range in CPO, due mainly to the fact that they have higher AAHOs. Research-oriented buildings had lower AAHOs, often fewer than 5 people, and thus their CPOs tend to be upwards of $480. Therefore, simply put, they are far more expensive to operate in terms of their benefits to each occupant. Many are also likely candidates for early closure in after-hours, but there is the major obstacle that research activities would be greatly hindered. A better solution would be to save as much as possible while leaving the buildings just barely hospitable for their sparse occupants.

Suggestion for Setback Times

Setting back of HVAC systems to a value that is still hospitable for the few occupants that are left inside later in after hours would result in 25% of the working-hours expenditure, according to Facilities Management. Assuming no current measure is in place, and that electricity and water bills would get lowered by the same amount, savings would be according to the last two columns in Table 2.

Table 2. 2007 Occupancy Taper-off, Suggestion for Setback Hours, and Potential Savings.

Building	6:15 p.m.	8:15 p.m.	10:15 p.m.	12:15 a.m.	Setback Time	Setback Hours	Savings*, ($)	Savings, (%)
Ryals	31	5	3	2	7:00 p.m.	98	18,129.58	43.75
BEC	352	209	32	21	10:00 p.m.	86	13,911.15	38.39
HUC	51	24	11	Closed	9:00 p.m.	90	14,716.29	40.18
Humanities	149	25	5	1	9:00 p.m.	90	6,887.36	40. 18
Hoehn	10	8	4	0	7:00 p.m.	98	3,685.61	43.75
UBOB	52	4	Closed	Closed	7:00 p.m.	98	7,903.03	43.75
Scrushy	12	2	2	2	7:00 p.m.	98	7,183.02	43.75

The setback hours used in the third-to-last column above call for full setback on weekends (never on those days does the total occupancy exceed 100 people in any building), setback from the given time until 5:00 a.m. on weekdays, and beginning at 7:00 p.m. on Friday night. In most instances, this would result in a roughly 40% decrease in the monthly bill, as seen above. As for the 2008 surveys, results were very similar, as is seen in Table 3.

Table 3. 2008 Occupancy Taper-off, Suggestion for Setback Hours, and Potential Savings.

Building	<6 p.m.	6–8 p.m.	8–10 p.m.	>10 p.m.	Setback Time, (p.m.)	Setback Hours	Savings, ($)	Saving, (%)
Lister Hill	69	53	47	24	10:00	86	7,970.55	38.39%
Peters	25	4	4	2	6:00	103	11,439.34	45.98%
Worrell	10	2	2	2	6:00	103	9,969.70	45.98%

CBSE data was missing; therefore its analysis is unavailable. Research facilities in particular would benefit the most, as a 46% savings could be realized by the suggested setback.

***Percentage of Rooms Fully Lit and Utility Cost per Gross Square Foot* (GSF)**

Up until now, all parameters have dealt with the occupancy of the rooms if they were occupied, but how about the rooms which were fully lit and unoccupied? The next phase of the study deals with rooms in that condition, and the average number of these rooms per survey per building is displayed below under the heading AAHL, which means "average after-hours lighting". This value is simply divided by the total number of rooms in the building to yield a percentage of fully lit rooms. This is displayed, among other things, in Table 4.

Table 4. Percent Fully Lit Rooms, Cost per GSF, and Energy Efficiency Scores.

Building	AAHL	Room	Fully Lit, (%)	Rank	Average $/GSF	Rank	CPO, ($)	Rank	Score
Ryals	33.45	312	10.72%	6	$0.41	8	546.77	9	23
BEC	52.64	305	17.26%	10	$0.25	3	31.51	1	14

HUC	19.82	310	6.39%	3	$0.25	4	118.96	4	11
Humanities	18.50	170	10.88%	7	$0.26	5	67.26	3	15
Hoehn	17.75	139	12.77%	8	$0.26	6	215.33	6	20
UBOB	12.90	70	18.43%	11	$0.42	9	141.19	5	25
Scrushy	7.52	194	3.88%	1	$0.21	2	778.61	11	14
Lister Hill	15.21	155	9.81%	5	$0.13	1	48.47	2	8
CBSE	18.90	270	7.00%	4	$0.49	10	607.37	10	24
Peters	16.90	296	5.71%	2	$0.29	7	228.56	7	16
Worrell	23.50	176	13.35%	9	$0.51	11	480.84	8	28

Unlike the previous analysis, the ranking seems to be independent of building usage, as the best is Scrushy, a nursing building, while the worst is UBOB, the criminal justice building. The other major classrooms and research facilities alternate, which seems to indicate a higher degree of bias based on each surveyor's judgment of what constituted a fully lit room. The utility cost per GSF is just that, and is a parameter which does not rely on any survey's results. It follows that, similar to CPO, it tends to favor libraries, administrative buildings, and classroom buildings, while research facilities perform poorly. However, that does not mean that there are no exceptions, as the Scrushy building, which is the second-best in terms of cost per gross square foot, is the worst in terms of after-hours CPO!

Summary and Conclusions

The final column in Table 4 in the previous section is a "score" of the utility efficiency of each building, based on three major criteria above:

1. After-hours cost per occupant;
2. Percentage of fully lit rooms; and
3. Cost per gross square foot.

All 11 buildings for which each parameter could be determined were ranked 1 to 11. The score was determined by summing the rankings for the percent fully lit rooms, the average cost/gross square foot, and the cost per occupant. The buildings with the lowest scores can be termed ideal models for how buildings should operate, while the ones with the highest scores appear to require some work. Lister Hill Library is the only building with a score less than 10, and HUC, which was determined by the team to be "an ideal model for how a building to operate" scored an 11. Five buildings had scores in the 20s: Ryals; UBOB; Hoehn; Scrushy; and Worrell. By this analysis, it appears these buildings could use some work in making them more utility-efficient.

Whether a building "scores" low or high, there are a few things that can be done to make it more energy efficient:

- Academic deans can pressure personnel to turn out lights when they are not in use;
- Replacing all incandescent lights with fluorescent lights, both straight and compact;
- Layered lighting in hallways and corridors for security;
- Motion/occupancy sensors in hallways and classrooms not currently equipped with such devices;
- Setback of HVAC systems to 25% of the daytime level in after-hours usage;
- Replacing washers and other parts in water systems which may lead to leaks; and
- Controlling water temperature in water fountains based on the building's occupancy.

If all of these measures were followed to regulate building resources, a potential savings of 50% or more could result.

References Cited

1. Birmingham Newschart, 2007. State Leads Nuclear Comeback. ***The Birmingham News***, (April 15).
2. Carnegie Mellon University. (2003). ***Environmental Decision Making, Science, and Technology*: *Science Notes: Chemistry of Fossil Fuels***. Retrieved August 3, 2008, from: http://telstar.ote.cmu.edu/environ/m3/s3/09fossil.shtml.
3. Energy Information Administration, (2008, May). *Greenhouse Gases, Climate Change, and Energy*. Retrieved August 3, 2008, from: http://www.eia.doe.gov/bookshelf/brochures/greenhouse/Chapter1.htm.
4. Jackins, G. A., 1983a. *Study of Energy Use and Conservation Opportunities for Sterne Library, University of Alabama in Birmingham*, prepared by George A. Jackins and Michael E. Scruggs, Energy Management Consultants, Birmingham, AL.
5. Jackins, G. A., 1983b. *Study of Energy Use and Conservation Opportunities for UAB Building #1, University of Alabama in Birmingham*, prepared by George A. Jackins, Michael E. Scruggs, Energy Management Consultants, Birmingham, AL.
6. Jackins, G. A., 1983c. *Study of Energy Use and Conservation Opportunities for UAB Building #2 Annex, University of Alabama in Birmingham*, prepared by George A. Jackins and Michael E. Scruggs, Energy Management Consultants, Birmingham, AL.
7. Jackins, G. A., 1983d. *Study of Energy Use and Conservation Opportunities for UAB Building #2, University of Alabama in Birmingham*, prepared by George A. Jackins and Michael E. Scruggs, Energy Management Consultants, Birmingham, AL.
8. Jackins, G. A., 1983e. *Study of Energy Use and Conservation Opportunities for UAB Building #3, University of Alabama in Birmingham*, prepared by George A. Jackins and Michael E. Scruggs, Energy Management Consultants, Birmingham, AL.
9. National Weather Service. (2008, January 8). *2007 Year In Review*. Retrieved August 3, 2008, from NOAA's National Weather Service: http://www.srh.noaa.gov/bmx/significant_events/2007/weather_review.php

A STUDY ON THE AFFECTED POPULATION SURROUNDING THE INDUSTRIES RELEASING OSHA CARCINOGENS IN JEFFERSON COUNTY USING TOXIC RELEASE INVENTORY

Shammi Rahman and Dr. Robert W. Peters. University of Alabama at Birmingham, Birmingham, AL

Abstract

The Toxics Release Inventory (TRI) is a publicly available EPA database that contains information on toxic chemical releases and waste management activities reported annually by certain industries as well as federal facilities.

The data obtained from the TRI can be used to determine the actual number of people living in the two mile zone of each facility releasing carcinogens in the Jefferson County. The study is conducted with the help of the data obtained from TRI. The information regarding the location of the Facilities releasing OSHA carcinogens in the Jefferson County is gathered from the TRI Explorer. There were 49 such facilities. Then with the help of GIS the information is incorporated with the population map of Jefferson County, Alabama, to see the actual number of people who live within the two mile area of the facilities.

The study also looks at the number of people of different age groups within the area. The population with age below 5 years and those above 65 are most susceptible to such carcinogens. Thus emphasis has been given particularly to these two groups. Variations in the different racial groups were studied, i.e., which racial groups dominate in the sphere of influence of the industries.

The limitation of the study is that the data uses only the OSHA Carcinogens which comprises of only 4 groups of chemicals: arsenic compounds; dioxin and dioxin-like compounds, chromium compounds, lead compounds, and polycyclic aromatic compounds (PACs). Further study on all carcinogens could benefit the locals and policy makers to establish reduction targets and measure progress toward those targets.

Potential Toxicity of Nanomaterials

Pickrell JA (1-Toxicology, Diagnostic Medicine/Pathobiology)
Dhakal K (1)
Erickson LE (2-Chemical Engineering)
Klabunde KJ (3-Chemistry)
Maghirang R (4-Biological and Agricultural Engineering)
Dhakal M (1; currently Infectious Disease, University of Missouri)
Van der Merwe D (1)
Oehme FW (1)

Kansas State University, Manhattan, KS

Nanoparticles (manufactured) or ultrafine particles (naturally produced in the environment; often from transportation sources) are particles with at least one dimension ≤ 0.1 microns (100 nanometers [nm]). Inhalation of potentially injurious particles small enough to travel to deep lung (fine particles - ≤ 2.5 microns [2,500 nm] or ultrafine- (nano-) particles) [7, 15] can have different outcomes. For example, one outcome might be that they cause no biomedical consequences. Alternatively, ultrafine or nanoparticles can cause and heal minor injuries in relatively healthy lung epithelial cells (LEC). Or, they may lead to chronic degenerative lung disease or even to lung cancer if the LEC are injured, or become injured [10a, 15, 17].

Induction and release of the cytokines interleukins 6 and 8 (IL6 and IL8) by airway cells in vitro has been used extensively to assess relative toxicity of occupational particles and particles occurring in nature [19], as soil particles and as particles originating from transportation vehicles. TRP channels (transmembrane receptors) established as primary sensors of thermal stimuli in the peripheral nervous system of mammals [1] are associated with pores and the release of cytokines such as IL-6 and IL-8 (Sabnis et al., 2008) [16]. Nano sized particles of the same nominal composition are useful for the induction of IL-6 and IL-8. The manufactured pure oxides were less potent than natural $PM_{2.5}$ particles whose source was soil or the positive controls. The potency for inducing IL-6 secretion by BEAS-2B did not correlate well with the generation of reactive oxygen species (ROS), possibly because they absorbed IL-6 on the high surface area silica and titanium dioxide particles. Measured IL-6 increased when BSA or fetal calf sera was added to the incubation mixture, suggesting that there was non-specific binding of IL-6 to surfaces > 1 order of magnitude above that of plastic on which they were cultured.

Particle artifacts from high surface areas need to be considered in experiments of this type. However, the study suggests that manufactured metal oxide nanoparticles are not highly toxic to lung cells compared to other environmental nanoparticles [19].

Exposure to air polluted with particles less than 2.5 micrometers (2,500 nm) in size was associated with epidemiologically adverse cardiopulmonary health consequences in humans [6]. Six healthy volunteers exposed to fine and ultrafine but non toxic magnesium oxide (MgO) particles. Exposure levels were: mean +/- standard deviation for exposure was ~ 4,138 +/- 2163 x mg/m^3. Ninety-eight (98%) percent of the particles were fine (< 2.5 micrometers, 2,500 nm) and 28% were ultrafine (< 100 nm). No differences were noted from controls, as indicated by bronchoalveolar lavage (BAL) inflammatory cell concentrations, interleukin concentrations (IL-1, IL-6, IL-8), tumor necrosis factor, pulmonary function or peripheral blood neutrophil concentrations [6].

In animal studies, rats exposed *in vivo* to very high levels (1.5-2 g/m^3) of nanoactive magnesium oxide, titanium dioxide or FAST ACT (combined MgO and TiO_2) for ~4 hours had no reductions of body weight by 3-14 days after initiation of exposure. They had no clinical signs or macroscopically visible pathologic changes on examination at necropsy at 14 days after initiation of exposure. Based on this evidence, manufactured magnesium, titanium or combined magnesium and titanium nanoparticles had no "in life" effect, at very high doses for short exposure times [5].

MgO particles, at least, may have been cleared relatively rapidly. In studies at Kansas State University [13], we have shown that non-toxic magnesium oxide (MgO), relatively insoluble in distilled water, is much more rapidly soluble in LEC tissue culture fluid (Hanks Balanced Salt Solution (HBSS) or Dulbecco's Modified Eagles Medium (DMEM); distilled water peak ~2-4 days; HBSS peak $\leq$ 1-3 hours). Peak dissolution speed and amount both increased ~ 2 fold (amount) or ~2 to 5+ fold (speed) with lung stimulant fluid containing bicarbonate relative to distilled water. Speed and amount of solubility were related directly to the bicarbonate concentration (a major human body buffer) in the dissolving media. Most likely, a more soluble carbonate compound (magnesium carbonate trihydrate; nesquehonite) was formed. It should be noted that solubility in these studies was defined as material that remained after micro-centrifugation at 13,200 x gravity for 20 minutes. Measurement of particle size by laser diffraction showed reduction in particle size from 100-1,000 nm (un-centrifuged) to ~40-100 nm (centrifuged) consistent in similar particle distributions with $\geq$ 99% removal as soluble material [13-14], within the reproducibility of the metal measurements.

Contrast this to titanium dioxide particles. Such particles had minimal solubility (1.5-2.1 ppm $\pm$ ~ 10-25%) at high (5X) relative and at low (1X) concentrations from distilled water, or media containing bicarbonate at different concentrations. Concentrations were significant above with a $p < 0.01$) 2 times out of 3; but, with a $p > 0.05$ 1 time out of 3, when corrected for 3 comparisons (Games, 1977). Measurement of particle size by laser diffraction showed reduction in particle size from 300-2,000 nm (un-centrifuged) to ~20-300 nm (centrifuged) consistent in similar particle distributions with $\geq$ 95% removal [14] as soluble material, which was within the reproducibility of the metal measurements. Carbonaceous particles which are relatively insoluble and would also be expected to persist. If sufficiently small these particles would be expected to go to the interstitial space and remain as a potentially injurious source. Alternatively, if they were not, they would most likely be cleared. Titanium dioxide particles have a tendency to aggregate in biological fluids [2, 8, 15]. Thus, TiO_2 particles have some probability of being phagocytized.

Lung responses are influenced by repeated injury, or persistence of ultrafine particles or nanoparticles. For example, mice inhaling urban traffic particles that were organic and would be expected to persist. When air was filtered successively by bag-, JFL-90- and HEPA-filters concentrations were 2.9 $\mu g/m^3$ showed no "in life" changes as indicated by no changes in lung and body weights, and lung/body weight ratios; respiratory physiology (pressure-volume curve) measurements; histopathology changes and morphometry changes [9]. When airborne particles were not filtered, concentrations were 16.8 $\mu g/m^3$. Comparisons were among unexposed controls, prenatal, post-natal or both (combined prenatal and post natal) exposures. Combined exposures had no differences in body weight, among the four groups. Although both levels are taken to be acceptable or nearly so by US-EPA standards ($PM_{2.5} < 15$ mg/m^3), post natal and combined groups had trends toward reduced lung/body weight ratios at 90 days exposure and significantly reduced surface to volume ratios when compared to pre natal (before birth) exposures or unexposed controls. Reduced surface to volume ratios in lungs may have reduced their ability to exchange oxygen into the capillaries. Thus, lungs of the mice exposed to higher levels had structural and histopathology changes relative to prenatal or unexposed controls [9]. Mice in the filtered air chamber had no detectable lung responses, suggesting the probable utility of HEPA masks.

Although nanoparticles commonly have a dimension of < 100 nm, on suspension in media agglomerates of nanoparticles are more common structures [10]. The authors exposed mice to relatively dispersed single walled carbon nanotubes (DSWCNT; mean diameter 0.69 micrometers, 690 nm; aggregated to small fine particles; ~10 μg/mouse lung; ~300 μg/kg mouse) by pharyngeal aspiration. Lung sections in sacrificed mice demonstrated transient inflammatory and neutrophil phases that resolved and were similar to that observed with larger agglomerates. There were no granulomatous lesions or epithelioid macrophages, but the average thickness of connective tissue in alveolar regions stained by Sirius red (collagen special stain) in cryo fixed sections increased from ~0.1 to 0.5 at 7 days- and 0.9 at 1-month-post exposure. Health and respiratory physiology appeared unchanged as indicated by body weight, lung volume, alveolar surface area, alveolar gas volume and alveolar tissue volume. Quantum dots attached to DSCWNT were not detected in alveolar macrophages. Quantum dots alone in exposures were avidly phagocytized by such macrophages. No differences were seen when compared with unlabelled DSWCNT. Gold-labelled- and quantum dot labeled-DSWCNT were rapidly incorporated into pulmonary interstitium, but didn't produce granulomatous lesions [10].

Less dispersed CNT can result in deposits ~ 15 micrometers (15,000 nm). Once in the lungs, the action of surfactant and protein may reduce the interactions between nanotubes and water and allow dispersion of these agglomerates. The dispersive action of alveolar lining fluid on nanoparticles has been demonstrated in vitro [10, 16a]. Gold and quantum dot labeling of less dispersed DSWCNT demonstrated that a significant fraction of aspirated DSWCNT material may exist at < 1 micrometer (1,000 nm) and can rapidly migrate into the alveolar interstitial spaces. These small aggregates were not a target for macrophage activity. The only detectable alteration was an increase in interstitial collagen as indicated by Sirius red staining of cryo-fixed tissue at 1 month after exposure. Carbon nanotubes were seen in lung tissue indicating their persistence [10].

Although total lung collagen was not measured, lung sections seemed to have very little persisting inflammation, less intense than animals in an earlier study who inhaled 50 nanocuries (nCi; 1,500 roentgen equivalent to man [rems]) of $^{238}PuO_2$ with diffuse interstitial pulmonary fibrosis similar to that shown in Mercer et al., 2008 spontaneously by 288 days after exposure (Pickrell et al., 1983) [11]. Similar plutonium exposures involved 8-12% of the projected histologic section's area with diffuse alveolar thickening [3]. Thus, our 1.5 to 2 fold increases in total lung collagen are ~ equivalent in extent to the diffuse alveolar thickening seen in collagen special stains of Mercer et al. [10]. If equivalent responses can be expected in human lungs following exposure to nanoparticles, this is good news, because they are probably negligible or reversible. Taken together data suggests that we may need to look to cellular and tissue injury caused by other components of polluted atmosphere, such as oxidant gases for injuries necessary to cause irreversible degenerative changes. The exact level of injury needed to cause irreversible degenerative changes in human lungs continues to be of considerable investigative interest [12].

Particles < 0.5 microns (500 nm) are cleared from lung with minimal efficiency by lung phagocytes or simply ignored and travel to the pulmonary interstitial space. Lack of clearance can be one source of persistence of the particles and continued injury. Relative solubility is a second source of persistence for poorly soluble particles [14, 15]. However, diffuse interstitial pulmonary fibrosis $\geq$ intensity and inflammation than DSWCNT, the most persistent of nanoparticles, is reversible anatomically, biochemically and physiologically [11].

Research supported by US Marine Corps Systems Commend - M-2 Corporation, NanoScale Corporation and Kansas State University.

References:

1. Caterina MJ, Leffler A, Malmberg AB, Martin WJ, Trafton J, Zeitz KR, Koltzenburg M, Basbaum AI Julius D: 1997. Impaired nociception and pain sensation in mice lacking the capsaicin receptor; Science 389; 816-824

2. Churg, A; Gilks, B; Dai, J; 1999; induction of fibrogenic mediators by fine and ultrafine titanium dioxide in rat tracheal explants; Am J Physiol; 277: L975-L982.

3. Diel JH: 1984; Research communication; Lovelace Inhalation Toxicology Research Institute, Albuquerque, NM.

4. Games PA: 1977. An improved t table for simultaneous control on g contrasts. J Amer Stat Assn 72 (359): 531-534.

5. Koper OB, Bergmann J, Klabunde KJ, Wilson CM, Pickrell JA: September 18, 2008. Research communication "Acute Inhalation of Nanoactive® Metal Oxides in rats". Kansas State University and NanoScale Corporation, Manhattan, KS.

6. Kuschner WG, Wong H, D'Allesandro A, Quinlan P, Blanc PD; 1997; Human pulmonary responses to experimental inhalation of high concentration fine and ultrafine magnesium oxide particles. Environ Health Perspect 105(11); 1234-1237.

7. Lin, C-C; Chen, S-J; Huang, K-L; Hwang, W-I; Chang-Chien. G-P; Lin, W-Y; 2005. Characteristics of metals in nano/ultrafine/fine/coarse particles collected beside a heavily trafficked road. Environ Sci Technol 39: 8113-8122.

8. Long, TC; Saleh N; Tilton, RD; Lowry, GV; and Veronesi, B; 2006; Environ Sci Technol. Titanium dioxide (P25) produces reactive oxygen species in immortalized brain microglia (BV2) implications for nanoparticle neurotoxicity; 40 (14): 4346-4352.

9. Mauad T, Rivero H, de Oliveira R, Lichtenfels A, Guimaraes E, de Andre A, Kasahara D, Bueno H, Saldiva P: 2008. Chronic exposure to ambient levels of urban particles affects mouse lung development. Am J Crit Care Med 178: 721-8.

10. Mercer RR, Scabilloni J, Wang L, Kisin E, Murray AR, Schwegler-Berry D, Shvedova AA, Castronova V; 2008; Alteration of deposition pattern and pulmonary response as a result of improved dispersion of aspirated single-walled carbon nanotubes in a mouse model; Am J Physiol Lung Cell Mol Physiol 294; L87-L97.

10a. Mroz, RM; Schins, RPF; Jimenez, LA; Drost, EM; Holownia, A; MacNee, W; Donaldson, K; 2008; Nanoparticle-driven DNA damage mimics irradiation-related carcinogenesis pathways; Eur Respir J 31: 241-251.

11. Pickrell, JA; Diel, JH; Slauson, DO; Halliwell, WH; Mauderly, JL; 1983; Radiation-induced pulmonary fibrosis resolves spontaneously if dense scars are not formed; Exp Mol Pathol 38(1): 22-32.

12. Pickrell JA, Abdel-Mageed AB: Radiation induced pulmonary fibrosis; In *Pulmonary Fibrosis*; Phan SH, Thrall RS eds. (Lenfant C, series ed; *Lung Biology in Health and Disease*) New York, NY Marcel Dekker Inc, **1995;** 80: 363-381.

13. Pickrell JA, Dhakal M, Castro SD, Gakhar G, Klabunde KJ, Erickson LE, Oehme FW: 2006. Comparative Solubility of Nanoparticles and Bulk Oxides of Magnesium in Water and Lung Simulant Fluids. AIChE, Nov 15, San Francisco, CA (Abstract and Extended Abstract AIChE DVD).

14. Pickrell JA: 2007. Biomedical responses and toxicity of nanoparticles; In Gupta R, editor; Veterinary Toxicology: Basic and Clinical Principles. Academic Press; NY; pp 305-313.

15. Pickrell JA, Erickson LE, Dhakal K, Klabunde KJ: 2009. Toxicity of Inhaled Nanomaterials, In Klabunde KJ and Richards R eds. Chemistry of Nanomaterials Acad Press, NY (In Press)

16. Sabnis AS, Shadid M, Yost GS, Reilly CA: 2008. Human lung epithelial cells express a functional cold sensing TRPM8 variant. Am J Respir Cell Mol Biol 39: 466-474.

16a. Sager T, Porter D, Robinson V, Lindsey G, Schwegler-Berry D, Castanova V: 2007; Improved method to disperse nanoparticles ofr in vitro and in vivo investigation of toxicity Nanotoxicology 2; 1-6.

17. Sydlik, U; Bierhals, K; Soufi, M; Abel, J; Schins, RP; Unfried, K; 2006; Ultrafine carbon particles induce apoptosis and proliferation in rat lung epithelial cells via specific signaling pathways both using EGF-R; Am J Physiol Lung Cell Mol Physiol 291 L725-L733.

18. Unfried, K; Sydlik, U; Weissenberg, A; Abel, J; 2008; Carbon nanoparticle-induced lung epithelial cell proliferation is mediated by receptor-dependent Akt activation. Am J Physiol Lung Cell Mol Physiol 294: L358-L367.

19. Veranth JM, Kaser EG, Veranth MM, Koch M, and Yost GS: 2007. Cytokine responses of human lung cells (BEAS-2B) treated with micron sized and nanoparticles of metal oxides compared to soil dusts. Particle and Fibre Toxicology 4-2: 1-18 of 18 (pages not for citation) 27 February 2007; doi10.1186/1743-8977-4-

53d Physiologically-Based Inhalation Dosimetry Modeling for Nanoparticles: Considerations of Activity, Age, and Gender

Pamela R. Shade and Panos G. Georgopoulos. Environmental and Occupational Health Sciences Institute, University of Medicine and Dentistry of New Jersey (UMDNJ)/Robert Wood Johnson Medical School (RWJMS), 170 Frelinghuysen Road, Piscataway, NJ 08854

The results of several epidemiological studies suggest that acute and chronic exposures to particulates in ambient air are associated with increases in respiratory and cardiovascular morbidity and mortality. Due to certain features (large surface area, high lung deposition efficiency, etc.) and their prevalence and persistence in ambient air, nanoparticles may contribute to these adverse health effects. Although ambient concentration is what is measured and regulated, the ultimate biological response is dependent upon the target tissue dose. Nanoparticles deposit primarily by diffusion and are more likely to deposit in the distal airways where gas exchange occurs.

Inhalation dosimetry plays a key role in determining the link between environmental exposure to airborne contaminants and observed human health effects. Due to the limited availability of human experimental data and the complex nature of exposure-dose-response scenarios, mathematical modeling is an important tool in studying the mechanisms which dictate the inhaled "dose." Numerous factors affect the inhaled dose, and available computational models must include an extensive amount of information regarding physiological and anatomical parameters.

A Matlab-based inhalation module was developed at the Computational Chemodynamics Laboratory (CCL) as part of the Modeling ENvironment for TOtal Risk (MENTOR) system. A flexible design allows for adaptation to a variety of exposure conditions and scenarios. This semi-empirical model accounts for variation in inhalation rate and delivered dose across individuals and populations due to differences in factors such as age, gender, and level of physical activity. Capturing the inter- and intra-individual variability of input parameters is accomplished primarily through the interaction of the module with databases of anatomical and physiological parameters.

Predictions from existing models are compared to one another and to available experimental data. Results indicate that the variation in human airway anatomy and respiration physiology among individuals of different ages and genders affects the inhaled dose of particulates. Results also suggest that the level of physical activity is an important factor in nanoparticle deposition, and that available models often underpredict the effects of exercise.

53e Carbon Nanotube Bacterial Cytotoxicity: Does the Type of Carbon Nanotubes Matter?

Seoktae Kang[1], Moshe Herzberg[2], Debora F. Rodrigues[1], and Menachem Elimelech[1]. (1) Department of Chemical Engineering, Environmental Engineering Program, Yale University, Mason Laboratory 204, 9 Hillhouse Ave., New Haven, CT 06520, (2) Department of Desalination and Water Treatment, Ben Gurion University of the Negev, Zuckerberg Institute for Water Research, Sede-Boqer Campus, Midreshet Ben Gurion, 84990, Israel

We investigate the bacterial toxicity of well-characterized carbon nanotubes (CNTs) by several independent techniques. Results demonstrate that single-walled carbon nanotubes (SWNTs) are much more toxic to bacteria (E. coli) than multi-walled carbon nanotubes (MWNTs). The likely main CNT-cytotoxicity mechanism is cell membrane damage by direct contact with CNTs. Gene expression data shows that in the presence of both MWNTs and SWNTs, E. coli express high levels of stress-related gene products, with the quantity and magnitude of expression being much higher in the presence of SWNTs. The enhanced bacterial toxicity of SWNTs may be attributed to: (1) a smaller nanotube diameter that facilitates partitioning and partial penetration of nanotubes into the cell wall, (2) a larger surface area for contact and interaction with the cell surface, and/or (3) unique chemical and electronic properties conveying greater chemical reactivity.

IN-SITU EVAPORATION OF WATER WITHIN ROCK PILES – A MODELING AND LABORATORY STUDY

Edward M. Trujillo and Paul S. Evans[1], University of Utah, Salt Lake City, UT, U.S.A.

Abstract

Acid rock drainage (ARD) is an environmental problem commonly faced by the mining industry. ARD occurs when sulfide-bearing rock, usually in the form of a large rock pile on the side of a mountain, is exposed to air and water and oxidizes to form sulfuric acid which can leach hazardous metal ions from various minerals. Laboratory tests are typically conducted using short columns packed with a crushed mining rock sample to determine a rock sample's potential to produce acid. The room temperature procedures that are typically involved in these tests are inherently non-isothermal because they involve a drying phase in which water evaporation occurs. We are investigating the processes that govern evaporation within a rock pile and the important parameters that describe these processes. A two-dimensional axially-symmetric model of the drying phase of the testing procedure has been developed by our research group. The model was developed using Darcy's law, heat transfer in porous media, and mass balance equations and was verified with an analytical solution under simplifying assumptions. The model was calibrated with experimental data from our laboratory for certain rock types. Hypothetical studies of the effects of heterogeneities in the form of inclined layers were also conducted.

Introduction

ARD (acid rock drainage) is a very complex process occurring in nature as well as in mining operations. It occurs when sulfide-bearing rock, usually in the form of a large rock pile on the side of a mountain or in a tailings impoundment, is exposed to air and water and oxidizes to form sulfuric acid which can leach hazardous metal ions from various minerals. The processes include biological as well as geochemical and physical mechanisms but the main ingredients are air (oxygen), water and a sulfide mineral. In most cases, air flow is the limiting factor, primarily in tailings. However, due to the heterogeneity of rock piles, convective air currents can develop deep within the pile due to temperature gradients caused by exothermic oxidation reactions and these air currents can evaporate moisture within the pile and then condense it in other, cooler, locations.

Typically, the internal structure of the pile is not known and calibration of mathematical models describing the rock pile over time is limited to several field measurements taken from a few well bores throughout the pile at various time periods. Mathematical models that have been used in the past range from very simple homogeneous diffusion models to complex transport models with heterogeneous zones and containing several geochemical reactions. One aspect that has not been studied sufficiently is the process of drying and redistribution of water within the pile for arid and semi-arid regions. Many piles contain highly permeable zones, typically at inclined angles, that facilitate an up flow of air caused by temperature gradients within the pile, and thus allow the transport of gases, such as oxygen, water vapor, and carbon dioxide, to be distributed unevenly throughout the

[1] Now with Frontline BioEnergy, LLC

pile. In addition, entering dry air can evaporate some of the water contained in the pore structure and then condense that water in other parts of the pile that are cooler or have different humidity levels. Since ARD is dependent to a large degree on water saturation, it is important to understand the redistribution of water and water vapor over time and space. The objective of this study was to understand the equations and mechanisms involved in evaporation of water within rock piles in the laboratory in order to develop better models for field-scale studies.

Laboratory tests are conducted using columns packed with a crushed mining rock sample to determine a rock sample's potential to produce acid [ASTM, 2000]. The procedure typically involves a weekly cycle that contains a three-day drying period, a three-day moisture period and a one-day leaching period. The room temperature procedures that are typically involved in these tests can be non-isothermal if they involve a convective drying phase in which water evaporation occurs. Perhaps the most applicable study on evaporation from surfaces is the one done by Wilson *et al.* (1997). They were able to illustrate the effect of capillary pressure on the total drying rate using a soil water characteristic curve. They, however, did not actually model evaporation. The evaporative fluxes were simply compared to the evaporative flux from a sample in which there was no capillary pressure. Viollaz and Suarez (1985) give a short review of mass transfer correlations used to predict evaporative fluxes under various conditions. They then use numerical algorithms to find solutions to the problem of drying shrinking particles. Silva *et al.* (2000) studied porous media with very small pores (<5 nm.) in which Darcy's law was deemed invalid. They reported a method of estimating flow rates through the use of an effective diffusivity which was a function of many parameters such as porosity, tortuosity and pore size. Because waste rock piles have much larger pores (>>5 nm.) it is not reasonable to use this approach.

Stacey and Udell (1997) studied evaporation in porous media. They developed an analytical solution for the speed of an evaporation wave (V_e) traveling through a partially saturated porous medium by transforming the one-dimensional mass balance into wave coordinates. Their isothermal solution (given below) assumes phase equilibrium and no migration of water due to capillary pressure gradients.

$$V_e = \frac{u_d \rho_{id}}{\phi\left(S_{gd}\rho_{id} + S_{ld}\rho_{ld}\right)} \qquad [1]$$

In Equation 1 the subscript *d* represents the downstream condition. u is the superficial velocity of the gas phase. ρ_i is the mass concentration of component i (the component being evaporated) and ρ_l is the liquid density. Equation 1 can be used to give a good estimate of the time it will take to completely dry an isothermal unsaturated sample. However, drying is rarely an isothermal process. Stacey and Udell also give analytical solutions for the temperature drop due to evaporation (ΔT) and the propagation speed (V_T) of the thermal wave. Their solutions were compared to numerical models and experimental data with reasonable agreement. They conclude that an isothermal treatment will almost always over predict evaporation rates in porous media.

Experimental Apparatus and Procedure

A typical humidity cell is constructed of Lexan or similar plastic material and is four inches in diameter and eight inches tall. The cell has a removable lid, which allows the rock sample to be placed onto a felt pad that is supported by a perforated plastic disk. Air flows from the bottom upwards through a side port below the perforated disk while the aqueous leachate flows out the bottom. Procedures typically follow that in ASTM standard methods [ASTM, 2000] however, in this project, temperature probes were added to measure transient temperatures within the humidity cell sample.

Four Omega brand PR-17 RTD probes were fitted to each cell to in order to measure temperature at different locations in the humidity cell during testing. According to the literature the probes give readings with an accuracy of ±0.3° C for a temperature range of -25° C to 85° C (Pace Scientific, 2005). The four RTD probes were inserted into the humidity cells, at heights of 0.8, 3.3, 5.7 and 8.0 cm from the top of the polypropylene felt. Cell F11 contained waste rock from a mining site in the western United States, while crushed marble was used for the control cell. An Opto22 control system was used to continuously record the temperatures measured by the RTD probes at predetermined intervals. Figure 1 shows the control cell packed with crushed marble. The picture shows the location of the temperature probes. Note that the top probe is located at the top surface of the rock material. Figure 2 shows a similar cell packed with an acid-producing rock sample (F11) from a mining site. Note that, in this cell, the top probe is about one inch above the rock surface, since the particle size and bulk density are different than that of the crushed marble. Only data taken with the bottom three probes were used for calibration of the model.

Figure 1 A humidity cell containing crushed marble, used as the control cell. Also shown are the locations of the RTD probes, which are approximately 1 inch apart.

Figure 2 A humidity cell containing sample F11, an acid-producing sample. Also shown are the locations of the RTD probes, which are approximately 1 inch apart.

Modeling

Comsol Multiphysics™ (Comsol, 2007), a finite element based computer program, was used to create the humidity cell models. As multi-physics software, Comsol has many modules which can be used to assist in modeling physical phenomena. Ambient conditions were assumed to be standard temperature and pressure. The thermal conductivities of air and water as functions of temperature were given by the expressions used by Eldredge (2005). Other thermal conductivities were assumed to change insignificantly over the temperature range of interest. The enthalpy of vaporization of water was also assumed to change very little over the temperature interval and was therefore treated as a constant. The Clausius Clapyeron equation was used to predict the saturation vapor pressure of water as a function of temperature. Water's normal boiling point was used for the reference condition. Additionally, the solid phase was assumed to be immobile. Several models were developed for the project:

- A 1D Isothermal model
- A 1D Adiabatic model
- A 1D Non-adiabatic model
- A 2D Non-adiabatic Homogeneous model
- A 2D Non-adiabatic Heterogeneous model .

For this paper we will only discuss the 1D isothermal model, the 1D non-adiabatic model and the 2D non-adiabatic heterogeneous model. P.S. Evans describes all these models in more detail [Evans, 2007]

1D Isothermal Evaporative Model

The isothermal case was run using two mass balances, one for liquid and one for vapor phase water. Concentration (*c*) of each species is determined by solving the continuity equation at each element.

$$\frac{\partial c}{\partial t} + \nabla \cdot (\vec{u}c - D\nabla c) = R \qquad [2]$$

Evaporation was accounted for using the reaction term (*R*). The evaporation rate was determined from the flux of water vapor from the liquid surface which was then multiplied by the specific surface area (*Ai*) of the medium. Although the liquid surface area will change with saturation, it was assumed have a constant value equal to that of uniform cubic face center packed spheres. A conceptual diagram of the 1D model is given in Figure 3 below.

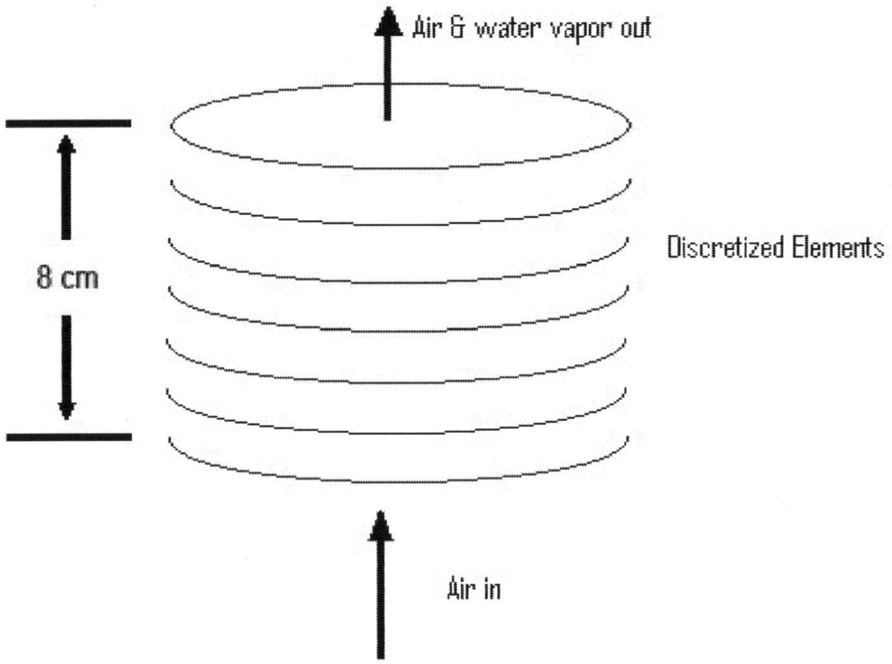

Figure 3 Conceptual diagram of the 1D evaporation model. Note, not all mesh elements have been shown.

The liquid phase velocity was assumed to be zero, while the vapor phase velocity was set to the superficial velocity of the entering air. This neglects a small increase in vapor phase velocity due to evaporation. This assumption was made because the equilibrium vapor pressure of water at the temperatures of interest is much less than one atmosphere. The liquid phase diffusivity was set to a very small number, but not zero. This gave stability to the numerical solution, while simulating the wicking that arises due to capillary action.

A zero flux boundary condition was given to the liquid mass balance at both boundaries. The concentration of the water vapor was set to zero at the inlet, and a convective flux boundary condition (no diffusive flux) was applied at the outlet. Table 1 provides a summary of the parameters used for the 1D isothermal evaporative model, which are applicable to Cell F11.

Table 1 Parameters used in the 1D isothermal evaporative model for Cell F11

Parameter	Symbol	Value	Units	Reference
Porosity	φ	0.36	–	Eldredge, 2005
Initial saturation	S_0	0.44	–	Klinker, 2006
Initial & ambient temperature	T_0	298	Kelvin	Klinker, 2006
Air flow rate	q_{air}	1	Lpm	Klinker, 2006
Inner diameter of humidity cell	D	10	cm	Eldredge, 2005
Height of sample in cell	h	8	cm	Klinker, 2006
Mass transfer Peclet Number	Pe_m	0.06	–	calculated
Lumped mass transfer coefficient	kA	100,000	s^{-1}	Bird, 2002

Figure 4 gives the results of the one dimensional isothermal model for the set of parameters given in Table 1. The lumped mass transfer coefficient (kA) is the product of the mass transfer coefficient and the specific surface area used to calculate total evaporative flux per unit volume using equation 2.19. Figure 4 gives the water saturation profile within the cell at $5x10^4$ seconds and shows a relatively sharp front with the rock sample essentially drying out completely after the front moves through. For easy comparison with the analytical solution (Equation 1) a vertical black line has been included to indicate the position of the evaporation front at $5x10^4$ seconds, as predicted by the analytical expression. It was found that the evaporation rate is insensitive to the mass transfer coefficient and the specific surface area above a threshold surface value. This indicates that, because the pore spaces are so small, the water vapor is able to come to equilibrium almost instantly. This is to be expected with such a small Peclet Number (0.06). This may not be the case for much larger rock particles or in situations with a much higher convective velocity.

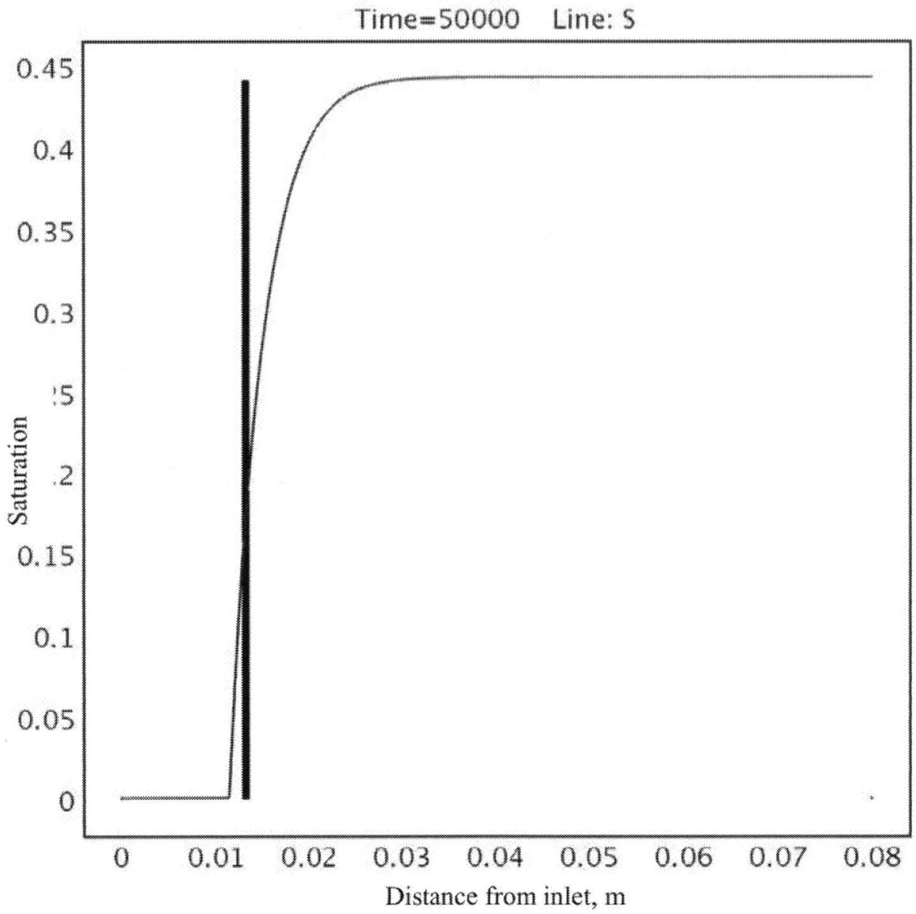

Figure 4 Saturation (Y axis) as a function of distance from inlet in meters (X axis) after 50,000 seconds as predicted by the 1D isothermal model. The vertical line indicates the position of the evaporation front according to the analytical solution (Equation 1).

1D Nonadiabatic Evaporative Model

In order to have a numerical model to compare to experimental results, a non-adiabatic model was developed to account for heat transfer with the surroundings. Due to the one-dimensional nature of this model some simplifications had to be made with regard to heat transfer in the radial direction. Incropera and DeWitt (2002) give the steady state solution to the radial heat transfer rate (q_r) as the following.

$$q_r = \frac{T - T_\infty}{\frac{\ln\left(r_i/r_r\right)}{2\pi L \lambda_r} + \frac{\ln\left(r_o/r_i\right)}{2\pi L \lambda_w} + \frac{1}{2\pi L r_o h}} \qquad [3]$$

In Equation 3, the subscripts r and w refer to properties of the rock and the humidity cell wall, respectively. r_i and r_o are the inner and outer radii of the cylinder and h is a convective heat transfer coefficient. The temperature of the surface of the cell was assumed to have a constant value equal to the ambient temperature (T_∞). This is the same as saying that the last resistance term in the denominator is much less than either of the other two terms. Additionally, the first resistance term vanishes under the assumption that the rock has no radial temperature gradient (it is a 1D problem). In order to be included in the volumetric heat generation term, q_r must be divided by the total volume of the porous medium. Thus, as might be expected, the volumetric heat generation due to radial conduction (Q_r) is independent of the length of the cell (L).

$$Q_r = \frac{2\lambda_w (T - T_\infty)}{r_i^2 \ln\left(r_o / r_i \right)} \quad [4]$$

The boundary conditions used were all the same as in the previous model, but with one exception. The temperature was not held constant at the inlet. Instead, a heat flux boundary condition was applied. The heat flux was determined from the enthalpy and flow rate of the incoming air. To model Cell F11, the non-adiabatic model used the same parameters as the adiabatic model, with the addition of the thermal conductivities of the humidity cell wall and the insulation used by Klinker (2006). These are given in Table 2. The control cell was also simulated using the 1D non-adiabatic model. The differences in the thermal properties of the control cell as compared with Cell F11 were insignificant. Therefore, when adjusting the model to represent the control cell, the only parameters that needed to be changed were the porosity and the initial saturation, which were 45% and 16%, respectively (Eldredge, 2005).

Table 2 – Parameters used in the 1D nonadiabatic evaporative model for Cell F11 and the Control Cell

Parameter	Symbol	Value	Units	Reference
Thermal conductivity of solids	k_s	2.5	$W\ m^{-1}\ K^{-1}$	Eldredge, 2005
Thermal conductivity of plexiglass	k_p	0.2	$W\ m^{-1}\ K^{-1}$	Incropera & DeWitt, 2002
Thermal conductivity of Insulation	k_i	0.08	$W\ m^{-1}\ K^{-1}$	Incropera & DeWitt, 2002

Figures 5 and 6 show the results of the one dimensional non-adiabatic model. The experimental data from Klinker (2006) have also been shown for comparison. The following two figures display the results of the 1D nonadiabatic model for each humidity cell along with the respective experimental data from Week One. The experimental data exhibits small fluctuations in temperature which are seen on both sets of data (F11 and Control). This is likely due to changes in ambient temperature.

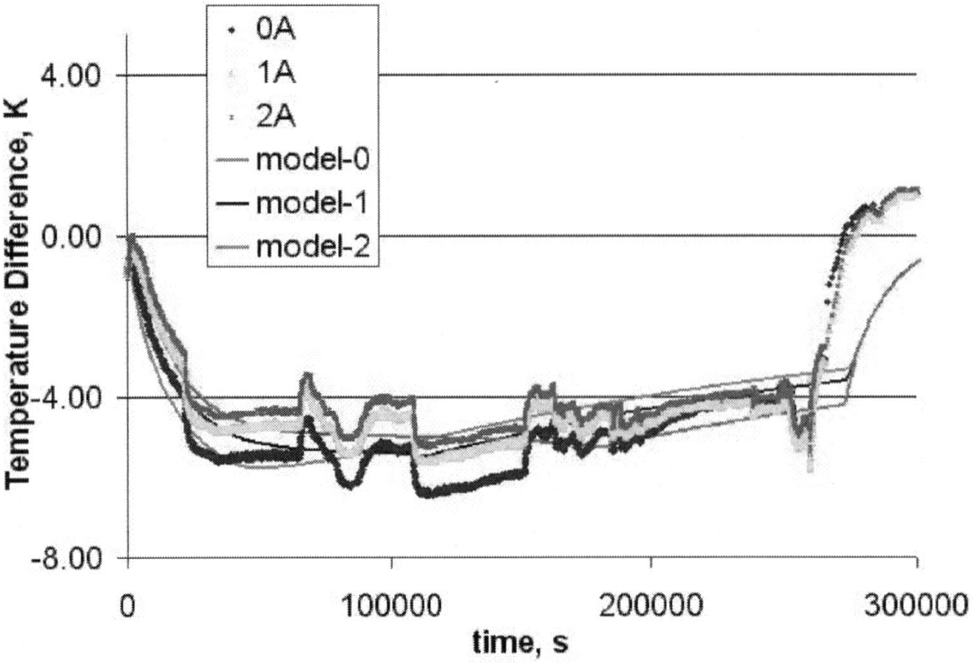

Figure 5 Experimentally determined and model predicted temperatures in humidity cell F11, Week One. 0A, 1A and 2A are the bottom, middle and top RTD probes respectively. Model points 0, 1 and 2 are the temperatures predicted by the 1D non-adiabatic model at the corresponding distances from the inlet of the humidity cell. Initial saturation was set at 44%.

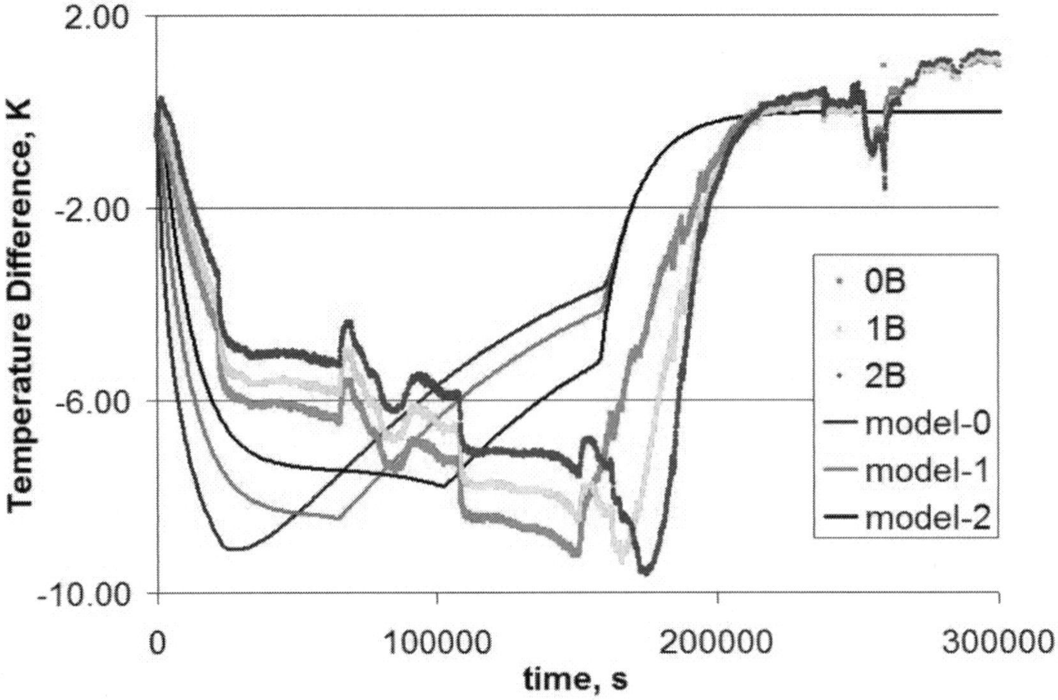

Figure 6 Experimentally measured and model predicted temperatures in the control cell, Week One. 0A, 1A and 2A are the bottom, middle and top RTD probes respectively. Model points 0, 1 and 2 are the temperatures predicted by the 1D non-adiabatic model at the corresponding distances from the inlet of the humidity cell. Initial saturation was set at 16% in the model.

It has been observed that the humidity cell does not always retain the same amount of water each week (Eldredge, 2005). There can also be settling from week to week. Because of this, it may be necessary to adjust the initial saturation and even sometimes the porosity, in order to match the experimental data. After making minor adjustments to the initial saturation, the non-adiabatic model was able to predict the temperature in the humidity cell with good agreement to the experimentally collected data. The best fit was obtained by raising the initial saturation of the Control Cell from 16% to 19%. The initial saturation of Cell F11 was lowered from 44% to 40%. It was not necessary to adjust the porosity for either of the humidity cells to obtain a good fit with the experimental data.

2D Heterogeneous Evaporative Model

A two-dimensional axis-symmetric model of the evaporation phase was also developed (see Figure 7). The mass balances used were the same as those already described for the 1D problem. The energy balance is essentially the same as well. The temperature of the wall of the humidity cell was held constant. A zero flux boundary condition was applied to both mass balance equations along the cell wall. All the parameters used in this model were the same as those of the 1D non-adiabatic model.

In the heterogeneous model an inclined layer was introduced to the domain in order to examine the effect of an irregular flow field (Figure 7). It should be noted that this layer, in cylindrical coordinates, is actually shaped like a hollow cone. Two heterogeneous scenarios were run. In the first one, the inclined layer was given a permeability four orders of magnitude higher than the surrounding rock. In the second, the permeabilities were switched such that the inclined layer had the lower permeability. For transient problems the following equation is solved.

$$[\beta(1-\phi)+\alpha\phi]\frac{\partial P}{\partial t}+\nabla\cdot\frac{k}{\mu}\nabla(P+\rho gz)=Q_s \quad . \qquad [5]$$

The solid was assumed to be incompressible (β = 0)and air was given the compressibility of an ideal gas (α = 1/P. The evaporation rate was used for the source term (Q_s). The parameters for the 2D heterogeneous evaporative model are the same as those for the 1D nonadiabatic model; also, this model requires additional parameters for Darcy's law which can be found in Table 3.

The boundary conditions used for this model were the same as those used for the homogeneous model. Additionally a constant velocity boundary condition was used at the inlet and a constant pressure boundary condition was used at the outlet for the momentum balance. A zero flux boundary condition was used on the cell wall.

Table 3 – Darcy's Law parameters used in the 2D heterogeneous models

Parameter	Symbol	Value	Units	Reference
High permeability	k	$1\text{x}10^{-5}$	m^2	Schwartz & Zhang, 2003
Low permeability	k	$1\text{x}10^{-9}$	m^2	Schwartz & Zhang, 2003
Viscosity of Air	μ_a	$1.8\text{x}10^{-5}$	Pa s	Incropera & DeWitt, 2002

With this geometry, it was observed at the four corners of the inclined layer that the energy balance predicted local temperature irregularities which did not appear to be physically realistic. This was likely due to a numerical effect of a single mesh element into which flow converges or diverges. No matter how fine the mesh, there would always be a single element in the vertex. This was remedied by "rounding the corners." After smoothing the mesh in this manner the hot and cold spots disappeared.

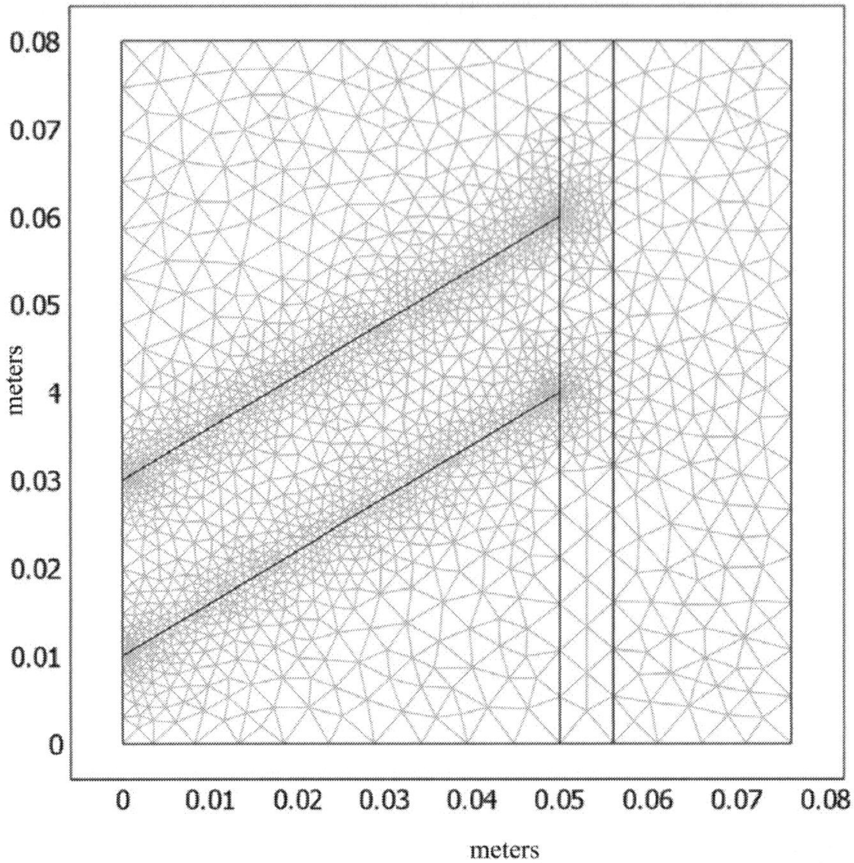

Figure 7 The finite element mesh used for the 2D heterogeneous model. Note the high density of mesh elements at the transition between zones of different permeability. The scale is in meters.

High permeability layer

Figure 8 shows a higher radial temperature gradient within the humidity cell than was exhibited by the homogeneous model (not shown). There is a difference in temperature of nearly two Kelvin from the center to the inner edge of the humidity cell. This is roughly double that of the homogeneous 2D evaporative model. This is due to the fact that the flow field is no longer uniform in the radial direction (see Figures 8 and 9). Figure 9 shows how the irregular flow field gives rise to an irregular evaporation front, which explains the higher radial temperature gradient.

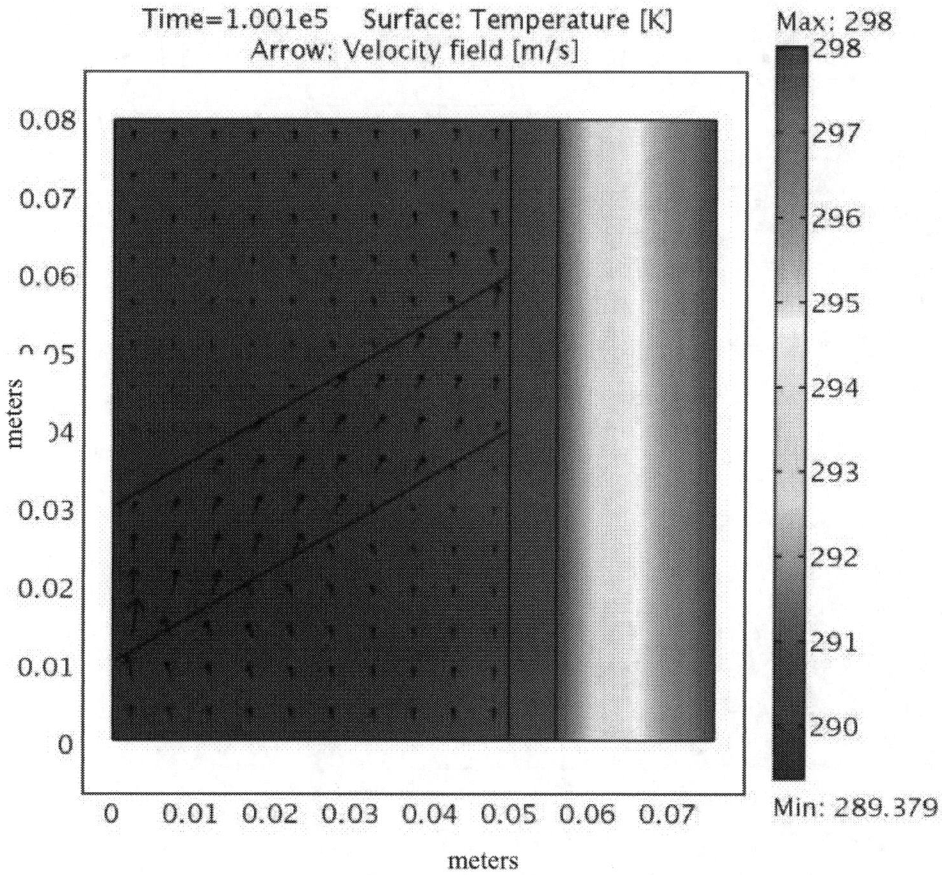

Figure 8 Temperature (color scale) and flow field (arrows) in the 2D humidity cell model with a layer of high permeability material after $1x10^5$ s. Note that the air flow through the high permeability layer is nearly parallel to its orientation. Also note the modest radial temperature gradient in the medium. The x and y axes are given in meters.

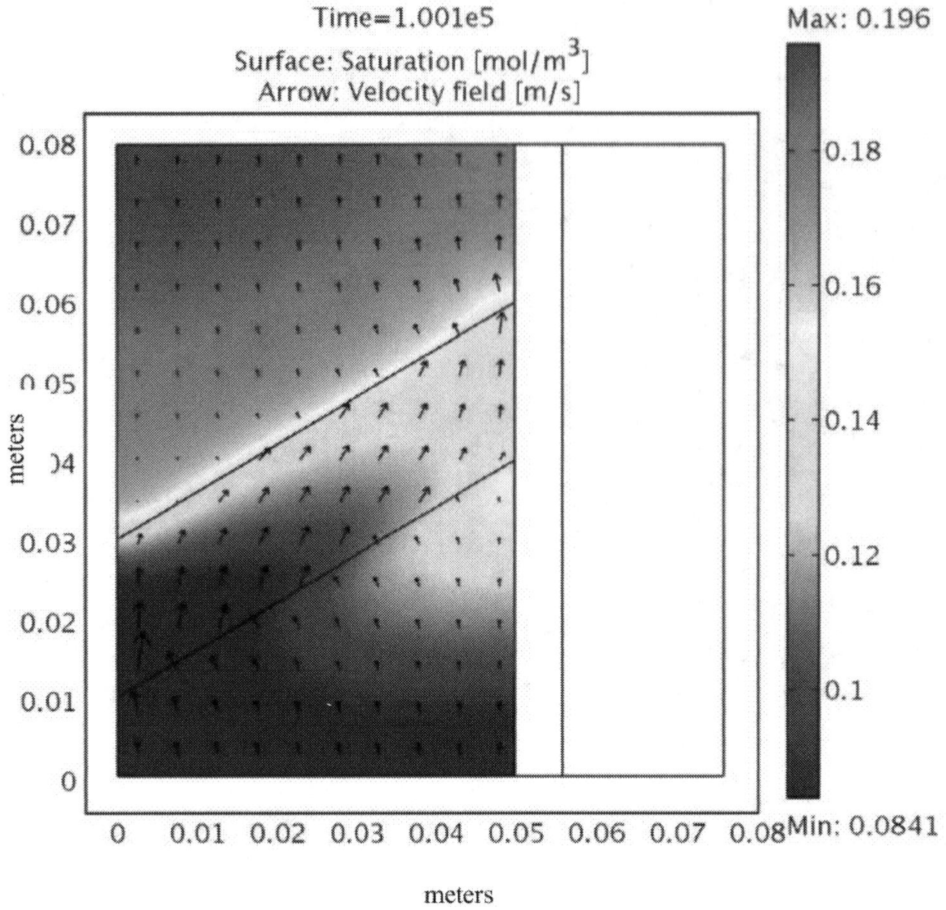

Figure 9 Saturation (color scale) and flow field (arrows) in the humidity cell with a high permeability layer after $1x10^5$ s. Note the irregular evaporation front. The x and y axes are given in meters.

Low permeability layer

The flow fields predicted by the model in the heterogeneous cases are what would be expected. The air moves through the low permeability layer in a direction perpendicular to the interface, while the air moves in a nearly parallel direction through the high permeability layer (Compare Figure 9 to 11). The radial temperature gradient within the humidity cell with the low permeability layer is roughly the same as that of the previous case. Figure 11 shows that the evaporation front is also irregular. However, this irregularity is not as pronounced as it was in the previous case (with the high permeability layer). The transient temperature profile in the low permeability layer 2D heterogeneous evaporative model appears to be roughly the same as with the high permeability layer.

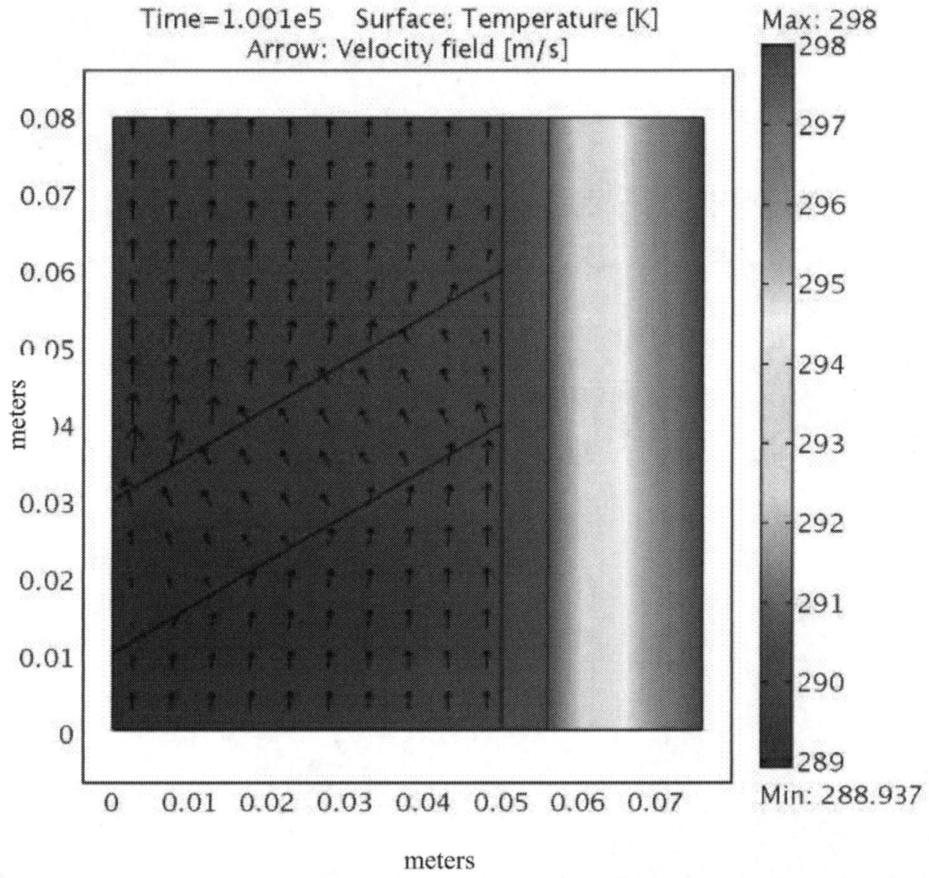

Figure 10 Temperature (color scale) and flow field (arrows) in the humidity cell with a layer of low permeability material after 1x10^5 s. Note that the air flow in the low permeability layer is roughly perpendicular to its orientation. Also note the radial temperature gradient. The x and y axes are both given in meters.

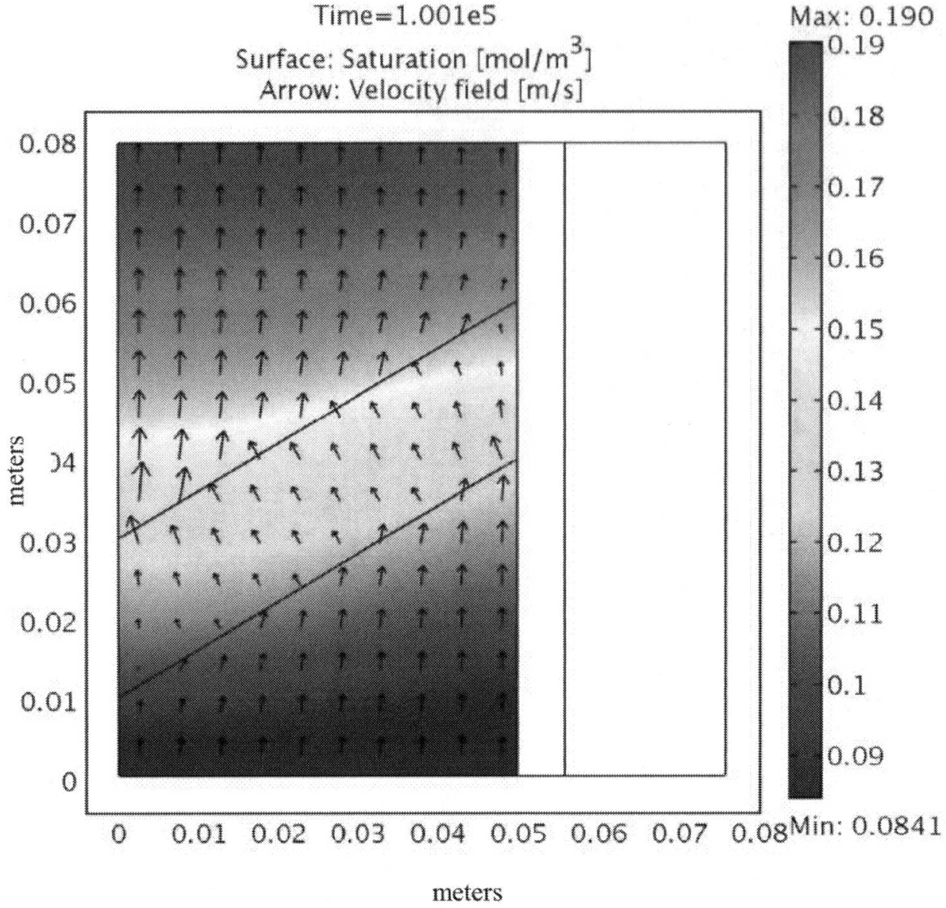

Figure 11 Saturation (color scale) and flow field (arrows) in the humidity cell with a low permeability layer after 1x10^5 s. Note the irregular evaporation front. The x and y axes are given in meters.

CONCLUSIONS AND FUTURE WORK

The one dimensional isothermal model shows excellent agreement with the analytical solution. The one-dimensional non-adiabatic evaporative model is capable of matching the temperature changes observed experimentally in the humidity cell. The model is relatively insensitive to the thermal conductivity of the composite rock material. It has also been found that the evaporation rate is insensitive to the mass transfer coefficient and the specific surface area above a threshold surface area. This indicates that, because the pore spaces are so small, the water vapor is able to come to equilibrium almost instantly. Therefore, for the purposes of modeling evaporation in crushed rock, it is not important to determine the mass transfer coefficient or the surface area of the porous medium with a high degree of accuracy. Heterogeneities in the porous media, however, can distort the flow patterns and can influence the amount and rate of evaporation.

ARD occurs over time scales of tens and hundreds of years. In continuation of this work, we plan to develop a model that will simulate all three phases of the humidity cell and include several geochemical and biological reactions that are involved in ARD over these time periods. We intend to compare our weekly simulations with not only the drying phase of the experimental procedure, but the leaching phase as well, so that we can compare the predicted leachate water quality with experimental data for a number of ions in hopes of generating kinetic rate constants. Thus, we hope to accurately capture the important aspects of humidity cell weathering that will help explain the weathering experienced in the field on a much larger scale and help predict the effects on the environment and rock pile stability.

Acknowledgement:

We would like to thank Chevron Mining, Inc. and the US Bureau of Land Management (Salt Lake City office) for their financial support. We would also like to thank Wes Eldredge and Andrew Klinker for assistance with the modeling and conducting the experiments.

REFERENCES

ASTM, "D 5744-96, Standard Test Method for Accelerated Weathering of Solid Materials Using a Modified Humidity Cell," in "Annual Book of ASTM Standards, 11.04". American Society for Testing and Materials, West Conschohocken, PA. (2000).

Bird, R.B.; Stewart, W.E.; Lightfoot, E.N., "Transport Phenomena," 2nd ed. John Wiley and Sons, New York (2002).

Comsol Group, http://comsol.com/, (Apr. 2007).

Eldredge, W.M., "Modeling Evaporation in Laboratory Rock Samples," M.S. Thesis, University of Utah: Chemical Engineering Dept., Salt Lake City, UT (2005).

Evans, P.S., "Modeling the Effects of Evaporation on Acid Rock Drainage in Mine Rock Samples," M.S. Thesis, University of Utah: Chemical Engineering Dept., Salt Lake City, UT (2007).

Incropera, F.P.; DeWitt, D.P., "Fundamentals of Heat and Mass Transfer," John Wiley and Sons, New York (2002).

Klinker, A.D., "Flow Effects on Temperature Controlled Humidity Cells," B.S. Thesis, University of Utah: Chemical Engineering Dept., Salt Lake City, UT (2006).

Pace Scientific, Staff Homepage, Temperature Probes, http://www.pace-sci.com/temperature-xr5.htm (July 8, 2005).

Stacey, D.A., and Udell, K.S., "Propagation of the Thermal and Evaporation Waves Through a Water-Wet Porous Media During Through-Flow Air Drying," Proceedings, Heat Transfer and Fluids Engineering, ASME (1997).

Schwartz, F.W.; Zhang, H., "Fundamentals of Groundwater," John Wiley and Sons, New York (2003).

Silva, M.A.; Kerkhof, P.J.A.M.; Coumans, W.J., "Estimation of Effective Diffusivity in Drying of Heterogeneous Porous Media," *Ind. Eng. Chem. Res.*, **39** (5) (2000).

Viollaz, P.; Suarez, C., "Drying of Shrinking Bodies," *AIChE Journal,* **31** (9) (1985).

Wilson, G.W.; Fredlund, D.G.; Barbour, S.L., "The Effect of Soil Suction on Evaporative Fluxes from Soil Surfaces." *Can. Geotech. J.*, **34** pp. 145-155 (1997).

86b Performance Targets for Batch Wastewater Treatment Operations

Benjamin J. Davis, UCLA Chemical Engineering Department, 5531 Boelter Hall, Los Angeles, CA 90095 and **Vasilios I. Manousiouthakis**, Chemical Engineering Department, UCLA,, 5531 Boelter Hall, Los Angeles, CA 90095-1592.

Evaluation of limits on the performance of reactors and reactor networks is crucial to the economic success of any chemical process. Consequently, the analysis and design of reactors and reactor networks has been one of the primary foci of process systems engineering research in the chemical and environmental engineering fields. The advancement of fundamental research in reactor network performance is an important exercise not only for the improvement of science education but also in the continued economic development of environmentally-friendly processes.

One method of identifying these performance limits for networks of batch reactors is through identification of the batch attainable region (BAR). Identifying BAR will provide us with the opportunity to quantify the fundamental limitations that may be imposed on any batch reactor network design by the underlying reaction kinetics, the provided initial conditions or the batch mode of operation itself. To consider the impact on a variety of objectives we will classify these objectives in two categories. First those whose value is uniquely correlated to each point of the BAR, and second those that do not have such a unique value at each point of the BAR. Characterization of the objectives which belong to one or the other category will be the subject of our research. It should be emphasized that the causal nature of the batch reactor network may play a crucial role in determining the category in which a particular objective belongs.

Synthesis of the batch reactor network that realizes some desirable characteristic will also be the subject of our research. Upon completion of the BAR construction, limits on the performance of all batch reactor network designs are established. Creating networks that can realize these performance limits will be the subject of our research. The forward algorithm may naturally construct networks that lead to particular locations in BAR. If however the desired characteristic is not simply a function of the BAR location, then linear programming approaches may be employed to identify a network with desirable characteristics. Even in this case however, identification of the BAR reduces the complexity of the underlying optimization by eliminating candidate network components whose "inlets" and / or "outlets" do not belong to BAR.

Moreno-Andrade et al. describe in their 2006 work a model for the batch treatment of wastewater containing 175 to 625 milligrams per liter of 4-chlorophenol. They used the dissolved oxygen concentration in the reaction vessel as an indicator of the rate of pollution removal by the biomass and used that control variable to optimize the reactor with respect to reaction time. We propose to incorporate this model in to the proposed batch attainable region formulation and to give performance limits on the operation of an optimal network of these batch treatment reactors.

Sensitivity Analysis of Adsorption Isotherms Subject to Measurement Noise in the Data

Karim K. Farhat, Mohamed N. Nounou, Ahmed Abdel-Wahab

Chemical Engineering Program, Texas A&M University at Qatar

Abstract

The analysis and design of many environmental and separation processes rely on the availability of accurate adsorption isotherms. These isotherms are estimated from measurements of the adsorption process variables. Unfortunately, these variables are usually contaminated with errors that affect the accuracy of their estimated parameters. Therefore, one objective of this work was to study the effect of measurement noise in the variables on the estimation accuracy of the Langmuir isotherm. In fact, Langmuir has three linearized forms, so it was sought to determine which out of these three forms would provide the most accurate estimated parameters. Another objective of this work was to estimate the isotherm parameters using the nonlinear Langmuir form using nonlinear optimization, and to compare the accuracy of the estimated parameters to the ones obtained using the most accurate linearized form. A third objective was to study the effect of measurement noise level on the accuracy of the Langmuir isotherm. As a result of this study, the following was found. One of the three linearized Langmuir forms provided the most accurate estimates. In fact, its accuracy was even comparable to that obtained by nonlinear optimization using the nonlinear isotherm. In addition, the estimation accuracy was more sensitive to the magnitude of the affinity constant than to the maximum amount of adsorbate in adsorbent; larger values of affinity constant result in higher estimation accuracy of both model parameters. Finally, it was confirmed that the higher the noise content in the variables, the larger the uncertainty of their estimation.

Keywords: Langmuir, Isotherms, Estimation, Sensitivity Analysis, Measurement Noise

1. Introduction

In the past few decades, a great deal of attention has been given to water pollution problems. For example, the removal of toxic heavy metals, such as Zinc, Nickel, and Lead from groundwater and wastewater has been approached by various studies, taking into consideration the serious implications that these pollutants can have on the health of humans and other living organisms. Consequently, many water purification methods have been developed and used to remediate consumable and waste water from those pollutants. These methods include chemical precipitation (Hence, 1998), reverse osmosis (Ning, 2002), electro dialysis, ion exchange, and finally adsorption, which is the focus of this proposal.

Adsorption is a mass transfer process, which involves the contact of solid called adsorbent with a fluid containing certain pollutants called adsorbate (Alkan and Dogan, 2001). These pollutants can be organic compounds, pathogens, and heavy metals, and their contact with the surface of the adsorbent results in permanent bonds, ensuring their removal from the fluid. The adsorption capacity depends on several factors, such as the adsorbent type, its surface area, and its internal porous structure. Additionally, since the attachment of the pollutant can be physical or chemical, the physical and chemical structures as well as the electrical charge of the adsorbent can significantly influence its interactions with the adsorbates, and thus the effectiveness of pollutant removal.

Adsorption processes are characterized by their kinetic and equilibrium isotherms. The adsorption isotherms specify the equilibrium surface concentration of the adsorbate as a function of its bulk concentration. Several mathematical models have been proposed to describe the equilibrium isotherms of adsorption. Some of the most popular models include Langmuir, Freundlich, Redlich-Peterson, and Sips. A summary of these isotherms is provided by Dabaybeh (2001). Even though most of these adsorption isotherms were derived based on some theoretical assumptions about the adsorption mechanism, they involve model parameters that need to be estimated from experimental measurements of the process variables. As an example, the Langmuir isotherm has the following form,

$$q_e = \frac{Q_c b C_e}{1 + b C_e} \qquad (1)$$

where, C_e is the equilibrium liquid phase concentration (mg/l), q_e is the equilibrium solid phase concentration (mg/g), Q_c is the maximum amount of adsorbate per unit weight of the adsorbent to form a complete monolayer, and b is a constant related to the affinity between the adsorbent and adsorbate. In the above Langmuir model, Q_c and b are model parameters to be estimated using measurements of the initial and equilibrium concentrations.

Unfortunately, measurements of the adsorption process variables are usually contaminated with noise or measurement errors due to random errors, human errors, or malfunctioning sensors. The presence of such measurement noise, especially in large amounts, can largely degrade the accuracy of the estimated isotherm parameters, which in turn limits the ability of the isotherm to accurately predict the process adsorption capacity. This is because most modeling techniques estimate the model parameters by minimizing some objective functions related to the prediction errors of the model output (q_e). Unfortunately, since the isotherm input and output variables are affected by measurement errors, only minimizing the output prediction errors may not lead to acceptable estimation.

Therefore, the objectives of this project are as follows:

i. Perform a sensitivity analysis to investigate the effect of measurement noise on the estimation accuracy of the Langmuir isotherms for all its linearized forms.

ii. Recommend the best linearized form for Langmuir parameter estimation.

iii. Assess the effect of measurement noise on the parameters estimated from the nonlinear Langmuir model and compare that to the effect on the parameters estimated using the most accurate linearized model.

iv. Assess the effects of different levels of noise on the accuracy of estimated isotherm parameters.

The rest of this paper is organized as follows. In Section 2, a summary of some related recent work is presented, followed by a description of isotherm estimation in Section 3. Then, the effects of measurement noise on the estimation accuracy of the Langmuir linearized forms and the nonlinear form are presented in Section 4, and the results of a sensitivity analysis for the estimation accuracy of model parameters are summarized in section 5. In section 6, the effect of noise content on the estimation accuracy of estimated isotherms is demonstrated for all Langmuir forms. Finally, some concluding remarks are presented in Section 7.

2. Literature Review

Several researchers have addressed the problem of assessing the accuracy of adsorption isotherms. For example, the concept of comparing adsorption isotherms and parameters and evaluating their accuracy by using different models has been discussed by Otun et. al. (2006) for the removal of selected metal ions by powdered egg shell, by Pikaar et. al. (2006) for the sorption of organic compounds to activated carbons, by Ridhika et. al. (2006) for the adsorptive removal of chlorophenols from aqueous solution by low cost adsorbent, and by Karahan et. al (2006) for the removal of boron from aqueous solution by clays and modified clays, and many others. However, these studies focused on determining the best fitting model only for the material, substance or process under study.

Other studies focused on determining the most suitable model to explain the adsorption process depending on the type of instruments used to get the experimental data. For example, Gormi et. Al. (2006) studied the estimation of adsorption-desorption models in the presence of noise in the gas sensors using Langmuir and Wolkenstein. Other studies focused on assessing the effect of some variables on the accuracy of the adsorption parameters. For example, the choice of column hold-up volume, range and density of the data point was found to have an impact on systematic errors in the measurement of adsorption isotherms by frontal analysis (Gritti et. al., 2005). This study showed that the concentration range within which the adsorption data are measured and the way the data points are distributed are important factors in error estimation. In another study of Gritti et. al. (2004), they showed that the fluctuations of the column temperature and the composition and the flow rate of the mobile phase affect the accuracy and precision of the adsorption isotherm parameters measured by dynamic HPLC methods.

Other studies performed statistical analysis on adsorption isotherms to determine the most accurate isotherm, i.e., isotherm model selection. For example, Joshi et. al. (2006) performed model based statistical analysis of adsorption equilibrium data. After comparing the parameter estimation by different linearized and non-linear adsorption models, it was shown that the Langmuir isotherm does not give a satisfying description of the considered experimental data, and that Freundlich isotherm provides the most accurate estimation for the liquid phase concentration range used in the experiment. However, the effect of noise on the accuracy of adsorption parameter estimation has been ignored.

It can be seen that the above studies dealt with analysis or assessment of adsorption isotherms with respect to model selection, effect of certain variables or instruments. In this work, however, the objective is to sturdy the effect of measurement noise on the estimation accuracy of adsorption isotherms. In particular, the focus will be on the Langmuir isotherm to investigate the best method of estimation using a linearized form and the nonlinear form itself.

3. Estimation of Adsorption Isotherms

3.1 Isotherm Model Formulation

Adsorption isotherms are usually nonlinear models relating the adsorption uptake (q_e) to the equilibrium concentration (C_e), which can be expressed as follows:

$$q_e = f\left(C_e, \theta\right) \tag{2}$$

where, θ is a vector of the isotherm parameters to be estimated empirically from experimental measurements. Sometimes, these nonlinear isotherms can be linearized by mathematical manipulations, and can be expressed in a linear form as follows:

$$g_1\left(q_e, C_e\right) = a_1\left(\theta\right) \times g_1\left(q_e, C_e\right) + a_2\left(\theta\right) \tag{3}$$

where, g_1 and g_2 are some functions of the uptake and equilibrium concentration, and a_1 and a_2 are the linearized isotherm parameters, which are functions of the model parameter vector, θ.

Given measurements of the initial concentration and equilibrium concentration data, $\{C_o(1) \;\; C_o(2) \;\; . \;\; . \;\; C_o(n)\}$ and $\{C_e(1) \;\; C_e(2) \;\; . \;\; . \;\; C_e(n)\}$, which are assumed to be contaminated with additive zero-mean Gaussian noise, i.e., $C_o = \tilde{C}_o + \varepsilon_o$ and

$C_e = \tilde{C}_e + \varepsilon_e$ where $\varepsilon_o \sim N(0,\sigma^2)$ and $\varepsilon_e \sim N(0,\sigma^2)$, it is desired to estimate the isotherm parameter vector, θ, that satisfies the relationship:

$$q_e(k) = f(C_e(k),\theta),\ k \in [1,n]. \tag{4}$$

Note that the equilibrium uptake is not measured and in calculated using the initial and equilibrium concentration data follows:

$$q_e = \frac{(C_o - C_e)V}{w} \tag{5}$$

where, V is the volume of the solution and w is the mass of the adsorbent.

The linearized form of equation (4) can be written similar to equation (3) as follows,

$$g_1(q_e(k),C_e(k)) = a_1(\theta)\times g_1(q_e(k),C_e(k)) + a_2(\theta),\ k \in [1,n]. \tag{6}$$

Defining: $g_1(k) \equiv g_1(q_e(k),C_e(k))$, $g_2(k) \equiv g_2(q_e(k),C_e(k))$, $a_1 \equiv a_1(\theta)$, and $a_2 \equiv a_2(\theta)$, equation (6) simplifies to,

$$g_1(k) = a_1 \times g_1(k) + a_2 \tag{7}$$

which can be expressed in matrix form as,

$$\begin{bmatrix} g_1(1) \\ g_1(2) \\ . \\ . \\ g_1(n) \end{bmatrix} = \begin{bmatrix} g_2(1) & 1 \\ g_2(2) & 1 \\ . & 1 \\ . & 1 \\ g_2(n) & 1 \end{bmatrix} \begin{bmatrix} a_1 \\ a_2 \end{bmatrix}. \tag{8}$$

Defining: $Y = \begin{bmatrix} g_1(1) \\ g_1(2) \\ . \\ . \\ g_1(n) \end{bmatrix}$, $X = \begin{bmatrix} g_2(1) & 1 \\ g_2(2) & 1 \\ . & 1 \\ . & 1 \\ g_2(n) & 1 \end{bmatrix}$, and $a = \begin{bmatrix} a_1 \\ a_2 \end{bmatrix}$, equation (8) reduces to,

$$Y = Xa. \tag{9}$$

3.2 Least Squares Isotherm Estimation

The model parameter vector, θ, can be computed using the estimated linearized isotherm parameters, a_1 and a_2. The linearized model parameters can be estimated using least squares regression by solving the following minimization problem,

$$\{\hat{a}\} = \arg\min_a (Y - Xa)^T (Y - Xa) \tag{10}$$

The above minimization problem has the following closed form solution,

$$\hat{a} = (X^T X)^{-1} X^T Y \tag{11}$$

Once the parameters a_1 and a_2 are estimated, the isotherm parameter vector θ can be computed using the relations, $a_1 \equiv a_1(\theta)$ and $a_2 \equiv a_2(\theta)$.

3.3 Estimation of Langmuir Isotherm Parameters

Consider the Langmuir isotherm shown in equation (1), where the parameters, Q_c and b, are to be estimated using measurements of C_e. Note that in this case, $\theta = [Q_c \quad b]$.

The Langmuir isotherm can be linearized three different ways as shown below:

$$\frac{1}{q_e} = \left(\frac{1}{Q_c b}\right)\frac{1}{C_e} + \frac{1}{Q_c} \qquad \text{(Langmuir 1)} \qquad (12)$$

$$\frac{C_e}{q_e} = \left(\frac{1}{Q_c}\right)C_e + \frac{1}{Q_c b} \qquad \text{(Langmuir 2)} \qquad (13)$$

$$\frac{q_e}{C_e} = -b q_e + Q_c b \qquad \text{(Langmuir 3)} \qquad (14)$$

As an example, for the second linearized form (Langmuir 2), $g_1 = \frac{C_e}{q_e}$, $g_2 = C_e$, $a_1 = \left(\frac{1}{Q_c}\right)$, and $a_2 = \frac{1}{Q_c b}$. Therefore, once estimates of the linearized isotherm, $\hat{a}_1$ and $\hat{a}_2$, are obtained using equation (11), estimates of the parameters Q_c and b can be computed as follows,

$$Q_c = \left(\frac{1}{a_1}\right), \text{ and } b = \frac{1}{a_2 Q_c}. \qquad (15)$$

It can be seen from equation (10), that the least squares estimation method relies on minimizing the prediction error of the model output, Y, when estimating the isotherm parameters. This is because it assumes that the input matrix, X, is noise-free. However, there is measurement noise in all variables, and therefore, the matrix, X is also noisy, which violates the basic assumption of this approach. Therefore, the presence of measurement noise in the data can greatly affect the estimation accuracy of estimated isotherms.

4. Effect of Noise on the Estimation Accuracy of the Langmuir Models

4.1 Effect of Noise on the Estimation Accuracy of the Linearized Langmuir Models

Since there are three different linearized forms for the Langmuir isotherms, it is expected to get three different estimates of the isotherm parameters. This is because the three forms minimize different objectives functions in the estimation of the isotherm parameters. Therefore, in this section, the objective is to study the effect of measurement noise on the estimation accuracy of the three linearized Langmuir forms, which will lead to the selection of the best form for estimating the isotherm parameters.

To perform this study, the Langmuir isotherm is used to generate data assuming Q_c =150, and b =0.15. These data are assumed to be noise-free. Then, the initial and equilibrium concentration data are contaminated with zero mean Gaussian noise of variance 0.5. Then, the noisy data are used to estimate the isotherm parameters using the three linearized forms as described in Section 3. To provide a statically meaningful results, a Monte Carlo simulation of 1000 realizations is performed, and the distributions of the estimated parameters are constructed for the three linearized forms as shown in Figures 1 and 2 for the parameters, Q_c and b, respectively.

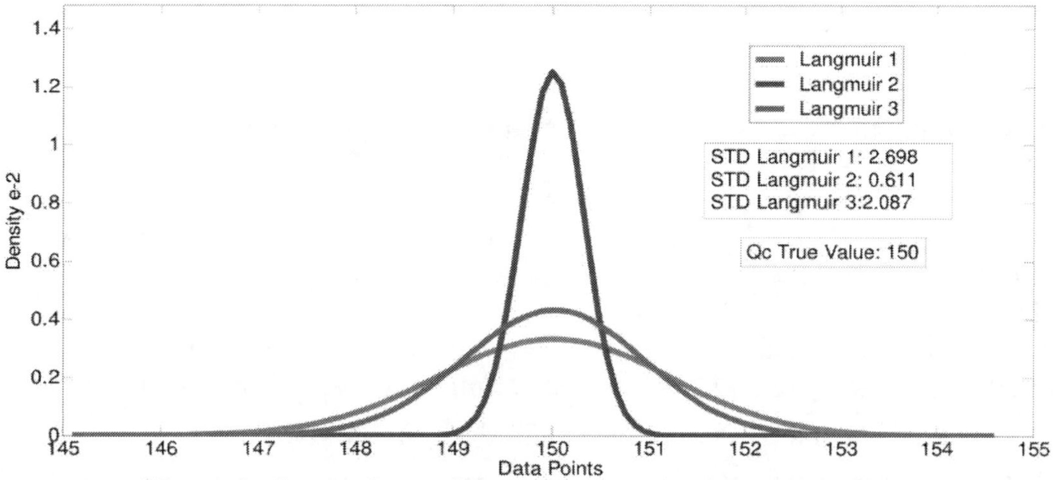

Figure 1. Comparison of the accuracy of estimated *Qc* using the three linearized Langmuir forms

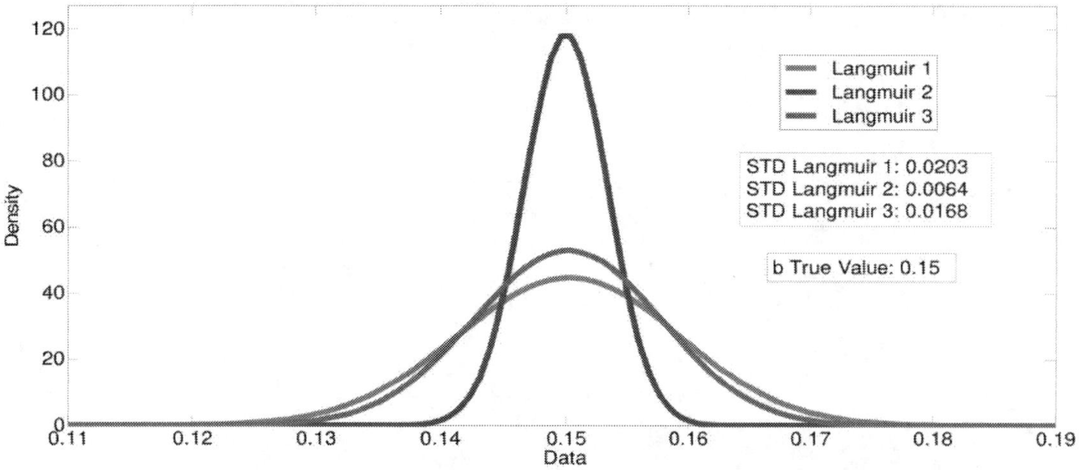

Figure 2. Comparison of the accuracy of estimated *b* using the three linearized Langmuir forms

Both Figures 1 and 2 show that the second linearized form of Langmuir provides the best estimated of the model parameters. This is also demonstrated by the standard deviations of the estimated parameters by the three forms as shown in Table 1.

Table 1. Standard deviations of the estimated parameters for the three linearized Langmuir forms

	Qc	*b*
Langmuir 1	2.698	0.0203
Langmuir 2	0.611	0.0064
Langmuir 3	2.087	0.0168

Therefore, the second linearized form of Langmuir, shown in equation 13, provides the most accurate estimates for the adsorption parameters.

4.2 Effect of Noise on the Estimation Accuracy of the Nonlinear Langmuir Model

In this section, the estimation accuracy of the best linearized form of the Langmuir isotherm (Langmuir 2) is compared to that based on the nonlinear model itself. The parameter estimates based on the nonlinear model are obtained using nonlinear optimization as follows:

i. A mesh of the possible Qc- b values is created as shown in Figure 3.
ii. At each node (i.e., $\{Qc,b\}$ pair), the values of the equilibrium uptake, q_e, is computed given the values of the equilibrium concentrations, C_e.
iii. The mean squares errors between the predicted q_e and the measured values are computed at each node.
iv. The isotherm parameter pair $\{Qc,b\}$ that provides the minimum mean squares error is selected as the optimum isotherm parameters.

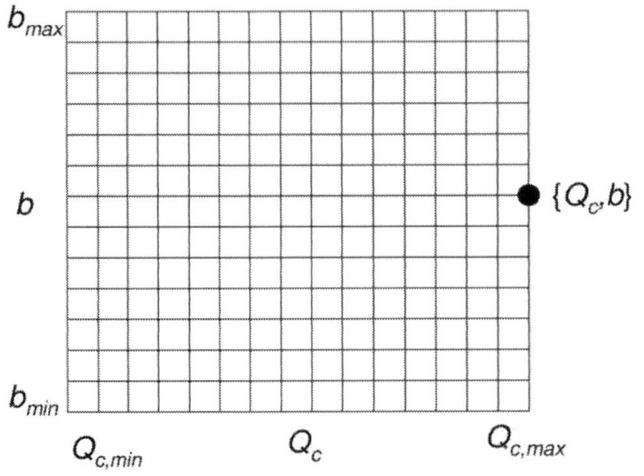

Figure 3. The mesh used to estimate the optimum $\{Qc,b\}$ pair based on the nonlinear Langmuir isotherm

To compare the accuracy of estimated isotherm parameters using Langmuir-2 and the nonlinear model, a Monte Carlo simulation is performed similar to the one described in Section 4.1, but for both the linearized and nonlinear forms and for different values of *Qc* and *b*, and the results are illustrated in Figures 4 and 5. Figures 4 and 5 show that the accuracies obtained using the second linearized isotherm form (Langmuir-2) and the nonlinear model, are comparable and differ for different magnitudes of *Qc* and *b*. For example, for small values of *b*, the nonlinear model provided slightly better estimates of the parameters, but for larger values of *b*, the linearized model provided better results.

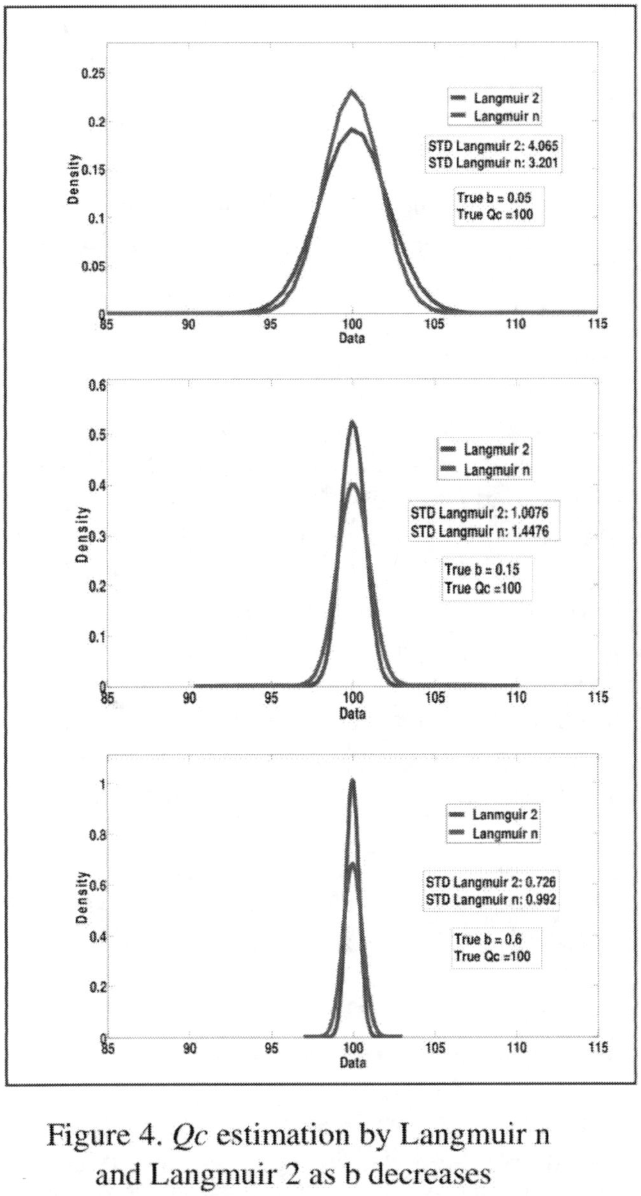

Figure 4. *Qc* estimation by Langmuir n and Langmuir 2 as b decreases

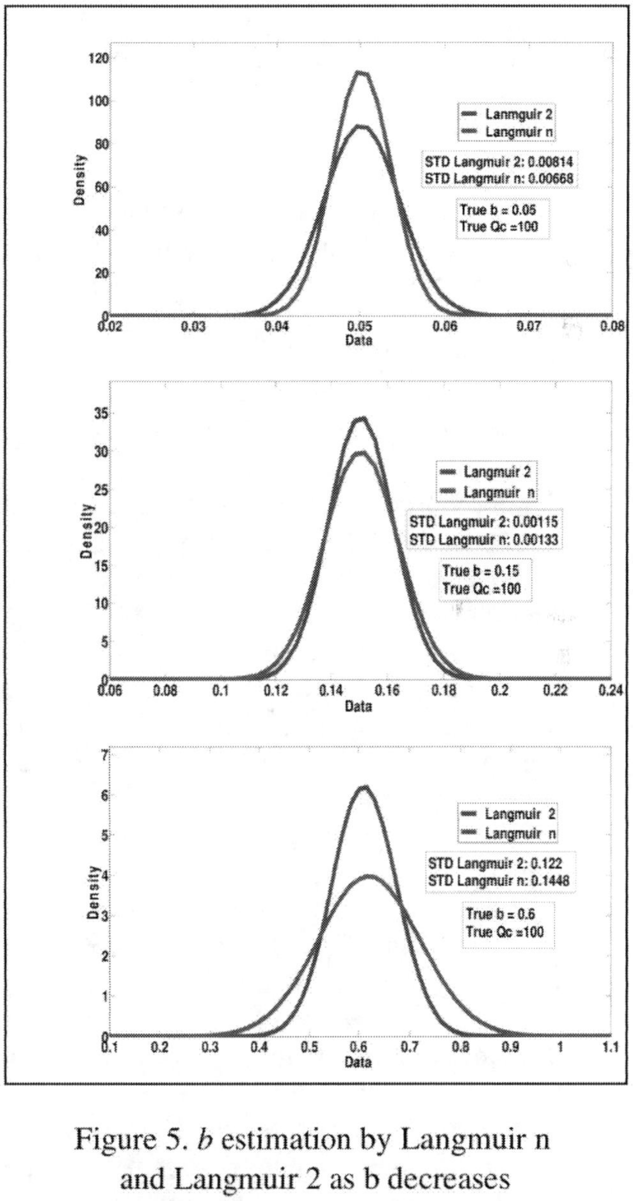

Figure 5. *b* estimation by Langmuir n and Langmuir 2 as b decreases

5. Sensitivity Analysis for the Estimation Accuracy of Model Parameters *Qc* and *b*

5.1 Estimation of *Qc* and *b* in Linearized Langmuir Models

The variation of *Qc* true value, at fixed *b* values, shows no significant effect on the distribution of *Qc* experimental data obtained from the three linearized models. On the other hand, increasing the true value of *b*, at *Qc* fixed values, shows that the standard deviation of data distribution for the three models decreases and the three models gives better and closer estimation. Also, the data fit according to Langmuir 3 and Langmuir 1 becomes increasingly more similar. In fact, the effect of *b* variation is most significant on Langmuir 1 estimation whose standard deviation changes significantly with the change in b value. Finally, increasing the true values of both parameters *Qc* and *b*, simultaneously, shows that the effect of the variation of *b* dominate the way the models' estimation for parameters changes; i.e. the results are very similar to those obtained in the case of *b* variation. A summary of the aforementioned results is demonstrated in Appendix A.

Equally true, the results of varying the parameters' magnitude on *b* estimation by the three linearized Langmuir models are the same as those on *Qc* estimation. A summary of those results is illustrated in Appendix B.

5.2 Estimation of *Qc* and *b* in Nonlinear Langmuir Model

Similar to the results of the linearized forms, the variation of *Qc* true value, at fixed *b* values, shows no significant effect is noticed on the distribution of *Qc* experimental data obtained from the two models. On the other hand, under very small *b* values, the nonlinear model Langmuir n gives better estimation for *Qc* parameter than the linear model Langmuir 2. As *b* increases, however, Langmuir 2 estimation improves and becomes more accurate than that of Langmuir n. Also, as *b* true value increases, it is noticed that the standard deviation of data distribution for the two models decreases and thus both models give better and closer estimation. Finally, increasing the true values of both parameters *Qc* and *b*, simultaneously, shows that the effect of the variation of *b* dominate the way the models' estimation for parameters changes; i.e. the results are very similar to those obtained in the case of *b* variation. A summary of the aforementioned results is demonstrated in Appendix C.

Again, the results of varying the parameters' magnitude on *b* estimation by the nonlinear Langmuir model are the same as those on *Qc* estimation. A summary of those results is illustrated in Appendix D.

6. Effect of Noise Content on the Estimation Accuracy of Estimated Isotherms

In this study, the effect of noise content in the measured variables on the estimation accuracy of the Langmuir isotherm parameters is investigated. To do that, the sensitivity analysis performed in Sections 4.1 and 4.2 are repeated for different noise levels (variances of 0.1, 0.5, and 0.9), and the results are illustrated in Figures 6 and 7. In these Figures, the line colors denote the following: red: Langmuir-1, blue: Langmur-2, green: Langmuir-3, grey: nonlinear model. As expected, these Figures show that for larger noise contents (higher noise variance), the uncertainty of the estimated parameters increases.

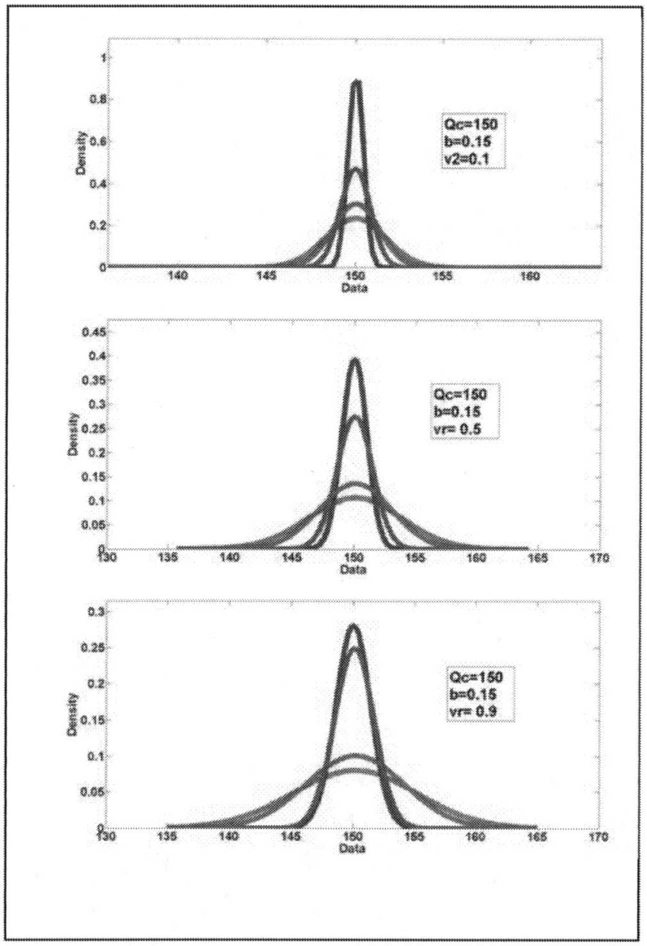

Figure 6: The effect of noise variation on the estimation of *Qc*

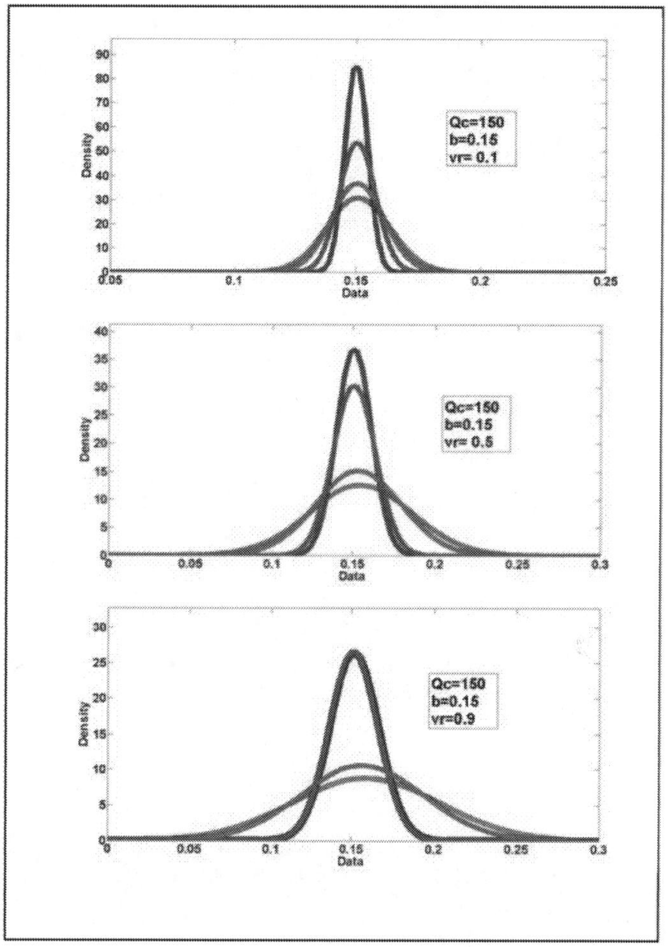

Figure 7: The effect of noise variation on the estimation of *b*

7. Conclusions

In this project, a statistical analysis of the Langmuir isotherm was preformed. It was found that out of three linearized isotherms forms of Langmuir, (Langmuir-2) provides the most accurate estimates of the isotherm parameters. In fact, Langmuir-2 provides comparable estimated to the ones obtained using the nonlinear isotherm form. This is an important finding because it is more computationally expensive to estimate the isotherm parameters by nonlinear optimization. It is also confirmed that the level of noise in the adsorption data degrades the estimation accuracy of estimated isotherm parameters.

Acknowledgement

The authors would like to gratefully acknowledge the support of the Qatar National Research Funds and Qatar Foundation to accomplish this work.

References

Alkan, M. and Dogan, M., J. Colloid Interface Sci, 243, 280 (2001).

Dabaybeh, M., M.S. Thesis, Jordan University of Science and Technology (2001).

Gormi, S., Seguin, J.L., Guerin, J., and Aguir, K., Sensors and Actuators B: Chemical, 119, 1, 351 (2006).

Gritti F. and Guiochon G., Journal of Chromatography A, 1097, 1-2, 98-115 (2005).

Gritti F. and Guiochon G., Journal of Chromatography A, 1043, 2, 159-170 (2004).

Hence, K.R., Water Environ. Res., 70, 1178 (1998).

Joshi, M., Kremling, A., and Seidel-Morgenstern, A., Chemical Engineering Science, 61, 23, 7805-7818 (2006).

Karahan, S., Yurdakoc, M., Seki, Y., Yurdakoc, K., Journal of Colloid and Interface Science, 293, 1, 36-42 (2006).

Ning, R.Y., Desalination, 143, 237 (2002).

Otun, J. A., Oke, I. A., Olarinoye, N. O., Adie, D. B., Okuofu, C. A, Journal **of** Applied Sciences, 6: 11, 2368-2376 (2006).

Pikaar, Ilje, Albert A. Koelmans and Paul C.M. van Noort, Chemosphere , 65, 11, 2343-2351 (2006).

Radhika, M., Palanivelu, K., Journal of Hazardous Materials, 138, 1, 116-124 (2006).

Appendix A

The variation of *Qc* estimation by the linearized Langmuir models as function of *Qc* and *b* true values

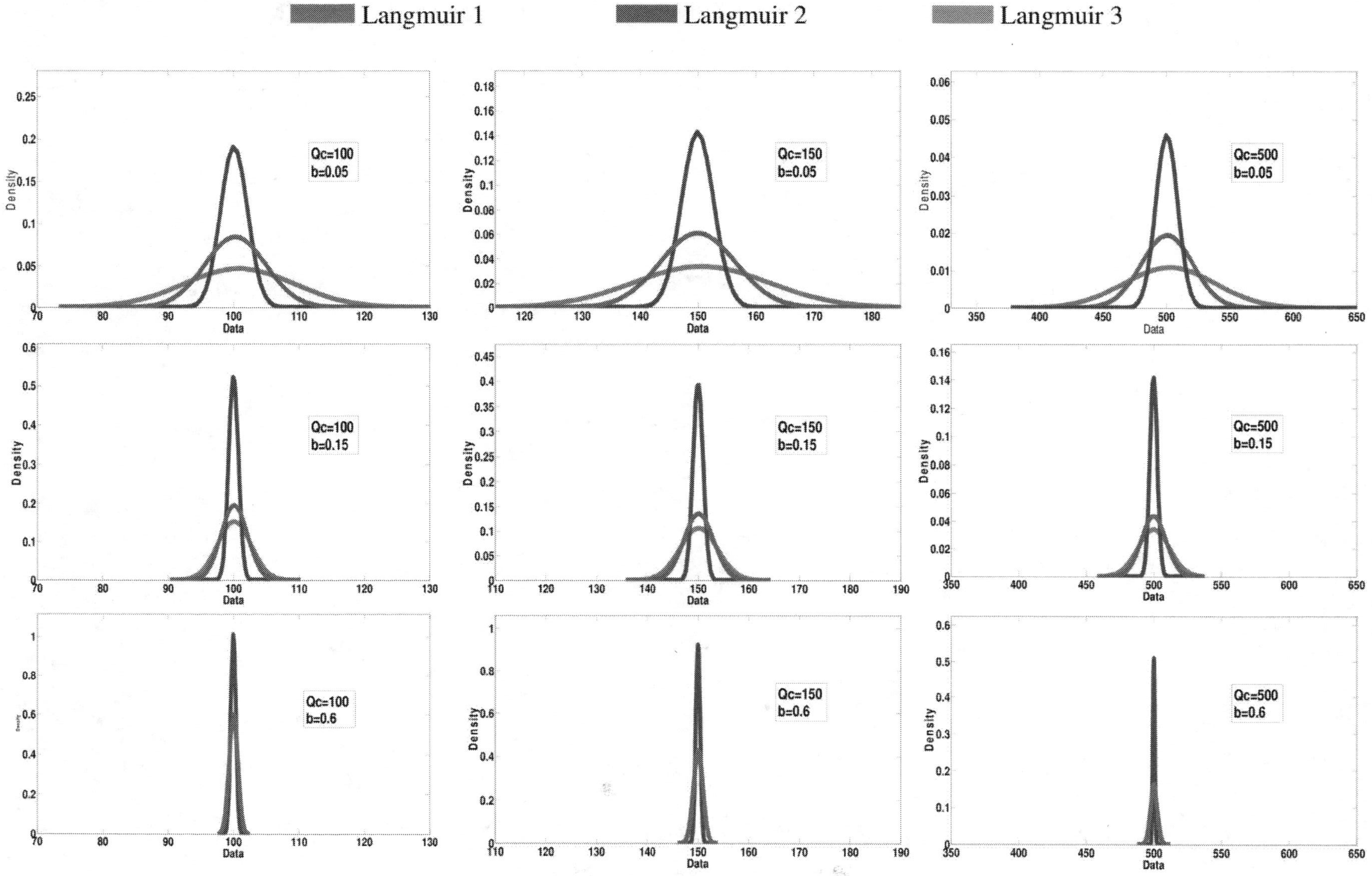

Appendix B

The variation of *b* estimation by the linearized Langmuir models as function of *Qc* and *b* true values

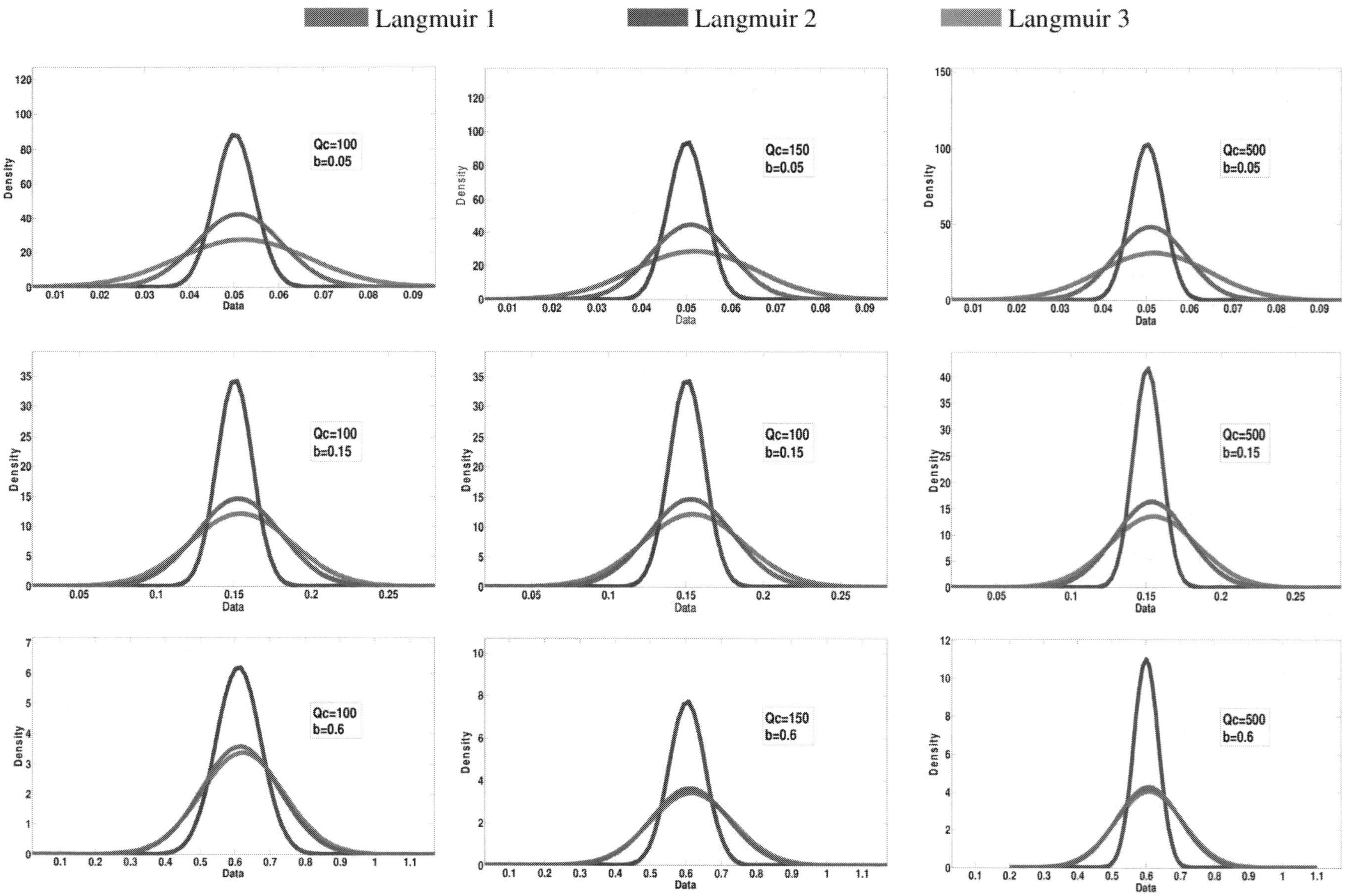

Appendix C

The variation of *Qc* estimation by Langmuir 2 and Langmuir n as function of *Qc* and *b* true values

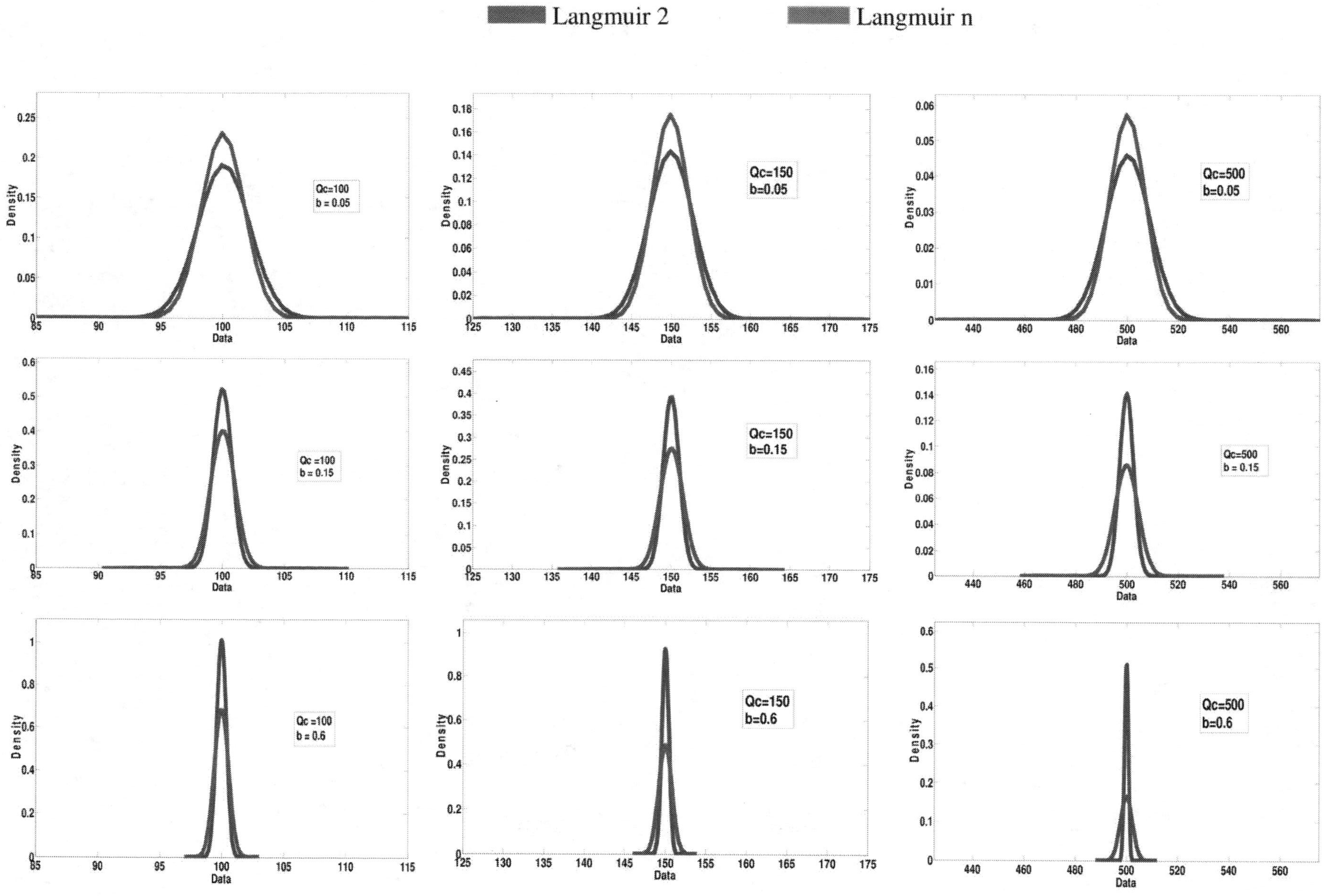

Appendix D

The variation of *b* estimation by Langmuir 2 and Langmuir n as function of *Qc* and *b* true values

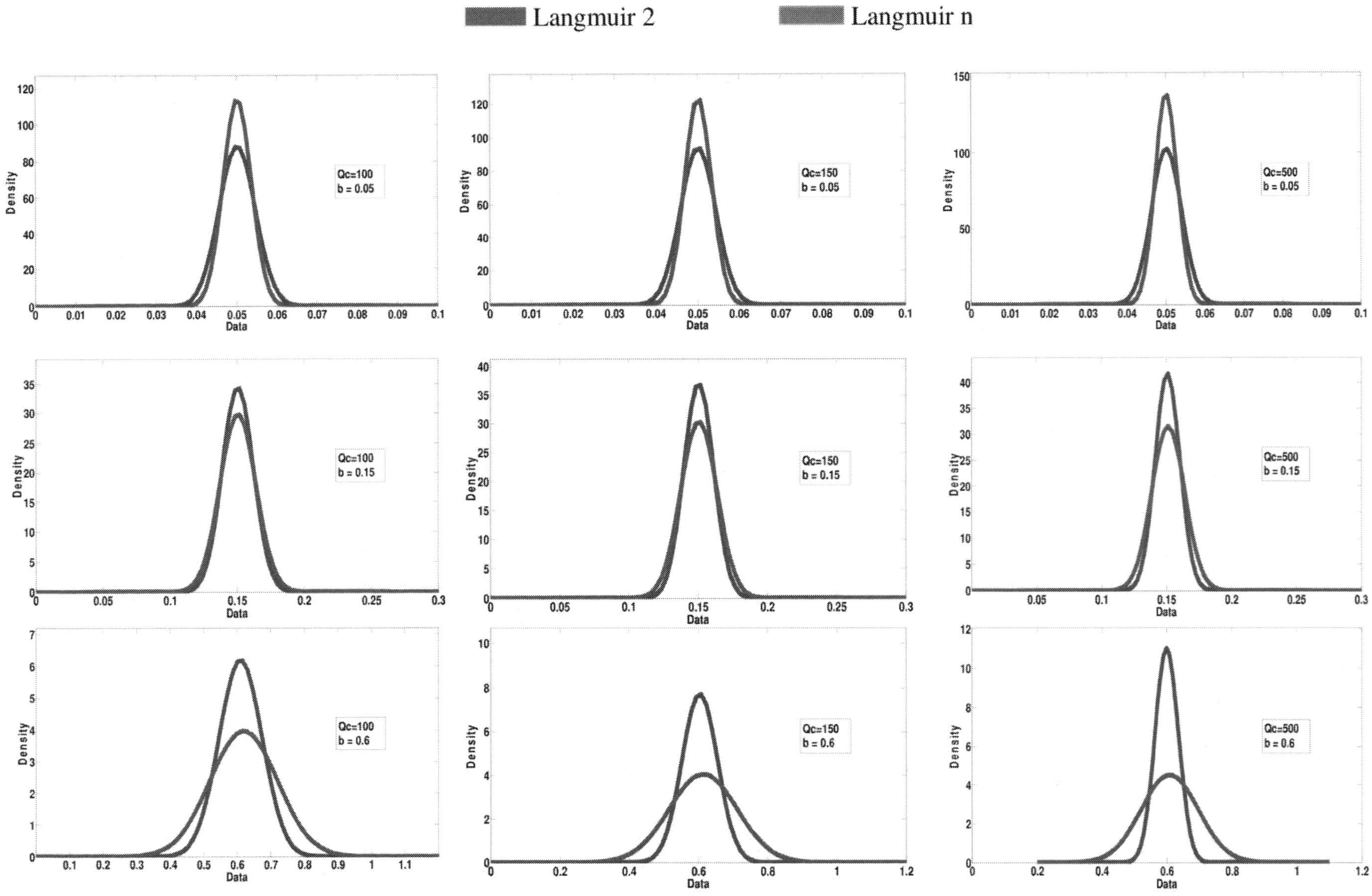

86d a Systematic Methodology for the Selection of Particles with Optimal Material Properties for Advanced Oxidation Process Applications

David M. Follansbee[1], Lealon L. Martin[1], John Paccione[2], and Joel L. Plawsky[1]. (1) Department of Chemical and Biological Engineering, Rensselaer Polytechnic Institute, 110 8th st, Troy, NY 12180, (2) Environmental Health Sciences, SUNY at Albany, School of Public Health, 547 River Street, Flanigan Square, Troy, NY 12180

The use of photocatalytic based materials for mineralization of organic compounds is a technique that has recently been receiving a lot of attention in the water treatment community for removal of organic contaminants from water. The photocatalytic process is such that when these catalyst absorb the appropriate amount of energy from photons (give by their respective band gap) an electron-hole pair forms and allows reduction and oxidation reactions to occur. This photocatalytic material has been employed in a wide variety of reacting systems from coated walls reactor to packed bed reactors, however, implementation of these catalysts in the various systems demands the functionality be such to accommodate the requirements of that given reactor. For example the catalyst particles chosen for a packed bed reactor need to be able to adsorb organic compounds as well as be photoactive. On the other hand the catalyst chosen for a slurry reactor needs to be highly photoactive and the adsorption properties are not a concern due to continuous mixing. The main properties of interest are physical properties such as adsorption, mechanical stability, light absorption, and photocatalytic activity. More often then not one photocatalytic material will not have all of the desired properties (whether it is adsorption, density, or simply particle size) and consequently these catalysts are composite material with a given catalyst support, adsorbent/zelolite, and photocatalyst. These catalyst particles need to be carefully screened for each different applications and many trials need to be performed to determine the optimal composite and the volume of each material rendering this process very costly and time consuming.

Recently the use of a novel water treatment process that utilizes draft tube transport and advanced oxidation has been posed to the community. This process is designed around the characteristics of a draft tube spouted moving bed reactor (DTSMB) in which the reactor consists of an annular moving packed bed (water purification region) and an internal draft tube for hydraulic transport of catalyst for regeneration. This system was studied for overall optimal design and operation of based on a utility cost objective function, however no investigation was placed on the design of the catalyst particle.

In this paper we would like to present a systematic approach used to study the effects of the bulk properties of the catalyst particles and in particular investigate the optimal selection of bulk properties to be used in the DTSMB system. We develop and present a theoretical model that captures the steady state design and operation of the DTSMB reacting system. We then define a problem statement and formulate it into a nonlinear mathematical program and outline a method to determine globally optimal catalyst properties. A case study is performed using water contaminated with reactive red (RR) to demonstrate the strength of the proposed methodology and discuss the selected catalyst properties. This catalyst selection process will give insight as to what photocatalyst, adsorbent, and support should be considered and the optimal loading of each when selecting catalysts.

86e Laboratory-Scale Fast Pyrolysis of Cca-Treated Wood Waste

Amy M. Parker, Dave C. Swalm School of Chemical Engineering, Mississippi State University, P.O. Box 9595, Mississippi State, MS 39762 and Mark Bricka, Chemical Engineering, Mississippi State University, P.O. Box 9595, Mississippi State, MS 39762.

The most widely used wood preservative since the early 1970's has been chromated copper arsenate (CCA), resulting in nearly 80% of all treated wood products in North America being treated with CCA. However, by late 2003 CCA-treated wood was restricted to industrial applications, leading to a considerable increase in the volume of CCA-treated wood slated for disposal. Landfilling was an acceptable means of discarding CCA-treated wood products until recently, which has lead to approximately 24 million tons of CCA waste. Problems associated with soil and groundwater contamination, caused by CCA-metals leaching from landfills, have generated the need for an effective and efficient disposal method for CCA-treated wood.

In this research, fast pyrolysis is evaluated as a means of recycling CCA-impregnated wood waste. A primary goal is to concentrate the CCA components in the bio-oil, for possible re-use in wood preservatives, while recovering the energy value of the wood. An easily modified lab-scale pyrolysis system, capable of operating in the desired temperature range under atmospheric and vacuum conditions, was designed. The system allows for the collection of each pyrolysis product so that complete mass balances on the metals can be performed, tracking the fate of the CCA metals. Initial testing was completed with the system in a batch reactor configuration. Experiments were performed in which the reactor was operated at temperatures between 400°C and 550°C using different carrier gas flow rates, to determine the optimal conditions needed to produce bio-oil rich in CCA-metals. Mass balances were performed by evaluating the concentrations of the metals in each pyrolysis product using ICP-AES analysis. This paper discusses the lab-scale pyrolysis system, characterization of the CCA-treated wood waste, results of the optimization experiments, and details of on-going research.

86f Pre- and Post-Combustion Carbon Capture Using Metal-Organic Framework Adsorbents

Craig M. Tenney, Tran D. Trinh, and **Christian M. Lastoskie**. Civil & Environmental Engineering, University of Michigan, 2350 Hayward, 2340 G.G. Brown Building, Ann Arbor, MI 48109-2125

The global warming impacts associated with anthropogenic CO2 emissions from fossil fuel combustion present an enduring problem to achieving energy sustainability. Although conservation measures and carbonless energy technologies may eventually stabilize CO2 emissions from stationary and mobile power sources, a continued reliance on coal, oil and natural gas in the near term necessitates development of strategies to reduce CO2 loadings into the atmosphere. Carbon capture and storage (CCS) is a promising concept in which CO2 is recovered from process gas streams, concentrated to a fraction suitable for pipeline transport, and sequestered in geologic formations or other storage media. Large, stationary point sources, such as fossil fuel electric power plants, are obvious targets for CCS. Improved adsorbents are needed however to reduce the high cost of separating CO2 from other process gases, particularly N2 in the case of (post-combustion) flue gas and H2 in the case of (pre-combustion) gasified coal or oil. Novel porous materials based on metal-organic framework (MOF) chemistries show promise for selective capture of CO2 from gas mixtures. In this paper, we report molecular simulation studies of mixed-gas CO2 adsorption on a copper-bipyridine MOF adsorbent that exhibits highly selective gated adsorption for CO2. Adsorption capacities and selectivities for CO2/H2 and CO2/N2 mixtures were studied via Monte Carlo simulations at 300, 350, and 400 K under realistic process pressures. Monte Carlo simulation results for isosteric heats of adsorption of CO2 and H2 are also presented.

86h Natural Gas Based Hydrogen Production with No Carbon Dioxide Emissions

Jorge Pena Lopez, Chemical Engineering Department, University of California at Los Angeles, 5531 Boelter Hall, Los Angeles, CA 90095 and Vasilios I. Manousiouthakis, Chemical Engineering Department, UCLA,, 5531 Boelter Hall, Los Angeles, CA 90095-1592.

The use of fossil fuels as a transportation energy source has driven large metropolitan areas around the world towards a deplorable air quality state. This urgent situation has motivated the development of technologies which aim to reduce the polluting emissions from automobiles. One viable option is the use of hydrogen as fuel, using hydrogen fuel cells to generate the energy in vehicles. Several options exist for hydrogen production; one of the most explored has been the reforming of natural gas given that methane contains potentially two hydrogen gas molecules. The main problem with conventional natural gas reformers is the emission of carbon dioxide, a greenhouse whose emissions have led to an increase in its concentration in the atmosphere. Transforming carbon dioxide into a useful product prevents its emission into the atmosphere, however an inadequate process design might largely increase the price of hydrogen with respect to gasoline, making it non-competitive.

In this work we present a hydrogen production process in which hydrogen and methanol are co-generated, while no carbon dioxide is emitted. Water and natural gas are fed simultaneously with oxygen in a partial oxidation reactor. The oxygen is separated from air in a pressure swing absorber process. The gases coming from the partial oxidation reactor are dried and mixed together and sent into a reverse shift-gas reactor to produce syngas that is then used in the production of methanol. In order to address and fulfill the energy requirements of the entire plant, heat and power integration techniques are applied into the process allowing us to maintain a self-powered process plant, reducing the carbon emissions to near zero and being economically feasible. We also show that there is an optimal operational and thermodynamical point for hydrogen production and power generation, and that on an operating cost basis, the proposed process is superior to that of the traditional steam reforming process.

Keywords: Hydrogen, Methanol, Partial oxidation, carbon dioxide

105a New Developments in Instrumentation for Characterizing Nano-Particles for Health Exposure Studies

Richard C. Flagan, California Institute of Technology, 1200 E. California Blvd. - MC210-41, Pasadena, CA 91125

Invited keynote paper - abstract pending

105b Spectroscopic and Microscopic Characterization of Ambient Carbonaceous Particulate

Randy L. Vander Wal[1], Victoria M. Bryg[1], and Michael D. Hays[2]. (1) M.S. 110-3, NCSER c/o NASA-Glenn Research Center, 21000 Brookpark Drive, Cleveland, OH 44135, (2) National Risk Management Research Laboratory, US EPA, MD E343-02, Research Triangle Park, NC 27711

Epidemiological studies have shown that inhalation of soot particles can cause pulmonary disease, cardiovascular damage and mortality. New findings show soot may be contributing to changes happening near the North Pole, such as accelerating melting of sea ice and snow and changing atmospheric temperatures. Soot surface chemistry may determine whether it can act as cloud nuclei, its participation in atmospheric heterogeneous reactions, thereby significantly affecting the atmospheric radiative balance. These properties are further modified by soot's nanostructure, along with its oxidation characteristics. With increasing knowledge of the detrimental health effects of soot and environmental impacts, identification of emission sources is of increasing importance

Electron microscopic and spectroscopic methods have great potential for carbonaceous aerosol characterization. High-resolution transmission electron microscopy (HRTEM) provides information of the carbon nanostructure (or relative lack thereof). Image quantification allows extraction of various statistical parameters including lamella length, separation distance and tortuosity. These may be summarized in the form of histograms. X-ray photoelectron spectroscopy (XPS) provides a means for probing the surface chemistry of materials. It provides not only information about the atomic composition of a sample but also information about the structure and oxidation state of the constituent elements.

Combustion produced soot is a highly variable material. Physically the nanostructure can range from amorphous to graphitic to fullerenic. Chemically nearly any element could be included, though the surface functional groups are predominantly oxygen-based. Results will be presented for analysis of HRTEM images of the physical nanostructure and XPS analysis of the chemical composition of soots collected from plant, institutional and residential scale oil-fired boilers, diesel engine, gas-turbine combustor (jet) and a wildfire. Physically soots from these emission source classes may be differentiated based on carbon lamella length, mean separation and tortuosity. Chemically these soots may also be distinguished by elemental composition, surface (oxygen) functional groups and carbon bonding.

Characterization of source (particulate) emission profiles is integral to formulation of source profiles towards development of receptor-based models and extraction of source contributions to monitored sites. Atmospheric and local environmental impacts can then be accurately assessed. Regulatory statutes may be specifically targeted. Human exposure and related health effects can be accurately evaluated.

105c Measurement of Particulate Matter within School Bus Passenger Compartments Under Realistic Operating Conditions: An Evaluation of Retrofit Devices

Robert P. Hesketh[1], David Martinez-Morett[1], Krishan K. Bhatia[2], and Anthony J. Marchese[3]. (1) Chemical Engineering, Rowan University, 201 Mullica Hill Rd, Glassboro, NJ 08028, (2) Mechanical Engineering, Rowan University, 201 Milluca Hill Road, Glassboro, NJ 08028, (3) Mechanical Engineering, Colorado State University, Fort Collins, CO 80523-1374

Recent studies have reported elevated levels of diesel particulate matter inside school bus passenger compartments. The elevated particulate levels have been attributed to the self-pollution from the school bus tailpipe and/or crankcase. Up to date there are still uncertainties as to the source of the particulate matter found in-cabin which can be attributed to the bus self-pollution as well as ambient pollution, neighboring vehicles, and re-entrained road dust. The issue has gained national attention because children are a particularly susceptible group to the adverse health effects from diesel particulate matter. The objectives of this study are to measure the concentrations of fine and ultrafine particles within the cabin of a school bus with and without retrofit technologies. To satisfy these objectives, mobile tests were conducted with a single school bus powered by an International DT466E engine on an outdoor test track at the Aberdeen Test Center in Aberdeen, MD. The study compared the ambient with the in-cabin pollutant levels measured during the operation of the bus following a realistic and repeatable test cycle developed using GPS data from New Jersey school bus routes. Particulate matter concentrations were measured using three Thermo Electron DataRAM-4 units, and three TSI P Trak ultrafine particle counters. Gaseous emissions (CO, CO2, HC, NOx) as well as pertinent engine parameters such as engine speed, fuel flow rate, engine oil temperature, and percent engine load were measured using the Sensors SEMTECH-D tailpipe emissions analyzer. The retrofit technologies used were a crankcase ventilation system for the crankcase emissions; a flow through filter, and a diesel particulate filter for tailpipe emissions control. Testing results indicate that operating the bus with the windows open results in only low levels of particulate matter (PM) during the cycle, but when operating with the windows closed a significant accumulation is observed. The combination of a closed crankcase filtration system with a tailpipe retrofit significantly improves the air quality within the cabin. This study suggests that current retrofit technologies can nearly eliminate diesel particulates from self-pollution inside school bus cabins.

105d Relationships Between Composition and Pulmonary Toxicity of Prototype Particles from Coal Combustion and Pyrolysis

William P. Linak[1], Seung-Hyun Cho[1], Jong-Ik Yoo[1], Audrey T. Turley[1], C. Andrew Miller[1], Jost O.L. Wendt[2], Frank E. Huggins[3], and M. Ian Gilmour[4]. (1) National Risk Management Research Laboratory, U.S. Environmental Protection Agency, 109 TW Alexander Dr, E305-01, Research Triangle Park, NC 27711, (2) Chemical Engineering, University of Utah, 3290 MEB, 50 S. Central Campus Dr., Salt Lake City, UT 84112, (3) University of Kentucky, 533 South Limestone Street, 103 S. J. Whalen Building, Lexington, KY 40506-0043, (4) National Health and Environmental Effects Research Laboratory, U.S. EPA, 109 T.W. Alexander Drive, Research Triangle Park, NC 27711

The hypothesis that health effects associated with coal combustion fly-ash particles are exacerbated by the simultaneous presence of iron and soot was tested through two sets of experiments. The first set created prototype particles from complete and partial combustion, or oxygen free pyrolysis of a high iron Illinois bituminous coal in an externally heated drop-tube furnace. The second experiment created prototype particles consisting of iron and soot in various concentrations from doped ethylene Burke-Schumann flames. Size-classified samples from the coal tests were separated into coarse (>2.5ìm), fine (0.5-2.5ìm) and ultrafine (<0.5ìm) fractions, and analyzed for total carbon, elemental composition, and detailed iron and sulfur speciation. In a similar manner, ultrafine particles from the ethylene flame tests were also analyzed for total carbon and elemental composition. Pulmonary inflammatory responses were determined after intratracheal aspiration of 100ìg samples in female CD1 mice. IL-6 and neutrophil responses were monitored as markers of inflammation. With carbon present, the coal data suggested that the ultrafine particles containing soot were more toxic than fine or coarse particles containing char, even though the iron and sulfur speciation varied only slightly with particle size. Iron and sulfur chemistry were, however, dependent on the extent of carbon burnout achieved. In the absence of carbon, ultrafine particles (high in bisulfates and semi-volatile alkali metals) were less toxic than the fine fraction (high in oxidized iron and sulfates). Iron-soot particles created from ethylene flames were more toxic than an equivalent physical mixture of iron oxide and soot, and the toxicity depended primarily on the soot concentration. However, taken as whole, these data do not support the notion that iron and soot interact to enhance pulmonary inflammatory responses.

105e Physicochemical Properties of Diesel Exhaust Particles Collected during Inhalation Exposure Studies

Seung-Hyun Cho[1], William P. Linak[1], Dennis G. Tabor[1], Charly J. King[2], Jost O.L. Wendt[3], Q. Todd Krantz[4], and M. Ian Gilmour[4]. (1) National Risk Management Research Laboratory, U.S. EPA, 109 T. W. Alexander Drive (MD-E305-01), Research Triangle Park, NC 27711, (2) ARCADIS U.S., Inc., 4915 Prospectus Drive, Suite F, Durham, NC 27713, (3) Department of Chemical Engineering & Institute for Clean and Secure Energy, University of Utah, 3290 MEB, 59 S. Central Campus Dr., Salt Lake City, UT 84112, (4) National Health and Environmental Effects Research Laboratory, U.S. EPA, 109 T.W. Alexander Drive, Research Triangle Park, NC 27711

Diesel exhaust and especially diesel exhaust particles (DEP) are a health concern because of their complex chemistry, small size, and ubiquitous presence in urban environments. As part of a research program designed to relate diesel exhaust properties to adverse health effects, a series of inhalation exposure studies were conducted to investigate the propensity of diesel exhaust to promote cardiopulmonary diseases, immune polarization, and trans-generational effects in mice. Three diesel engines were utilized to generate diesel exhaust for whole animal exposures. These include a four cylinder 31 kW non-catalyst equipped engine and air compressor (CDEP), an eight cylinder 134 kW mid-1990s vintage pickup truck engine equipped with an oxidation catalyst and eddy current dynamometer (TDEP), and a one cylinder 4.8 kW non-catalyst equipped engine and generator (GDEP). The CDEP and TDEP engines were operated at approximately 25% of their maximum engine loads at speeds of 1700 and 2500 rpm, respectively. These operating conditions were chosen to be representative of typical equipment operation, and for TDEP, to imitate steady-state highway driving. The GDEP engine was operated at approximately 80% of its maximum load at 3600 rpm. This also was intended to be representative of how a small generator might be used. During the exposures, gas/particle concentrations and particle size distributions were monitored in real-time. DEP was collected at engine exhaust and inhalation chamber locations using filters, baghouses, and/or an electrostatic precipitator, and analyzed for particle mass concentration, organic and elemental carbon (OC-EC), inorganic (non-EC) elements, and organic components. Dichloromethane-extracted organic material (EOM) was sequentially sub-fractionated through a silica column using four organic solvents of increasing polarity: hexane, 50:50 hexane:dichloromethane, dichloromethane, and methanol. The EOMs and the sub-fractions were also determined gravimetrically. The exhaust particle mass concentration of the CDEP was approximately 11 times greater than either the TDEP or GDEP. This also corresponded to higher amounts of quantifiable polycyclic aromatic hydrocarbons (PAHs) in the CDEP samples (>400 μ g/g) compared to the TDEP and GDEP samples (<100 μ g/g). Geometric mean number diameters of the CDEP was slightly larger (84-102 nm) than the TDEP and GDEP (48-74 nm). However, with the exception of PAH species, other organic chemical components were similar for all three engines. Sulfur, chlorine, zinc, calcium, and iron were the most abundant inorganic elements, comprising <5% of the mass. OC/EC mass ratios ranged from 0.2-2.0. EOM comprised 10-40% of the DEP mass, and partitioned at 54-77% and 8-13% between the least-polar (hexane) and most-polar (methanol) solvents, respectively. Less than 2.5% of mass of all DEP samples was identified and quantified as specific organic compounds, and 84-98% of the organics identified were alkanes or organic acids. The overall characteristics placed the three EPA-generated DEP samples between the NIST SRM2975 and Japanese (Sagai) samples, which are commonly used as reference samples but have drastically different properties from each other (This abstract does not reflect EPA policy.).

SEARCHing for Answers

Rick D. Saylor[1], *Eric S. Edgerton*[2] *and Benjamin E. Hartsell*[3]

[1] *Atmospheric Research & Analysis, Inc., Snellville, GA*
[2] *Atmospheric Research & Analysis, Inc., Cary, NC*
[3] *Atmospheric Research & Analysis, Inc., Plano, TX*

Introduction

Many studies have demonstrated associations between atmospheric particulate matter (PM) with aerodynamic diameter less than 2.5 µm ($PM_{2.5}$) and health effects (1, 2). In particular, evidence has arisen that hourly increases in $PM_{2.5}$ levels may lead to the onset of myocardial infarctions within a few hours after exposure (3). As a result of these health concerns, the U.S. Environmental Protection Agency promulgated regulations on the maximum $PM_{2.5}$ concentrations to which the general public should be exposed. More recently, studies have suggested that not only the total concentration is important, but knowledge of the abundance of the individual chemical constituents of $PM_{2.5}$ may be linked with particular health endpoints (4-6). Furthermore, in the past decade, the study of atmospheric PM has intensified due to the identification of its potential impacts on climate change (7, 8).

In response to these concerns, the Southeastern Aerosol Research and Characterization Study (SEARCH) was implemented in 1998-1999 in an effort to provide extensive long-term data on the sources and chemical characteristics of atmospheric particulate matter (PM), in particular for the southeastern United States with its distinctive meteorological and demographical characteristics. The SEARCH study's objectives include: (i) development of a long-term chemical climatology of PM; (ii) advancement of the understanding of PM composition, variability and sources; (iii) establishment of a platform for the development and testing of new measurement methods related to PM; and, (iv) collaboration with states, federal agencies and others in carrying out special studies at selected network sites.

SEARCH is unique in the breadth, high time resolution and longevity of trace atmospheric measurements that have been made at eight sites across the southeastern U. S. At four pairs of urban/rural or urban/suburban monitoring sites located in Atlanta, GA (JST and YRK), Birmingham, AL (BHM and CTR), Gulfport, MS (GFP and OAK), and Pensacola, FL (PNS and OLF), (see Figure 1) meteorological (temperature, pressure, wind speed and direction, humidity, solar insolation), trace gas (O_3, SO_2, CO, NO, NO_2, NO_y, HNO_3, NH_3), fine PM ($PM_{2.5}$ mass and composition), and coarse PM data ($PM_{10-2.5}$ mass and composition) have been routinely collected since the network's inception. The SEARCH dataset provides a rich repository for the investigation of many scientific questions concerning the sources, formation and behavior of atmospheric PM. This paper presents a broad overview of selected results from investigations that have been undertaken to date with SEARCH data. Details of the network including descriptions of measurement methods, data adjustment techniques, and quality control procedures can be found in various published articles (9-13) and on the Atmospheric Research & Analysis, Inc., website (http://www.atmospheric-research.com).

Results and Discussion

SEARCH PM measurements include both integrated 24-hr filter-based $PM_{2.5}$ and $PM_{10-2.5}$ mass and composition as well as semi-continuous (5 min resolution) $PM_{2.5}$ mass and composition. Figure 2 presents typical composition data for filter-based $PM_{2.5}$ from the JST site in midtown Atlanta and Figure 3 presents monthly variation of this data at all sites in the network. "Best Estimate" data is a reconstruction of $PM_{2.5}$ mass and composition taking into account nitrate, ammonium and organics that volatilize from the primary filter after initial collection. Organic matter comprises the largest fraction of $PM_{2.5}$ Best Estimate (BE) mass at each site (4 – 7 $\mu g/m^3$), accounting for 30–36% of the total. In combination with black carbon, which by itself is a fairly small contributor to total mass (4-10%), carbonaceous matter as a whole makes up 35-45% of total $PM_{2.5}$ BE mass across SEARCH. Sulfate accounts for 25-30% of total BE mass, with concentrations falling within a fairly narrow range (3.5 – 4.8 $\mu g/m^3$). Ammonium concentrations account for 9-12% of total $PM_{2.5}$ BE mass (1 – 2 $\mu g/m^3$). Ratios of

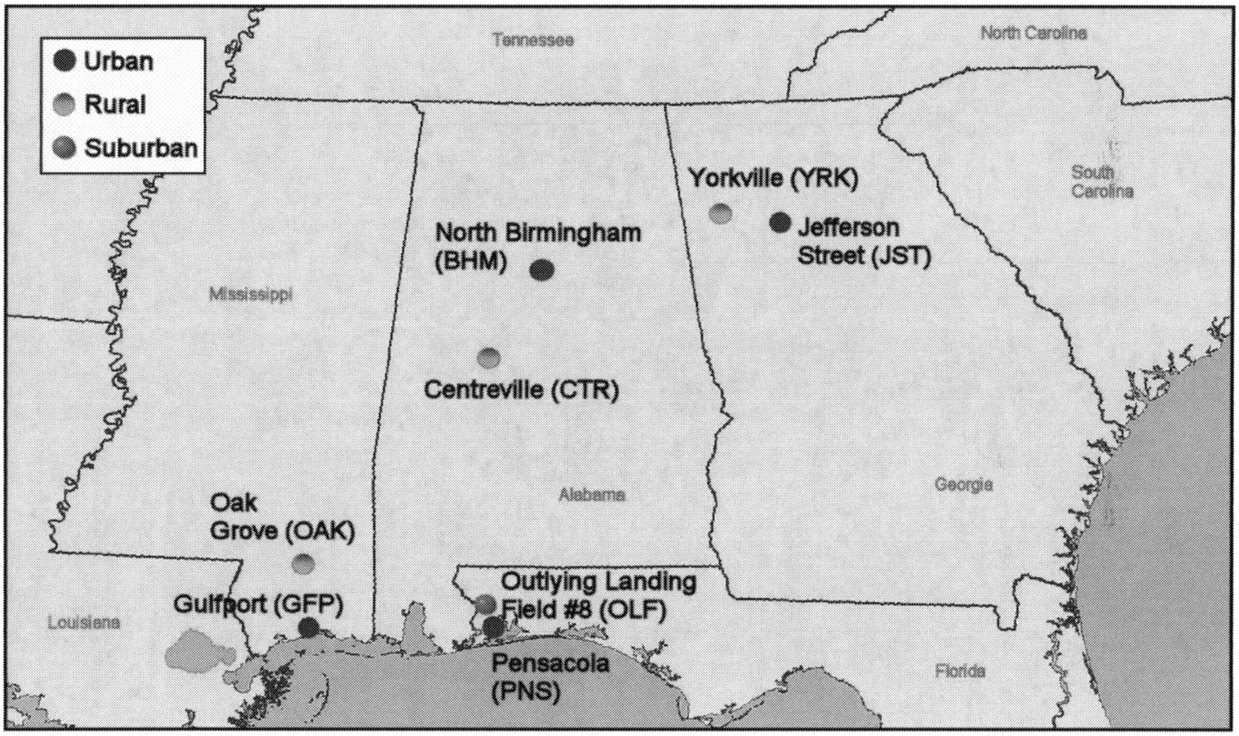

Figure 1. Location of SEARCH monitoring sites in the southeastern U. S.

($[SO_4^{2-}] + [NO_3^-])/[NH_4^+]$ (as equivalents) increase from 1.1 to 1.3 from north to south across the network and the degree of neutralization as defined by the ratio ($[NH_4^+] - [NO_3^-])/[SO_4^{2-}]$ (as equivalents) decreases from 0.92 to 0.73-0.78, possibly indicating a reduction in available ammonia as one goes from the continental interior towards the Gulf of Mexico. Nitrate concentrations are low (0.4 – 1.0 $\mu g/m^3$) and account for less than 6% of total $PM_{2.5}$ BE mass. Major metal oxides (MMOs) contribute less than 5% to total $PM_{2.5}$ BE mass at all sites. "Other" is a derived quantity considered to be a combination of species not measured, particle-bound water, analytical error, and uncertainty in the assumed form of PM components. Aside from particle-bound water perhaps the most important of this latter group is the molecular form of organic carbon which comprises OM, where a constant factor of 1.4 is used. The "Other" category accounts for 10-19% of the total $PM_{2.5}$ BE mass (2.0 – 2.5 μg/m^3), which is a lesser but not inconsequential portion of the total mass.

Figure 3 presents monthly average $PM_{2.5}$ Best Estimate mass and composition for all SEARCH sites. Monthly-averaged Best Estimate mass concentrations exhibit a distinct seasonality with higher concentrations during the warmer months, generally peaking in July-August. This seasonality is more pronounced at the more northerly sites JST, YRK, BHM, and CTR than at more southerly locations. In fact, data for the southern sites show virtually no summertime enhancement of $PM_{2.5}$ compared to other seasons. Sulfate is a primary driver of the observed seasonality, with values 2.2-3.9 times larger in July-August than in December-January. The larger sulfate concentrations during warm months are likely due to increased photochemical activity and/or increased precursor emissions during summer as a result of peak electricity generation during this time. OM, in contrast, shows a relatively flat pattern,

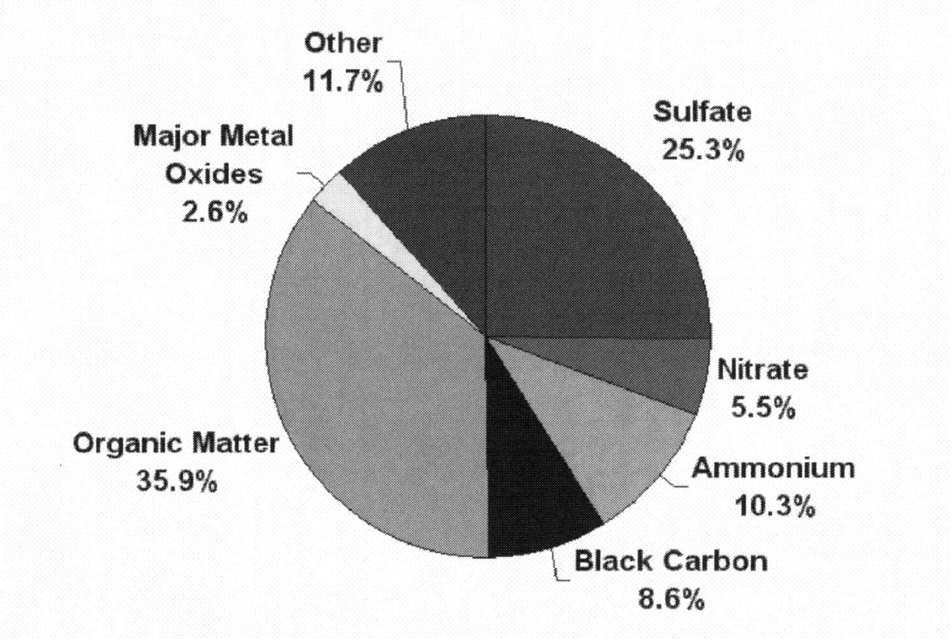

Figure 2. Best Estimate $PM_{2.5}$ mass and composition from SEARCH site JST in midtown Atlanta, GA.

possibly reflecting substitution of secondary production during the warm (growing) season with primary production (e.g., wood smoke) during the cold season. Nitrate concentrations peak during the cool months of December-January and reach minima during July-August, consistent with the thermodynamics of gas/particle partitioning of ammonium nitrate. At all sites, MMO concentrations peak during April-August and reach their lowest values during December-January. Reasons for this pattern are unclear, but could be related to rainfall and/or soil moisture and/or long-range transport of African or Asian dust.

Deployment of continuous PM analyzers in the SEARCH network began in 1998 and continues today as new technologies are developed. Measurement of $PM_{2.5}$ mass is performed using a dried, 30°C tapered element oscillating microbalance (TEOM). Light scattering by nephelometry complements the TEOM. Measurements of major constituents include: (a) SO_4^{2-} via reduction to SO_2; (b) NH_4^+ and NO_3^- via respective catalytic oxidation and

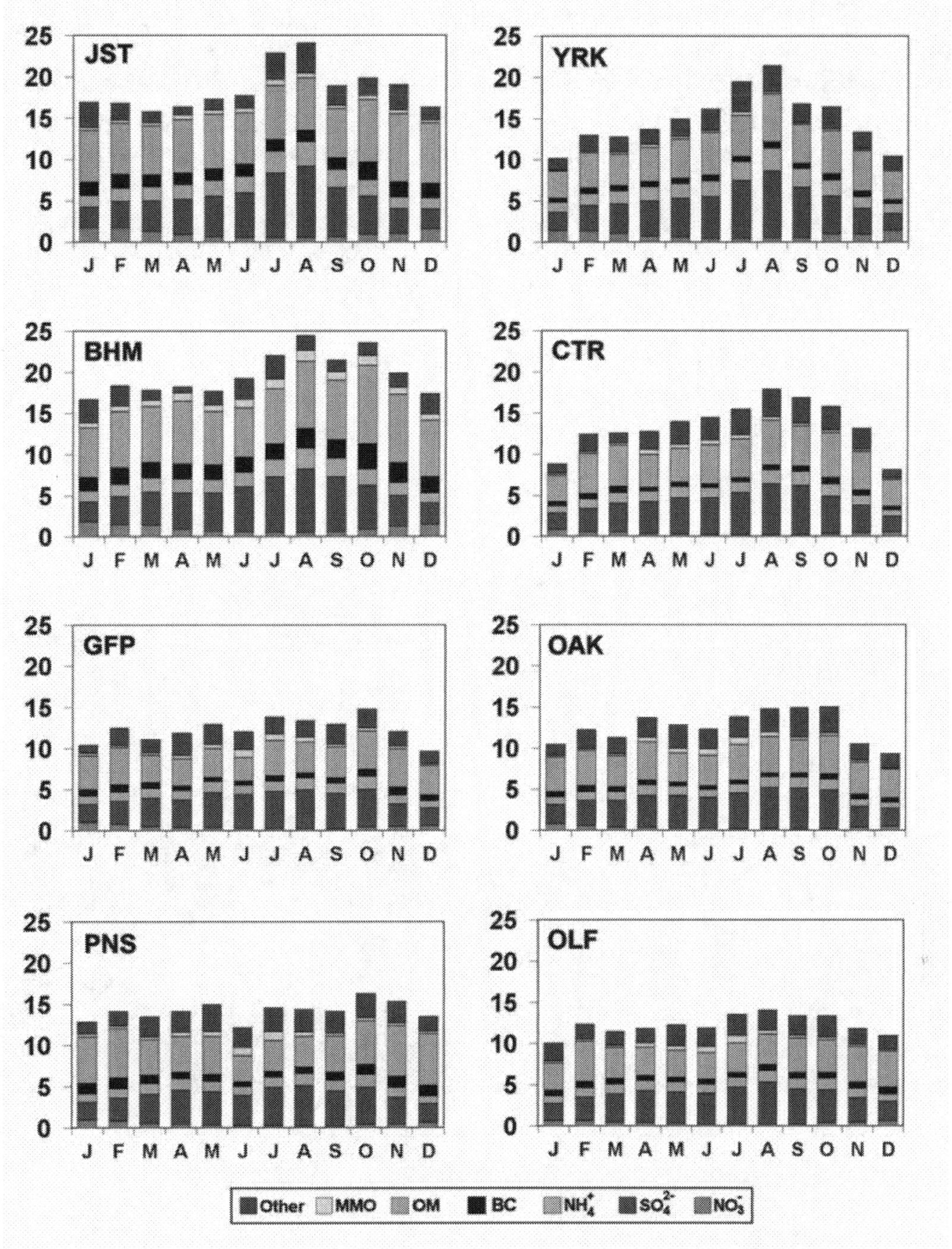

Figure 3. Best Estimate $PM_{2.5}$ mass and composition from filter-based measurements at SEARCH sites.

reduction to NO, (c) black carbon (BC) by optical absorption, (d) total carbon (TC) by combustion to CO_2, and (e) organic carbon by difference between the latter two measurements. These continuous measurements are intended to be long-term, and in conjunction with co-measured meteorological and trace gas data, offer opportunities for: (1) investigating sources and chemical and physical dynamics of $PM_{2.5}$; (2) evaluating chemical transport and transformation models; and; (3) assessing the effectiveness of emissions reduction programs.

Figure 4 presents hourly mass and composition data from a high-$PM_{2.5}$ episode that occurred in the southeast during June-July, 2002. At the Atlanta site (JST), a multi-day buildup of $PM_{2.5}$ mass began on June 29 and culminated in a peak near 60 µg/m^3 during the afternoon of July 2. Sulfate concentrations are seen to be the driving influence in this mass increase, rising from a value near 1 µg/m^3 on June 29 to > 30 µg/m^3 during the mass concentration maximum on July 2. Other components, most notably OM and BC, also contributed to the total mass increase, but sulfate dominated the change, going from < 10% of the total mass on June 29 to over 50% on July 2. Using high-temporal resolution mass and composition data such as this allows detailed analysis of the factors contributing to high pollution episodes such as the one presented here.

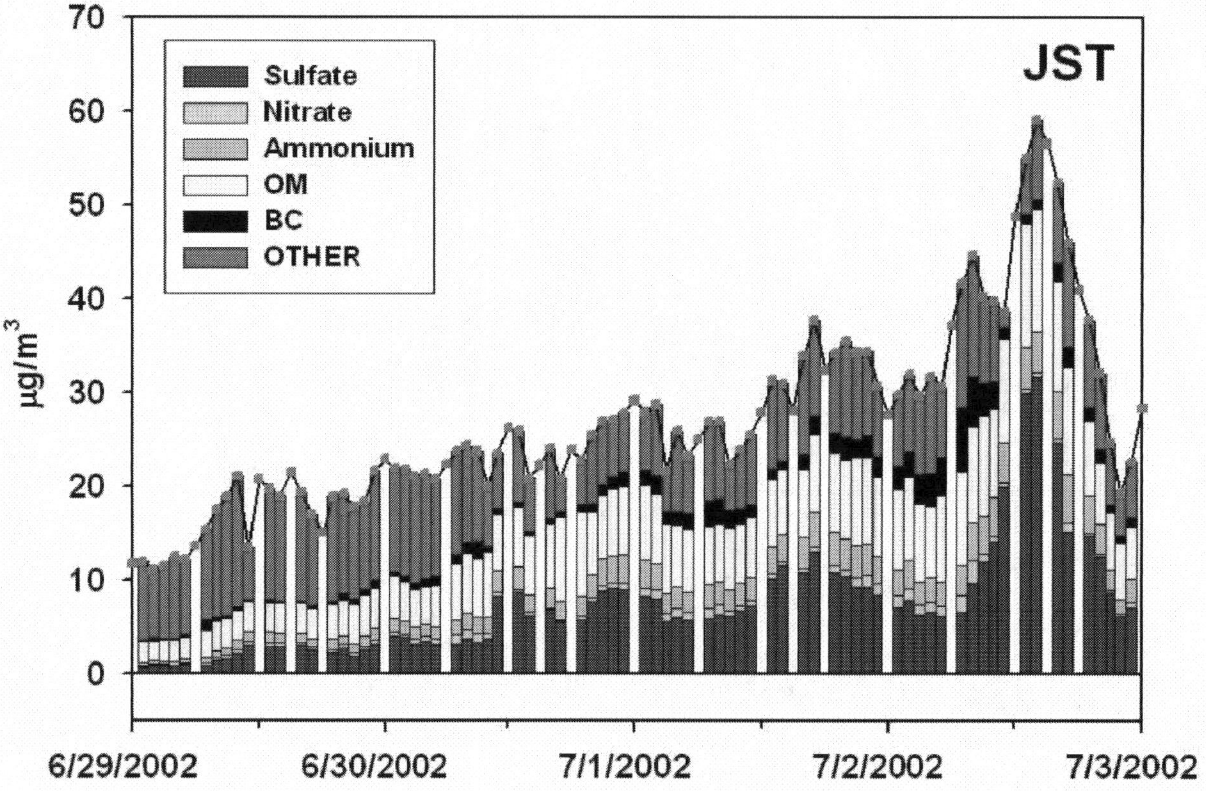

Figure 4. Hourly $PM_{2.5}$ mass and composition at SEARCH site JST during June-July 2002.

Controlled burning activities are prevalent in the southeastern U. S. during winter months and are used by various government agencies to control vegetation accumulation, mitigate disease and insect infestations, and manage habitat for endangered species. Smoke plumes from these fires are regularly observed at SEARCH sites during these times. The measurements made at the sites present a unique opportunity to characterize the chemical nature of these plumes and provide an initial gauge of the overall effect that these activities have on the air quality of the southeastern region. An example of a smoke plume interception at OAK (Oak Grove, MS) in 2001 is shown in Figure 5. CO, NO_y, and $PM_{2.5}$ total carbon (TC) concentrations all increased sharply on two separate occasions during late January and early February. The first episode, January 25-27, occurred as the wind direction swung from west-northwest to east to south over January 25 and 26. CO and TC values are seen to follow each other closely over the episode and CO:NO_y molar ratios in the plume are 25-35, quite different from typical urban plume ratios < 10. O_3 mixing ratios through the episode were not substantially higher than before or after the passage of the plume and SO_2 values were slightly elevated. The second episode occurred February 3-6, again as wind direction shifted from west-northwest through the east to the south on February 3, and again on February 4. Once again, CO and $PM_{2.5}$ TC tracked each other closely, CO:NO_y molar ratios obtained similar values as in the first episode, and O_3 increased somewhat as the episode continued. In both of these plumes, $PM_{2.5}$ TC concentrations reached very high values ranging from 15 $\mu g/m^3$ to nearly 80 $\mu g/m^3$. As additional evidence that these plumes represent fires, satellite fire detection data for this time period was obtained from the U. S. Forest Service Remote Sensing Applications Center website. Several fires were detected in southern Mississippi, western Alabama and eastern Louisiana during this time period with the closest occurring 53 km west of the OAK site, 64 km southwest of the site, and 54 km north-northwest of the site.

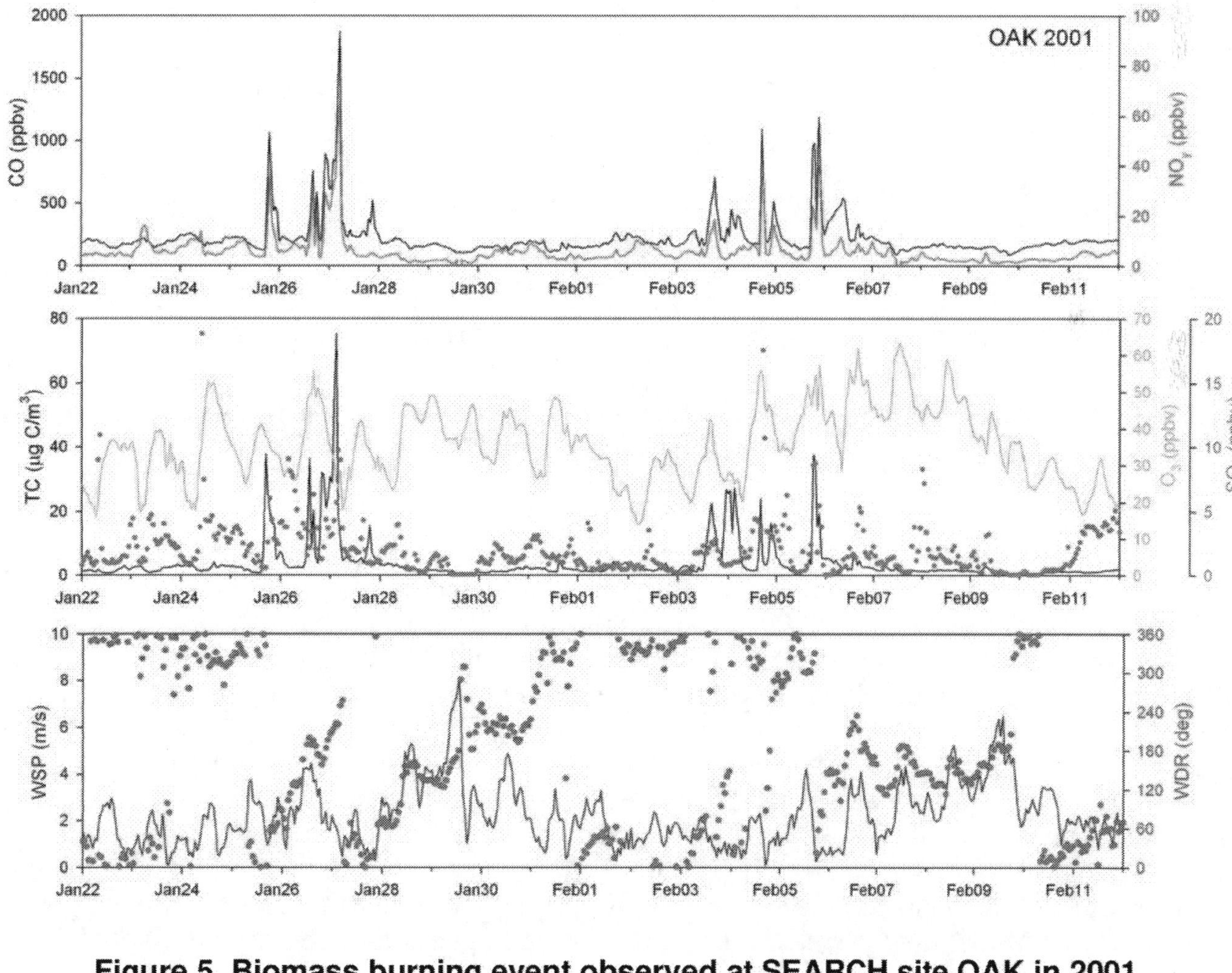

Figure 5. Biomass burning event observed at SEARCH site OAK in 2001.

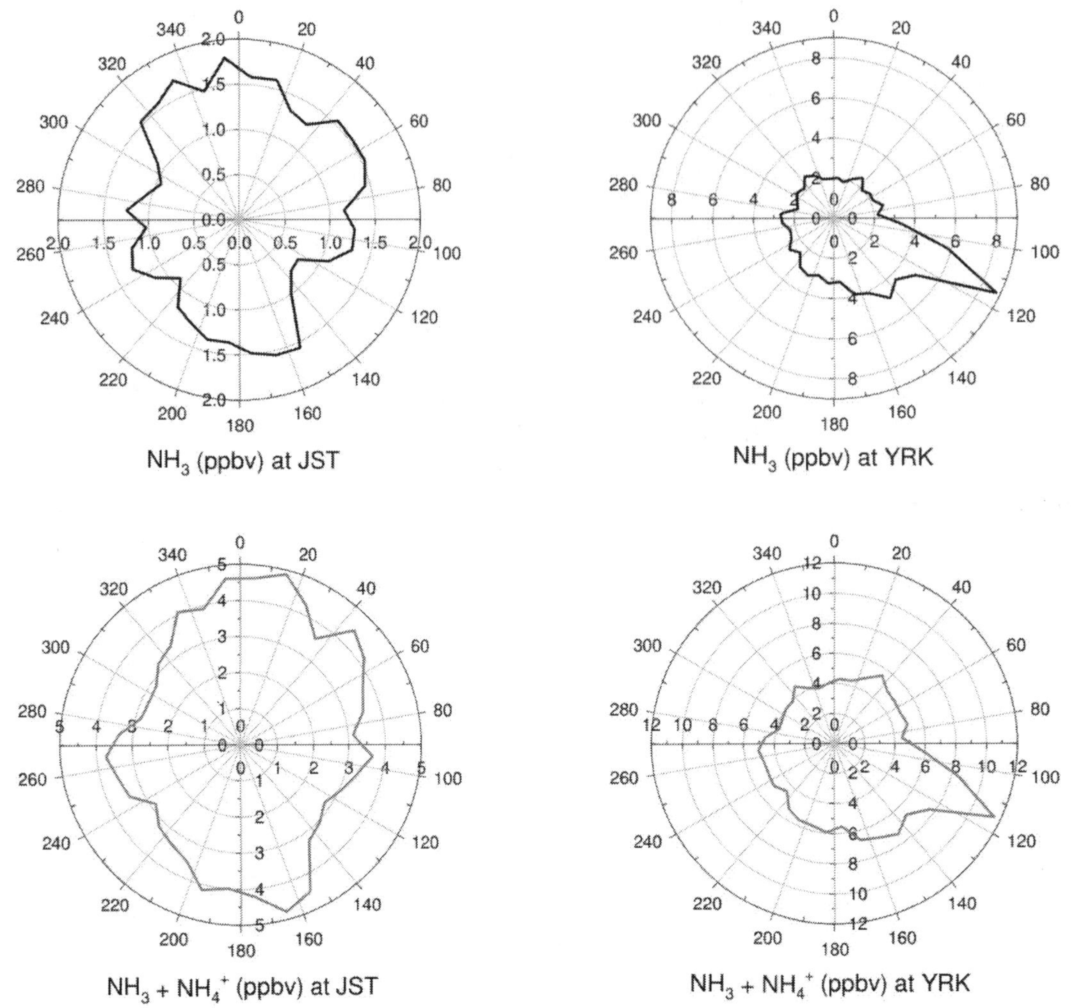

Figure 6. Measurements of gaseous and total ammonia as a function of wind direction at SEARCH sites JST and YRK.

In an effort to better characterize the dynamics of NH_3 in the atmosphere of the southeastern U. S., continuous NH_3 measurements were implemented at three SEARCH sites in 2007, an industrial-residential site located at Jefferson Street in midtown Atlanta, Georgia (JST), a rural-agricultural site located near Yorkville, Georgia (YRK), approximately 55 km west-northwest of the Jefferson Street, Atlanta site, and a rural-forested site near Oak Grove, Mississippi (not reported here). Figure 6 presents NH_3 and total ammonia ($NH_x = NH_3 + NH_4^+$) as a function of wind direction. The mixing ratio averages presented in these plots are the means of all hourly-average values falling into 10° wind direction bins, and are plotted at the midpoint of each bin. At JST there is a slight tendency for larger NH_3 and NH_x values when winds are generally from the 350°-20° and 150°-170° sectors, which may be related to motor vehicle exhaust. A trucking facility is located < 100 m due north of the JST site, while a parking lot and a moderately-traveled urban two-lane road lies within 150 m to the south.

In contrast, the wind direction plots for YRK exhibit a markedly different tendency. In the 110-150° sector, two large peaks occur for both NH_3 and NH_x and are substantially larger than the means for all other wind sectors. In fact, excluding the quadrant 90-180°, the mixing ratio values for NH_3 and NH_x are only slightly larger at YRK than at JST. YRK is located in a mixed agricultural and forested landscape and is situated in a pasture with nearby poultry-rearing houses, the two closest of which are about 1.3 km from the YRK site. Two more distant facilities are 2.6-3.0 km away. All of these poultry houses lay within the directional sector of 100°-160° with respect to the YRK site, and the two major peaks in the NH_3 and NH_x plots of Figure 6 (at 115° and 145°) correspond very closely to the direction of the two closest broiler facilities. Of the 40 hourly-average NH_3 values > 20 ppbv measured at YRK during July through December 2007, 27 (67.5%) have hourly-average wind directions from within the 110-150° sector and 39 (97.5%) fall within the 90-180° sector. Similarly, for the 167 hourly-average NH_3 values > 10 ppbv, 90 (53.9%) have wind directions within the 110-150° sector and 146 (87.4%) fall within the 90-180° sector. It therefore seems very likely that the high average values and large variability of NH_3 and NH_x observed at YRK are due directly to emissions from the nearby poultry facilities. Back-trajectory analyses of selected episodes of high NH_3 further confirm this hypothesis.

Summary

By the end of 2008, more than ten years of data will have been collected at eight continuously operating sites in SEARCH. The comprehensive nature and rigorous quality assurance of the data collected at each SEARCH location provides a rich repository for scientific investigations of atmospheric chemistry and air quality processes, for both gaseous species and PM. SEARCH is slated to continue operation through 2011, with possible further extension beyond that time. All data collected as part of SEARCH is freely available from Atmospheric Research & Analysis, Inc.'s website (http://www.atmospheric-research.com). Special data requests can be sent to Eric Edgerton (eedgerton@atmospheric-research.com).

References

1. Laden, F., Neas, L. M., Dockery, D. W., Schwartz, J. (2000) Association of fine particulate matter from different sources with daily mortality in six U. S. Cities. *Environ. Health Perspectives*, 108, 941-947.
2. Somers, C. M., McCarry, B. E., Malek, F., Quinn, J. S. (2004) Reduction of particulate air pollution lowers the risk of heritable mutations in mice. *Science*, 304, 1008-1010.
3. Peters, A., Dockery, D. W., Muller, J. E., Mittleman, M. A. (2001) Increased particulate air pollution and the triggering of myocardial infarction. *Circulation*, 103, 2810-2815.
4. Peel, J. L., Tolbert, P. E., Klein, M., Metzger, K. B., Flanders, W. D., Todd, K., Mulholland, J. A., Knox, T., Ryan, P. B., Frumkin, H. (2005) Ambient air pollution and respiratory emergency department visits. *Epidemiology*, 16, 164-174.
5. Metzger, K. B., Tolbert, P. E., Klein, M., Peel, J. L., Flanders, W. D., Todd, K., Mulholland, J. A., Ryan, P. B., Frumkin, H. (2004) Ambient air pollution and cardiovascular emergency Department Visits. *Epidemiology*, 15, 46-56.
6. Klemm, R. J., Lipfert, F. W., Wyzga, R. E., Gust, C. (2004) Daily mortality and air pollution in Atlanta: Two years of data from ARIES. *Inhalation Toxicology*, 16, Supplement 1, 131-141.

7. Charlson, R. J., Lovelock, J. E., Andreae, M. O., Warren, S. G. (1987) Oceanic phytoplankton, atmospheric sulfur, cloud albedo and climate. *Nature*, 326, 655-661.
8. Intergovernmental Panel on Climate Change (2001) *Climate Change 2001: The Scientific Basis*, Cambridge University Press, Cambridge, UK.
9. Hansen, D. A., Edgerton, E. S., Hartsell, B. E., Jansen, J. J., Kandasamy, N., Hidy, G. M., Blanchard, C. L. (2003) The Southeastern Aerosol Research and Characterization Study: Part 1 – Overview. *J. Air & Waste Manage. Assoc.*, 53, 1460-1471.
10. Edgerton, E. S., B. E. Hartsell, R. D. Saylor, J. J. Jansen, D. A. Hansen, and G. M. Hidy (2005) The Southeastern Aerosol Research and Characterization Study: Part II. Filter-based measurements of fine and coarse particulate matter mass and composition. *J. Air & Waste Manage. Assoc.*, 55, 1527-1542.
11. Edgerton, E. S., B. E. Hartsell, R. D. Saylor, J. J. Jansen, D. A. Hansen, and G. M. Hidy (2006) The Southeastern Aerosol Research and Characterization Study: Part III. Continuous measurements of $PM_{2.5}$ mass and composition. *J. Air & Waste Manage. Assoc.*, 56, 1325-1341.
12. Edgerton, E. S., R. D. Saylor, B. E. Hartsell, J. J. Jansen, and D. A. Hansen (2007) Ammonia and ammonium measurements from the southeastern United States. *Atmos. Environ.*, 41, 3339-3351.
13. Edgerton, E. S., G. S. Casuccio, R. D. Saylor, T. L. Lersch, B. E. Hartsell, and J. J. Jansen (2008) Measurements of $PM_{10-2.5}$ organic, elemental and carbonate carbon in the southeastern U. S. *J. Air & Waste Manage. Assoc.*, in press.

Acknowledgements

SEARCH and the work presented here are sponsored by the Southern Company and the Electric Power Research Institute.

MULTI-POLLUTANT PLUME-IN-GRID MODELING

Prakash Karamchandani and Krish Vijayaraghavan
Atmospheric & Environmental Research, Inc. (AER), San Ramon, CA, USA

1. INTRODUCTION

One-atmosphere three-dimensional (3-D) grid models are now being widely used to predict the impacts of emission controls on the atmospheric concentrations and deposition of pollutants such as ozone (O_3), fine particulate matter ($PM_{2.5}$), mercury (Hg) and other air toxics. Such a grid-based approach necessarily averages emissions within the volume of the grid cell where they are released. This averaging process may be appropriate for sources that are more or less uniformly distributed at the spatial resolution of the grid system. However, it may lead to significant errors for sources that have a spatial dimension much smaller than that of the grid system. For example, stack emissions lead to plumes that initially have a dimension of tens of meters, whereas the horizontal resolution in grid-based air quality models is typically several kilometers in urban applications and up to about 100 km in regional applications. This artificial dilution of stack emissions leads to (1) lower concentrations of plume material, (2) unrealistic concentrations upwind of the stack, (3) incorrect chemical reaction rates due to the misrepresentation of the plume chemical concentrations and turbulent diffusion, and (4) incorrect representation of the transport of the emitted chemicals.

Plume-in-Grid (PiG) modeling has been demonstrated to be an effective approach to resolve sub-grid scale effects associated with discrete sources (e.g., Seigneur et al., 1983; Sillman et al., 1990; Kumar and Russell, 1996; Gillani and Godowitch, 1999; Karamchandani et al., 2002; 2006a; Godowitch, 2004). In this approach, the errors associated with the grid-averaging of stack emissions can be eliminated by using a subgrid-scale representation of stack plumes that is imbedded in the 3-D grid system of the air quality model.

This paper describes the development and evolution of one such PiG model that was originally developed and applied for ozone (Karamchandani et al., 2002; Vijayaraghavan et al., 2006) and then extended to particulate matter (PM) (Karamchandani et al., 2006a), mercury (Karamchandani et al., 2006b; 2006c), and more recently, to investigate the sub-grid scale impacts of air toxics emissions from roadways (Karamchandani et al., 2008a). We present results from the application of the model for a number of case studies to illustrate the differences in results from traditional grid modeling versus PiG modeling, and the implication of these differences.

2. PLUME-IN-GRID MODEL

The PiG model presented here consists of a reactive plume model that is embedded within a three-dimensional grid model. The 3-D grid model, also referred to as the "host" model, is based on the U.S. EPA Community Multiscale Air Quality (CMAQ) model (Byun and Schere, 2006). The embedded reactive plume model is adapted from the Second-Order Closure Integrated puff model with CHEMistry (SCICHEM) (Karamchandani et al., 2000). The resulting PiG model is referred to as CMAQ with Advanced Plume Treatment or CMAQ-APT.

In SCICHEM, the reactive plume component of CMAQ-APT, plume transport and dispersion are simulated using a second-order closure approach to solve the turbulent diffusion equations (Sykes et al., 1993; Sykes and Henn, 1995). The plume is represented by a myriad of three-dimensional puffs that are advected and dispersed according to the local micrometeorological characteristics. Each puff has a Gaussian representation of the concentrations of emitted inert species. The overall plume, however, can have any spatial distribution of these concentrations, since it consists of a multitude of puffs that are independently affected by the transport and dispersion characteristics of the atmosphere. SCIPUFF can simulate the effect of wind shear since individual puffs will evolve according to their respective locations in an inhomogeneous velocity field. As puffs grow larger, they may encompass a volume that cannot be considered homogenous in terms of the meteorological variables. A puff splitting algorithm accounts for such conditions by dividing puffs that have become too large into a number of smaller puffs. Conversely, puffs may overlap significantly, thereby leading to an excessive computational burden. A puff-merging algorithm allows individual puffs that are affected by the same (or very similar) micro-scale meteorology to combine into a single puff. Also, the effects of buoyancy on plume rise and initial dispersion are simulated by solving the conservation equations for mass, heat, and momentum.

The formulation of nonlinear chemical kinetics within the puff framework is described by Karamchandani et al. (2000). Chemical species concentrations in the puffs are treated as perturbations from the background concentrations. The chemical reactions within the puffs are simulated using a general framework that allows any chemical kinetic mechanism to be treated. The puff chemical mechanism is the same as the host grid model mechanism for consistency. For example, typical applications of CMAQ-APT for ozone (Karamchandani et al., 2002) and fine particulate matter (Karamchandani et al., 2006a) have used the Carbon-Bond Mechanism (CBM-IV) in both SCICHEM and CMAQ. Another feature of SCICHEM is that it allows the option of explicitly simulating the effect of turbulence on chemical kinetics for selected reactions. This effect is more pronounced near the stack (Karamchandani et al., 2000) and requires additional computational time for its simulation. More details on the SCICHEM model formulation and its evaluation with plume data from the 1995 Southern Oxidants Study (SOS) in Nashville/Middle Tennessee are presented elsewhere (Karamchandani et al., 2000).

The formulation of PM processes in SCICHEM follows that of the host model and has been described by Karamchandani et al. (2006a). Two versions of CMAQ-APT are available based on the treatment of aerosols in the host and plume models. The first version uses the AERO3 module of CMAQ for the treatment of aerosols and is referred to as CMAQ-AERO3-APT. The second version uses the Model of Aerosol Dynamics, Reaction, Ionization and Dissolution (MADRID) (Zhang et al., 2004) for aerosol treatment and is referred to as CMAQ-MADRID-APT. The primary differences between the two aerosol modules are in their representation of particle size distribution (sectional in MADRID and modal in AERO3) and in their treatment of the formation of secondary organic aerosols (SOA). Both aerosol treatments use the ISOROPPIA module of Nenes et al. (1999) for the thermodynamics of inorganic species.

The consistency between the host model and reactive plume model is extended to their treatment of heterogeneous chemistry on aerosol particles and cloud droplets as well as aqueous-phase chemistry in clouds and removal by dry and wet deposition.

The treatment for mercury in CMAQ-AERO3-APT and CMAQ-MADRID-APT (Karamchandani et al., 2006b) is based on the chemical mechanism and deposition algorithms of Seigneur et al. (2004). Additionally, an empirical reaction is used to represent the potential reduction of Hg^{II} to Hg^{0} in coal-fired power plant plumes (Lohman et al., 2006); the kinetics of this reaction is taken to be proportional to the SO_2 gas-phase concentration and significantly affects Hg^{II} concentrations in plumes but has little effect in the background (Seigneur et al., 2006). Another recent improvement to the treatment of mercury processes in the host and reactive plume models is the addition of modules to simulate the adsorption of reactive gaseous mercury (Hg^{II}) to atmospheric particulate matter (Vijayaraghavan et al., 2008).

The coupling of SCICHEM and CMAQ follows the established protocols for incorporating new science modules into CMAQ. SCICHEM is invoked in CMAQ-APT by a single subroutine call similar to the invocation of any other physical or chemical process module in the host model. Like the other process modules in CMAQ, all relevant information related to the emissions and the dynamic state of the atmosphere required by SCICHEM are accessed directly from the input files via the Input/Output Applications Programming Interface (I/O API) (Coats et al., 1993); only the three-dimensional concentration fields are shared directly between the host model and SCICHEM. On input to SCICHEM, these concentrations serve as the background (ambient) concentrations for SCICHEM calculations. On output from SCICHEM, these concentrations are updated whenever plume-to-grid transfer occurs and are returned to the host model. The transfer of puff material to the 3-D grid system (referred to as "puff dumping") is triggered when the puff size is commensurate with the grid cell size. As an option, a chemical dumping criterion may be selected for situations where the physical criterion may result in premature transfer of the plume material to the grid (Karamchandani et al., 2002). However, for most applications, the physical dumping criterion is sufficient.

The computational overhead associated with PiG modeling can make it impractical to use for long-term simulations with a large number of point sources treated explicitly with the plume component of the model. To overcome this limitation, we have recently parallelized the PiG code (Karamchandani et al., 2008b) to exploit the widespread availability of multi-processor workstations and workstation clusters that are commonly used today for air quality model simulations. A fully parallel PiG code allows efficient utilization of all the available compute cycles in these modern computer systems.

3. MODEL APPLICATIONS

Applications for Ozone

The first version of the PiG model described above was applied to examine the effects of the PiG formulation on model predictions of ozone concentrations in the vicinity of and downwind of large power plants in the northeastern U.S (Karamchandani et al., 2002). This early version of the model did not include a treatment for aerosols, aqueous-phase chemistry or mercury. The model was applied to a domain covering the northeastern U.S. for a five-day episode from 11 to 15 July 1995. The grid had a horizontal resolution of 12 km and 30 large point sources were selected for PiG treatment. The model was evaluated with available observations.

The simulation results showed that the use of the PiG model had a significant effect on the spatial patterns of ozone (O_3) and nitric acid (HNO_3) surface concentrations downwind of the sources considered for PiG treatment to distances of 100 to 200 km. Surface O_3 concentrations from the PiG simulation showed both decrements and increments with respect to the base simulation (a simulation with all sources treated with the grid model). The maximum decrement was about 80 ppb, while the maximum increment was about 77 ppb.

Most of the surface O_3 decrements were associated with lower production of O_3 immediately downwind of the point sources in the PiG simulation relative to the base simulation. Furthermore, downwind O_3 production was generally delayed in the PiG simulation as compared to the base simulation. The large increments in surface O_3 concentrations in the PiG simulation relative to the base simulation were primarily associated with the higher titration of background surface O_3 in the base simulation near and upwind of the point sources, particularly in VOC-limited environments. When the PiG treatment was used, the plume NO remained aloft for longer distances and was not immediately available for titrating surface O_3 concentrations near the source. The smaller increments (typically less than 10 ppb) in the PiG simulation relative to the base simulations were associated with both higher titration of existing surface O_3 in the Base simulation and delayed production of O_3 further downwind in the PiG simulation, as the plume NO_x was transported and exposed to a NO_x-limited environment.

For surface HNO_3 concentrations, the differences between the PiG and base simulations range from a maximum decrement of 24 ppb to a maximum increment of 9 ppb. The PiG simulation generally predicted lower surface HNO_3 concentrations than the base simulation over a large portion of the modeling domain. Over the entire domain and episode, the PiG treatment resulted in a decrease in HNO_3 mass of about 3%.

Vijayaraghavan et al. (2006) applied the PiG model to simulate O_3 and HNO_3 formation during a four-day July/August 2000 episode in central California. The top ten NO_x emitting plants in the Central California Ozone Study (CCOS) domain were selected for explicit plume treatment. Their results were qualitatively similar to those of Karamchandani et al. (2002) for the northeastern U.S. However, the magnitudes of the differences between the PiG and base simulations were smaller in the central California application than in the northeastern U.S. simulation. This can be explained by the relatively smaller magnitude of NO_x emissions from large point sources in California as compared to the eastern United States.

Application for Fine Particulate Matter ($PM_{2.5}$)

The PiG model for ozone was extended by Karamchandani et al. (2006a) to include in the plume component state-of-the-science treatments of aerosol chemistry and dynamics as well as aqueous chemistry that were consistent with the treatments used in the host grid model. The model was applied to the southeastern United States with a 12 km horizontal grid resolution for the months of January and July 2002. Emissions from fourteen coal-fired power plants (CFPPs) in Alabama, Georgia, Florida and Mississippi were explicitly simulated with the PiG treatment.

The model evaluation included both comparisons with routine measurements of $PM_{2.5}$ components from the Interagency Monitoring of Protected Visual Environments (IMPROVE) network, the Clean Air Status and Trends Network (CASTNet), the Southeastern Aerosol Research and Characterization study (SEARCH) network. In addition, the SEARCH network measurements also included more detailed observations corresponding to plume events from specific upwind power plants. This special database provided an opportunity to conduct a refined evaluation of the PiG treatment. The comparisons of measured peak plume increments of SO_2 concentrations for plume events at the SEARCH monitoring sites during January and July 2002 with the corresponding peaks from the base simulation (i.e., all sources treated with the grid model) and PiG simulations showed that, for a large majority (nearly 90%) of the cases, the PiG simulation provided a better representation of the plume events than the base simulation.

The results showed that the use of the PiG model had a significant effect on the spatial patterns of particulate sulfate and total inorganic nitrate (gas-phase nitric acid + particulate phase nitrate) concentrations. The effects were largest in the four states containing the 14 power plants selected for PiG treatment. However, differences in sulfate and total inorganic nitrate concentrations were also predicted in some of the surrounding states. Using a PiG treatment resulted in a reduction in the contributions of the 14 CFPPs to sulfate and total inorganic nitrate concentrations over most of the regions impacted by these sources. In July 2002, using the PiG treatment reduced the calculated contributions of the power plants to sulfate concentrations by 5% to 40% in most of Alabama, north-western Georgia, south-eastern Mississippi and north-western Florida and by up to 5% in eastern Tennessee and most of the Carolinas. The maximum CFPP contributions in the PiG simulation to monthly-average sulfate and total inorganic nitrate concentrations in July 2002 were about 41% and 45% lower respectively than the corresponding values in the base simulation.

These differences between the base and PiG simulation results are due to the more correct treatment of elevated point source emissions and the near-field transport and chemistry of these emissions in the PiG simulation. The purely gridded approach used in the base simulation cannot resolve these subgrid-scale phenomena and overestimates the oxidation rates of SO_2 and NO_x emissions from the power plants. Over the entire modeling domain, about 77% of the SO_2 emissions from the power plants were oxidized to sulfate in July 2002 in the base simulation, as compared to 69% in the PiG simulation, a difference of about 10%.

This application was one of the first studies to use a PiG treatment for $PM_{2.5}$ and the results from this study showed that it is important to use a PiG treatment for emissions from large elevated point sources to accurately capture the subgrid-scale features of the transport and chemistry of the plumes from these sources, so that the impacts of these sources on downwind $PM_{2.5}$ sulfate and nitrate concentrations can be correctly simulated.

Applications for Mercury (Hg)

Karamchandani et al. (2006b) subsequently extended the PiG model for ozone and $PM_{2.5}$ to include a treatment for mercury. The model was applied to a domain covering the southeastern U.S. with a 12-km horizontal resolution for the entire year 2002. The vertical grid structure consisted of 19 layers from the surface to the tropopause with finer resolution near the surface (e.g., the surface layer is about 35 m deep). The meteorological fields for the air quality modeling simulations were obtained from a prognostic simulation conducted with the non-hydrostatic meteorological model, MM5. Forty coal-fired power plants in the 12 km resolution domain with the highest emissions of SO_2, NO_x and Hg were selected for explicit PiG treatment.

Two simulations with the PiG model were conducted for the model performance evaluation. One simulation included an empirical reaction between Hg^{II} and SO_2 to represent the potential reduction of Hg^{II} to Hg^0 in coal-fired power plant plumes, while the other simulation does not include this reaction. As shown in a previous study (Karamchandani et al., 2006c), the incorporation of this reaction provides significantly better agreement with Hg concentration measurements downwind of power plants as compared to simulations without this reaction. The results from the two simulations were compared with data from national and regional ambient and deposition monitoring networks, as well as with measurements of ambient Hg concentrations downwind of several coal-fired power plants during plume events.

Vijayaraghavan et al. (2008) applied the PiG model for 2001 over a domain covering the U.S. at 36 km horizontal resolution. The top thirty Hg emitting power plants in the U.S. were selected for explicit plume treatment. The version of the PiG model used in this application included a treatment of the gas-phase adsorption of reactive gaseous mercury (RGM) on atmospheric particulate matter. The PiG treatment resulted in improved performance for Hg wet deposition over a purely Eulerian grid-based model, partial correction of over-predictions of wet deposition downwind of coal-fired power plants in the northeastern U.S., and in decreases of approximately 10% in simulated dry and wet deposition over large parts of the eastern U.S. with larger decreases near the plants selected for PiG treatment.

Application for Nitrogen Deposition

The PiG model was also used to estimate the decrease in atmospheric nitrogen deposition in Escambia Bay and its watershed in Florida and southern Alabama due to planned emissions controls of NO_x and SO_2 at a nearby coal-fired power plant (Vijayaraghavan et al., 2007). The model was applied for 2002 to a domain covering the southeastern USA and centered on Alabama (AL) and Georgia (GA), the ALGA domain, with a horizontal resolution of 12 km. The use of PiG treatment resulted in less simulated dry deposition of atmospheric nitrogen than a purely gridded model, approximately half as much in the immediate vicinity of the plant and by 10% over the Escambia Bay watershed. This difference is due to the correct treatment in the PiG model of the depletion of oxidants by NO in the early stages of plume development in NO_x-rich plumes.

Application for Sub-Grid Scale Modeling of Air Toxics Concentrations Near Roadways

The PiG model was recently adapted by Karamchandani et al. (2008a) to develop a prototype model that could be used to simulate near-roadway concentrations due to mobile emissions from roadways. The motivation for the study was the increasing concern about population exposure to hazardous air pollutants (HAPs), especially in the vicinity of the sources of these pollutants. For example, measurements of toxic air pollution levels near a busy freeway in Los Angeles showed that exposures near the freeway were up to 10 times greater than those at background locations and dropped to background levels within 300 m downwind of the freeway (Zhu et al., 2002).

The model was applied to a busy interstate highway in New York City. The model was able to successfully capture the observed spatial variability in exposure levels from near the source to several hundreds of meters from the source.

4. SUMMARY

We have described the evolution of a multi-pollutant plume-in-grid model from its initial development for ozone applications to a model that can be used to calculate air quality impacts for a suite of pollutants including fine particulate matter, mercury, and other air toxics. Newer developments include the parallelization of the model code to improve its computational efficiency and the application (currently ongoing) of the parallelized code to conduct annual base simulations and emission scenario simulations with over 150 point sources explicitly treated with the plume component of the model.

Applications of the PiG model show that the model can predict significantly different air quality impacts of emissions from large point sources than the traditional gridded approach. These differences are due to the more realistic treatment of the transport and chemistry of the plumes from these sources in the PiG model.

5. ACKNOWLEDGMENTS

The PiG model described here was funded by several sources including EPRI, Palo Alto, CA who funded the model development and several applications, the Southern Company, Birmingham, AL, who funded the model application for $PM_{2.5}$ and Hg to the southeastern United States, and AER, Inc. who funded the development of the roadway prototype of the model. We also acknowledge the contributions of several organizations for providing some of the modeling data sets used in the model applications, such as the Georgia Environmental Protection Division (GEPD) and the Visibility Improvement State and Tribal Association of the Southeast (VISTAS), and Atmospheric Research and Analysis, Inc. (ARA) for providing mercury and PM data for the model evaluation from the Southeastern Aerosol Research and Characterization study (SEARCH) network. Finally, we acknowledge the contribution of scientists at L-3 COM, who collaborated with us in several aspects of the model development.

6. REFERENCES

Byun, D., and K. L. Schere, 2006. Review of governing equations, computational algorithms, and other components of the Models-3 Community Multiscale Air Quality (CMAQ) modeling system, *Appl. Mech. Revs.*, **59**, 51–77.

Coats, C. J., A. Hanna, D. Hwang, and D. W. Byun, 1993. Model engineering concepts for air quality models in an integrated environmental modeling system, *AWMA Specialty Conference on Regional Photochemical Measurement and Modeling Studies*, November 8-12, San Diego, CA.

Gillani N. V. and J. M. Godowitch, 1999. Plume-in-grid treatment of major point source emissions, Chapter 9, in: *Science algorithms of the EPA Models-3 Community Multiscale Air Quality (CMAQ) modeling system, EPA-600/R-99/030*, U. S. Environmental Protection Agency, Washington, D.C.

Godowitch, J., 2004. Simulations of aerosols and photochemical species with the CMAQ plume-in-grid modeling system, *3rd Annual CMAS Models-3 User's Conference*, October 18-20, Chapel Hill, NC.

Karamchandani, P., L. Santos, I. Sykes, Y. Zhang, C. Tonne, and C. Seigneur, 2000. Development and evaluation of a state-of-the-science reactive plume model, *Environ. Sci. Technol.*, **34**, 870–880.

Karamchandani, P., C. Seigneur, K. Vijayaraghavan, and S.-Y. Wu, 2002. Development and application of a state-of-the-science plume-in-grid model, *J. Geophys. Res.*, **107**, 4403-4415.

Karamchandani, P., K. Vijayaraghavan, S.-Y. Chen, C. Seigneur and E.S. Edgerton, 2006a. Plume-in-grid modeling for particulate matter, *Atmos. Environ.*, **40**, 7280-7297.

Karamchandani, P., K. Vijayaraghavan, S.-Y. Chen and C. Seigneur, 2006b. Plume-in-grid modeling for PM and mercury, *5th Annual CMAS Conference*, October 16-18, Chapel Hill, NC.

Karamchandani, P., K. Vijayaraghavan and C. Seigneur, 2006c. Detailed treatment of power plant plumes in the regional modeling of atmospheric mercury, *8th International Conference on Mercury as a Global Pollutant*, August 6-11, Madison, WI,.

Karamchandani, P., K. Lohman and C. Seigneur, 2008a. Using a sub-grid scale modeling approach to simulate the transport and fate of toxic air pollutants, *Environ. Fluid Mech.*, in press, doi:10.1007/s10652-008-9097-0.

Karamchandani, P., K. Vijayaraghavan, S.-Y. Chen and C. Seigneur, 2008b. Development and application of parallel plume-in-grid models, *7th Annual CMAS Conference*, October 6-8, Chapel Hill, NC.

Kumar, N., and A. G. Russell, 1996. Development of a computationally efficient, reactive subgrid-scale plume model and the impact in the northeastern United States using increasing levels of chemical detail, *J. Geophys. Res.*, **101**, 16737–16744.

Lohman, K., C. Seigneur, E. Edgerton, and J. Jansen, 2006. Modeling mercury in power plant plumes, *Environ. Sci. Technol.*, **40**, 3848-3854.

Nenes, A., S. N. Pandis and C. Pilinis, 1999. Continued development and testing of a new thermodynamic aerosol module for urban and regional air quality models, *Atmos. Environ.*, **33**, 1553-1560.

Seigneur, C., T. W. Tesche, P. M. Roth, and M. K. Liu, 1983. On the treatment of point source emissions in urban air quality modeling, *Atmos. Environ.*, **17**, 1655–1676.

Seigneur, C., K. Vijayaraghavan, K. Lohman, P. Karamchandani and C. Scott, 2004. Global source attribution for mercury deposition in the United States, *Environ. Sci. Technol.*, **38**, 555-569.

Seigneur, C., K. Vijayaraghavan and K. Lohman, 2006. Atmospheric mercury chemistry: sensitivity of global model simulations to chemical reactions, *J. Geophys. Res.*, **111**, D22306, doi:10.1029/2005JD006780.

Sillman, S., J. A. Logan, and S. C. Wofsy, 1990. A regional scale model for ozone in the United States with subgrid representation of urban and power plant plumes, *J. Geophys. Res.*, **95**, 5731–5748.

Sykes, R. I., S. F. Parker, D. S. Henn, and W. S. Lewellen, 1993. Numerical simulation of ANATEX tracer data using a turbulence closure model for long-range dispersion, *J. Appl. Meteor.*, **32**, 929–947.

Sykes, R. I., and D.S. Henn, 1995. Representation of velocity gradient effects in a Gaussian puff model, *J. Appl. Meteor.*, **34**, 2715–2723.

Vijayraghavan, K., P. Karamchandani and C. Seigneur, 2006. Plume-in-grid modeling of summer air pollution in Central California, *Atmos. Environ.*, **40**, 5097-5109.

Vijayaraghavan, K., R.T. Balmori, S.-Y. Chen, P. Karamchandani and C. Seigneur, 2007. Modeling of atmospheric nitrogen deposition to the Escambia Bay and watershed in the southeastern United States, *6th Annual CMAS Conference*, October 1-3, Chapel Hill, NC.

Vijayaraghavan, K., P. Karamchandani, C. Seigneur, R. Balmori, S.-Y. Chen, 2008. Plume-in-grid modeling of atmospheric mercury, *J. Geophys. Res.*, in press.

Zhang, Y., B. Pun, K. Vijayaraghavan, S.-Y. Wu, C. Seigneur, S. Pandis, M. Jacobson, A. Nenes and J.H. Seinfeld, 2004: Development and application of the Model of Aerosol Dynamics, Reaction, Ionization and Dissolution, *J. Geophys. Res.*, **109**, D01202, doi:10.1029/2003JD003501.

Zhu, Y., W.C. Hinds, S. Kim, S. Shen and C. Sioutas, 2002. Study of ultrafine particles near

a major highway with heavy-duty diesel traffic,
Atmos. Environ., **36**, 4323-4335.

121c Hygroscopic Properties of Microstructured Aerosols

Asit K. Ray, University of Kentucky, 161 F. Paul Anderson Tower, Lexington, KY 40506-0046 and Harry H. Hunter Jr., Chemical and Materials Engineering, University of Kentucky, 161 F. Paul Anderson Tower, Lexington, KY 40506-0046.

Physical and chemical processes related to aerosols are influenced by the microstructure of the individual particles, and in most cases, the region near the surface of a particle dictates its interactions with surrounding gas phase molecules. For example, surface properties dictate condensation of water vapour on cloud nuclei, and the water cycle depends on whether the surface of the particles is hydrophilic or hydrophobic. In addition, the presence of a layer or coating on a particle can significantly influence its physical and chemical transformations. For instance, although black carbon (BC) is hydrophobic, the presence of layers of hygroscopic materials, such as OC, sulphates, and sea salts, on BC particles can significantly increase the water absorbing capacity of BC-containing aerosol. One important factor that dictates cloud coverage is the water content of aerosols. The water content depends on the relative humidity, and deliquescence and crystallization points. The physical state of aerosols (i. e., solid particles or liquid droplets) is dictated by the deliquescence and crystallization humidities and the hysteresis loop. We have very limited understanding of the hygroscopic properties of microstructured aerosols, such as BC-containing aerosols.

In view of the situation we have examined hygroscopic properties of microstructured aerosols. In this study we present results on deliquescence, growth and recrystallization characteristics of inorganic salt particles coated with water-insoluble layers, as well as soot particles coated with inorganic salts. An electrodynamic balance was used to suspend single charged particles, and a suspended particle was then exposed to an air stream whose relative humidity was adjusted in a controlled manner. The mass of the particle was determined from the d.c. voltage required to maintain the particle at the centre of the balance. The deliquescence and recrystallization points were measured from the jump in the balancing d. c. voltage as well as from the light scattering signals. For examining salt particles coated with water insoluble layers, a droplet generated from a homogeneous solution, for example, of NaCl, DOP, water and ethanol, containing known amounts of NaCl and DOP, was suspended in the balance. The droplet was exposed to a dry air stream for about 20 minutes, and during this period the water and ethanol evaporated leaving a particle containing known proportion of solid NaCl and liquid DOP. After drying, the relative humidity of the air stream was altered in steps. For generating soot particles coated with inorganic salts, we dispersed a known amount of soot particles in a given amount of an aqueous solution that was saturated with a salt. We obtained a particle of soot aggregates coated with salt when a droplet generated from this mixture was dried.

The results obtained from a pure NaCl particle with the results from a NaCl particle coated with a DOP layer. For the pure NaCl particle the observed deliquescence relative humidity of 75.5% and recrystallization humidity of 45% agree well with the values reported in the literature. The results from the coated particle show that the particle gradually absorbs water as the relative humidity is increased from 0 to 24%, where the water content reaches about 36.5 wt% on a DOP-free basis. Then the water content remains constant up to 75% humidity, where the water content rises sharply similar to the deliquescence of a pure NaCl particle. During the period of lowering of humidity, NaCl remains in the solution at very low humidities. For soot particles coated with inorganic salts we observed that deliquescence relative humidities increase slightly, but the water content after the deliquescence is similar to pure salt particles.

A COMPUTATIONALLY EFFICIENT MODEL FOR SIMULATING AEROSOL INTERACTIONS AND CHEMISTRY (MOSAIC)

Rahul A. Zaveri,[1] Richard C. Easter,[1] Jerome D. Fast,[1] and Leonard K. Peters[2]
[1]Pacific Northwest National Laboratory, Richland, WA
[2]Global Laboratory Operations, Battelle, Richland, WA

Introduction

Atmospheric aerosols are ubiquitous suspended particulate matter that can range from a few nanometers up to a few microns in size. In large urban areas and mega-cities, anthropogenic aerosols are implicated in many human health related problems [Gauderman et al., 2000; Osornio-Vargas et al., 2003; Pope et al., 2004]. On regional and global scales, these as well as other naturally occurring aerosols directly influence the Earth's radiation balance by scattering and absorption of radiation, and indirectly through their impact on cloud microphysical properties and amount [Intergovernmental Panel on Climate Change, 2001]. The role of aerosols in climate forcing is, therefore, a critical factor in climate change assessment, as well as an essential element in advancing the state of the art in climate modeling. As a result, there is an urgent need for developing accurate yet computationally efficient models of aerosol chemistry and microphysics for use in air quality and climate models. This paper describes a new Model for Simulating Aerosol Interactions and Chemistry (MOSAIC), with a special focus on addressing the long-standing issues in solving the dynamic partitioning of semi-volatile inorganic gases (HNO_3, HCl, and NH_3) to size-distributed atmospheric aerosol particles.

MOSAIC Framework

MOSAIC is made highly modular to allow easy coupling between various chemical and microphysical processes. It treats all the major aerosol species important at urban, regional, and global scales, which include sulfate (SULF = SO_4^{2-} + HSO_4^-), methanesulfonate (CH_3SO_3), nitrate (NO_3), chloride (Cl), carbonate (CO_3), ammonium (NH_4), sodium (Na), calcium (Ca), black carbon (BC), primary organic mass (OC), and liquid water (W). Other unspecified inorganic species such as silica (SiO_2), other inert minerals, and trace metals are lumped together as "other inorganic mass" (OIN). The gas-phase species that are allowed to partition to the particle-phase include H_2SO_4, HNO_3, HCl, NH_3, and MSA (methanesulfonic acid). Work is currently underway to include secondary organic aerosols (SOA).

The urban to global scale trace gas photochemistry in MOSAIC is presently modeled with the "lumped-structure" photochemical mechanism CBM-Z [Zaveri and Peters, 1999], which affords a reasonable tradeoff between accuracy and computational efficiency. A condensed dimethylsulfide (DMS) photooxidation mechanism [Zaveri, 1997] has also been added to simulate the temperature dependent formation of SO_2, H_2SO_4, and MSA in the marine environment. The augmented CBM-Z scheme consists of 67 prognostic species and 164 reactions.

In the present study, MOSAIC is implemented in the sectional framework where the aerosol size distribution is divided into discrete size bins. Each bin is assumed to be internally mixed so that all particles within a bin have the same chemical composition, while particles in different bins are externally mixed. The number of bins is flexible and can be specified by the user. Both mass and number are simulated for each bin. Particle growth or shrinkage resulting from the dynamic gas-particle partitioning of trace gases (H_2SO_4, CH_3SO_3H, HNO_3, HCl, NH_3, and eventually secondary organic species) is first calculated in a Lagrangian manner. Transfer of particles between bins is then calculated using the (default) two-moment approach of Simmel and Wurzler [2006] or the moving section approach of Jacobson [1997]. Aerosol coagulation is calculated using the algorithm of Jacobson et al. [1994] with a Brownian coagulation kernel. H_2SO_4-H_2O homogeneous nucleation is modeled with Wexler et al. [1994] parameterization.

The thermodynamics module in MOSAIC is specially designed for use in dynamic gas-particle partitioning aerosol models, and is both accurate and computational efficient. It consists of the recently developed Multicomponent Taylor Expansion Method (MTEM) to calculate the activity coefficients in aqueous atmospheric aerosols [Zaveri et al., 2005a] and a computationally efficient Multicomponent Equilibrium Solver for Aerosols (MESA) to compute the intra-particle solid-liquid phase equilibrium [Zaveri et al., 2005b].

Gas-particle partitioning of semi-volatile species is a highly dynamic and competitive process, and plays an important role in the continuous evolution of atmospheric aerosols and the associated physical and chemical properties relevant for the climate forcing and air quality issues. The coupled ordinary differential equations (ODE) for dynamic gas-particle mass transfer are extremely stiff, and the available numerical techniques are either inaccurate, very expensive, or produce oscillatory solutions [e.g., Capaldo et al., 2000; Pilinis et al., 2000; Koo et al., 2003; Jacobson, 2005]. These limitations are overcome in MOSAIC with a new dynamic gas-particle partitioning algorithm called ASTEM, which is briefly described below.

Dynamic Gas-Particle Mass Transfer Algorithm: ASTEM

ASTEM stands for Adaptive Step Time-split Euler Method. It is specially designed to take advantage of several prominent characteristics of the atmospheric gas-particle partitioning problem to systematically reduce the stiffness while still maintaining the overall fidelity of the numerical solution. The first step in reducing the stiffness involves separating the non-volatile gases from semi-volatile ones in the numerical solver. Because gases such as H_2SO_4 and MSA are non-volatile, their solution is relatively straightforward and does not depend on the phase state of the particles in different bins. On the other hand, mass transfer of semi-volatile HNO_3, HCl, and NH_3 is rather complex and depends greatly on the phase-state of the aerosols. Thus, ASTEM consists of two parts which are time-split. In the first part, ASTEM analytically integrates the condensation of non-volatile gases H_2SO_4 and MSA for all the bins over a user-specified time splitting interval, h_{ASTEM}, typically set at 5 min. In the second part, it numerically integrates condensation and/or evaporation of HNO_3, HCl, and NH_3 for all the bins over h_{ASTEM} using adaptive time steps, h. The two parts are described below.

Part 1: Condensation of H_2SO_4 and MSA

Since H_2SO_4 and MSA are non-volatile, it is reasonable to assume their equilibrium surface concentrations to be zero. Furthermore, by assuming that the mass transfer coefficients of H_2SO_4 and MSA for all the bins remain constant over h_{ASTEM}, the condensation of these non-volatile gases reduces to a simple first order process. If H_2SO_4 and MSA condense on particles containing any one or mixture of salts such as $CaCO_3$, $Ca(NO_3)_2$, $CaCl_2$, $NaNO_3$, and NaCl, then the equivalent amounts of CO_2, HCl, and HNO_3 gases are evaporated due to the displacement reactions. However, if any of these salts are not present in the pre-existing aerosol bins, then some NH_3 is allowed to condense so that the aerosols do not artificially become "sulfate rich" before HNO_3, HCl, and additional NH_3 are condensed or evaporated in the next part. Once the new particle-phase sulfate, CH_3SO_3, and NH_4 concentrations are computed, the internal solid-liquid phase equilibrium in each size bin is updated with the thermodynamic module MESA.

Part 2: Condensation/Evaporation of HNO_3, HCl, and NH_3

The gas-particle mass transfer rates of HNO_3, HCl, and NH_3 greatly depend on whether the particles are completely solid, completely liquid, or mixed phase. It is then necessary to explicitly track the aerosol species concentrations in both the solid and liquid phases during each time step taken to integrate HNO_3, HCl, and NH_3. Formulae for computing the gas-particle mass transfer fluxes are described in detail in Zaveri et al. [2008]. The numerical solution for the dynamic mass transfer consists of a combination of semi-implicit and explicit Euler methods for the gas-liquid and gas-solid systems, respectively.

The ASTEM algorithm also includes a new concept of "dynamic pH" and an adaptive time-stepping scheme were developed to further reduce the stiffness in liquid particles, and thereby allow the solver to take long time steps (~100 s) and still produce smooth and accurate solutions over the entire relative humidity range. Dynamic pH is a function of equilibrium constants, mass transfer coefficients, and the gas- and particle-phase concentrations of all the involved species, and is equal to equilibrium pH at steady state. This approach provides for the first time a more accurate calculation of aerosol pH in gas-aerosol systems at non-equilibrium compared to other formulations that are based on intra-particle thermodynamic equilibrium only. The dynamic pH approach not only eliminates the spurious oscillations typically seen in numerical solutions that are based on equilibrium pH [e.g., Jacobson, 2005], but also significantly speeds up the numerical solution.

Results and Discussion

We first evaluate MOSAIC in the box-model format with a focus on validating the new ASTEM algorithm. We then evaluate MOSAIC within PNNL's 3-D chemical transport model PEGASUS using the 1987 Southern California Air Quality Study (SCAQS) dataset.

Box-Model Results

MOSAIC/ASTEM is evaluated here using 10 idealized test cases which represent different inorganic gas-aerosol systems commonly found in the urban, rural, and marine troposphere [Zaveri et al., 2008]. The time-dependent gas and aerosol concentrations are verified against those computed with a more rigorous but computationally expensive version of the model that uses LSODES (Livermore Solver for Ordinary Differential Equations with general Sparse Jacobian matrix) [Hindmarsh, 1983] instead of ASTEM for directly integrating the gas-particle partitioning equations. These test cases consist of monodisperse (i.e., single-bin) aerosols so that the steady-state MOSAIC results can also be validated against the highly accurate equilibrium model AIM, which serves as another independent benchmark for accuracy [Wexler and Clegg, 2002]. We also compare the steady-state MOSAIC results and the CPU time requirements for these test cases with those obtained with the computationally efficient equilibrium model ISORROPIA [Nenes et al., 1998, 1999].

Figure 1 shows the time evolution of gas and particle concentrations for cases 1-3 at low RH where the given aerosols are completely solid. Figure 2 shows the results for cases 4-7 at high RH where the given aerosols are completely liquid. Figure 3 shows the results for cases 8-10 at moderate RH where the given aerosols contain both the solid and liquid phases. In all cases, MOSAIC/ASTEM predictions (lines) are in excellent agreement with the MOSAIC/LSODES results (filled circles), and the final steady-state solutions matched the AIM equilibrium results (filled squares) almost exactly. On the other hand, the ISORROPIA equilibrium results (filled triangles) show significant deviations from MOSAIC and AIM predictions in solid and mixed-phase test cases. ISORROPIA results were in very good agreement with MOSAIC and AIM for most of liquid cases. However, H^+ ion concentrations predicted by ISORROPIA had large errors compared to MOSAIC and AIM. Finally, despite using long time steps of 100 s at all times, MOSAIC/ASTEM produced smooth and accurate solutions in all the cases.

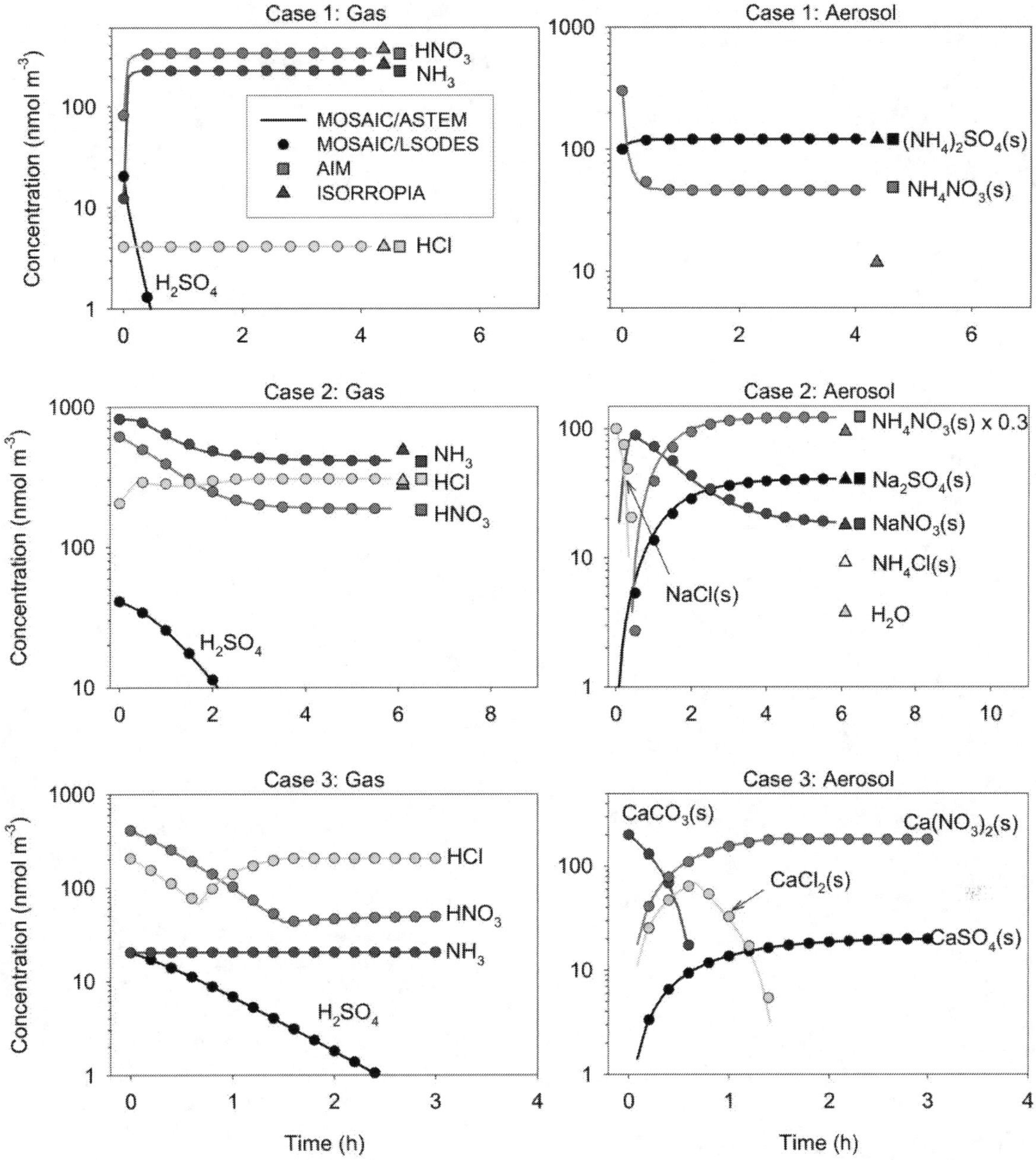

Figure 1. Comparison of the predicted gas and aerosol species from MOSAIC/ASTEM (lines), MOSAIC/LSODES (filled circles), equilibrium ISORROPIA (filled triangles), and equilibrium AIM Model III (filled squares) for completely solid aerosol test cases 1-3. Note that AIM predictions are not included for case 10 which contains calcium.

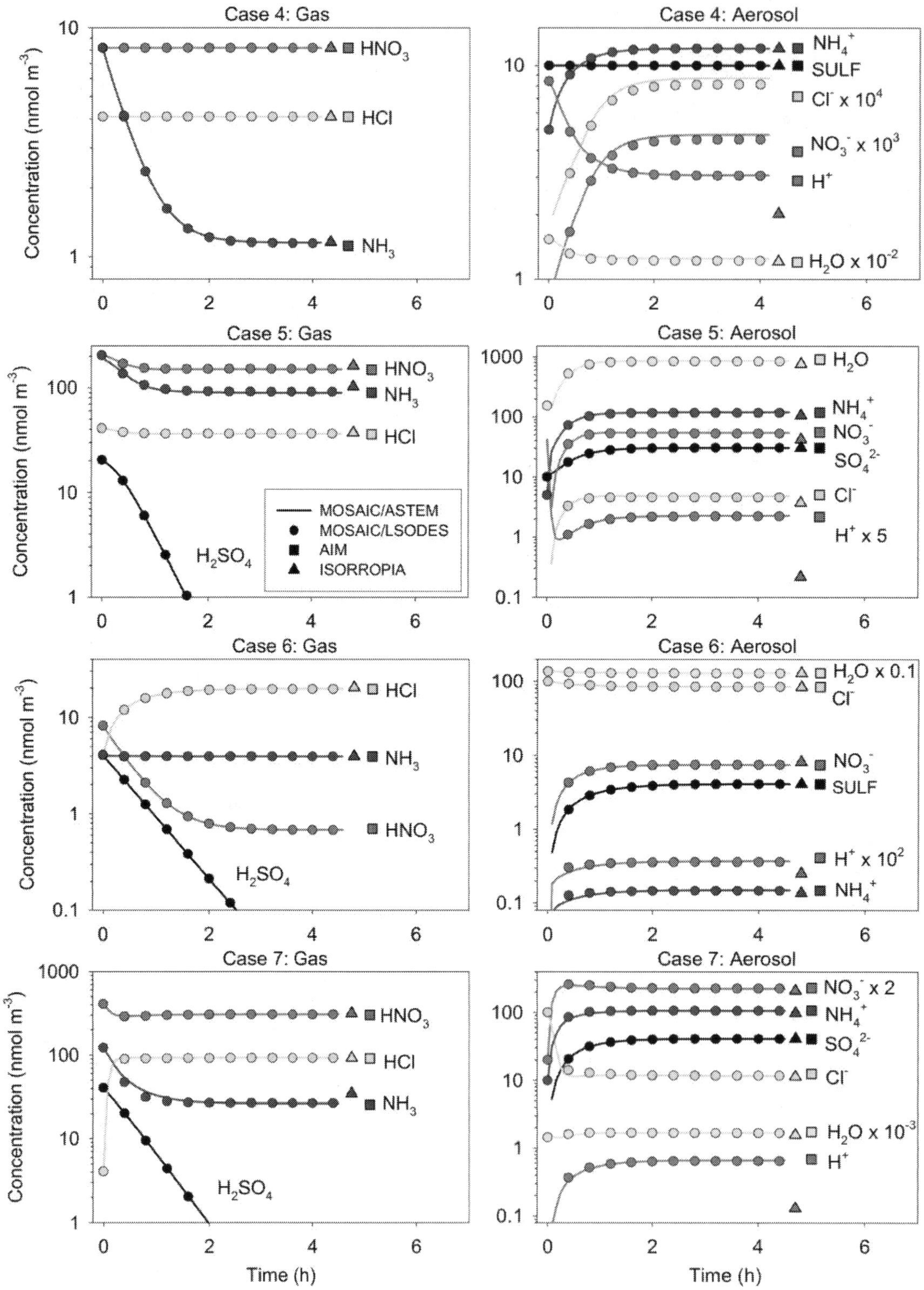

Figure 2. Comparison of the predicted gas and aerosol species from MOSAIC/ASTEM (lines), MOSAIC/LSODES (filled circles), equilibrium ISORROPIA (filled triangles), and equilibrium AIM Model III (filled squares) for completely liquid aerosol test cases 4-7.

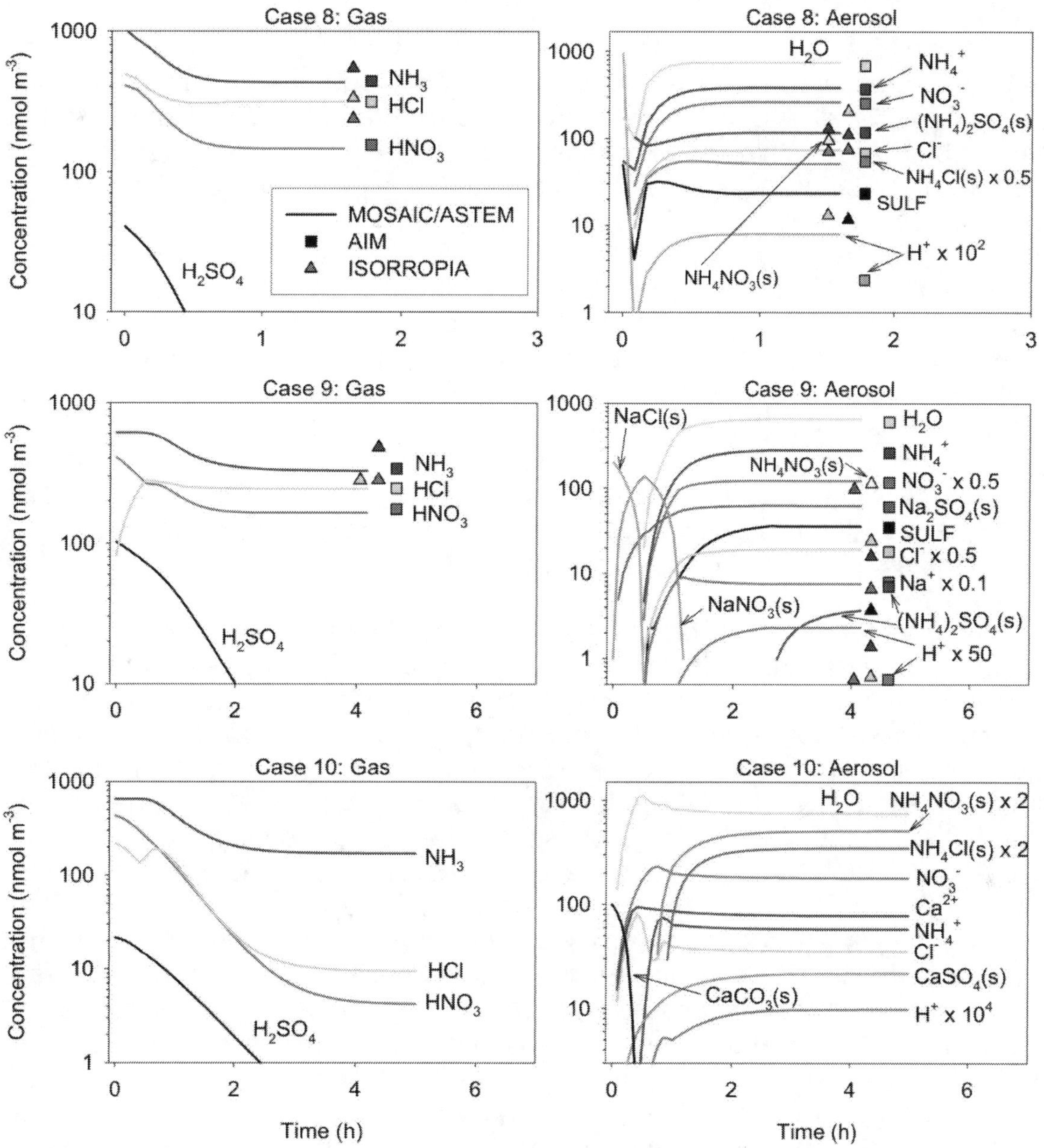

Figure 3. Comparison of the predicted gas and aerosol species from MOSAIC/ASTEM (lines), equilibrium ISORROPIA (filled triangles), and equilibrium AIM Model III (filled squares) for mixed-phase aerosol test cases 8-10. Note that ISORROPIA and AIM predictions are not included for case 10 which contains calcium. LSODES had difficulty converging for the mixed-phase cases, and therefore MOSAIC/LSODES results are not shown in the comparisons.

3-D Model Results

The above box-model tests and analysis verifies and validates the accuracy and robustness of the new thermodynamic and gas-particle partitioning modules in MOSAIC under widely different chemical and RH conditions. we now evaluate MOSAIC within a comprehensive 3-D Eulerian chemical transport model, with the primary goal of estimating its computational efficiency under more realistic ambient conditions and continuously changing meteorological controls. To this end, MOSAIC was implemented in PNNL's Eulerian Gas and Aerosol Scalable Unified System (PEGASUS) [Fast et al., 2002] and applied to simulate trace gas and aerosol evolution in the Los Angeles basin between 08 UTC 26 August and 06 UTC 29 August during the 1987 Southern California Air Quality Study (SCAQS). The model domain was resolved with 78 x 28 x 31 grid points in the X, Y, and Z dimensions, respectively, which gives a total of 67,704 grid cells. The horizontal grid resolution was 5 X 5 km^2 while the vertical grid was non-uniformly spaced. Both the transport-chemistry time splitting interval and ASTEM time splitting interval were set at 300 s, and the maximum allowable internal time step in ASTEM was restricted to 100 s.

Pollutants emitted in downtown Los Angeles were transported primarily from west to east during each afternoon for 3-day period simulated here. Because of photochemical production of secondary aerosol species, the highest particulate mass concentrations were usually measured at the downwind Claremont and Riverside sites located ~50 km and ~75 km east of downtown Los Angeles, respectively. Model performance at the central Los Angeles and Claremont sites only are therefore chosen in the present evaluation for brevity. The predicted diurnal and spatial variations in ozone and particulate matter (not shown) were consistent with the observations, with the peak values east of Los Angeles in the late afternoon. Figure 4 shows comparisons of the predicted (line) and observed (filled circles) diurnal variations in several key gas and particulate (PM2.5) species concentrations at the central Los Angeles and Claremont sites. The predicted values are averages of the values in the grid cell closest to the site and the eight horizontally-adjacent grid cells. The gray shading represents the range of the lowest and the highest values predicted in these nine grid cells. The predicted (average) and observed O_3 diurnal profiles are in excellent agreement, and the small differences between the minimum and maximum values in the nine grid cells indicate relatively weak horizontal gradients in O_3 concentrations at the two sites. On the other hand, relatively larger differences in the predicted minimum and maximum concentrations are seen for various aerosol species, indicating sharp gradients for aerosols at the two sites. As a result the predicted aerosol concentrations are sensitive to small errors in winds and the associated meteorological transport at these and other observation sites. Nevertheless, the predicted average concentrations for the various gas and particulate species agree quite well with the observations at the central Los Angeles and Claremont sites. The model tends to overpredict the concentrations of NH_3 and HNO_3 in the gas phase as well as particulate NH_4 and NO_3 at Claremont, although the

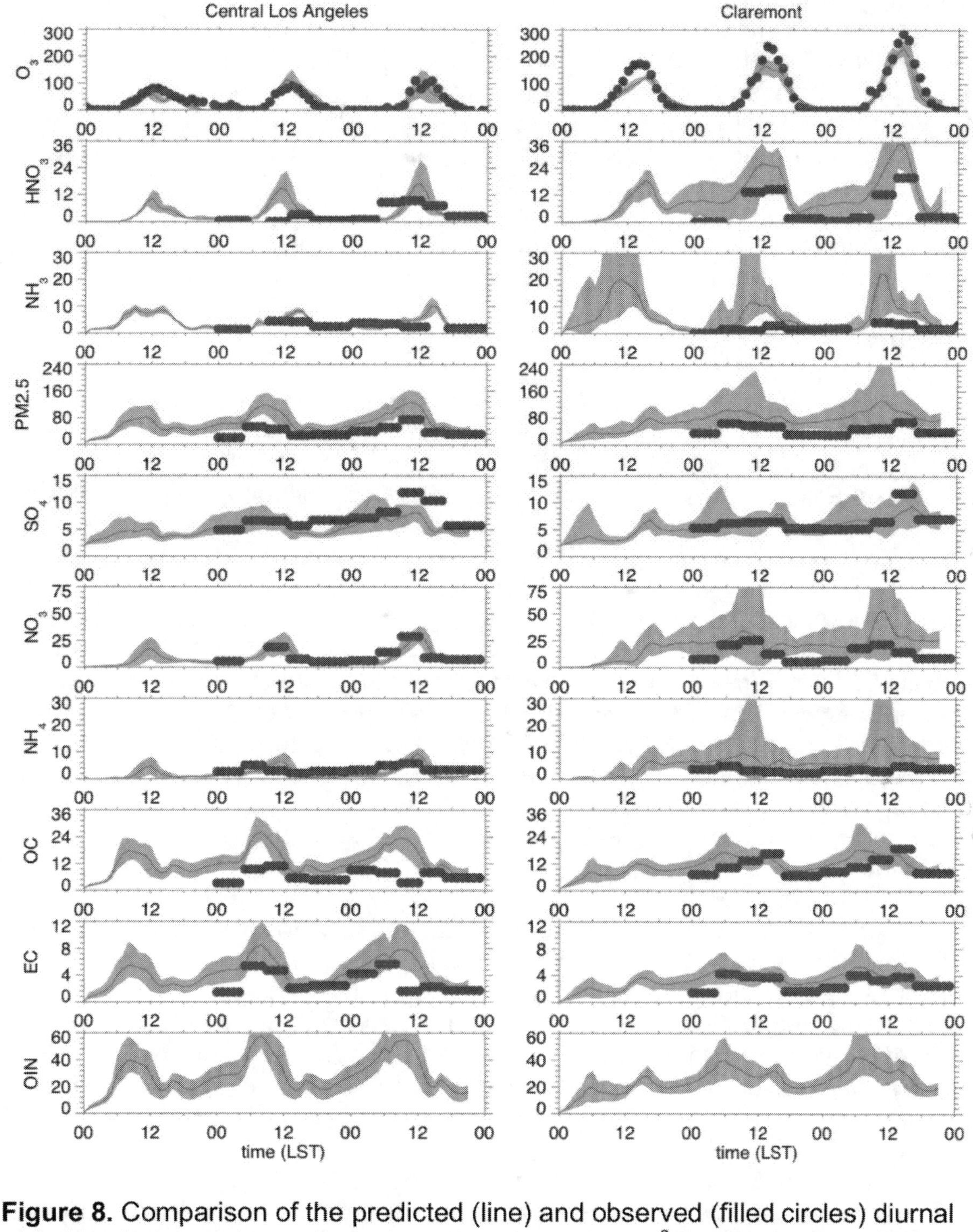

Figure 8. Comparison of the predicted (line) and observed (filled circles) diurnal variations in several key gas (ppbv) and aerosol ($\mu g\ m^{-3}$) species concentrations at the central Los Angeles and Claremont sites. The lines represent average of the values in the grid cell closest to the site and the eight horizontally-adjacent grid cells. The gray shading represents the range of the lowest and the highest values predicted in these nine grid cells.

minimum of the nine grid cells compared well with the observations. This suggests that the discrepancies between model predictions and observations at Claremont are likely due errors/uncertainties in transport and mixing and/or in the emission inventory for NO_x and NH_3. The predicted PM2.5 mass was somewhat higher than observed at Los Angeles as a result of high organic carbon and other inorganic mass, while predicted PM2.5 was higher than observed at Claremont as a result of nitrate and other inorganic mass. Unfortunately, there were no measurements of other inorganic mass to directly evaluate the model for this species. The reason predicted organic carbon was similar to the measurements even though MOSAIC does not presently treat SOA is that emissions of organic carbon were increased from the original estimates as in Zhang et al. [2004].

Computational Efficiency

The CPU timing tests for the box-model tests showed that MOSAIC/ASTEM is two orders of magnitude faster than MOSAIC/LSODES. MOSAIC's efficiency was also determined by comparing its CPU speed with that of ISORROPIA, which is considered to be one of the fastest equilibrium models currently available [Ansari and Pandis, 1999]. Since MOSAIC is a fully dynamic model, we compare its average CPU time requirement per size bin per 5 min integration interval (typical 3-D model time step) to the CPU time requirement per single, bulk equilibrium calculation with ISORROPIA. The codes were tested with three different compilers (Portland Group Fortran 77/90, PathScale Fortran 77/90, and Intel Fortran 77/90) on two different Linux workstation platforms (3.0 GHz Intel Xeon and 3.0 GHz Intel EM64T processors). The CPU time requirements for MOSAIC and ISORROPIA were found to be similar for most of the cases tested in this study. However, for some test cases, MOSAIC was found to be up to 20 times faster ISORROPIA. At the same time, MOSAIC was also found to be much more accurate than ISORROPIA under these conditions as discussed earlier.

For the 3-D model test, the 3-day Los Angeles Basin simulation took ~1 CPU day. The average CPU time per grid cell per hour was ~20 ms of which 12 ms (60% of the total CPU time) was taken up by the dynamic gas-particle partitioning, thermodynamic equilibrium, and sectional growth calculations in MOSAIC. This corresponds to an average CPU time of 125 μs per bin per 5 min, which is comparable to the range of CPU times seen for the box-model tests cases.

While more comprehensive comparisons of accuracy and efficiency of MOSAIC with other dynamic models are warranted in the future, the present study demonstrates that MOSAIC is extremely efficient without sacrificing accuracy, and is therefore highly attractive for use in large scale, 3-D air quality and aerosol models.

Summary

We have described the various chemical, thermodynamic, and microphysical processes currently represented in the new aerosol model MOSAIC (Model for Simulating Aerosol Interactions and Chemistry), with a special focus on addressing the long-standing problems in dynamic gas-particle partitioning of semi-volatile inorganic species. The coupled ordinary differential equations describing the dynamic gas-particle mass transfer process are stiff and have proved to be extremely difficult to solve – the numerical techniques available in the literature are either computationally too expensive or produce oscillatory and/or inaccurate steady-state results. These problems were overcome in MOSAIC with a new dynamic gas-particle partitioning solver ASTEM (Adaptive Step Time-split Euler Method). The ASTEM algorithm reduces the stiffness in the ODEs by first analytically solving the condensation of all the non-volatile gases (H_2SO_4 and CH_3SO_3H) over a time-splitting interval, which is typically set at ~300 s. This is then followed by a numerical solution for the dynamic mass transfer of semi-volatile gases (HNO_3, HCl, and NH_3) over the same time interval with a combination of semi-implicit and explicit Euler methods for the gas-liquid and gas-solid systems, respectively. A new concept of "dynamic pH" and an adaptive time-stepping scheme were developed to further reduce the stiffness in liquid particles, and thereby allow the solver to take long time steps (~100 s) and still produce smooth and accurate solutions over the entire relative humidity range.

MOSAIC was evaluated for several test cases representing different gas-aerosol systems commonly found in the troposphere. Performance of the new ASTEM solver was verified against a benchmark version of MOSAIC that uses the stiff ODE solver LSODES to rigorously integrate the dynamic mass transfer of all the species simultaneously without the numerical approximations used in ASTEM. The time-varying predictions with MOSAIC/ASTEM and MOSAIC/LSODES were found to be in excellent agreement for all the cases (monodisperse and polydisperse aerosols) tested under low and high RH conditions, thus validating the ASTEM algorithm. Furthermore, the final (steady-state) MOSAIC results were in excellent agreement with those obtained with the benchmark equilibrium model AIM for the monodisperse aerosol test cases at low, moderate, and high RH. On the other hand, aerosol composition, concentration, water content, and pH predicted by the computationally efficient equilibrium model ISORROPIA were also in very good agreement with AIM and MOSAIC for most of the high RH test cases, but relatively larger errors were seen under low and moderate RH conditions. Furthermore, the CPU times required for dynamic solutions by MOSAIC per size bin per 5 min integration interval (typical 3-D model time step) were similar to the CPU times required for single, bulk equilibrium solutions by ISORROPIA, although in some instances MOSAIC took ~20 times less CPU time than ISORROPIA. For the polydisperse aerosol test cases, MOSAIC/ASTEM was found to be 100-300 times faster than MOSAIC/LSODES.

An eight-bin version of MOSAIC was then applied and evaluated within Pacific Northwest National Laboratory's offline 3-D chemical transport model PEGASUS for southern California area using the historic 1987 SCAQS dataset. The primary goal of this exercise was to estimate MOSAIC's computational efficiency under more realistic conditions. Simulated diurnal gas and aerosol variations at the central Los Angeles and downwind Claremont sites compared well with the observations. The total CPU time per grid cell per hour averaged over the entire simulation was 20 ms (on a 3.0 GHz Intel Xeon processor), ~60% of which was used up by aerosol calculations in MOSAIC. These results show that MOSAIC is extremely efficient without compromising accuracy, and is therefore highly attractive for use in 3-D air quality and large scale aerosol models.

Acknowledgements

The authors thank A. Nenes of Georgia Tech, Atlanta, for providing a copy of the ISORROPIA code and assisting in comparing its performance with MOSAIC. Funding for this research was provided by the U.S. Department of Energy (DOE) under the auspices of the Atmospheric Science Program of the Office of Biological and Environmental Research, the NASA Earth Science Enterprise under grant NAGW 3367, and Pacific Northwest National Laboratory (PNNL) Laboratory Directed Research and Development (LDRD) program. Pacific Northwest National Laboratory is operated for the U.S. Department of Energy by Battelle Memorial Institute under contract DE-AC06-76RLO 1830.

References

1. Capaldo, K. P., C. Pilinis, and S. N. Pandis (2000), A computationally efficient hybrid approach for dynamic gas/aerosol transfer in air quality models, *Atmos. Environ. 34*, 3617–3627.
2. Fast, J. D., R. A. Zaveri, X. Bian, E. G. Chapman, and R. C. Easter (2002), Effect of regional-scale transport on oxidants in the vicinity of Philadelphia during the 1999 NE-OPS field campaign, *J. Geophys. Res.*, *107*(D16), 4307, doi:10.1029/2001JD000980.
3. Gauderman W. J., R. McConnell, F. Gilliland, S. London, D. Thomas, E. Avol, H. Vora, K. Berhane, E. B. Rappaport, F. Lurmann, H. G. Margolis, and J. Peters (2000), Association between air pollution and lung function growth in southern California children, *Am. J. Respir. Crit. Care Med.*, *162*, 1383-1390.
4. Hindmarsh, A. C. (1983), ODEPACK, a systematized collection of ODE solvers. In *Scientific Computing*, edited by R. S. Stepleman et al. North-Holland, Amsterdam, pp. 55-64.
5. Intergovernmental Panel on Climate Change (2001), *Climate Change 2001: The Scientific Basis*, Cambridge Univ. Press, New York. (Available at http://www.grida.no/climate/ipcc_tar/)

6. Jacobson, M.Z. (1997), Development and application of a new air pollution modeling system – II. Aerosol module structure and design, *Atmos. Environ.*, *31*, 131-144.
7. Jacobson, M. Z., A solution to the problem of nonequilibrium acid/base gas-particle transfer at long time step, *Aerosol Science and Technology*, *39*, 92–103, 2005.
8. Jacobson, M. Z., R. P. Turco, E. J. Jensen, and O. B. Toon (1994), Modeling coagulation among particles of different composition and size, *Atmos. Environ.*, *28*, 1327-1338.
9. Koo, B., T. M. Gaydos, and S. N. Pandis (2003), Evaluation of the equilibrium, dynamic, and hybrid aerosol modeling approaches, *Aerosol. Sci. Technol.*, *37*, 53–64.
10. Nenes, A., S. N. Pandis, and C. Pilinis (1998), ISORROPIA: A new thermodynamic equilibrium model for multiphase multicomponent inorganic aerosols, *Aquatic Geochem.*, *4*, 123–152.
11. Nenes, A., C. Pilinis, and S. N. Pandis (1999), Continued development and testing of a new thermodynamic aerosol module for urban and regional air quality models, *Atmos. Environ.*, *33*, 1553–1560.
12. Osornio-Vargas, A. R., J. C. Bonner, E. Alfaro-Moreno, L. Martinez, C. Garcia-Cuellar, S. P. Rosales, J. Miranda, and I. Rosas (2003), Proinflammatory and cytotoxic effects of Mexico City air pollution particulate matter *in vitro* are dependent on particle size and composition, *Environ. Health Perspectives*, *111*, 1289–1293.
13. Pilinis C., K. P. Capaldo, A. Nenes, and S. N. Pandis (2000), MADM – A New Multicomponent Aerosol Dynamics Model, *Aerosol Sci. Technol., 32*, 482–502.
14. Pope C. A., R. T. Burnett, G. D. Thurston, M. J. M. Thun, E. Calle, D. Krewski, and J. J. Godleski (2004), Cardiovascular mortality and long term exposure to particulate air pollution, *Circulation*, *109*, 71-77.
15. Wexler, A. S., F. W. Lurmann, and J. H. Seinfeld (1994), Modelling urban and regional aerosols – I. Model development, *Atmos. Environ.*, *28*, 531-546.
16. Wexler A. S., and S. L. Clegg (2002), Atmospheric aerosol models for systems including the ions H^+, NH_4^+, Na^+, SO_4^{2-}, NO_3^-, Cl^-, Br^-, and H_2O, *J. Geophys. Res.*, *107*(D14), 4207, doi:10.1029/2001JD000451.
17. Zaveri, R. A. (1997), Development and evaluation of a comprehensive tropospheric chemistry model for regional and global applications, Ph.D. thesis, Va. Polytech. Inst. and State Univ., Blacksburg, Va.
18. Zaveri, R. A., and L. K. Peters (1999), A new lumped structure photochemical mechanism for large-scale applications, *J. Geophys. Res.*, *104*, 30387-30415.
19. Zaveri, R. A., R. C. Easter, and A. S. Wexler (2005a), A new method for multicomponent activity coefficients of electrolytes in aqueous atmospheric aerosols, *J. Geophys. Res.*, *110*, D02201, doi:10.1029/2004JD004681.

20. Zaveri, R. A., R. C. Easter, and L. K. Peters (2005b), A computationally efficient Multicomponent Equilibrium Solver for Aerosols (MESA), *J. Geophys. Res. 110*, D24203, doi:10.1029/2004JD005618.
21. Zaveri, R. A., R. C. Easter, J. D. Fast, and L. K. Peters (2008), Model for Simulating Aerosol Interactions and Chemistry (MOSAIC), *J. Geophys. Res.,* 113, D13204, doi:10.1029/2007JD008782.
22. Zhang, Y., B. Pun, K. Vijayaraghavan, S.-Y. Wu, C. Seigneur, S. N. Pandis, M. Z. Jacobson, A. Nenes, and J. H. Seinfeld (2004), Development and application of the Model of Aerosol Dynamics, Reaction, Ionization, and Dissolution (MADRID), *J. Geophys. Res.*, *109*, D01202, doi:10.1029/2003JD003501.

121e Effect of Temporal Averaging of Vertical Eddy Diffusivity on the Forecast Quality of Surface Zone Concentration of the National Air Quality Forecast Capability

Pius C. Lee[1], Youhua Tang[1], Jeff McQueen[1], Hochun Huang[1], Sarah Lu[1], Marina Tsidulko[1], Rohit Mathur[2], Jon Pleim[2], Tanya Otte[2], George Pouliot[2], Ken Schere[2], and Paula Davidson[3]. (1) Emc/ncep/noaa, NOAA, 5200 Auth Rd, Camp Springs, MD 20746, (2) Atmospheric Modeling Division, NOAA, US EPA, Rtp, NC 27711, (3) Nws/noaa, Office of Science and Technology, 1327 East West Hwy, Silver Spring, MD 20910

Air Quality forecast models are employed to provide numerical guidance for forecasters to issue timely ozone and particulate matter concentration forecast pertinent to human health exposures to the communities they serve. The National Centers for Environmental Prediction (NCEP), National Weather Service Weather Research and Forecasting Non-hydrostatic Mesoscale Model (WRF/NMM) has been coupled with the EPA Community Multi-scale Air Quality (CMAQ) model to form the National Air Quality Forecast Capability (NAQFC). It provides such guidance for the majority part of the U.S. since 2006. This capability first provided forecast coverage over the northeastern U.S. in 2004 with WRF predecessor, the Eta model, as the meteorological driver. Both the earlier and the current capabilities are off line systems. Coupling of the turbulent mixing processes of the meteorological driver and CMAQ has been a challenge. Both Eta and WRF/NMM use the Mellor-Yamada-Janjic (MYJ) turbulence mixing scheme in the boundary layer. The option of directly using MYJ provided vertical eddy diffusivity to calculate turbulent mixing of the atmospheric constituents in CMAQ has been tested. This study investigated the effect of various temporal averaging treatments of vertical eddy diffusivity in redistributing ozone in the boundary layer. Model predicted concentration structures of ozone resulting from these various temporal treatments of vertical eddy diffusivity have been studied and verified using AIRNOW, an EPA surface ozone concentration network, and ozone-sonde data.

SOLVENT EXTRACTION OF LOW GRADE COALS FOR CLEAN LIQUID FUELS

Elliot B. Kennel, Mayuri Mukka, Alfred H. Stiller and John W. Zondlo
West Virginia University
Department of Chemical Engineering
PO Box 6102 Morgantown WV 26506-6102
304-293-2111 x2423
Elliot.kennel@mail.wvu.edu

Abstract

Solvent extraction of bituminous coals has been used as a means of coproducing clean liquid transportation fuels as well as solid fuels for gasification. Coal solvents are created by first hydrogenating coal tar distillate fractions to the level of a fraction of a percent and then simply dissolving the coal in the hydrogenated solvent. This enables bituminous coal to enter the liquid phase under conditions of high temperature (above 400 °C). The pressure is controlled by the vapor pressure of the solvent and the cracked coal. Once liquefied, mineral matter can be removed via centrifugation, and the resultant superheavy crude can be processed to make pitches, cokes and lighter products.

Lower rank coals such as sub-bituminous coal and lignite are desirable feedstocks for this process due to their low cost and high aliphatic character, which can result in superior transportation fuels. However, these advantages are partially offset by the disadvantages of high moisture content and high ash content which typically accompany lignite and sub-bituminous coals.[1] In particular, ash content of approximately 20% is problematic because centrifugation might not succeed in decreasing the ash content of the extract. Much of the liquid product would be contained in the nominal tails rather than in the separated liquid centrate.

In order to overcome this, a more complete liquid separation can be accomplished by vacuum distillation. Mineral matter is further heat treated to produce a value-added slag product. Solids separation can be over 90% effective using this technique depending upon the degree to which coal molecules are broken down during the solvent extraction process.

Concept

Solvent extraction is an evolutionary way by which criteria emissions can be reduced during the production process while replacing processes such as co-production of coal tar from metallurgical grade coke production, as a means of obtaining liquid products from coal. Coal tar results from the condensation of vapors emitted from coal as it is coked in a metallurgical coke oven. However, non-recovery ovens, in which the coal tar is consumed in order to fuel the metallugical grade cokemaking, offer superior environmental performance, and thus receive additional consideration as the process of choice in the modern environmentally conscious era. Hence, supplies of coal tar are decreasing even as demand for it is increasing.[2]

Because coal tar is an aromatic liquid, aromatic chemicals from it are more easily produced rather than aliphatic chemicals preferred for transportation fuels and the like. This is not to say, however, that coal liquids are inherently useless for fuels. Other aromatic heavy liquids, such as Athabasca Oil Sands crudes and Venezuelan Orinoco crudes, might have been considered unacceptable for the fuels industry a generation ago, but are now routinely handled in state-of-the-art North American refineries.[3] Were it not for the fact that established markets exist for chemicals derived from coal tars, no doubt coal tar would be increasingly utilized as a feedstock for transportation fuels, as the technology certainly exists to upgrade such hydrocarbons.

In the case of solvent extraction processes, the present authors advocate the creation of heavy aromatic coal-derived crudes from coal as a simpler and less expensive process than the creation of synthetic sweet light crudes.

This process involves dissolving coal in a commodity solvent such as decant oil or coal tar distillate, usually with the addition of hydrogen-rich diluent in order to increase the total solubility of the coal, and to decrease the viscosity. The total amount of hydrogenation in the solution can be as low as a fraction of a percent by mass, compared to the level obtained in commodity solvents (e.g., coal tar distillate).

In this way, it is possible to dissolve up to about 90% of the coal feedstock as determined on a dry ash-free basis. Both bituminous as well as sub-bituminous rank coals have been successfully trialed in this way. The result is slurry containing dissolved coal as well as undissolved mineral matter and fixed carbon. The solid phase is then removed via centrifugation, resulting in a heavy, aromatic synthetic hydrocarbon liquid.

The role of hydrogen was historically viewed as being "donated" to solid molecules in the coal in order to create lower molecular weight species and render them more fluid.[4] In the current application, however, it is recognized that the amount of hydrogen is insufficient to cause fluidity by transfer to molecules in the coal structure. Rather, hydrogen is seen as an agent that causes the breakup of polymerized molecules in the coal. Hence a single hydrogen atom can break up many hydrocarbon molecules. This somewhat surprising issue is supported by examination of Tables 1-3, in which it is seen that the mass ratio of hydrogen to carbon is actually lower for binder pitch (i.e., the product), than either the coal or coal tar distillate solvent.

Table 1. Elemental Analysis of Bituminous Coal (Kittanning).

Element	Average	% Rel. S. D.
Nitrogen%	1.7	2.2
Carbon%	77.9	1.1
Hydrogen%	5.3	1.0
Sulfur%	1.8	22.9
H/C ratio	0.07	

Table 2. Elemental Analysis of Coal Tar Distillate (Solvent/Diluent).

Component Name	Average	% Rel. S. D.
Nitrogen %	1.1	1.2
Carbon %	92.7	0.2
Hydrogen %	6.2	0.05
Sulfur %	0.3	6.4
H/C ratio	0.07	

Table 3. Elemental Analysis of Binder Pitch (110 °C Softening Point).

Component Name	Average	% Rel. S. D.
Nitrogen%	1.2	3.8
Carbon%	93.3	0.05
Hydrogen%	4.3	0.6
Sulfur%	0.3	2.5
H/C	0.05	

Thus basic principles suggest that additional hydrogen is not needed to convert coal to binder pitch, but in fact a small amount of excess hydrogen might be generated. In practice, however, a certain amount of hydrogen is lost due to methane generation or entrainment of liquid products in the centrifuge tails, requiring that coal feedstocks be supplemented by addition of a makeup solvent/diluent. The environmental consequence is that hydrogen is consumed only in small amounts (less than one mass percent of the product) to generate a pitch product, implying that carbon generation during hydrogen production is also very small. A zero-hydrogen process is also potentially achievable.

Generation of thermal energy for processing is another requirement that could be associated with generation of carbon dioxide. This would have to be evaluated on a case-by-case basis depending on the particular method used to generate thermal energy, and the associated capture and sequestration strategy associated with each case.

The refining process of the resultant coal extract solution would presumably be conceptually similar to the processes used for upgrading petroleum-based crudes such as Orinoco Venezuelan or Athabasca Oil Sands. Extensive hydrotreating would be necessary in order to remove sulfur and nitrogen compounds, as well as to enhance aliphatic content and to lower the molecular weight distribution.

Fourier Transform InfraRed (FTIR) spectroscopy has been used to observe enhancements in the aliphatic content of treated solvents. This is based on the donor solvent model, in which naphthalene acquires four donatable hydrogen atoms via the hydrotreatment process, resulting in the formation of tetrahydronaphthalene (Tetralin®). Thus hydrogen is added to the system in an aliphatic state, as shown in the figures below.

The ratio of the aromatic peak area to the aliphatic peak area in the FTIR spectrum is used as an indicator of the effectiveness of the process. However, the modifications introduced in the current research effort have complicated the situation. Because coal tar distillation cuts have been substituted for pure Tetralin®, the donor solvent reaction may take place in several aliphatic compounds and not just in Tetralin®. Conversely, other aromatic-to-aliphatic conversions may occur without resulting in donatable hydrogen.

For this reason, Nuclear Magnetic Resonance (NMR) is the preferred technique to distinguish between donatable and non-donatable hydrogen. In other words, separate measurements identify hydrogen as aromatic hydrogen, aliphatic donatable hydrogen and aliphatic non-donatable hydrogen.

Although FTIR is not definitive in this respect, it is very fast and suitable for making measurements in a matter of a few minutes. Hence an algorithm was developed for determining the success of hydrotreatment based on the ratio of FTIR peak areas corresponding to aromatic hydrogen and to aliphatic hydrogen. The basic technique is to define a peak area by truncating the peak, and subtracting the background as illustrated by the calculations below.

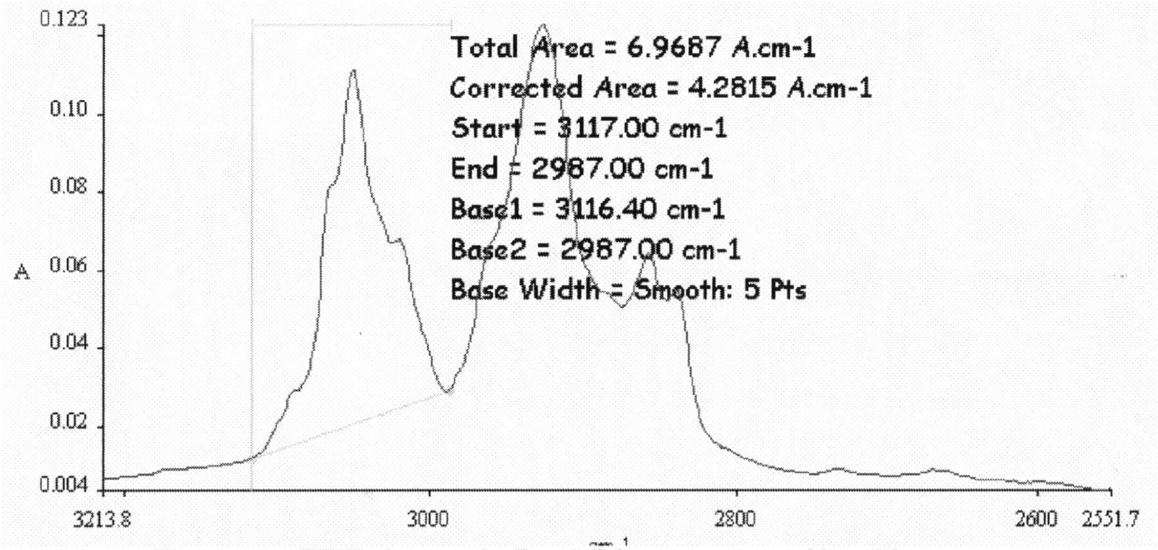

Figure 1. FTIR Aromatic Peak Measurement Algorithm.

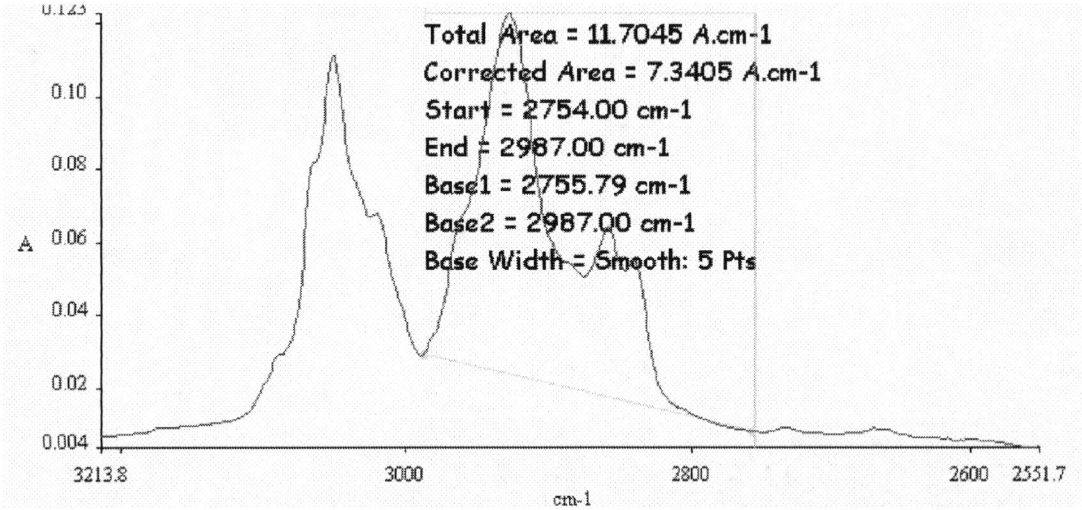

Figure 2. Sample FTIR Aliphatic Peak Measurement Calculation.

To correlate FTIR measurements with the actual amount of donatable hydrogen absorbed, and to verify that results obtained using the Perkin-Elmer FTIR can be duplicated on other FTIR machines, different calibration standards were examined using both instruments and compared. Typical values for aromatic to aliphatic ratio are about 2.0 before the reaction and 0.5 to 1.0 after the reaction is carried out as estimated by FTIR, consistent with the notion that aromatic content is decreased as the reaction is carried out. Again however, this does not prove that hydrogen donation is actually the primary reaction that causes a phase transition. Depolymerization may be at least as significant in many processes.

FTIR studies verify that the digestion process tends to result in reduced aromatic content, although coal derived fuels are still higher in aromatic content than desired by fuels producers, suggesting that blends with highly aliphatic liquids may be desirable.

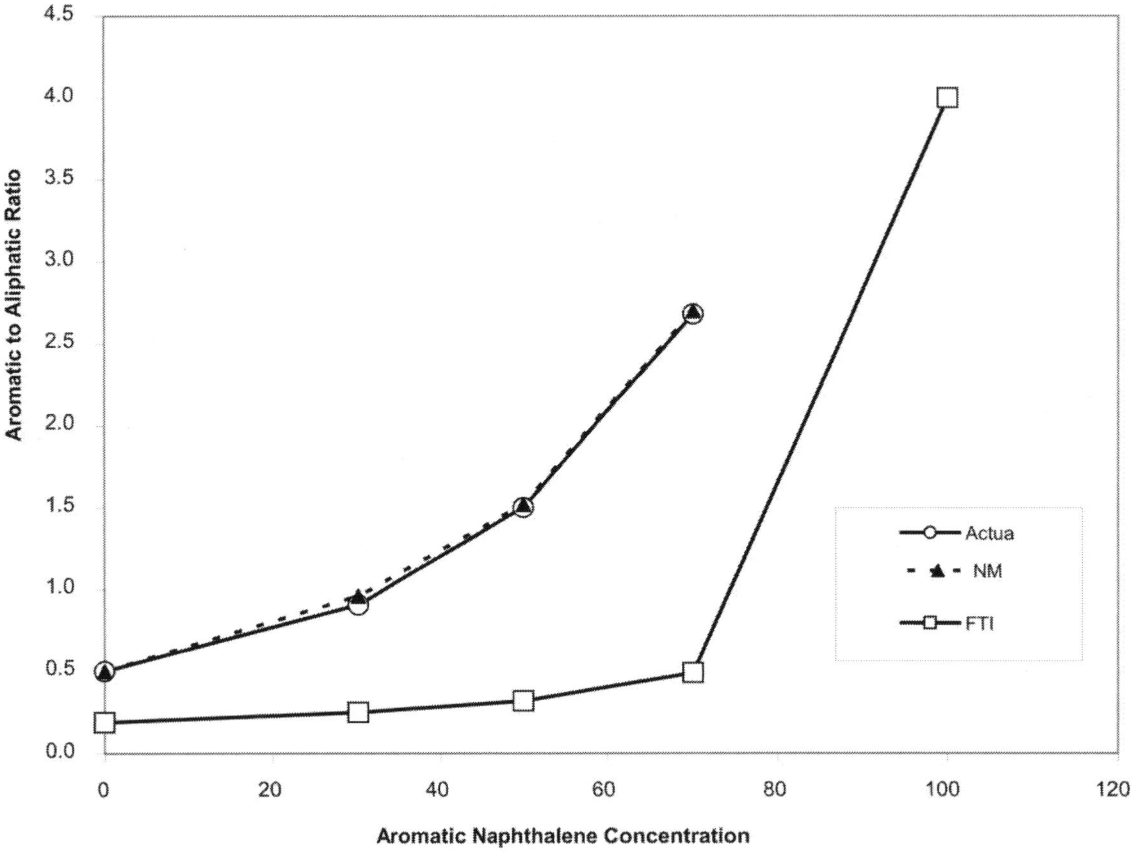

Figure 3. Aromatic to aliphatic ratio as estimated by FTIR and NMR in a solution of naphthalene and tetralin.

A logical extension of this effort would be to create extracts from lower rank coals from the western US and Canada. One reason is that lower rank coals may offer enhanced aliphatic content and reduced aromatic content. This may be a disadvantage from the standpoint of commercial pitch production since pitches require aromaticity in order to create well-ordered structural carbon for use in the metals smelting industry. However, from the standpoint of fuels production, lower aromatic content is an advantage from the standpoints of handling, combustion rate kinetics, etc.

Blends of coal-derived liquids with other fuels such as biofuels may be desirable in order to achieve proposed standards on carbon footprint. In addition, other parameters such as aromatic-to-aliphatic ratio, combustion rate constant, specific enthalpy content, lubricity, cold viscosity etc. might be effectively modified by combining coal derived fuels with other fuels having desirable environmental characteristics. As an example, a blend of 85% ethanol with 15% coal derived liquids of the appropriate molecular weight would likely combust cleanly since the aromatic content of the coal derived liquids would be outweighed by the nonaromatic ethanol. At the present time, an agreed-upon carbon accounting methodology has not yet been developed, however.

Summary

The results suggest that fuels produced from solvent extraction of coal can be "greener" than conventional fuels if not perfectly so. Direct liquefaction processes do not produce CO_2 directly although anytime energy or hydrogen is consumed, the processes used to generate those commodities need to be considered for their environmental footprint. Combinations of coal-derived fuels with fuels with exceptional environmental characteristics such as biofuels may offer the best overall combination of environmental and other performance metrics.

Acknowledgement

The support of the US Department of Energy, National Energy Technology Laboratory, Contract DE-FC26-06NT42761, is gratefully acknowledged. The support of the Kingwood Mining Company, which provided the coal for this effort, is also gratefully acknowledged.

References

[1] Kouichi Miura et al., Dewatering of coal through solvent extraction, Kyoto University, Fuel 81 (2002) 1417-1422.

[2] K. Fitzgerald et al., "Pitch Quality and Supply Trends," Sixth Annual Carbon Conference, Great Lakes Carbon Corporation, The Woodlands, Texas, 27 September 2000).

[3] S. Sayles, "Unconventional Crude Upgrading Challenges," Petroleum Technology Quarterly, Vol 12; No 5, pp. 43-48, 2007.

[4] S. Lee, J. G. Speight, S. K. Loyalka (editors), *Handbook of Alternative Fuel Technologies*, CRC Press, New York, 2007.

Effects of Quench Rate, NO, and Quartz Surface Area on Gas Phase Oxidation of Mercury by Bromine

Brydger Cauch
Reaction Engineering International, Salt Lake City, UT, USA

Constance L. Senior
Reaction Engineering International, Salt Lake City, UT, USA

Geoffrey D. Silcox
University of Utah, Salt Lake City, UT, USA

JoAnn S. Lighty
University of Utah, Salt Lake City, UT, USA

Abstract

Oxidized mercury species may be formed in combustion systems through gas-phase reactions between elemental mercury and halogens, such as chorine or bromine. Injection of bromine compounds or brominated sorbents have been shown to affect mercury speciation and removal in coal-fired power plants. This study examines how bromine species affect mercury oxidation in the gas-phase. Experiments were conducted in a bench-scale, laminar, methane fired (300 W), quartz-lined reactor in which gas composition (Br_2, NO_x), quench rate, and quartz surface area were varied. Speciated mercury was measured using a wet conditioning system and continuous emissions monitor (CEM). Oxidation levels ranged from 13% to 80% with a quench rate of 450 K/s and bromine concentrations ranging from 3.5 to 41 ppmv equivalent HBr. A lower quench rate (220 K/s) increased oxidation by about one third. Most of the experiments were conducted with 30 ppm NO (dry). Mercury oxidation was unaffected by the presence of 500 ppmv NO. The surface area was increased by inserting a bundle of thin-walled quartz tubes. The insert decreases the reactor residence time by about 5 %. Tripling the interior, quartz surface area of the reactor from 1000 cm^2 to 3000 cm^2 did not appreciably affect the extent of oxidation. The experimental results were compared to detailed kinetic modeling predictions.

Introduction

The recent U.S. EPA Clean Air Mercury Rule (CAMR) (1) has put limits on mercury emissions from U.S. coal-fired utilities. The related Clean Air Interstate Rule or CAIR (*2*) specifies reductions in NO_x and SO_2 emissions, which are projected to result in a 40% reduction in mercury emissions in the utility sector by 2010, the start of Phase I of the CAMR. The CAIR mercury reduction is assumed to result from existing technologies, primarily as a co-benefit of reducing other pollutants such as NO_x, and SO_2, and will affect 28 eastern states and the District of Columbia. A similar plan has been introduced in Canada, with emissions caps in 2010 that vary from province to province, and the possibility of more stringent reductions by 2018 (*3*). Mercury-specific technology will be required in the U.S. to achieve more stringent reductions in mercury emissions mandated by 2018 under the Clean Air Mercury Rule. CAIR and CAMR are currently in the midst of legal battles, and have been vacated by the DC circuit court. As of December, 2007, nineteen states had passed their own mercury regulations, some incorporating parts of CAMR (*4*). The timing and extent of federal mercury regulations are in question, but states are going ahead with their own regulations.

In the last several years, demonstrations of a number of different mercury control technologies have been performed in the US on commercial coal-fired power plants in anticipation of the control of mercury emissions from electricity generating plants, as promulgated by CAMR. Activated carbon injection is one technology that is considered close to commercial application. Wet flue gas desulphurization (FGD) scrubbers are also effective for removing mercury if there is sufficient oxidized mercury in the flue gas at the FGD inlet. The ability to oxidize elemental mercury or to make activated carbon more effective at removing mercury from coal-fired power plants could substantially lower the cost of compliance; thus, a better understanding of the role of flue gas composition, boiler conditions, and homogeneous oxidation processes in combustion gases is of great importance.

Mercury exists in the elemental form (Hg^0) in the high temperature regions of the coal-fired boiler. As the flue gas cools, complex reactions transform the elemental mercury into gaseous oxidized forms (Hg^{2+}) (*5*) and particulate bound mercury (Hg_p). The amount of Hg^0 which is converted to Hg^{2+} and Hg_p depends on several factors including the quench rate, the flue-gas composition, and the amount and properties of fly ash.

Mercury speciation and capture is affected by halogens. Full-scale demonstrations have shown that bromine is effective for mitigating mercury emissions. Vosteen et al. (*6*) have patented one method for introducing bromine species into the combustion system, involving the addition of bromine containing compounds to the furnace or flue-gas. This method has been licensed in the U.S. by ALSTOM under the name KNX™ additive. The addition of bromine compounds has been shown to increase oxidation and mercury capture at numerous power plants firing various coal types (*7*-12). The addition of bromine compounds to the flue gas of boilers in which activated carbon is injected has been shown to increase oxidation and enhance the activated carbon mercury capture (*8-10), (12), (*13). Various brominated sorbents have also been shown to be effective for decreasing mercury emissions (*8-10), (13*-17).

We have performed a study of homogeneous mercury oxidation in a methane fired furnace. Mercury and bromine (Br_2) were added through the flame. When bromine is injected through the flame, Br, Br_2, and HBr should all be present at various temperatures in the reactor. Impingers were used to speciate and condition the flue gas upstream of a mercury continuous emission monitor (CEM). A KCl impinger, similar to the Ontario Hydro method, was treated with $Na_2S_2O_3$ and used to remove oxidized mercury in order to allow the CEM to quantify elemental mercury in the flue-gas.

Mercury-bromine reactions have not been studied under combustion conditions using time-temperature profiles representative of a power plant. Laboratory studies have focused on specific individual reactions, not overall speciation, and most experiments have been performed at atmospheric conditions. Ariya et al. (*18*) and Donohoue et al. (*19*) studied the reaction between Hg and Br under atmospheric conditions in Pyrex flasks using photolysis. The reaction rate found by Donohoue et al. was nine times slower than that determined by Ariya et al. There is obvious discrepancy in the calculated homogeneous rates even when performing similar experiments. Liu et al. (*20*) studied the reaction between Hg and Br_2 under flue gas conditions in Pyrex flasks. Arrhenius Law constants were found for temperatures between 298 K and 365 K. Although the gaseous mixture contained flue gas constituents, the higher temperatures seen by flue gas is a power plant could not be obtained due to the melting point of the halocarbon wax used to coat the Pyrex flasks and prevent wall reactions. In this paper, we report experimental data that were collected using actual flue gas and temperature profiles which cover the wide range of temperatures seen in power plants.

Experimental

The reactor used in this study was developed to study gas-phase mercury oxidation and is described in detail in Fry et al. (*21*). The reactor is a 47 mm-ID quartz glass tube, 132 cm in length. Five separately controlled temperature regions allow the time-temperature profile of the reactor to be manipulated to create various quench rates. The quench rates used in these experiments are representative of industrial conditions and allow adequate time at temperatures necessary for mercury oxidation reactions. The temperature profiles used are shown in Figure 1. The low temperature region (around 350°C) represents flue gas temperatures through air pollution control devices.

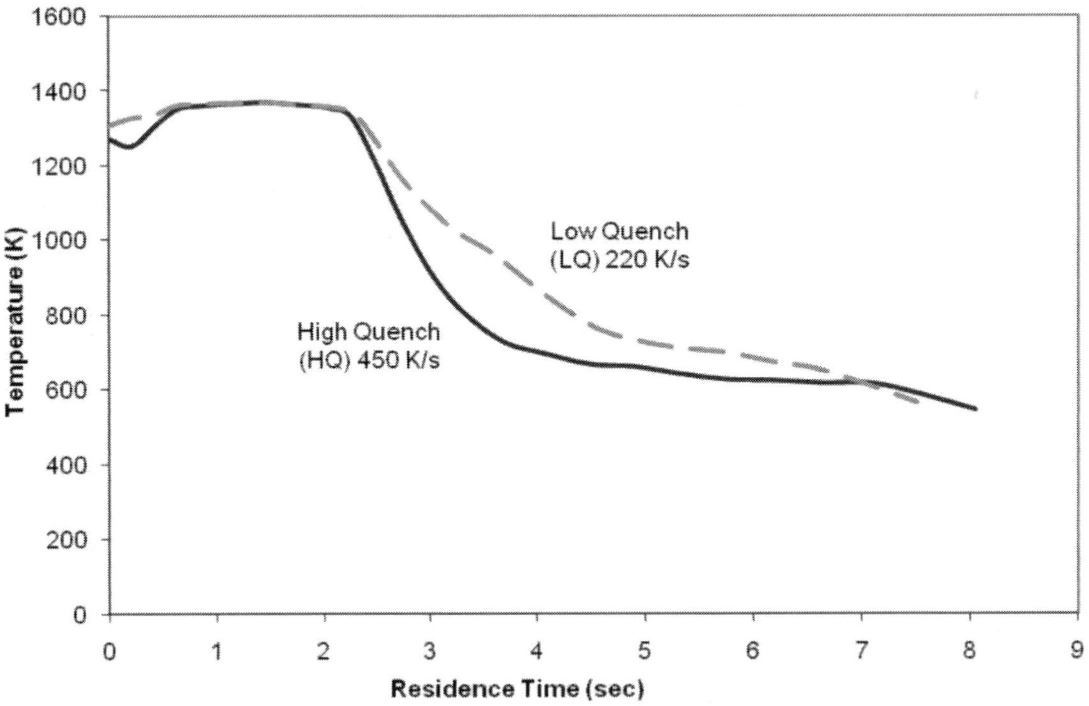

Figure 1: Time-temperature profiles used in this study

A radical pool representative of real combustion systems was created in the reactor by introducing all species through the flame of a 300 W methane-fired burner. A PS Analytical Mercury Calibration Gas Generator or "CavKit" was the source of elemental mercury. The reactor gas was analyzed for mercury, O_2, and NO. For mercury analysis, two sample streams were taken from the outlet of the reactor and introduced into a wet-chemical conditioning system. Mercury speciation was accomplished by passing one stream through impinger solutions containing $KCl/Na_2S_2O_3$ (to remove oxidized mercury while preventing oxidation of elemental mercury in solution) and NaOH, and the other stream through impingers containing $SnCl_2$ (to convert all mercury to the elemental form) and NaOH. A Tekran 2537A mercury analyzer coupled with an intermittent 4-port sampler was used to measure the elemental mercury concentration of each stream. At the start of each experiment, a mass balance on mercury was closed through the reactor with a flame, before any oxidant was introduced.

Bromine was introduced using a 3000 ppm Br_2 in air calibration gas. The flow of bromine was controlled by a stainless steel micro-needle valve. A Gilian Gilibrator automated bubble flow column was used to measure the flowrate before and after introducing bromine into the reactor. Bromine concentration was then calculated by mass balance.

The effect of NO was also investigated. Pure NO was used as the source for these experiments. The flows in the reactor were adjusted to maintain a constant 6 standard liters per minute through the reactor and 1.5% O_2 in the dry exhaust gas. The NO concentration was continuously measured after the conditioning system.

For some experiments with bromine, the quartz reactor surface area was increased. The reactor normally has 1000 cm^2 of surface area. This value was increased to 3000 cm^2 by inserting a bundle of thin walled quartz tubes. The bundle was made of seven tubes arranged in a hexagonal configuration. Each tube was 60 cm in length, 14 mm-OD, and 12 mm-ID. The tube bundle decreased the residence time in the reactor by only 5%.

Results and Discussion

Baseline oxidation experiments with bromine were performed with a high quench rate (450 K/s). The concentrations of O_2 and NO were 1.5% (dry) and 30 ppm (dry) respectively. Bromine concentrations ranged from 6.7 to 41.4 ppmv equivalent HBr (wet). Results from these experiments are shown in Figure 2. Oxidation ranged from 18% to 78%. Increasing the bromine concentration significantly increased mercury oxidation in the reactor. Error ranged from 2.2-6.3% based on the oxidation through the reactor during the initial mass balance with no bromine.

Several operating conditions were changed individually to see the effect they had on homogeneous mercury oxidation by bromine. First, the lower quench rate (220 K/s) was applied, keeping all other operating conditions the same. These results are compared to the high quench results in Figure 2. Decreasing the quench rate increased the extent of oxidation by about one third for similar bromine concentrations. The effect of quench rate may be due to bromine radical chemistry. HBr, Br_2, and Br are present in the system at various ratios depending on temperature (*22*). Second, at the high quench rate, 500 NO (dry) was added. The NO concentration in the flue gas was continuously monitored, after a slip stream was removed for mercury analysis. The increased NO concentration had no discernable impact on oxidation, as shown in Figure 3. This is in contrast to modeling predictions presented by Senior et al. (*23*) and Niksa et al., (*24*) which predicted decreased oxidation with increased NO concentration. It is unclear why these modeling predictions do not match the experimental data. Perhaps the detailed kinetics involving NO_x and bromine species used in the modeling are incorrect. Lastly, the quartz surface area in the reactor was increased from 1000 cm^2 to 3000 cm^2. The increased surface area, again at 30 ppm NO, slightly increased oxidation, as shown in Figure 4. This surface area effect needs to be investigated further to ensure accurate extents of oxidation are being reported. Varying the surface area further will yield a correlation which may be extrapolated to zero surface area. This will isolate the gas-phase reactions from any possible surface area reactions.

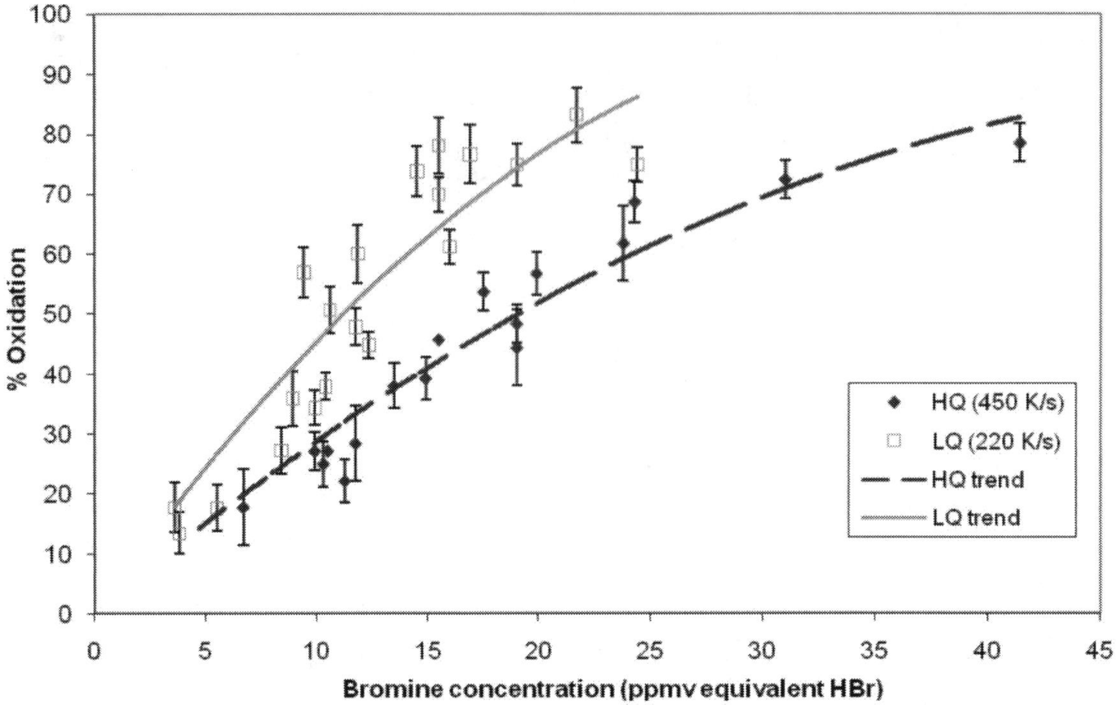

Figure 2: Comparison of homogeneous mercury oxidation by bromine at high and low quench rates (The lines are drawn to indicate trends and are not based on model predictions, error ranges from 2.2-6.3%)

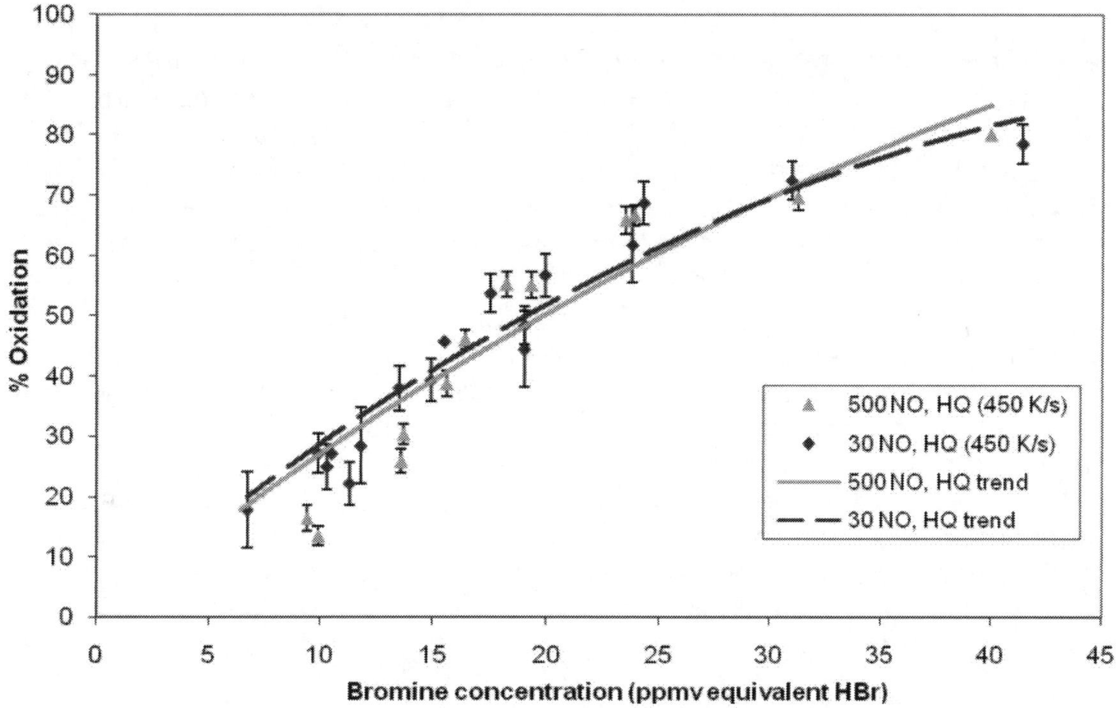

Figure 3: Effect of NO concentration on homogeneous mercury oxidation by bromine

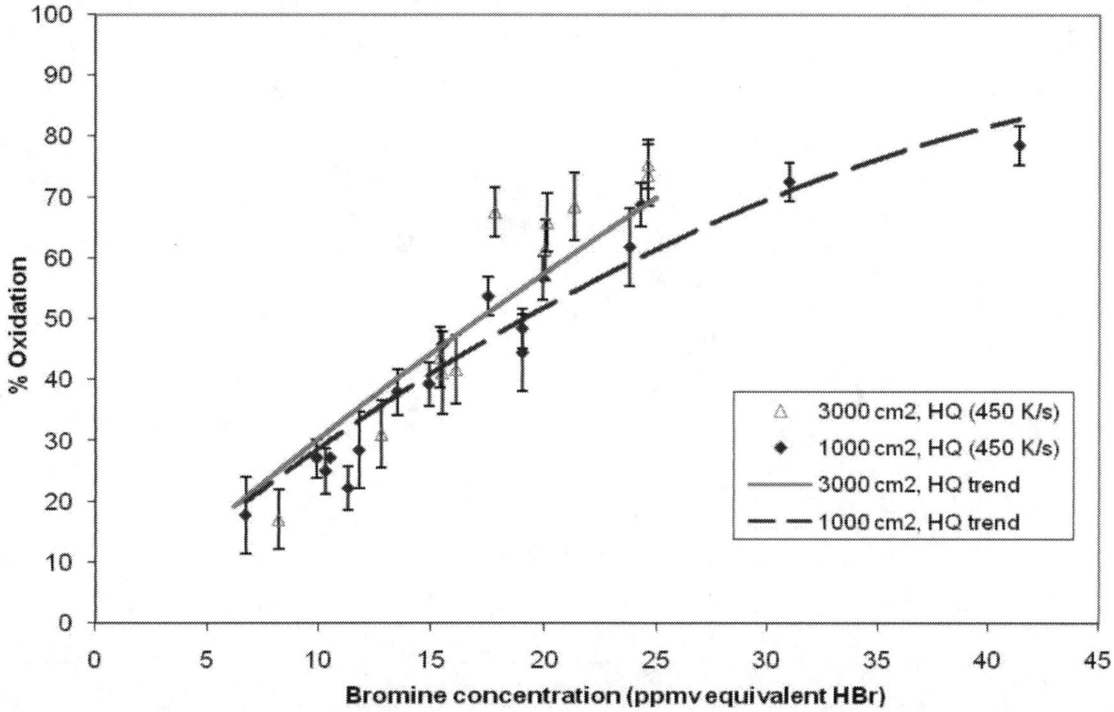

Figure 4: Effect of increased surface area on homogeneous mercury oxidation by bromine

Kinetic Modeling

The Hg-Br experiments were modeled using the Reaction Engineering Kinetics Solver (REKS). This program is similar to ChemKin. The model includes sub-mechanisms for hydrocarbon, mercury, bromine, SO_2, and NO_x chemistry. The bromine mechanism was put together using NIST Halon chemistry. The kinetic parameters for a simple, global Hg-Br mechanism were found by fitting the experimental data. The reactions used and the corresponding pre-exponential factors and activation energies are shown in Table 1. The model predictions are compared to the experimental data in Figure 5. Reactions 1 and 2 were selected because they have been previously found to be the main reaction pathway for mercury oxidation by chlorine. Reaction 3 was added later after the dependence on quench rate could not be matched using reactions 1 and 2 only. While the shape and general effect of quench rate (lower quench rate gives larger oxidation) are predicted, the magnitude of the effect of quench rate is under-predicted. The separation between the two quench rates needs to be larger to predict the experimental data correctly. Because a more global mechanism is being used and multiple parameters are being changed, several parameter combinations will results in an appropriate fitting of the experimental data. The effect of quench rate is most likely caused by differences in bromine radical chemistry at different temperatures.

Table 1: Hg-Br mechanism used in modeling

Reaction	A	Ea
1) Hg + Br + M = HgBr + M	1.0E16 cm^6/mol^2-sec	0 cal/mol
2) HgBr + Br_2 = $HgBr_2$ + Br	2.0E16 cm^3/mol-sec	1000 cal/mol
3) HgBr + HBr = $HgBr_2$ + H	2.7E14 cm^3/mol-sec	1000 cal/mol

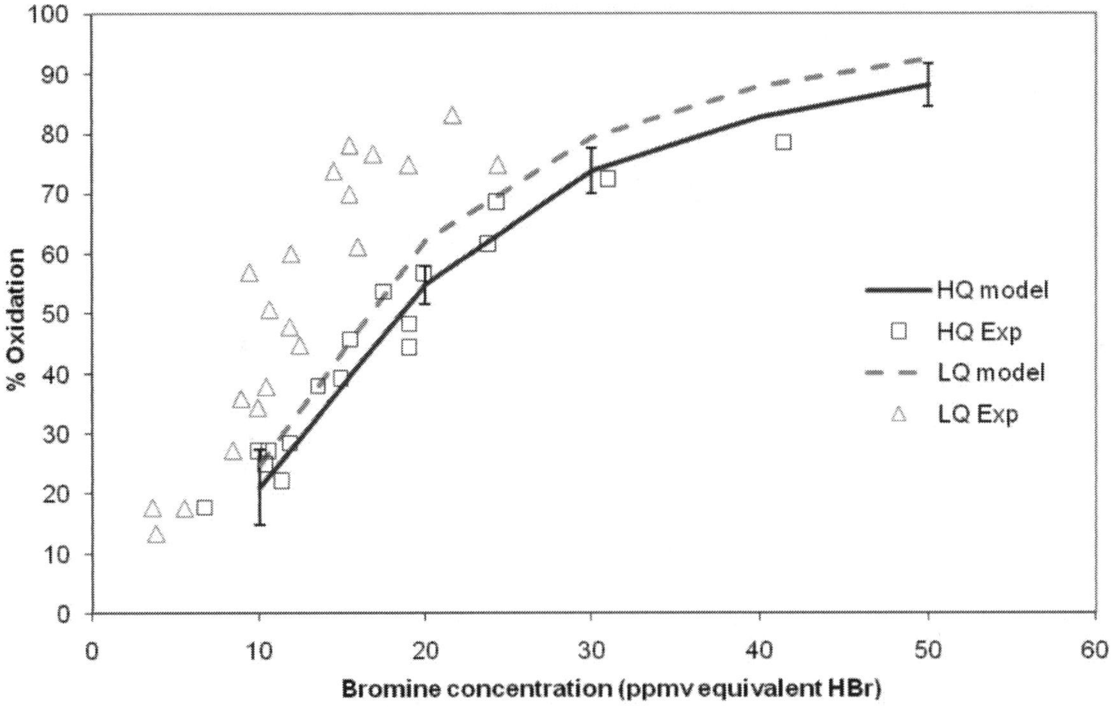

Figure 5: Comparison of model predictions and experimental data

Conclusions

In summary, bench-scale experiments were performed in a laminar, methane fired (300 W), quartz-lined reactor in which gas composition (Br_2, NO_x), quench rate, and quartz surface area were varied. Baseline oxidation with bromine using a quench rate of 450 K/s and 30 ppm NO (dry) showed oxidation of Hg^0 varied from 18% to 78% with bromine concentrations ranging from 6.7 to 41.4 ppmv (equivalent HBr wet). Lowering the quench rate to 220 K/s increased oxidation by approximately one third. Increasing the NO concentration to 500 ppm had negligible impact on oxidation. Tripling the quartz surface area slightly increased oxidation. These results indicate that a large fraction of elemental mercury in the gas phase is oxidized homogeneously by bromine species. A simple Hg-Br reaction mechanism was developed which approximates the experimental data.

Acknowledgments

This paper was prepared with the support of the U.S. Department of Energy, under Award No. DE-FG26-06NT42713. However, any opinions, findings, conclusions, or recommendations expressed herein are those of the author(s) and do not necessarily reflect the views of the DOE.

References

(1) U.S. EPA 2005a, Clean Air Mercury Rule, http://www.epa.gov/air/mercuryrule/ (accessed Aug 29, 2007).
(2) U.S. EPA 2005b, Clean Air Interstate Rule, http://www.epa.gov/interstateairquality/ (accessed Aug 29, 2007).
(3) Canadian Council of Ministers of the Environment (CCME), 2005, Canada-wide standards for mercury emissions from coal-fired electric power generation plants [draft proposal June 2005], 5 p. http://www.ccme.ca/assets/pdf/canada_wide_standards_hgepg.pdf (accessed November, 2005).
(4) National Association of State Clean Air Agencies, State Mercury Programs for Utilities, December 4, 2007, http://www.4cleanair.org/Documents/StateTable.pdf (accessed May 7, 2008).
(5) Senior, C. L.; Sarofim, A. F.; Zeng, T.; Helble, J. J.; and Mamani-Paco, R. Gas-phase transformations of mercury in coal-fired power plants. *Fuel. Proc. Technol.* **2000**, *63 (2-3)*, 197-213.
(6) Vosteen, B.; Beyer, J.; Bonkhofer, T.-G.; Fleth, O.; Wieland, A.; Pohontsch, A.; Kanefke, R.; Standau, E.; Mueller, C.; Nolte, M.; Koeser, H. Process for removing mercury from flue gases, U.S. Patent 6,878,358, April 12, 2005.
(7) Berry, M.; Dombrowski, K,; Richardson, C.; Chang, R.; Borders, E.; Vosteen, B. Mercury control evaluation of calcium bromide injection into a PRB-fired furnace with an SCR. *In Proceedings of the International Conference on Air Quality VI*, Arlington, VA, September 24-27, **2007**.
(8) Sjostrom, S. Evaluation of sorbent injection for mercury control. Topical report for: Sunflower Electric's Holcomb Station, DOE Cooperative Agreement No. DE-FC26-03NT41986, June 2005.

(9) Sjostrom, S. Evaluation of sorbent injection for mercury control. Topical report for: AmerenUE's Meramec Station Unit 2, DOE Cooperative Agreement No. DE-FC26-03NT41986, Sept 2005.
(10) Sjostrom, S. Evaluation of sorbent injection for mercury control. Topical report for: Basin Electric Power Cooperative's Laramie River Station, DOE Cooperative Agreement No. DE-FC26-03NT41986, January 2006.
(11) Richardson, C.; Dombrowski, K.; Chang, R. Mercury control evaluation of halogen injection into coal-fired furnaces. Presentation at Electric Utility Environmental Conference, Tucson, AZ, January 22-25, **2006**.
(12) Benson, S. A.; Holmes, M. J. ; McCollar, D. P. ; MacKenzie, J. M.; Crocker, C.; Galbreath, K.; Richardson, C.; Dombrowski, K. Large-scale mercury control technology testing for lignite-fired utilities- oxidation systems for wet FGD. Presentation at 2006 DOE-NETL Mercury Control Technology Conference, Pittsburgh, PA, December 11-13, **2006.**
(13) Holmes, M. et al. Enhancing carbon reactivity for mercury control in coal-fired power plants: results from Leland Olds, Stanton, and Antelope Valley Stations. Presentation at 2006 DOE-NETL Mercury Control Technology Conference, Pittsburgh, PA, December 11-13, **2006.**
(14) Richardson, C.; Dombrowski, K.; Orr, D. Sorbent injection for small ESP mercury control in low sulfur eastern bituminous coal flue gas. Final report, DOE NETL Cooperative Agreement DE-FC26-03NT41987, URS Group Inc., Austin, TX, March 2007.
(15) Dombrowski, K.; Padilla, J.; Richardson, C.; Fisher, K.; Campbell, T.; Chang, R.; Eckberg, C.; Hudspeth, J.; Pletcher, S. Evaluation of low-ash impact sorbent injection technologies at a Texas lignite/PRB fired power plant. In *Proceeding of the International Conference on Air Quality VI*, Arlington, VA, September 24-27, **2007**.
(16) Landreth, R.; Nelson, S.; Liu, X.; Tang, Z.; Donochod, D. R.; Kulpinski, K.; Wanninger, K.; Brickett, L. Additional B-PAC™, H-PAC™, and C-PAC™ trial results. In *Proceedings of the International Conference on Air Quality VI*, Arlington, VA, September 24-27, **2007**.
(17) Landreth, R.; Nelson, S.; Liu, X.; Tang, Z.; Miller, J. ; Brickett, L. Brominated sorbents for small cold-side ESPs, hot-side ESPs, and fly ash use in concrete. Presentation at 2007 DOE-NETL Mercury Control Technology Conference, Pittsburgh, PA, December 11-13, **2007.**
(18) Ariya, P. A.; Khalizov, A.; Gidas, A. Reactions of gaseous mercury with atomic and molecular halogens: kinetics, product studies, and atmospheric implications. *J. Phys. Chem. A.* **2002**, *106*, 7310-7320.
(19) Donohoue, D L.; Bauer, D.; Cossairt, B.; Hynes, A. J. Temperature and pressure dependent rate coefficients for the reaction of Hg with Br and the reaction of Br with Br: A pulsed laser photolysis-pulsed laser induced fluorescence study. *J. Phys. Chem. A,* **2006**, *110 (21)*, 6623-6632.
(20) Liu, S.; Yan, N.; Liu, Z.; Qu, Z.; Wang, H. P.; Chang, S.; Miller, C. Using bromine gas to enhance mercury removal from flue gas of coal-fired power plants. *Environ. Sci. Technol.* **2007**, *41 (4)*, 1405-1412.
(21) Fry, A.; Cauch, B.; Silcox, G. D.; Lighty, J. S.; Senior, C. L. Experimental evaluation of the effects of quench rate and quartz surface area on homogeneous mercury oxidation. In *31st International Symposium on Combustion*, Heidelberg, Germany, August, **2006**.
(22) Senior, C.; Fry, A.; Montgomery, C.; Sarofim, A.; Wendt, J.; Silcox, G.; Lighty, J.; Bozzelli, J. Gas-phase reactions of mercury and halogens in combustion environments. Presentation at the Air & Waste Management 100th Annual Meeting and Exhibition, Pittsburgh, PA, June 26-28, **2007**.

(23) Senior, C.; Fry, A.; Montgomery, C.; Silcox, G.; Cauch, B.; Bozzelli, J. Impact of halogen injection on mercury speciation and capture in coal-fired power plants. In *Proceedings of the International Conference on Air Quality VI*, Arlington, VA, September 24-27, **2007.**

(24) Niksa, S.; Naik, C. V.; Berry, M. S.; Monroe, L. Interpreting enhanced Hg oxidation with Br addition at Plant Miller. In *Proceedings of the International Conference on Air Quality VI*, Arlington, VA, September 24-27, **2007.**

Bench- and Pilot-Scale Studies on Mercury Removal by Potassium Iodide in Coal-Fired Flue Gas

Ying Li, Michael Daukoru, Achariya Suriyawong, Pratim Biswas

Aerosol and Air Quality Research Laboratory, Department of Energy, Environmental and Chemical Engineering, Washington University in St. Louis, St. Louis, MO 63130, USA

Introduction

Mercury (Hg) is a toxic air pollutant with coal-fired utility plants being the largest anthropogenic Hg emission source in the United States.[1] The extent that Hg can be removed from power plants using conventional air pollution control devices (APCDs) is significantly affected by Hg speciation in flue gas.[2] Oxidized Hg (Hg^{2+}) is preferable to elemental Hg (Hg^0) in that Hg^{2+} is less volatile and soluble in water and thus is easier to be captured through wet scrubbers and/or electrostatic precipitators. Chlorine is the major halogen species in coal and it plays a very important role in both homogeneous and heterogeneous Hg^0 oxidation in flue gas.[3, 4] Other halogen species have also demonstrated their abilities to oxidize Hg^0.[5, 6] It has also been reported that activated carbon impregnated with sulfur, chlorine, and iodine has greater ability of capturing Hg compared with untreated carbon.[3, 7, 8] Particularly, iodine impregnated activated carbon showed superior capacity of Hg capture.[7, 9] Another advantage of using iodine promoted sorbent is the lower volatility of HgI_2 (higher boiling point) compared to other mercury halides, which enhances the stability of the spent sorbent. On the other hand, injection of carbon based sorbents can compromise fly ash properties due to the residual carbon content, and thus, the operation cost will be increased since the fly ash is no longer sellable. Therefore, non-carbon based sorbent with an iodine promoter is preferred.

A simple form of iodine promoter is metal iodide, e.g. KI. I_2 sublimates at a relatively low temperature (113 °C) and it is more expensive than KI, and thus, I_2 is not an appropriate candidate for direct injection. KI is soluble in water, and it has been reported that an acidic KI solution was effective in capturing gas-phase Hg.[10] To the best of our knowledge, no one has directly tested KI as a Hg sorbent in coal-fired combustors. Although it is believed that Hg can be oxidized by I_2 forming HgI_2,[5, 11] the interactions between Hg and iodide (e.g. KI) are not clear in the literature. The purpose of this study was to investigate the effectiveness of Hg^0 removal by KI in a coal-fired flue gas and to explore the reaction mechanisms. Feasibility studies were first carried out in bench-scale systems using air as a carrier gas, and the factors that affect Hg^0 removal efficiency were investigated. Guided by the bench-scale findings, pilot-scale tests were then conducted in a pulverized coal combustor. This finding is important to the development of KI-based sorbents for Hg control in full-scale coal-fired power plants.

Experimental

Description of Bench-Scale Tests

The schematic diagram of the bench-scale experimental system is shown in Figure 1. Hg^0 vapor was introduced to the system by passing N_2 through a liquid Hg^0 reservoir. Two types of reactors were used: a packed-bed reactor (PBR) and an aerosol flow reactor (AFR). In the PBR system (Figure 1a), a certain amount of granular or powder KI (>99.6%, Mallinckrodt Chemicals) was packed in a glass tube (10 mm ID), which was placed inside a tubular furnace (Thermolyne, Type 21100). In the AFR system (Figure 1b), KI aerosols were generated in an atomizer and introduced to the system through atomization. The KI particles mixed with the Hg^0 vapor and then entered a ceramic tube (ID = 1.9 cm; L = 50 cm) which was heated by the tubular furnace. The size distributions of the KI particles before and after the furnace were monitored by a Scanning Mobility Particle Sizer (SMPS, TSI Inc., St. Paul, MN, USA). For both PBR and AFR systems, Hg^0 concentration at the furnace outlet was measured by an online Hg analyzer (Model RA915+, OhioLumex Co.). Since the Hg analyzer only detects Hg^0, an EPA Method 5 type sampling train developed by Hedrick et al.[10] was used to measure both Hg^0 and Hg^{2+} concentrations in the gas phase. The impinger solutions were then analyzed by inductively coupled plasma mass spectrometry (ICP-MS). The experimental conditions for bench-scale studies are summarized in Table 1.

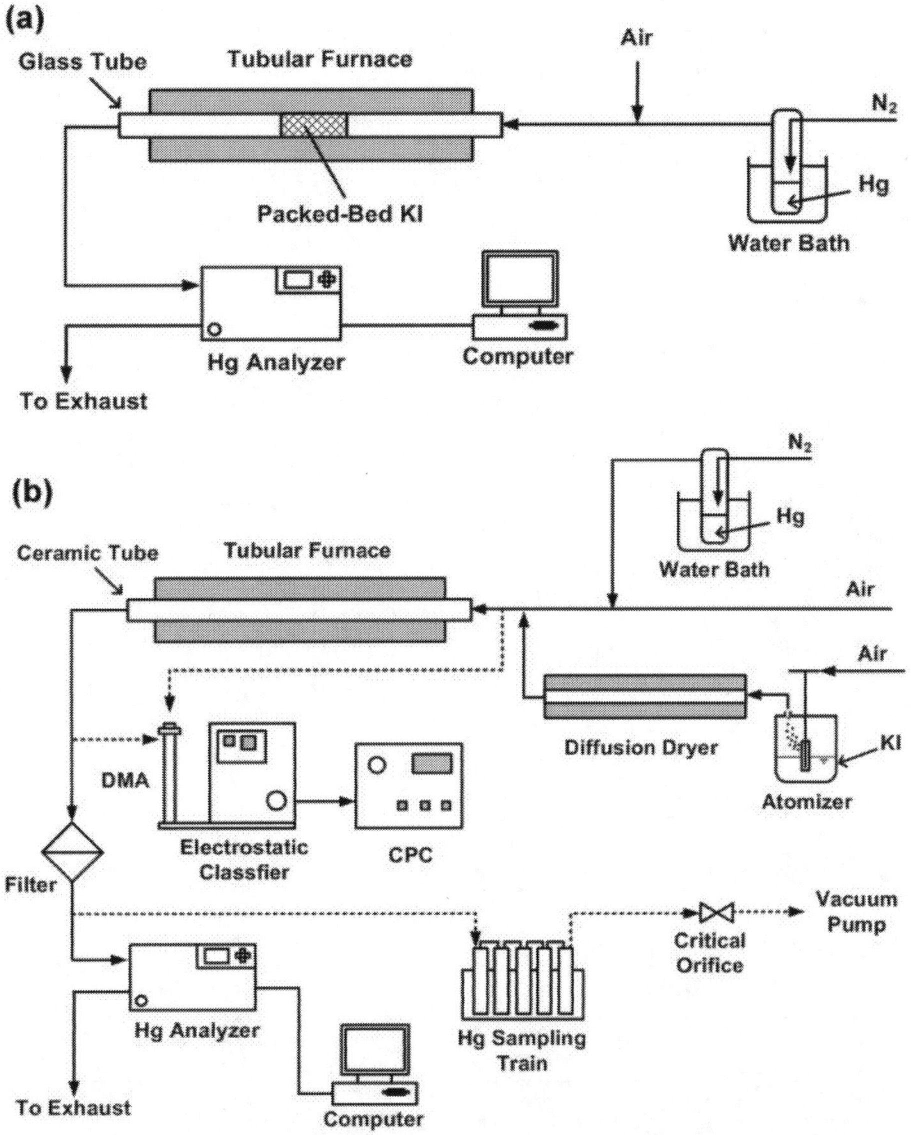

Figure 1. Schematic diagram of bench-scale experimental system: (a) packed-bed reactor (PBR), (b) aerosol flow reactor (AFR).

Table 1. Summary of bench-scale experimental conditions

PBR system				
Inlet Hg conc. (ppb)	Amount of KI (g)	Furnace Temp. (°C)	Gas flow rate (l min^{-1})	Residence time in packed-bed (s)
10.5 ± 0.5	2.0 (granule)	25 ~500	1.0	0.21
10.5 ± 0.5	0.5 (powder)	25 ~500	1.0	0.033
AFR system				
Inlet Hg conc. (ppb)	KI/Hg (molar ratio)	Furnace Temp. (°C)	Gas flow rate (l min^{-1})	Residence time in reactor (s)
13.5 ± 0.5	15, 30, 60, 600	200 ~ 1100	1.5, 3.0, 4.5	5.8, 2.9, 1.9

Description of Pilot-Scale Tests

The pilot-scale experiments were conducted by burning Powder River Basin (PRB) subbituminous coal in a 160 kW facility at the Energy and Environmental Research Center (EERC) at the University of North Dakota. The schematic drawing of the pilot-scale facility is shown in Figure 2. Coal feeding rate remained constantly at 27 kg/hr and the flue gas flow rate was 130 scfm or 210 Nm^3/hr. Hg speciation and concentration were measured at the ESP outlet using a continuous mercury monitor (CMM) (PSA Sir Galahad). KI was fed to the system in two ways as shown in Figure 2: (A) KI powder mixed with coal and (B) aqueous KI solution sprayed into flue gas. The KI feeding strategies and feeding rates are listed in Table 2.

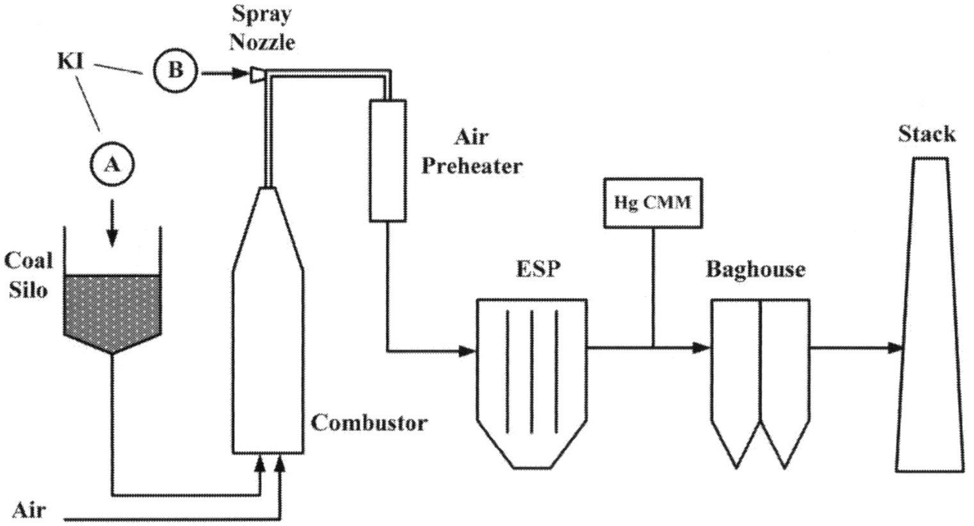

Figure 2. Schematic diagram of the 160 kW pilot-scale facility with KI injection

Table 2. Summary of pilot-scale experimental conditions

	KI powder mixed in coal (ppm_w)	KI solution sprayed (ml/hr)	KI feeding rate (g/hr)	KI conc. in flue gas (mg/m^3)
KI powder	235		6.4	31
	389		10.6	51
	777		21.2	101
KI solution (1.4 M)		200	46.5	221
		800	185.9	885

Results and Discussion

Results of Bench-Scale Tests

The PBR experiments were conducted using 2.0 g granular KI and 0.5 g powdered KI, respectively. The results are shown in Figure 3. For 2.0 g granular KI with air as the carrier gas, no Hg^0 removal was observed when the temperature was below 100 °C. When the temperature increased over 100 °C, a slight removal of Hg^0 (< 7%) was observed. A sharp increase in Hg^0 removal occurred at around 200 °C and reached 100% when the temperature was higher than 320 °C. This positive temperature dependence suggests that Hg^0 removal involves chemical reaction and certain activation energy is required. When N_2 (99.99%)was used as the carrier gas, a similar shape of curve was observed except that the sharp increase in Hg^0 removal occurred at a higher temperature (330 °C) and 100% Hg^0 removal was reached at above 400 °C. The result indicated that O_2 can promote Hg^0 removal by KI. When 0.5 g of powder KI was tested in air, the Hg^0 removal curve shifted to the lower temperature region, even though the amount of powder KI was less than the granular KI used. This is likely due to the better contact of Hg^0 with KI powder which has much smaller size and higher total surface area.

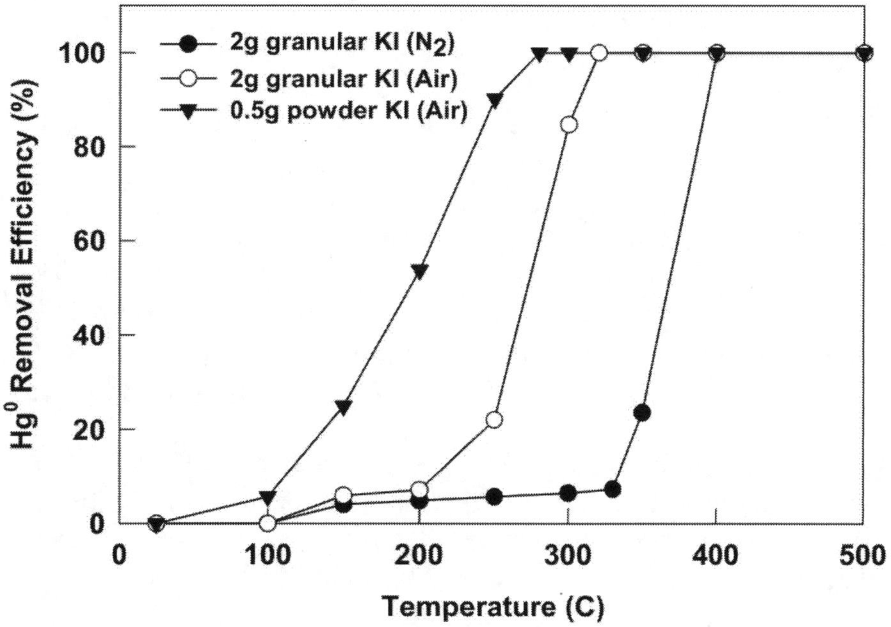

Figure 3. Hg^0 removal by KI in PBR as a function of temperature.

Experiments in the AFR system investigated the effectiveness of KI aerosols on Hg removal by varying the KI concentration and residence time in the temperature range of 200 to 1100 °C. The mass concentration of KI aerosols introduced to the gas stream was varied by changing the concentration of KI in the solution in the atomizer. The residence time was controlled by varying the total gas flow rate. Figure 4 shows the results when the residence time was 5.8 s. The curves of Hg^0 removal as a function of temperature followed a similar trend as observed in the PBR tests, i.e. Hg^0 removal became notable above a certain temperature and then increased with increasing temperature and finally reached a plateau at high temperatures. A higher temperature was needed to achieve observable Hg^0 removal at a lower KI concentration (i.e. smaller KI/Hg molar ratio). When the molar ratio of KI/Hg decreased subsequently from 600 to 60, 30, and 15, the temperature above which Hg^0

removal took place increased from 300 to 350, 380, and 400 °C, respectively. In the plateau region (500 – 1100 °C) where Hg^0 removal was relatively stable, a higher KI concentration also resulted in higher Hg^0 removal efficiency. A maximum of 98% and 100% Hg^0 removal was achieved at KI/Hg = 60 and 600, respectively. As the residence time decreased from 5.8 s to 2.9 s and 1.9 s (results not shown), the Hg^0 removal efficiency decreased accordingly. The Hg speciation analysis did not detect gaseous Hg^{2+} at the reactor outlet. Thus, it is believed that all the Hg^0 reduced in the gas-phase across the reactor was converted to particulate phase, possibly collected on the filter downstream.

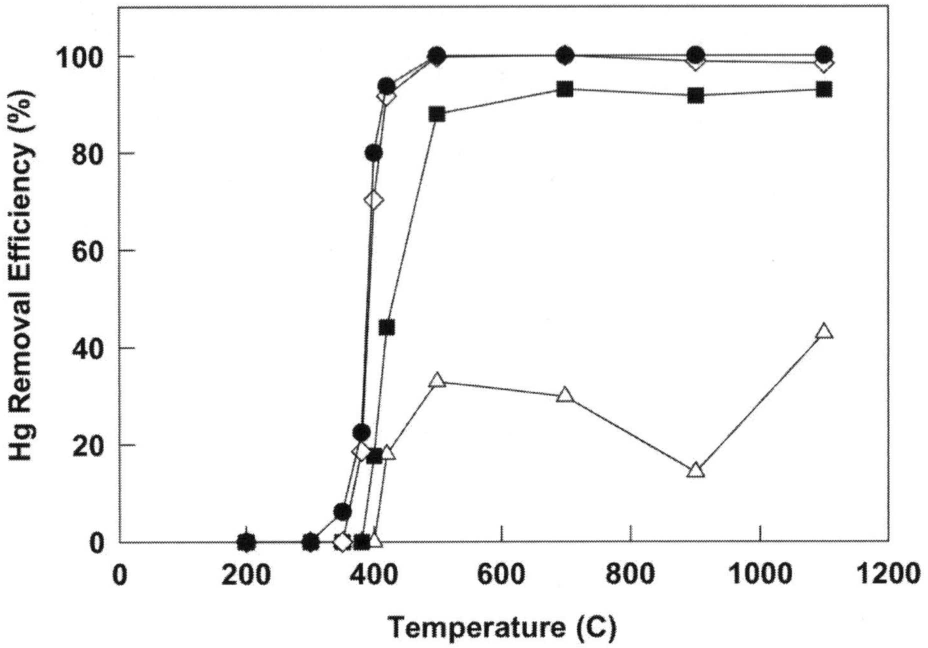

Figure 4. Hg removal efficiency as a function of temperature; resident time = 5.8s. Legend: △ KI/Hg (molar ratio) = 15, ■ KI/Hg = 30, ◇ KI/Hg = 60, ● KI/Hg = 600.

It should be noted that even the highest KI injection rate (KI/Hg = 600) tested in this study was actually very low, compared to the full-scale injection rates of other types of sorbents. It was reported that by using activated carbon a carbon-to-Hg mass ratio of 2,000 – 15,000 was required to achieve 25% to 95% removal of Hg in coal-fired power plants.[3] Typical flue gas residence time from coal-fired boiler furnace exit to ESP inlet is 3 – 5 s. At a comparable time scale, this study showed that a KI/Hg mass ratio of 500 (molar ratio = 600) achieved 100% and 90% Hg^0 removal at a residence time of 5.8 and 2.9 s, respectively. This demonstrates that KI is a highly effective sorbent for Hg^0 removal.

Reaction Mechanisms

In this study, a starch-iodine test was carried out to verify whether I_2 was generated and participated in the removal of Hg^0. It is known that I_2 dissolved in aqueous KI solution (forming I_3^- ion) reacts with starch producing a deep black-blue color. Neither I^- nor I_2 alone will lead to the color change. In this study, a white-color solution consisting of 0.3% w/w starch and 0.01 M KI was used as the reagent, and an impinger containing 15 ml of such solution was connected to the outlet of the PBR that was packed with 0.5 g KI powder. Air or N_2 (99.99%) was passed through the PBR at 1.0 l/min without feed of Hg^0. If gas-phase I_2 was produced from the KI powder in the PBR, it would be captured in the impinger and color change should be observed.

When air was the carrier gas and the furnace was operated at 25 °C, no color change was observed for a collection time of 120 min. When the furnace temperature increased to 300 °C, a pink color was observable in 120 min, probably due to the trace amount of I_2 captured from the gas-phase. When the furnace temperature further increased to 450 °C, the solution turned to blackish blue in one min, which clearly indicated the presence of I_2. The tests were then repeated using N_2 as the carrier gas. In this case, no color change was observed at 300 °C in 120 min but a pink color at 450 °C in 3 min, very likely due to the impurity O_2 (<100 ppm) in the N_2 cylinder. The results of starch-iodine tests indicated that I_2 vapor was produced from the oxidation of KI by O_2.

$$O_2 + 4KI \rightarrow 2I_2 + 2K_2O \qquad (1)$$

Accordingly, the removal of Hg^0 was due to the oxidation of Hg^0 by I_2 forming HgI_2.

$$Hg + I_2 \rightarrow HgI_2 \qquad (2)$$

This reaction mechanism was further verified by examining the number concentration and particle size distributions of the KI aerosols (without feed of Hg^0) at the AFR outlet measured by SMPS. The KI aerosol concentration was observed to decrease above 300 °C and dropped to the background level (no KI) at 600~700 °C. Because the melting point of KI is 681 °C, the decrease in the particle number concentration below this temperature is very likely due to oxidation of the fine KI particles to I_2 vapor. The melting and boiling points of HgI_2 are 259 °C and 354 °C, respectively. Considering that 100% Hg^0 removal was achieved when the AFR was operated at above 500 °C and no gaseous Hg^{2+} species was detected, the HgI_2 produced in the furnace should stay in gas phase at that high temperature and then possibly deposited along the tubing or adsorbed on unreacted KI particles at lower temperatures downstream of the AFR (most likely on the filter), forming $KHgI_3$ or K_2HgI_4.

$$HgI_2 + KI \rightarrow KHgI_3 \qquad (3)$$

$$HgI_2 + 2KI \rightarrow K_2HgI_4 \qquad (4)$$

Results of Pilot-Scale Tests

Since the bench-scale results showed that high temperatures promote Hg^0 removal by KI, KI was introduced into two high temperature zones in the pilot-scale combustor. Figure 5a shows the effect on Hg^0 removal by adding KI powder to coal and co-combusted at over 1400 °C. Without KI addition, the baseline Hg^T and Hg^0 concentrations at the ESP outlet were around 5.0 μg/m^3 and no Hg^{2+} was in the gas phase. Soon after 235 ppm$_w$ (6.4 g/hr) KI powder was added to the coal, the Hg^T concentration decreased to around 3.6 μg/m^3. Temporarily cutting off the KI powder feed caused the Hg^T concentration to increase to the baseline level. When the KI powder was added again at 389 ppm$_w$, the Hg^T concentration decreased to 2.4 μg/m^3. Measurement of Hg^0 concentration at this condition showed the same level as Hg^T, indicating that no Hg^{2+} existed in the gas phase which agrees with that in the bench-scale experiments. It is very likely that the reaction product, HgI_2, was condensed in particulate phase or captured on the fly ash as the flue gas cools down. Increasing the KI powder mixing ratio further to 777 ppm$_w$ reduced Hg^T concentration to an average level of 1.9 μg/m^3. Finally, the KI powder feed was cut off and the Hg^T concentration increased but at a slower rate toward the baseline level. Figure 5b shows the effect of spraying KI solution to the flue gas where the temperature was around 900 °C. The average baseline Hg^T and Hg^0 concentrations were again equal to each other at around 5.8 μg/m^3. When spraying KI solution equivalent to 46.5 g/hr of KI, the Hg^T concentration gradually decreased to 4.1 μg/m^3. Increasing the spraying rate by four times to 185.9 g/hr, the Hg^T concentration slightly decreased to 3.5 μg/m^3. The average Hg^0 concentration at this condition was also close to Hg^T concentration.

Overall, the Hg removal efficiency increases with increasing KI feeding rate. 61% and 53% Hg removal were achieved at the feeding rates of 21.2 g/hr and 10.6 g/hr KI powder, respectively. It clearly shows that mixing KI powder with coal was more effective than spraying an aqueous KI solution into the flue gas. One important reason why co-combusting KI powder with coal was superior to spraying KI is because of the higher temperature and longer residence time, both of which favor the production of I_2.

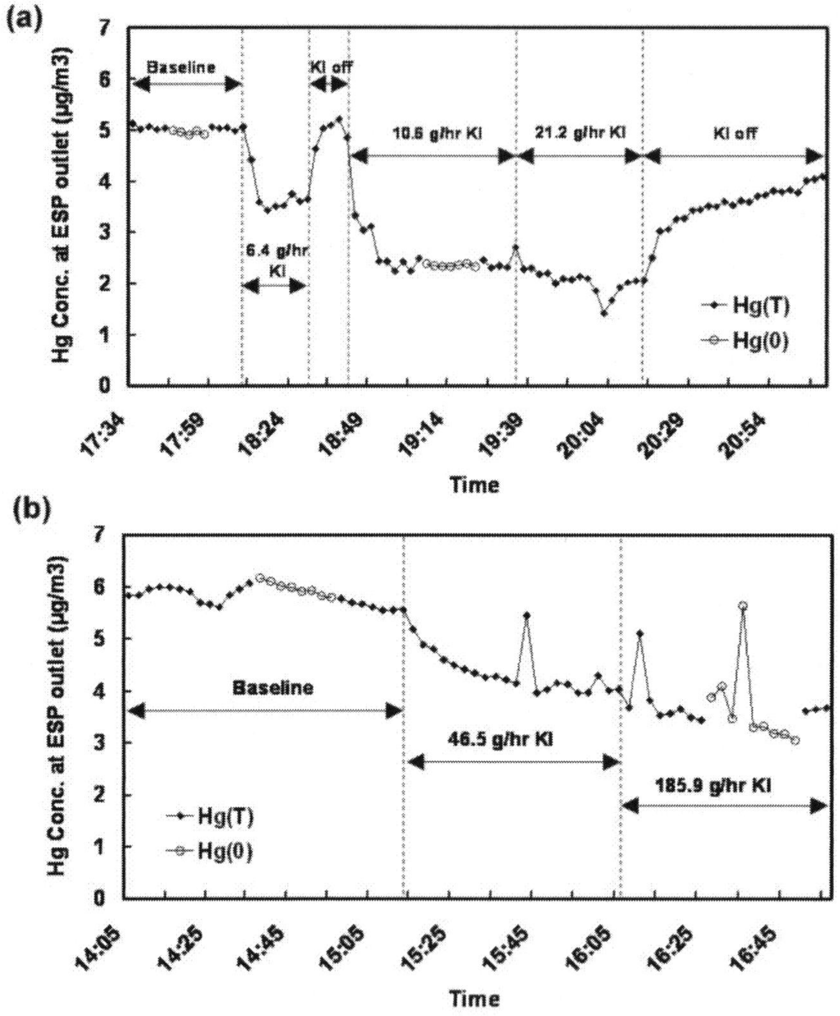

Figure 5. Hg concentration at ESP outlet under the condition of (a) KI powder mixed with coal (b) KI solution sprayed into flue gas.

Conclusions

It has been demonstrated that KI is an effective reagent for Hg^0 removal in both bench- and pilot-scale experiments. In the bench-scale PBR and AFR experiments, a positive temperature dependence of Hg^0 removal by KI was observed. In the PBR, 100% Hg removal was observed for either 2 g granular and 0.5 g powder KI at temperatures higher than approximately 300 °C. In the AFR, at a 5.8 s residence time and temperature above 500 °C, 98% and 100% Hg^0 removal was achieved at KI/Hg molar ratios of 60 and 600, respectively. The starch-iodine tests identified the formation of I_2 vapor due to oxidation of solid KI by O_2 at elevated temperatures, which resulted in the observed Hg^0 removal with HgI_2 being the most likely reaction product. Higher KI concentrations and longer residence time in the reactor also enhanced Hg^0 removal efficiency. Guided by the bench-scale results, pilot-scale experiments were conducted in a 160 kW pulverized coal combustor. First, KI powder was mixed with coal and fed to the combustor (>1400 °C); second, KI was sprayed to the flue gas where the temperature was approximately 900 °C. Hg removal efficiency increased with increasing feeding rate of KI. Mixing KI powder with coal resulted in a higher Hg removal efficiency than spraying KI solution to the flue gas, possibly due to the higher temperature of the combustion zone and the longer residence time of KI in the flue gas that favor the generation of I_2. No gaseous Hg^{2+} species was measured in the flue gas, indicating that the oxidation product HgI_2 was captured on the fly ash and removed by the ESP.

Acknowledgements

This work was supported partially by grants from the USDOE grant number DE-FGZ6-05NT42531 and Ameren UE, St. Louis, through contract number W-0001-06 Missouri Department of Natural Resources. We thank Dr. Ye Zhuang, UND-EERC and his staff for helping performing the pilot-scale experiments.

References

1. USEPA *Mercury Study Report to Congress*; EPA-452/R-97-003; US EPA: 1997.
2. Romero, C. E.; Li, Y.; Bilirgen, H.; Sarunac, N.; Levy, E. K., Modification of boiler operating conditions for mercury emissions reductions in coal-fired utility boilers. *Fuel* **2006,** 85, 204-212.
3. Pavlish, J. H.; Sondreal, E. A.; Mann, M. D.; Olson, E. S.; Galbreath, K. C.; Laudal, D. L.; Benson, S. A., Status review of mercury control options for coal-fired power plants. *Fuel Process. Technol.* **2003,** 82, (2-3), 89-165.
4. Presto, A. A.; Granite, E. J., Survey of Catalysts for Oxidation of Mercury in Flue Gas. *Environ. Sci. Technol.* **2006,** 40, 5601-5609.
5. Cao, Y.; Gao, Z.; Zhu, J.; Wang, Q.; Huang, Y.; Chiu, C.; Parker, B.; Chu, P.; Pan, W. P., Impacts of halogen additions on mercury oxidation, in a slipstream selective catalyst reduction (SCR), reactor when burning sub-bituminous coal. *Environ. Sci. Technol.* **2008,** 42, (1), 256-261.
6. Liu, S. H.; Yan, N. Q.; Liu, Z. R.; Qu, Z.; Wang, P.; Chang, S. G.; Miller, C., Using bromine gas to enhance mercury removal from flue gas of coal-fired power plants. *Environ. Sci. Technol.* **2007,** 41, (4), 1405-1412.
7. Granite, E. J.; Pennline, H. W.; Hargis, R. A., Novel sorbents for mercury removal from flue gas. *Ind. Eng. Chem. Res.* **2000,** 39, (4), 1020-1029.
8. Lee, S. F.; Seo, Y. C.; Jurng, J.; Lee, T. G., Removal of gas-phase elemental mercury by iodine- and chlorine-impregnated activated carbons. *Atmos. Environ.* **2004,** 38, (29), 4887-4893.
9. Miller, S. J.; Dunham, G. E.; Olson, E. S.; Brown, T. D., Flue gas effects on a carbon-based mercury sorbent. *Fuel Process. Technol.* **2000,** 65, 343-363.
10. Hedrick, E.; Lee, T. G.; Biswas, P.; Zhuang, Y., The development of iodine based impinger solutions for the efficient capture of Hg^0 using direct injection nebulization - Inductively coupled plasma mass spectrometry analysis. *Environ. Sci. Technol.* **2001,** 35, (18), 3764-3773.
11. *Sorbents for mercury removal from flue gas*; Report DOE/FETC/TR-98-01; U.S. Department of Energy: Pittsburgh, PA, Jan 1998.
12. Ponpon, J. P.; Sieskind, M.; Amann, M.; Bentz, A.; Corbu, C., Characterization of the HgI_2 surface layer after KI etching. *Nuclear Instruments & Methods in Physics Research A* **1996,** 380, (1-2), 112-116.

THE EFFECTS OF THE RENEWABLE PORTFOLIO STANDARDS IN THE STATE OF ALABAMA

Candace Watson and Dr. Robert W. Peters, University of Alabama at Birmingham
Birmingham, AL

Introduction

The major concerns throughout the world are global warming and climate change. Presently, many studies are being conducted to determine the catalyst responsible for these phenomena. Research is showing that we need to reduce the amount of carbon dioxide (CO_2) emitted into the atmosphere to correct or improve the problem. For 10,000 years CO_2 concentrations in the atmosphere have been consistent at 275 parts per million (ppm), in the past 50 years this level has risen to 378 ppm. If we continue with our current actions this level could possibly reach 840 ppm, by the end of the century[3].

Fossil-fuel-fired energy generation is the single greatest source of air pollution in the United States, in which power plants are the leading source of CO_2 emissions – the primary contributor the global warming. Currently in the United States, eighty-five percent of the electricity generation originates from the fossil fuels coal, oil and natural gas. The extraction and burning of these fuels contributes to global warming, chronic human health conditions, and degrades valuable land and water resources. The use of these fuels presents another problem, they are finite resources, meaning they cannot be replaced and will eventually run out. These facts inform us that it is time to find another primary resource to meet our current and future energy demands. The resource or resources selected need to be clean, cheap and comparable to that of fossil fuels. A standard has been established requiring electrical utility companies to replace a certain percentage of fossil fuel generated energy with renewable resources by a given date, known as the *Renewable Portfolio Standard* (RPS). Over half of the states in the country have already adopted this standard and are reaping its numerous benefits.

There are several renewable resources available to generate energy, the most common resources are biomass, geothermal, hydropower, landfill gas, solar, and wind. All of these resources excluding hydropower are accepted under the standard. Hydropower has been in use for so long and is almost at capacity that the standard does not allow states to incorporate hydropower in the percentage of renewable resources. Each state is encouraged to achieve the goal of replacing 20% of the finite fuel-generated energy with renewable resources by the year 2020. It is expected that this standard will significantly reduce the amount of carbon dioxide emitted into the atmosphere, which will improve overall air quality.

Why Consider Renewable Resources?

Renewable energy provides several benefits in the areas of economy, environmental benefits, risk management, energy assurance, reliability and public demand. States that import fuel from other locations for energy production lose vital financial resources because they leave the state and the local economy. The labor intensity of renewable resources can lead to a strong employment market in the

---

[3] http://www.architecture2030.org/

areas of construction, maintenance, manufacturing, and operation. Implementing local resources could potentially increase the local economy by reinvesting the dollars saved from renewables back into neighborhood projects. Implementing local biomass and wind farms can revitalize rural communities by providing new cash crops for farmers and new tax revenues.

Economics

Renewable energy resources can provide economical benefits to energy supplier's resources. During peak availability of renewable resources, utility plants can switch from costly purchased fuels to solar or wind sources, thus leading to less fuel purchased for energy generation and possibly a storage that can be used during peak energy demands.

Typically, energy efficient technologies can save money immediately and have short paybacks. Reducing costly fuel generated energy can lower operating costs and make industry, merchants and farms more economically competitive. Renewable technologies minimize the need to build new and expensive power plants to keep up with the current and future energy demands.

Environmental Impact

Renewable energy technologies improve the quality of the atmosphere by reducing carbon dioxide emitted into the atmosphere by fossil fuel-burned energy production. Solar and wind energy are emission free at the point of use and emissions from biomass are significantly lower than comparable conventional fuels. Biomass has the potential to be emission free with a closed carbon cycle. While the biomass is being combusted for energy production, new biomass crops are in the growing process taking in the carbon dioxide emitted. Renewable energy technologies have minimal impact on water resources in that they do not discharge harmful pollutants into surface water and toxins do not seep to groundwater. Minimal environmental impact will also lead to increased air quality resulting in improved human health.

Risk Management

Risk management and energy assurance go hand in hand. Renewable energy technologies contribute to the energy investment portfolio by providing diversification to evade against unplanned events that threaten delivery and affordability of energy. Unlike a central power plant, renewable energy supply sources are distributed in several locations thus reducing its vulnerability to security threats, natural disaster, and human error. Renewable energy can also offset fluctuating fuel costs for consumers, because many of the resources are free; when fuel prices rise renewable energy can be produced and used until fuel price decrease. Energy assurance provides the same benefits including that of when there is a peak in energy demand renewable resources can supplement fuel to keep prices down.

Reliability

Renewable resources are more reliable than fossil fuels. As this may seem contradictory, because certain resources are not always available, it is very easy to predict and plan their presence. When resources are in abundance, they can be captured, stored, and then used during times when conventional resources are unavailable. In combination with electricity generation, renewable energy can assist energy suppliers with stretching conventional fuels to mitigate the risks of interruptions during peak energy demands. The most reliable kilowatt is one that does not need to be generated, transmitted, distributed and stored. During California's energy crisis in 2001, energy efficiency and energy-conserving behavior increased system reliability by reducing the demand on the system. These methods also proved to be cost effective to taxpayers and ratepayers that invested $1.3 billion in energy efficiency upgrades, compared to the $2 billion to $20 billion in potential losses from rolling blackouts, which were avoided.

Public Interest

Studies indicate that the public would rather invest in environmentally clean energy and are willing to pay extra for "green" power. Research shows that green power decreases as conventional fuel increases. Surveys that have been conducted across the country in many demographic categories show that the public is interested in environmentally friendly energy production and view it as being fiscally sensible and environmentally responsible.

> *"United States oil and natural gas production has been on the decline since 1970's and global oil and natural gas reserves are limited (the static lifetime of global oil reserves is 42 years and natural gas only 63 years). With the global energy demand increasing, coal is becoming the fossil fuel of choice. The United States, Russia, China and Australia and India have plenty of it and it is cheap (and dirty) .Clean coal technology is decades away as is capturing and storing CO_2, and they are costly[5]."*

Renewable Energy Technologies

Renewable technologies tap into the ever-present energy that exist around us and turn it into a useful form. The movement of wind and water, the heat and light from the sun, heat in the ground, and carbohydrates in plants – are all natural resources that can supply our energy needs in a sustainable way. The capturing, storing, and processing of each of these resources will convert them into energy to supply power where needed. The most common types of renewable energy are biomass, geothermal, hydropower, landfill gas, solar and wind.

Biomass

Biomass is any organic material made from plants or animals. Domestic biomass resources include agricultural and forestry residues, municipal solid waste, industrial waste, and terrestrial and aquatic plants specifically grown for energy production. Some of the biomass producing resources includes the following: energy crops, trees, grasses, other crops, biomass residues, oil plants, forestry, agriculture and cities.

Energy crops, as known as "power crops," are can be grown on farms, in large quantities, similar to food crops. Trees and grasses prove to be the best crops for energy production. Some varieties of trees that are cut off close to the ground can grow back. This process called coppicing, allows trees to be harvested for 3 to 8 years for 20 to 30 years before replanting. Thin stemmed perennial grasses, switch grass, big bluestem and other native varieties grow quickly in many parts of the country and can be harvested for up to 10 years before replanting.

Oil plants are plants that produce oil that can be processed into energy. Examples are soybean and sunflower plants. This concept is similar to the idea of processing the oils from corn and producing energy. A different type of oil crop with great promise for the future is microalgae. These tiny aquatic plants have a potential to grow quickly in hot, shallow, saline water found in some lakes located in the

---

[5] http://www1.eere.energy.gov

desert southwest. Some research has shown that carbon dioxide can be used to stimulate algae growth. The algae can be converted into various ranges of fuels.

Biomass residues are the leftover wastes after plants have been used for a specific purpose. Forestry, agriculture and manufacturing industries generate large volumes of plant and animal waste. City waste, from garbage and sewage is another source for biomass energy. People generate biomass waste in many forms including "urban wood waste" (such as shipping pallets and leftover construction wood), the biodegradable portion of garbage (such as paper, food, leather and yard waste), and gas released in landfills because of waste decomposition. Sewage is another biomass resource. Some sewage treatment plants capture the off gas of methane from sewage and burn it for heat and power, reducing air pollution and emissions of global warming gases.

Forestry wastes are the largest source of heat and electricity, since lumber, pulp and paper mills use them to power their factories. After timber harvesting, treetops and branches are typically left behind and are the largest source of forestry waste. Some of the waste is left behind to provide essential nutrients and provide habitats for birds and mammals, but the rest could be used to generate electricity. Other sources of wood waste are sawdust and bark from sawmills, shavings from furniture manufacturing, and organic sludge (or "liquor") from paper and pulp mills.

Biomass is solar energy stored in organic matter. The process of photosynthesis uses the sun's energy to convert carbohydrates (sugars, starches, and cellulose), as trees and plants grow. Carbohydrates are the organic compounds that form biomass. When plants die, the process of decay releases the energy stored in carbohydrates and discharges carbon dioxide back into the atmosphere. Biomass is considered a renewable resource because the growth of new plants and trees replenishes the supply.

One of the major benefits to biomass energy production is it causes no net increase in carbon dioxide emissions into the atmosphere. As trees and plants grow, they remove carbon dioxide from the atmosphere through photosynthesis. In addition, using biomass for energy generation is a way to dispose of waste materials that otherwise would create environmental risks.

Biomass resources provide approximately 3% of all energy consumed in the United States. In 2002, biomass supplied about 47% of all renewable energy consumed in the United States. Electric generation from biomass (excluding municipal solid waste) represents about 11% of all generation from renewable resources in the United States. Biomass supplied more energy throughout the United States in 2002 than other renewable resources, including hydroelectric. Providing almost six times the energy of geothermal, solar, and wind energy resources, biomass has a high potential to meet the increases needs of the world's energy demand[4].

Geothermal

Geothermal energy is the conversion of hot water or steam from deep beneath the earth's surface into energy. The resources of geothermal energy are volcanoes, hot springs, fumaroles, pools of boiling mud, and geysers. A volcano is a fissure of earth's crust through which molten lava and gases erupt. There are three types of geothermal plants used to generate electricity, in which the type is based on depth, temperature and quality of water and steam in the area. The three types of plants are dry steam, flash steam, and binary-cycle.

The dry steam geothermal plant is where steam (no water) shoots from the well and passes through a rock catcher and then a turbine. Dry steam fields are very rare. Flash steams geothermal

---

[4] http://www.ucsusa.org

plants utilize water over 182°C from geothermal reservoirs. The water below ground is well above boiling point of water, at sea level, but remains in liquid state due to the high pressure. As the water is pumped from the reservoir to the power plant, the drop in pressure causes the water to convert or "flash" into steam to power the turbine. Any unused water, not flashed into steam, is re-injected back into the reservoir for reuse. Flash steam geothermal power plants are most common type of geothermal power plants in operation today.

Water used in binary-cycle geothermal power plants is typically cooler than the water used in flash steam plants, ranging from 107 to 182°C. The hot fluid from the geothermal reservoir passes through a heat exchanger, which transfers heat to a separate pipe containing fluids with a much lower boiling point. These fluids are usually iso-butane and iso-pentane, vaporized to power the turbine. The cooled water is returned to the reservoir. Some of the additional benefits of using a binary-cycle plant are lower operating cost, increased efficiency, no emissions of excess gas, able to utilize lower temperature reservoirs (which are most common), and many future geothermal power plants that are planned for construction are binary-cycle plants, offering more examples to learn from.

Hydropower

Hydropower generation is simply taking the mechanical energy present in moving or falling water, turning it into kinetic energy and using the kinetic energy in a useful form. The harnessing of moving water to provide electricity prevents the building of new dams that have a major impact on wildlife and water quality. Since water is 832 times denser than air, water features such as tides, waves and ocean currents, which represent a powerful, highly concentrated and clean-energy resource.

Solar Energy

Solar energy is free and exhaustible. As a viable alternative to the fossil fuels that continuously pollute our air and water, threaten human health and contribute to global warming, solar energy is a vast and clean energy technology. Solar energy is the largest supporter of life on earth and is the main ingredient for all energy used. Plants utilize the sun's energy for growth, which can be burned as biomass fuel or if left in the swamp and compressed underground for millions of years, in the form of coal and oil. Heat from the sun causes temperature difference in areas, producing wind that can power turbines. Water evaporates because of the sun, falls on high elevations, and rushes down to the sea, spinning hydroelectric turbines as it passes. Solar energy in the renewable portfolio standard refers to the exploitation of the sun's energy into electricity, lighting and heating.

The amount of energy that is absorbed in the earth from the sun is abundant. Energy collected from 20 days of sunshine is equal to the total amount of energy stored in oil, coal and natural gas reserves. Outside of the Earth's atmosphere, the sun's energy contains about 1,300 watts per square meter. Inside of the Earth's atmosphere, it has decreased to 1,000 watts per square meter, at noon on a cloudless day. Averaged over the entire surface of the planet, 24 hours per day, each square meter collects the equivalent of almost one barrel of oil per year or 4.2 kilowatt-hours of energy per day.

Wind Energy

Harnessing wind energy is one of the cleanest and most sustainable energy generating technology. Wind energy generation does not contribute harmful greenhouse gas emissions into the atmosphere and does not contribute to global warming. Wind power is also one of the most abundant and cost-effective energy generating technologies available, which makes them an increasingly viable resource. In 2005, wind energy produced 11,500 megawatts (MW) of electricity. In the United States, 2,431 MW were installed in 2005, providing enough energy to power 650,000 typical homes. Wind turbines collect wind energy. Modern turbines generally have two or three blades placed upwind on the tower it sits on. Inside of the wind turbine, motion is turned to energy and that energy used for a useful purpose.

Alabama's Potential for Renewable Energy Production

Alabama's potential for renewable energy includes biomass, hydropower, wind and solar. The potential of wind power in the state of Alabama is possible on a small-scale project, typically individual buildings and residential structures. There is a high potential for hydropower energy generation in Alabama, but this resources has already been used to the maximum capacity. Alabama has a high potential for biomass energy production. With 22.5 million acres of timberland forests in the state, biomass proves to be the state's most reliable renewable resource. Solar generated energy has a potential in the southeastern region in the state for both flat-plate collectors and concentrating collectors[6].

Alabama currently has 18 natural gas power plants, 14 hydroelectric power plants, 9 coal powered plants and 2 nuclear power plants. It is important to evaluate the existing electricity infrastructure to determine if these existing plants can support renewable energy technology. In addition to evaluating the existing plants, the impacts on the state's resources must also be evaluated.

Impact on the State of Alabama

If the state of Alabama were to implement the renewable portfolio standard, requiring it to replace 15% of finite resources with renewable resources, excluding hydropower, the state of Alabama would have to install: 6,600 wind turbines, each collecting 2MW of power, occupying 500 square miles of land, which is equivalent to the city of Phoenix, Arizona; 190 square miles of solar panels, an area equivalent to the city of Columbus, Georgia; or 5,500 square miles of switch grass, an area equivalent to the state of Connecticut. The implementation of the use of renewable resources would require a massive impact on existing land. Biomass energy is the choice resource for Alabama. Several obstacles are involved with the use of biomass energy[2].

Potential Obstacles with the Use of Biomass Energy

Biomass is typically grown in remote areas throughout the state, thus having to be transported to the power plants. The transporting of this material could essentially contribute to CO_2 emissions producing a negative impact on the atmosphere. Straw and forest residue biomass typically has a low volumetric density compared to its converse fossil fuels. A large variety of biomass materials have different chemical compositions resulting in different processing techniques. This can create problems when retrofitting existing power plants to support renewable energy technologies. This issue forces the state to select biomass materials with similar chemical compositions to simplify the conversion of existing power plants. The processing techniques of biomass materials cause a high ash content, which leads to corrosion and agglutination of machinery. It requires additional maintenance to keep equipment running properly. The most common vulnerability of biomass is the socio-economic aspect, that is using crops or land that could be used for human food sources for energy. A plan must be designed to avoid disrupting the food production in the United States.

---

[6] http://www.forestry.state.al.us/forest_facts.aspx

[2] Presentation given by Thomas W. Johnson, Research Program Manager at Southern Company Birmingham.

Case Study: Germany

Germany is the forerunning country in renewable energy production. In 2005, 20.2% of the electricity produced in Germany originated from renewable resources. Germany's renewable energy output was 26.5 tetra-watt (TWh) hours compared to 21.5 TWh of hydropower. Power production from biomass materials has reached an all time high in Germany at 13.2 TWh, and has increasing possibilities. Photovoltaic and geothermal energy still play a marginal role in Germany, but is continuing to increase from the year 2000. Germany is well on their way to meeting the goal to replace 20% of their fossil fuel generated energy with renewable resources, by the year 2020.

Another growing sector in Germany is heating from renewable energy sources. In 2005, space heating from renewable resources was 5.4%, with an annual production of 76.5 TWh from biomass resources. Germany uses renewable resources to produce heat using traditional wood burning combined with modern methods of wood-pellet heating systems. Solar thermal heating using rooftop collectors and other surface collectors are growing rapidly. The amount of heat produced from solar technology has more than doubled since the year 2000, and since 1990, it has increased more than twentyfold.

Fossil fuels still dominate energy production in Germany, in 2005 accounting for approximately 82% of the energy demands. In addition to fossil fuel use, nuclear power is responsible for 12% of the electric power generated. Renewable energy sources supply 3.6% of the primary energy consumption, increasing from the mid 1990's at about 1.5%. More than 50% of the energy produced from renewable resources is generated from biomass materials; wind energy contributes 14.6% and waterpower 11.9%.

Germany's motivation to reduce reliance on fossil fuel sources is politically driven. Over the past ten years, public, legal, and economic framework have been established to give renewable energy a chance to be implemented in the energy market, in spite of their relatively high power-generating costs. Several energy credit programs have been established which says that power generated from renewable sources can be sold to the public power grid preferentially, and will receive a guaranteed price. The costs are covered by adjusting the price of power sold.

The prices established for renewable–source power are created on a decreasing scale according to source and specific requirements of the individual energy carriers. These resources are graded on a declining scale (decreasing from year to year), to make renewable energy technology more competitive with other energy producing resources. This can only be accomplished with the implementation of temporary subsides, such as tax credits and incentive programs. During this period, renewable-resource energy will have a chance to prove its efficiency and reliability in practice, as well as demonstrate economic viability.

Germany is at the stage that is proving that renewable energy resources are reliable, efficient and economically viable, with an increasing potential over time. Based on the increase of use of renewable energy over the years, Germany could possibly be the first country to meet the 20% goal by the year 2020, and possibly might surpass this percentage soon thereafter. Germany's implementation demonstrates renewable energy can be our next electric fuel source.

Germany's use of renewable technologies to produce energy illustrates that renewable resources are capable of producing a sufficient amount of energy for the growing energy demand. The amount of energy produced from renewable resources in Germany has surpassed the amount of energy produced from hydropower. This demonstrates that there are alternative energy resources available to offset the amount of fossil fuels that are burned to produce energy[1].

---

[1] Wengenmayr, Roland. *Renewable Energy: Sustainable Energy Concepts for the Future.* Weinheim: Wiley-VCH, 2008.

Conclusions

At this point in the project, research indicates that Alabama has a potential use for biomass, solar, and wind power for energy generation. Solar and wind power generation is on a smaller scale, but any portion of fossil fuel-powered energy that can be replaced with a renewable resource makes a positive contribution to the ultimate goal. It is very possible for Alabama to replace at least 15% of the fossil-fuel generated energy with renewable resources, primary utilizing biomass technologies. Based on the success of Germany's renewable energy production, renewable materials are a viable resource to produce energy.

In order to determine a recommendation for Alabama, several factors need to be addressed to create a plan for Alabama to incorporate renewable energy generation. The potential vulnerabilities of renewable resources need to be researched and a plan must be created to avoid possible interruptions in electricity production. If biomass materials are the primary resource for energy production, a plan that does not disrupt food production will also have to be implemented to separate energy-producing materials from food supplies. An inventory of all the electricity plants in the state will have to be evaluated to determine how they can adapt to support renewable energy technologies. Weather and agricultural data from at least the past ten years will need to be analyzed to determine if the state can support renewable energy material production. Investigating the potential benefits of landfill gas technologies will determine if this is a possible resource for Alabama to take advantage of.

References Cited

1. Wengenmayr, Roland. *Renewable Energy: Sustainable Energy Concepts for the Future.* Weinheim: Wiley-VCH, 2008.
2. Presentation given by Thomas W. Johnson, Research Program Manager at Southern Company Birmingham.
3. http://www.architecture2030.org/
4. http://www.ucsusa.org
5. http://www1.eere.energy.gov
6. http://www.forestry.state.al.us/forest_facts.aspx

123e Enhancement of the Desulfurization Performances of Waste Cement Particles by Acid Treatment

Jiawei Wu[1], **Akihiro Yamasaki**[2], Atsushi Iizuka[3], Kazukiyo Kumagai[1], and Yukio Yanagisawa[4]. (1) Institute of Environmental Studies, Graduate School of Frontier Science, The University of Tokyo, 5-1-5 Kashiwanoha, Kashiwa-shi, Chiba 277-8563, Tokyo, Japan, (2) Department of Materials and Life Science, Seikei University, 3-1 Kichijoji-kitamachi 3-chome, Musashino, Japan, (3) Institue of Environmental Studies, Graduate School of Frontier Science, The University of Tokyo, 5-1-5 Kashiwanoha, Kashiwa-shi, Chiba 277-8563, Tokyo, Japan, (4) Institute of Envronmental Studies, Graduate School of Frontier Science, The University of Tokyo, 5-1-5 Kashiwa-no-ha, Kashiwa, Chiba, 277-8563, Japan

Effect of acid treatment on the desulphurization performances of waste cement particles were examined with a laboratory-scale experimental apparatus based on the weight change of the sample exposed to a gas flow containing SO2. The waste cement particles, a byproduct of recycling process of aggregates from waste concrete, were pre-treated with three kinds of acid, namely, carbonic acid, hydrochloric acid and acetic acid. The desulfurization rate of the waste cement particles was enhanced by the acid treatment; especially, the calcium utilization ratio was significantly improved. In some cases, the desulfurization rate per unit weight of the sample was higher than the original sample. These results can be attributed to the extension of the specific surface area of the particles, and consequently, the ratio of the calcium content active for the desulfurization was increased by the acid treatment. Based on the experimental results, cost estimation of the dry desulfurization process was conducted, and it was found that the desulfurization cost by using the waste cement as a dry sorbent is about 1/10 to 1/6 of that for the conventional method by using limestone. Therefore, the desulfurization process with the waste cement is highly competitive in terms of the cost.

A PRACTICAL ROUTE FOR GREENING OF COAL

Li Zhou, Yu Zhang, Chunfeng Liu, Yan Sun, Yaping Zhou
High Pressure Adsorption laboratory
Tianin University, Tianjin, China

Introduction

Global surface temperature anomaly yields serious consequences in the inhabited environment, which is considered responsible for the rising of sea levels, shrinkage of snow-covered areas, shortage of fresh water, and the disasters caused by abnormal weather. Accumulation of CO_2 in the atmosphere is considered responsible for the global temperature anomaly, and industrial activities have accelerated the CO_2 accumulation because of the combustion of fossil fuels. The rate of economic development is much faster in the early years of the 21st century than ever before. Therefore, the rate of CO_2 accumulation in the atmosphere is definitely much faster than past. Although coal is not clean, it is indispensable energy source of the world for quite a long time[1] as shown in Fig. 1. Therefore, more development of economy will result in heavier pollution of environment. However, complete greening of coal will radically solve environmental problems, and the fast knot between economic development and the environmental pollution is unlocked.

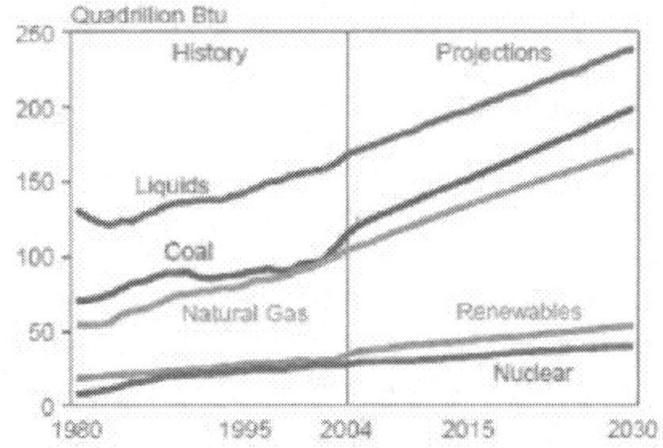

Fig. 1. World marketed energy use by fuel type for 1980-2030.

Complete greening of coal is different from any clean coal technology available presently. The latter cannot reduce CO_2 emission because the fuel still contains the element of carbon. The former changes coal to hydrogen, therefore, stops CO_2 emission. The Hymelt Process [2] is a typical technology of changing coal to hydrogen presently available. Coal is decomposed under the catalysis of molten iron and a gas mixture containing hydrogen is obtained. Some carbon is also produced in decomposition and the carbon is dissolved in molten iron. Oxygen reacts with molten iron in another reactor, and the dissolved carbon was taken out of iron yielding CO-rich gas. The process operates at temperatures as high as 1400-1800 ^{o}C. Such technologies feature in the fact that coal is the source of hydrogen, therefore, hydrogen is not the unique fuel obtained. The HydroMax process [3,4] produces pure hydrogen, but uses molten iron to react with steam and uses solid carbon to reduce iron oxide back to pure metal. The latter operation is just a typical metal smelting process. Another water-splitting process basing on the Fe_3O_4/FeO cycle [5,6] releases oxygen at an elevated temperature to realize the cycle. Extra energy is paid for keeping operations cycling between two temperatures, and the huge thermal capacity of real reactors renders the cycle a very slow process.

Proposed Strategy

The process

In the complete greening technology of coal, coal is assumed to convert to inorganic carbon. Hydrogen is produced in a cycle consists of three reactions:

1. Oxidation of FeO by steam at about 800°C. The product is pure hydrogen and Fe_3O_4/Fe_2O_3.

$$H_2O + 3\ FeO = H_2 + Fe_3O_4 \quad (1)$$

Or

$$H_2O + 2\ FeO = H_2 + Fe_2O_3 \quad (1a)$$

2. CO is used to reduce Fe_3O_4/Fe_2O_3 at same temperature. The product is FeO and CO_2.

$$Fe_3O_4 + CO = 3\ FeO + CO_2 \quad (2)$$

Or

$$Fe_2O_3 + CO = 2\ FeO + CO_2 \quad (2a)$$

3. The above produced CO_2 reacts with carbon at about 900°C generating the CO required in Reaction 2.

$$C + CO_2 = 2CO \quad (3)$$

Summing up reactions (1-3), a general reaction is obtained:

$$H_2O + C = H_2 + CO \quad (4)$$

Theoretical energy balance

To evaluate the viability of a novel process, theoretical energy gain (TEG) of the process was defined as the ratio of theoretically releasable energy by the process products over the theoretically invested energy in order to acquire the products. A process must not be energetically feasible if the TEG is less than unity. The theoretical energy input and output is calculated on the basis of general reaction and formation enthalpies at 298 K.[7] The energy input of the process is: (-110.54)-(-285.83) = 175.29 kJ. The theoretically releasable energy of the products is determined from the combustion reactions:

$$H_2 + 1/2O_2 = H_2O$$

$$CO + 1/2O_2 = CO_2$$

The enthalpy change of the above reactions is -285.83 and –282.97 (-393.51+110.54) kJ, respectively. The energy gain is, therefore, (285.83+282.97)/175.29 = 3.24. The energy input required by the whole process, 175.29 kJ, is only 62% of that releasable by CO (282.97 kJ), therefore, the process is energetically more than autarky.

Experimental proof

Practical feasibility of the process was experimentally proven.[8] Pure hydrogen could be continuously produced in a process shown schematically in Fig. 2. In order to increase the reaction interface, iron oxide was dispersed on the surface of silica SBA-15. Composition of the effluent stream from a reactor was analyzed with a mass- spectrograph. Pure hydrogen was produced in the oxidation period, and it was really obtained as is shown in Fig. 3.The reduction period is very important for the continuous production of pure hydrogen, though the period does not produce hydrogen. The reduction extent of iron oxides must be under control to avoid the product of metal iron since fresh iron is very tiny particulate and easily carried out from the reaction zone. The temperature and atmosphere of the reduction must be carefully tuned. Both oxidation and reduction reactions were kept at 800 °C, and a proper ratio of CO vs. CO_2 was maintained in the reduction period. Carbon is consumed as the source of carbon monoxide, but not as reductant to favor the control of reduction extent and to keep the hydrogen pure. The production rate of hydrogen will be stable in so doing as shown in Fig. 4. Avoidance of forming metal iron avoids the formation of iron carbide because metal iron is the prerequisite condition of forming iron carbide for temperatures higher than 628 °C.[9] Iron carbide can hardly form at 800 °C even H_2 joined the reducing gas.[8,10]

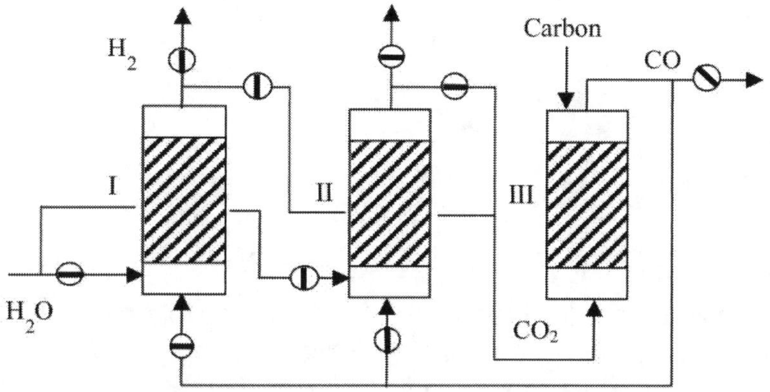

Fig. 2. A process for continuous operation. Reactors I and II are packed with iron oxide support and Reactor III is packed with carbon.

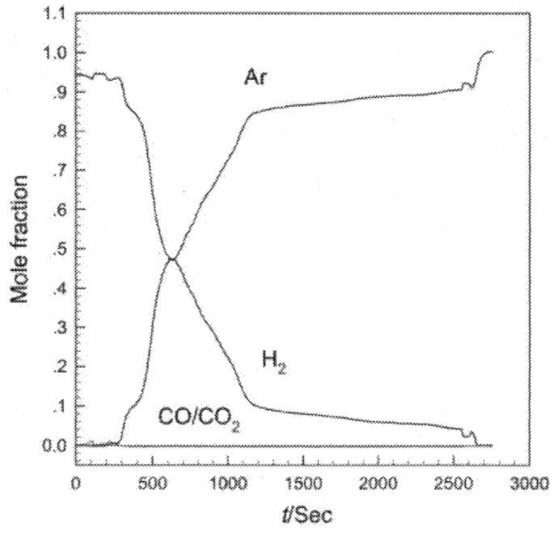

Fig. 3. Composition of the effluent stream during oxidation period (Ar is carrier gas).

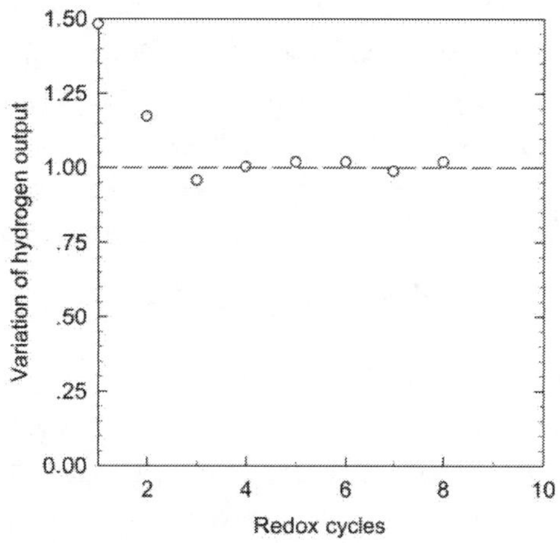

Fig. 4. Stability of hydrogen production in consecutive reduction/oxidation cycles using a mixture of CO and CO_2 as reducing gas.

Scenario of green energy unit

To satisfy the energy requirement of the hydrogen production process, 62% of the produced CO is burnt, and CO_2 is generated. Aside from CO_2, N_2 is also enriched in the flue gas of the combustion. Therefore, CO_2 and N_2 are by-products of the process if a separation operation is applied for flue gas. The 38% of produced CO is output as another by-product. CO is one of the initiating chemicals of many green synthesis processes and has eight times higher market value than hydrogen. Therefore, the process is economically profitable. The combined processing units may look like that shown in Fig. 5.

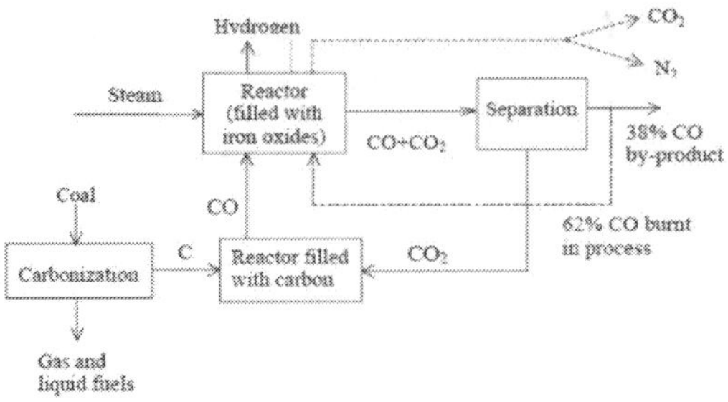

Fig. 5. Combined processing units.

There are three green ways to dispose the produced CO_2. One is to inject CO_2 into some oil fields aiming to increase oil recovery.[6] Another is to inject CO_2 into the reservoirs of natural gas and five times higher output was observed[11] while CO_2 is concealed in the reservoir. The third one is to inject CO_2 into the sediments of flammable ice to replace CH_4 as the guest gas of hydrates.[12] Therefore, there would be not CO_2 emitting into atmosphere.

The process is technically simple, and there is not any technical barrier towards industrialization. Coal is not the exclusive source of carbon, but is the one a national economy can rely on. The carbonization technology of coal is already matured, though the advanced technology has not been widely applied. More liquid and gaseous fuels are produced in the process of coal carbonization if supercritical extraction is utilized,[13] and the ever increasing petroleum price would subject a heavy impact if the proposed energy strategy be seriously considered. Coal bears the largest ratio of carbon to hydrogen compared to liquid or gaseous fuels, and the decrease of CO_2 emission is more than proportional if coal is replaced with hydrogen. A fast improvement in the world environment is definitely promising.

Acknowledgements

The persistent support of the National Natural Science Foundation of China for the author's research interests is greatly appreciated.

References

1. International Energy Outlook 2007, web site www.eia.doe.gov/oiaf/ieo/index.html.
2. T.D. Trowbridge, D.P. Malone (2001), "The Hymelt Process - Potential for revolutionizing the hydrogen industry," [325d] Annual AIChE Meeting, Nov. 4-9, At Louis, Missouri, USA.

3. http://www.alchemix.net. "Alchemix Corporation Hydrogen Technology."
4. J.K. Kindig, R.R. Odle, T.E. Weyand, B.R. Davis (2002), "Method and apparatus for the production of hydrogen gas," US Patent Application Publication Sep.12, Pub. No. US 2002/0127178A1.
5. F. Lemort, P. Charvin, C. Lafon, M. Romnicianu (2006), "Technological and chemical assessment of various thermochemical cycles: From the UT3 cycle up to the two steps iron oxide cycle," *Int J Hydrogen Energy*, 31, 2063-75.
6. Ministry of Petroleum and Energy, Norway (2007), "Use of CO_2 to increase oil recovery," web site http://www.regjeringen.no/ published August 20.
7. J.A. Dean (1999), Lange's Handbook of Chemistry, 15th Ed. New York, McGraw-Hill.
8. Y.P. Zhou, C.X. Jia, Y. Zhang, C.F. Liu, D.S. Huang, Y. Sun, L. Zhou (2008), "Experimental Study on a New Process of Producing Hydrogen in Consumption of Water and Coal," *AIChE J*, 54 (5) 1388-1395.
9. L.K. Liang, Y.C. Che (1990), "*Metallurgy Thermodynamics and Dynamics* (in Chinese)," Shenyang, Northeast Institute of Technology Press.
10. D.Y. Wang, Y. Min, C.J. Liu, P.Y. Shi, M.F. Jiang (2007), "Study on iron carbide production with CO-CO_2-H_2 mixing gas," (in Chinese) *The Chinese J Process Eng.* 7, 332-336.
11. C.M. Oldenburg, K. Pruess, S.M. Benson (2001), "Process modeling of CO_2 injection into natural gas reservoir for carbon sequestration and enhanced gas recovery," *Energy Fuels*, 15, 293-298.
12. J.H. Yoon, T. Kawamura, Y. Yamamoto, T. Komai (2004), "Transformation of Methane Hydrate to Carbon Dioxide Hydrate: In Situ Raman Spectroscopic Observations," *J Phys Chem A*, 108, 5057-5059.
13. S.C. Guo (1991), *Chemical Technology of Coal*, Beijing, Metallurgical Press.

123g High Temperature Adsorption of Mercury on Non-Carbon Based Sorbents

Sung Jun Lee, Chemical Engineering, ICSE, University of Utah, 50 S. Central Campus Dr., MEB Rm. 3290, Salt Lake City, UT 84112, **Jost O.L. Wendt**, Department of Chemical Engineering & Institute for Clean and Secure Energy, University of Utah, 3290 MEB, 59 S. Central Campus Dr., Salt Lake City, UT 84112, and Joep Biermann, MinPlus, BV., Arnhem, Netherlands.

A bench scale disperse phase, entrained-flow reactor, was employed to investigate elemental mercury (Hg) adsorption on non-carbon based sorbents. The major parameters varied for the Hg adsorption test were reaction temperature, sorbent feeding rate, gas compositions, and mineral composition of sorbents. The sorbent receiving the greatest attention was MinPlus, an engineered sorbent produced from a process which uses paper recycling wastes as a feedstock. Hg adsorption tests were performed of a temperature range of 600 - 1000°C and at various sorbent feeding rates (< 7g/hr). Since the Hg adsorption was occurred at high temperatures and the adsorption efficiency increased with increasing temperature, the reaction between sorbents and Hg should be considered to be chemisorption. Overall Hg capture was achieved at over 90% at 900-1000 °C, and could be divided into two mechanism; in-flight capture and capture through reactor wall effects involving deposited particles.

Surface analysis of spent sorbent (XRD, SEM-EDS) showed that high temperature mineral transitions, and the ensuing sorbent compositions could enhance the Hg sorption mechanism (in-flight and wall effect). Gas compositions such as CO2, H2O, HCl and O2 could have a significant effect on Hg adsorption, which appears to require an oxidation step. Furthethermore, interactions with types of fly ash also could affect the Hg adsorption efficiency, where bituminous coal derived fly ash showed higher Hg capture than that from mixtures involving PRB (Powder River Basin) fly ash.

Optimization of nanostructred smart biosensors incorporated in a fouling resistant ultrafiltration membrane used for water treatment

Amr Zaky[1], Guang Cai[1], Cyndee Gruden[1], Colleen Gorey[2], and Isabel Escobar[2]
[1]Department of Civil Engineering,
[2]Department of Chemical and Environmental Engineering
The University of Toledo
Toledo, OH

Membrane replacement is the largest operating cost in water separation applications. Irreversible fouling and permeate flux decline occur due to macromolecular adsorption of organic matter (e.g., bacteria) on the membrane surface, which cannot be removed by backflushing or backpulsing. The presence of microorganisms in feed water can further exacerbate fouling due to the accumulation of microorganisms onto the membrane surface and on the feed spacer between the envelopes, or biofouling. Accurate prediction of membrane performance requires early detection of bacteria in membrane-based water treatment systems. A novel temperature-activated cellulose acetate ultrafiltration membrane in a cross flow configuration has been developed. A stimuli-responsive polymer brush (hydroxypropyl cellulose) was attached to the membrane surface and coupled with antibodies to detect potential biofouling species, mycobacteria. It is hypothesized that by continuously triggering the phase transition via stimulus response, the non-equilibrium movement of the polymer brush would offer better protection of the membrane surface than at equilibrium, thereby reducing fouling. In addition, it is anticipated that the movement of the polymer brush will improve the performance of the sensor in the presence of foulants (organics). The smart biosensor recovery was measured following temperature activation (above 50°C) of the polymer brushes. Recovery was significantly decreased from 15% to 8%. Ongoing experiments involve alternative methods of polymer brush activation including using a low intensity ultrasound probe ($\approx 1\ W/cm^2$) to increase biosensor recovery efficiencies. A combination of the fouling resistant layer and early detection method could increase the potential for membrane use in a wider range of water separation applications.

141b Determining the Extent and Influence of Flagella, Injection Concentration, and Solution Chemistry In *Salmonella* Transport

Berat Z. Haznedaroglu, Chemical and Environmental Engineering, University of California, Riverside, 900 University Ave, Bourns Hall A242, Riverside, CA 92521 and Sharon L. Walker, Department of Chemical and Environmental Engineering, University of California, Riverside, Riverside, CA 92521.

This study focuses on the transport of Salmonella, a bacterial pathogen, and the extent to which mobility, cell injection concentration, and solution chemistry in groundwater. The goal was to determine how flagellar components, experimental solution chemistry and cell concentration affect its fate. For this purpose, Salmonella enterica subspecies enterica serovars Gallinarum (wild type, non-flagellated) and Typhimurium (wild type and dysfunctional flagellated mutant strain) were injected into a packed bed column with aquifer sand over a range of solution ionic strength (IS) and cell concentration simulating groundwater environments. Initial transport experiments conducted with non-flagellated Salmonella strains showed trends that cannot be explained with classic Derjaguin-Landau-Verwey-Overbeek (DLVO) theory alone (Derjaguin and Landau; 1941). Therefore, comprehensive cell surface characterization techniques including size, surface charge density, extracellular polymeric substances, electrophoretic mobility, and hydrophobicity analyses were conducted to explain observed transport trends. Electrophoretic mobility calculations, in agreement with titration analysis and column experiment results revealed that non-flagellated Salmonella was less electronegative than other previously studied non-pathogenic bacteria (Walker et al., 2004; Walker et al., 2005). Packed bed column experiments allow for calculation of transport distances and these distances were found to be up to 33 meters in low ionic strength conditions at our tested flow rate and injection concentration values. Results to be presented at the American Institute of Chemical Engineers (AIChE) 2008 Annual Meeting will include the relative transport behavior currently under investigation of flagellated and dysfunctional flagellated Salmonella spp. with the non-motile Salmonella strain. The findings of this study contribute to the understanding of fate and transport of important pathogens in groundwater and seek to establish the relative role of biological (flagellar components) and physicochemical (IS and concentration) phenomena.

References:

1. Derjaguin, B. V., and L. Landau. 1941. Acta Physicochim U.S.S.R. 14:300.

2. Walker, S. L., J. A. Redman, and M. Elimelech. 2004. Role of cell surface lipopolysaccharides in Escherichia coli K12 adhesion and transport. Langmuir 20:7736-7746.

3. Walker, S. L. 2005. The role of nutrient presence on the adhesion kinetics of Burkholderia cepacia G4g and ENV435g. Colloids and Surfaces B-Biointerfaces 45:181-188.

141c Biomedical Imaging Gold Nanoparticles: Interactions with Humic Substances

Vasanta L. Pallem[1], Holly A. Stretz[1], and Martha J. M. Wells[2]. (1) Department of Chemical Engineering, Tennessee Technological University, Prescott Hall Rm 311, Stadium Drive, Cookeville, TN 38505, (2) Center for the Management, Utilization and Protection of Water Resources, Tennessee Technological University, P.O. Box 5033, Cookeville, TN 38505

The interaction of biomedical imaging gold nanoparticles with environmental entities like humic substances was investigated applying UV-Vis absorbance and fluorescence spectroscopy. Gold nanoparticles have potential applications in drug delivery and biomedical imaging; however their fate in the environment is relatively unknown. The interactions of gold nanoparticles with environmental entities like humic substances present in natural aquatic systems will play a vital role in their fate and transformation. Humic substances (HS) are fluorescent in nature, giving us an interesting tool to characterize the interaction between these natural polyelectrolytes and model water soluble gold nanoparticles. When gold nanoparticles were exposed to HS, the fluorescence map (excitation/emission/intensity) of the humic substances was observed to be quenched. This quenching exhibited concentration dependent behavior. The conclusion was that the gold nanoparticles are associating with these water soluble polyelectrolytes closely enough that energy could be transferred between the different species. However, no precipitation was observed, indicating that gold nanoparticles might be mobilized in surface waters due to such interactions. The aggregation behavior of the gold nanoparticles/HS was also studied using UV-Vis absorbance. Absorbance of the gold nanoparticles was observed at 519 nm in the control and this response shifted to longer wavelengths (650-800 nm) as the concentration of HS increased, indicating that the gold nanoparticles in the presence of humic substances were aggregating. This behavior is best observed as a difference in the spectra of the gold nanoparticles/HS and the spectra of the HS alone.

141d Interaction of Fullerene (C_{60}) Nanoparticles with Humic Acid and Alginate Coated Silica Surfaces: Implications for Fate and Transport

Kai Loon Chen, Department of Geography and Environmental Engineering, Johns Hopkins University, Baltimore, MD 21218 and Menachem Elimelech, Department of Chemical Engineering, Yale University, Mason Lab, 313A, 9 Hillhouse Avenue, New Haven, CT 06520.

Buckminsterfullerene C_{60} is a nanomaterial with potential applications in the fields of optics, electronics, and biomedical engineering. With the impending widespread utilization of fullerene in industrial and consumer products, fullerene pollution of natural waters is inevitable. In aqueous solutions, hydrophobic fullerene molecules bind strongly together to form negatively charged fullerene nanoparticles. To better understand the fate and transport of these nanoparticles in natural aquatic systems, we employ the quartz crystal microbalance (QCM) to derive the deposition kinetics of fullerene (C_{60}) nanoparticles onto bare silica surfaces and surfaces pre-coated with humic acid and alginate. The deposition behavior of these nanoparticles is investigated over a range of monovalent (NaCl) and divalent ($CaCl_2$) concentrations. The deposition kinetics of fullerene nanoparticles onto bare silica surface are shown to be controlled by electrostatic interactions, consistent with the classical DLVO theory. The presence of dissolved humic acid and alginate in the solution leads to the adsorption of the macromolecules onto fullerene nanoparticles, resulting in significantly slower deposition rate due to steric repulsion. Pre-coating the silica surfaces with humic acid and alginate exerts similar steric influences in the presence of NaCl. At lower NaCl concentration, the increase in surface roughness, specifically for the alginate-coated surface, may result in faster initial deposition kinetics followed by slower deposition in the later stages. In the presence of $CaCl_2$, the deposition kinetics of fullerene nanoparticles onto both humic acid- and alginate-coated surfaces are relatively high due to the macromolecules undergoing complex formation with Ca^{2+} which reduces the surface charge and steric influences of the layers.

141e Environmental Remediation through Double Sequestration of Pollutants by Surfactant Templating Mesoporous Materials

Jia Zhou[1], Jingjing Zhan[1], Grace Tan[1], Vijay T. John[1], Gary L. McPherson[2], JiBao He[3], and Vladimir L. Kolesnichenko[4]. (1) Department of Chemical and Biomolecular Engineering, Tulane University, New Orleans, LA 70118, (2) Department of Chemistry, Tulane University, New Orleans, LA 70118, (3) Coordinated Instruments Facility, Tulane University, New Orleans, LA 70118, (4) Department of Chemistry, Xavier University of Louisiana, 1 Drexel Drive, New Orleans, LA 70125

Many ubiquitous pollutants in aqueous systems (chlorinated hydrocarbons, phenols, etc.) can be removed quite efficiently by entrapment in surfactant micelles through the well-studied membrane enhanced ultrafiltration process (MEUF). It is the common assumption that spherical micelles swell upon uptake of the pollutants. We show that uptake of specific pollutants (particularly phenols) results in a dramatic shape change of micelles from spherical to wormlike and vesicular structures. We have developed a facile follow up method where the entrapped species in surfactant microstructures becomes encapsulated in mesoporous silicas. The pollutant is therefore highly concentrated with a volume reduction of several orders in magnitude, and converted to solid materials for easy disposal. But additionally, the process results in the generation of functional mesoporous materials. The fundamentals of such entrapment, encapsulation are studied through a combination of small angle neutron scattering, cryo-transmission electron microscopy and X-ray diffraction. There are remarkable modulations in micelle characteristics upon entrapment, with transitions from spherical micelles to wormlike micelles and thence to spherical vesicles and enventually to wormlike vesicles. These structures which are then used as templates for ceramic synthesis, create modifications in ceramic mesoporosity which are correlated to the characteristics of self-assembly.

141f Filtration Mechanisms of Single-Walled Carbon Nanotubes In Porous Media

Deb P. Jaisi[1], **Navid Saleh**[2], Ruth E. Blake[1], and Menachem Elimelech[3]. (1) Geology and Geophysics, Yale University, 221 Whitney Avenue, New Haven, CT 06511, (2) Chemical Engineering, Yale University, 9 Hillhouse Ave, Room 300, New Haven, CT 06511, (3) Department of Chemical Engineering, Yale University, Mason Lab, 313A, 9 Hillhouse Avenue, New Haven, CT 06520

Increased production and use of carbon nanotubes (CNTs) may result in the introduction of CNTs to soils and ultimately into groundwater systems. Deposition of CNTs onto surfaces is key to their fate, bioavailability, and reactivity in aquatic systems. This study evaluates the transport and deposition behavior of functionalized single-walled carbon nanotubes (SWNTs) in a well-defined porous media comprising quartz sand. Our results show that increasing solution ionic strength results in increased SWNT deposition (filtration). This observation is consistent with conventional colloidal stability theories, thereby suggesting that physicochemical filtration plays an important role in SWNT transport. However, the relatively insignificant change of SWNT filtration at low ionic strengths (≤3.0 mM KCl) and the incomplete breakthrough of SWNTs in deionized water indicate that physical straining also plays an important role in the capture of SWNTs within the packed sand bed. It is proposed that SWNT shape and structure, particularly the very large aspect ratio and highly bundled (aggregated) state in aqueous solutions, contribute considerably to straining in flow through porous media. We conclude that both physicochemical filtration and straining play a role at low ionic strength (<3.0 mM), while physicochemical filtration is the dominant mechanism of SWNT filtration at higher ionic strengths.

141g Particle and Bacterial Deposition: Role of Gravity

Gexin Chen, Yongsuk Hong, and Sharon Walker. Chemical & Environmental Engineering, University of California, Riverside, Bourns Hall B365, 900 University Avenue, Riverside, CA 92521

The role of gravitational forces on the deposition of 0.5, 1.1, 1.83 μ m carboxylate-modified polystyrene latex microspheres and bacterium *Escherichia coli* D21g has been evaluated using a parallel plate flow chamber system. This experimental system utilized an inverted and an upright optical microscopes attached with image-capturing devices to directly observe and determine the deposition kinetics onto glass surfaces located at the top and bottom of the flow chamber where gravity was observed to minimize and enhance colloid deposition, respectively. Deposition kinetics was quantified at two ionic strengths (0.01 and 0.1M KCl) under electrostatically unfavorable and favorable attachment conditions and at two fluid velocities (0.13 mm/s and 6.6 mm/s), simulating the range of velocities from groundwater flow to rapid granular filtration. Comparing the particle deposition kinetics onto the top and bottom surfaces under identical flowing exposure time, fluid chemistries and hydrodynamic conditions, results showed that significant differences were observed between two surfaces, suggesting that gravity was a significant driving force for the initial stages of colloidal deposition. The only exception to this was 0.5 μ m particles at the high fluid velocity conditions, at which deposition onto top and bottom surfaces were virtually identical. To further understand the influence of gravitational forces on particle deposition in the paral lel plate flow chamber system, the experimental results were compared with both bacterial deposition rates and theoretical calculations of particle deposition on surfaces for diffusion-limited transport or by sedimentation. Finally, numerical simulations were conducted to further interpret and confirm our experimental data. The results of the experimental and theoretical studies will presented and the implications for groundwater quality and industrial applications will be discussed.

141h Surface Properties of Metal and Metal Oxide Nanoparticles In Aqueous Environments and Their Toxicity

John M. Pettibone, Department of Chemical and Biochemical Engineering, University of Iowa, 4122 Seamans Center, Iowa City, IA 52242 and Vicki H. Grassian, Chemistry and Chemical and Biochemical Engineering, University of Iowa, 244 IATL, Iowa City, IA 52242.

Metal and metal oxide nanoparticles make up a large portion of the growing nanotechnology commercial market. Because these materials are being used in commercial products and the use of nanomaterials is expected to grow, it is paramount that the fate, transport, toxicity and surface chemistry of these materials be understood. In this study, commercially manufactured metal and metal oxide particles have been well-characterized and their physicochemical properties investigated using a wide range of techniques and methods. The surface chemistry, toxicity and properties of titanium dioxide nanoparticles of several sizes (5, 18 and 32 nm) were investigated. Titanium dioxide nanoparticle surface reactivity and toxicity was compared as a function of primary particle size, pH and aggregation state. For metal nanoparticles (Fe, Ag and Cu), dissolution studies in aqueous environments and simulated biological environments were also examined. These laboratory experiments can be used to better understand the environmental fate and the toxicity of these materials.

148a A Review of Computational Methods for Predicting Environmental Transport and Fate

David A. Gallagher, CAChe Research, 13690 SW Otter Lane, Beaverton, OR 97008

The ability to predict the environmental transport and fate of candidate compounds and products, particularly before they are synthesized, can save significant laboratory time and expense, simply by eliminating those with potentially undesirable properties early in the development cycle. Although, R&D cost savings are important, the associated benefits of shortening time-to-market and minimizing environmental risk, offer the greatest potential for long-term profitability.

Transport and fate in the environment correlate closely with a compound's physical and chemical properties, and most properties can now be predicted from just molecular structure information alone using a variety of computational methods. The latest generation of computational tools makes it relatively simple to pre-screen large libraries of hypothetical candidate compounds for environmental transport and fate, even on a regular Windows-based computer.

The author will review various methods for the prediction of environmental transport and fate with examples covering aquatic fate, atmospheric fate, persistence, atmospheric lifetimes, and soil remediation rates. The approaches will be illustrated with both commercial and free tools, based on quantum chemistry, statistical thermodynamics, quantitative structure-property relationships (QSAR), and related methods.

148b Estimating the Transient Source Strength of Soil Fumigants for Use In Air Dispersion Calculations

Steven A. Cryer, R&D Operational Excellence, Dow AgroSciences, 9330 Zionsville Rd, Indianapolis, IN 46268 and Ian J. Van Wesenbeeck, R&D Reg Labs, Dow AgroSciences, 9330 Zionsville Rd, Indianapolis, IN 46268.

Soil fumigants are compounds with high vapor pressures and are often applied to agricultural land to control nematode populations and crop diseases. Lowering fumigant losses to the atmosphere improves air quality and is now mandated by the California Department of Pesticide Regulations (CDPR) for the state of California. Field studies help quantify the transient flux profile observed under typical agronomic conditions. Unfortunately, field trials only represent a small proportion of the near semi-infinite parameter combinations of environmental, agronomic, and meteorological conditions. Therefore, soil physics modeling is used to expand field knowledge and provide estimates of transient source strength for use in air dispersion modeling. The USDA model CHAIN_2D simulates chemical transport in variable saturated/unsaturated porous media. This model has largely been underutilized by the agricultural community due to lack of pre and post processing tools, slow CPU execution times, and lack of simulation of physical constraints often associated with agriculture. However, CHAIN_2D has been modified to make amenable to soil fumigation practices, with automated input/output and finite grid generation capabilities. Significant modifications include additions of a tarp boundary condition proposed by Wang et al (1998, 1999), along with the ability to simulate tarp tearing and tarp removal at user specified time intervals post application, and coupling to air dispersion modeling. A new air/soil boundary condition can be specified that ties meteorological conditions to the resistance to mass transfer at the air/soil interface, or the default and original stagnant (constant) boundary layer thickness of CHAIN_2D can be assumed. This tool, developed using Microsoft Excel with custom VBA programming, allows easy exploration of different best management practices (BMP's). A sensitivity analysis for CHAIN_2D, comparison to field observations for the fumigant 1-3-dichloropropene, and several BMPs examples illustrating the impact of source strength and resulting near-field air concentrations are provided.

148c Self-Organizing Classification of Ecotoxicological and Molecular Information as a Proxy for Identifying Persistent Bioaccumulative and Toxic Substances

Gabriela Espinosa, Departament d'Enginyeria Química, Universitat Rovira i Virgili, Av. dels Països Catalans 26, Tarragona, 43007, Spain, Rallo Robert, Departmament d'Enginyeria Informàtica i Matemàtiques, Universitat Rovira i Virgili, Av. dels Països Catalans 26, Tarragona, 43007, Spain, and Giralt Francesc, Grup de Fenomens de Transport Departament d'Enginyeria Quimica, Universitat Rovira i Virgili, Av. dels Països Catalans 26, Tarragona, 43007, Spain.

The exploration and characterization of the chemical space of REACH compounds is the first step towards the development of new computerized tools which integrate chemical domain knowledge with physicochemical, biological and regulatory information. Environmental risk assessment takes advantage of the so-called proxy indicators which provide reliable information for specific end-points by using simpler and less costly measurements and calculations. In the context of REACH, the development of these proxies requires the establishment of relationships between PBT profiling data and the corresponding molecular information. The identification of the most representative subsets of chemicals within each PBT and vPvB profile contributes to the design and implementation of optimal screening procedures which facilitate the risk assessment process.

In the current work the screening of structural and property profiles for substances of concern such as PBT and vPvB has been performed. The modes of toxic action (MOA) were used as categorical indicators of effects of chemicals in the environment and similar effect patterns were identified by cluster analysis using self organizing maps. Molecular information was added to facilitate the above clustering process on the basis that similar compounds share a common mode of action (chemical reactivity).

Web Page: www.etseq.urv.es/fet/1.html

167a Carbon-Metal Redox Reaction: Its Implications In Particle Toxicity and Air Pollution Control

Bing Guo, Department of Mechanical Engineering, Texas A&M University, College Station, TX 77843

Elemental carbon can reduce transition metal ions in aqueous solutions. This redox reaction may have significant effect on particle toxicity due to the valence state dependence of the metal's ability to induce oxidative stress. Results from some studies have suggested a synergistic effect between carbon and transition metals on the toxicity of particulate matter. The observed synergistic effect may be due to the carbon-metal redox reaction. Because of the ubiquitous abundance of both elemental carbon and transition metals in ambient particulate matter, understanding the implication of carbon-metal redox reaction in particle toxicity is of great importance to developing air pollution control strategies.

In this study we investigated the redox reaction between various elemental carbon nanoparticles and ferric ions in an acidic aqueous solution at 40 and 80 ° C. The objective was to determine the amount of Fe^{3+} reduced to Fe^{2+} (reduction yield) in batch mode reactions, as a function of reaction time, temperature, and the type of elemental carbon particles. We found that the reduction yield reached a steady-state value after about 1 hour of reaction, regardless of particle type and temperature. The carbon mass based reduction yield was dependent on the type of elemental carbon, as well as on the temperature. The carbon particle surface area based reduction yield showed weak dependence on the particle type at 40 ° C, but at 80 ° C it was significantly particle type dependent. X-ray photoelectron analysis of the carbon particles before and after the reaction suggested that the surface functional groups of the carbon particles were involved in the redox reaction.

In conclusion, the preliminary data have shown that elemental carbon of various origins can reduce ferric ions in aqueous solutions. This reaction appears to be dependent on the surface functional groups of the elemental carbon particles. Further investigation is needed to confirm the hypothesized biological effect of the redox reaction between elemental carbon and transition metals.

167b Ultra-Fine PM Derived from Fullerene-Like Carbon In Electrostatic Precipitator Fly Ash

Uschi M. Graham[1], Alan Dozier[2], Rajesh A. Khatri[1], Michael Tseng[3], James C. Hower[1], and Burtron H. Davis[1]. (1) Center for Applied Energy Research, University of Kentucky, 2540 Research Park Dr., Lexington, KY 40511, (2) Electron Microscopy Center, University of Kentucky, ASTeCC Building, Lexington, KY 40506, (3) Department of Medicine, University of Louisville, Lousville, 40292

Recently, Particulate Matter (PM) pollution received major attention based on scientific discoveries, health concerns, as well as climate change debates. Especially, the area of organic PM (primary and secondary) has many uncertainties that need further exploration. To help trace and determine the different source materials and points of origin of ultra small carbon particulate matter in the atmosphere, this study investigated fly ash-based ultra-fine PM carbon material (nanocarbons) derived from coal-fired power plants. The investigation only addressed the nature of the nanocarbons in fly ash from a high volatile bituminous coal although others are planned for future study. The nanocarbons observed are associated with fly ash captured and collected within coal combustion pollution-control systems' electrostatic precipitators (ESP units). The study addresses (a) the spatial relationship of ultra-fine (< 10 nm) carbons with respect to the glassy ash spheres, (b) describes variations in the structural characteristics of the nanocarbons, and (c) analyzes the composition of nanosized metal inclusions in the ultra-small carbon source. The most significant observation was the presence of carbon-based nanocoatings on the surface of individual fly ash particles. The carbon nanocoatings are extremely porous and fragile. This delicate appearance of the nanocarbon deposits on the Al-Si glassy particles suggests that fractions of the nanocarbon coatings may become liberated during ash management (either in or before ESPs or after collection) which could affect the size of the nanocarbon fraction, cause separation of the nanocarbons from the bulk fly ash, and contribute nanocarbons in form of aerosol to atmosphere. Given the seemingly delicate association of the nanocarbons with the glassy fly ash, it is suggested that this fraction could be a contributor to ultra-fine PM soot in the atmosphere and the associated trace elements released by power plants. A secondary organic aerosol formation via detachment of nanocarbons from coal ash material may be an important contributor to the overall ambient organic aerosol mass.

The ultra fine PM carbon in fly ash was distinguished as either nanoscale sooty or graphitic fullerene-like carbons. They are well below the optical limit of resolution and, furthermore, their graphitic nature has not been described in earlier studies. This ultra-fine PM carbon represents the carbon source in fly ash previously unaccounted for in optical microscopic investigations. The nanocarbons (10's nm) are delicately deposited on and interlaced between larger aluminosilicate (Al-Si) glassy fly ash spheres. They form an ultra-thin layer on the coarser inorganic fly ash particles and emerge as nanocoatings covering the glassy spheres in a halo or shell-like fashion. Nanocarbons can be traced as an ultra-thin halo around the majority of Al-Si glass fly ash spheres (carbon nanocoating). The carbon shells/ nanocoatings are vastly porous and consist of agglomerated soot particles, each exhibiting concentric ring structures. High-resolution transmission electron microscopy (HRTEM), scanning transmission electron microscopy (STEM), and electron energy-loss spectroscopy (EELS) investigations of a high-carbon fly ash with relatively high concentrations of As (10^2-10^3 μ g/g), Se (10^2 μ g/g), and Hg (1-3 μ g/g) demonstrated that nanocarbons with both soot-like structures and fullerene-like carbons contain finely dispersed nm-sized metal inclusions. The elemental associations of the metal inclusions follows the pattern Fe >>Ca>Ti>Pb>As>Ni>Co>Hg, with varying amounts of Se also present. The study showed a majority of Al-Si glassy spheres to have a carbon-based nanocoating or at least some fraction of the surface coated. The presence of the carbon nanocoatings has not been taken into account during past routine petrographic studies of fly ash materials. The delicate nature of the carbon deposited on the Al-Si glass was confirmed by the observed damage to the structure by the electron beam. This suggests that processes, such as the transport of the fly ash in the turbulent flue gas, could potentially separate a fraction of the fullerene-like carbon, with the included trace elements, from the Al-Si glass. If this happens in or before the electrostatic precipitators (ESP), the relatively coarser glass will be captured by the ESP (and ESP's routinely capture over 99% of the fly ash) and the finer carbons will escape into the atmosphere. Since the thickness of the carbon nanocoating varies among different Al-Si spheres in fly ashes captured in ESPs and can be irregular on the surface of individual spheres, such nanocarbon deposits appear to obstruct the smoothness and, hence, fluidity of the ash particles. HRTEM provides a superior resolution of the carbon nanostructure while STEM imaging emphasizes the difference between the low-atomic-number carbon and the inorganic entities captured or engulfed by the nanocarbons. The EELS spectra of selected metal inclusions in the nanocarbon agglomerates illustrate that a majority of the nanosized metals are iron-rich. The results are based on the presence of the FeL3 (715 eV) and FeL2 (720 eV) peaks in the EELS spectra. Elemental analysis indicated an association of Hg with the nanocarbon. Arsenic, Se, Pb, Co, and traces of Ti and Ba are often associated with Fe-rich particles within the nanocarbon deposits. A detailed high resolution STEM- X-ray map, with the corresponding STEM image, of a carbon-rich nanocluster indicates that As follows Fe and S signatures.

Temperatures within the ESP array range from about 200 °C at the entry to the first row to <150 °C at the exit of the third row. For both electrostatic precipitators (ESP) and fabric filters (FF), it is often observed that there is a relation between the collection point of the fly ash (ESP row) and the concentration of trace elements with the concentration of volatile elements, such as Pb, As, and Se, generally increasing with a decrease in temperature at the collection point and a decrease in ash particle size. Coarser particles are preferentially collected in the first row of an ESP while finer, higher surface-area particles are trapped at the third row. The combination of decreasing temperature and increasing surface area of fly ash particles leads to enhanced concentrations of volatile trace elements, but the concentration of trace elements in the fly ash is also strongly related to the concentration of elements in the feed coal.

The HRTEM investigation revealed abundant fullerene concentration over the nanocarbon agglomerate surfaces and the result is not surprising given the fact that the nanocarbons were formed in a combustion environment. Individual fullerene molecules range in size which is typically observed in samples produced from a fullerene-forming flame. In addition, some nanocarbons in the current study exhibit a fullerene-like structure over larger areas, suggesting that the growth mechanisms of the nanocarbons involved fullerene formation. The typical morphology of fullerene-like soot in the nanocarbons on Al-Si spheres shows chain-like assembly of primary fullerene particles, often exhibiting multiple shell structures that are interlinked and result in the appearance of strongly bent graphene sheets or ribbons and double shells. The key parameters required for fullerene synthesis on the nano fly ash carbons are suggested to depend in part on temperature and concentrations of H_2O vapor and CO_2 since those noticeably affect the amount of fullerenes, as well as their size formed under experimental conditions on carbon supports. Future research should address the nature of the nanocarbons in fly ashes derived from these ranks and, very importantly, also the reactivity of such nanocarbons, as this may play important roles in pulmonary and other health effects.

Hydrocarbons Adsorbed to Combustion-Derived PM Localize to Lipid Droplets and Activate Oxidative Stress Response Genes In Respiratory Cells (128910)

Arthur L. Penn, Gleeson Murphy*[1], Rodney L. Rouse*[2] and William W. Polk**

**Comparative Biomedical Sciences, School of Veterinary Medicine, Louisiana State University, Skip Bertman Drive, Baton Rouge, LA 70803. [1] Present address: U.S. Army Veterinary Medical Corps, Analytical Toxicology Division, USAMRICD, Aberdeen, MD, 21010-5400. [2] Present address: U.S. Food and Drug Administration, CDER, OPS, OTR, Department of Applied Pharmacology Research, White Oak Campus, Silver Spring, MD 20993.*

Airborne particles, especially those arising from combustion and industrial processes, are of increasing environmental and public health concern due to a) their contribution to ambient pollution and b) the toxic responses they elicit, especially in people already at risk for cardiovascular and respiratory diseases. The varied toxic responses have been attributed both to physical and chemical characteristics of the particles (**1,2**).

Among physical characteristics, particle size is inversely associated with particle toxicity, at least in part, because the smaller the particle, the more likely it is to penetrate deeply into the lungs. In addition, smaller particles provide greater available surface area for adsorption of toxic agents than do larger particles of equivalent mass.

There is growing evidence of deleterious health effects associated with inhalation of particulate matter (PM) in the fine (<2.5 μ) and ultrafine (<0.1 μ) size ranges (**3**). Cigarette smoke and diesel exhaust (DEP) are among the most common, and intensively studied, sources of complex particulates. Organic chemicals adsorbed to DEP elicit pro-inflammatory and oxidative stress responses in respiratory system cells following DEP exposure both *in vivo* and *in vitro* (**4-9**). Cigarette smoke, whose contributions to major diseases, including lung cancer and cardiovascular disease, have been documented exhaustively for over a half century, contains > 4,500 chemical components, of which at least 100, including benzo(a)pyrene [B(a)P] and other carcinogens, are polynuclear aromatic hydrocarbons (PAHs; **10**). Organic chemical-rich products of incomplete combustion (soots) also are released to the atmosphere in large quantities in petrochemical-related processes, e.g., during routine flaring of volatiles. In addition, sabotage and accidents involving pipelines and refineries, pose major risks both in the U.S. and abroad.

We have been investigating the *in vivo* and *in vitro* responses to exposure to butadiene soot (BDS), which serves as both a model mixture and real-life example of PAH-rich, combustion-derived nanoparticles (**11-14**). 1,3-butadiene, an aliphatic hydrocarbon by-product of petroleum refining, is a "top 40" U. S. chemical and is the starting material for virtually all commercially produced rubber. Early, during combustion of petrochemical volatiles, hydrocarbon free radicals are formed. These can fuse, ultimately giving rise to PAHs, which can aggregate into nanoparticles and then into extended, branched-chain soots (**15**). Generation as well as physical and chemical characterization of the branched-chain BDS, composed of 20-50 nm, PAH-rich, carbonaceous particles, has been described (**11**). The characteristic blue fluorescence of PAHs allows the time-and dose-dependent uptake of BDS by cells to be followed and recorded (**11-13**).

Human broncho-epithelial cells exposed to BDS develop blue fluorescence, which over time becomes localized in discrete cytoplasmic vesicles (**11**). Following BDS exposure, these cells display the same profile of extractable PAHs as the parent BDS. The fluorescence does not develop if the cells are exposed to carbon black instead of BDS, or if the BDS is extracted with organic solvents before the soot particles are presented to the cells.

The experiments described here were designed to identify the cytoplasmic compartment(s) within which BDS-associated fluorescence is localized and to determine whether there is concomitant activation of xenobiotic metabolism pathways known to potentiate PAH toxicity. Complete details of these studies have recently been published (**13**).

Methods

Cell Culture. Cells (1.5×10^6) of the BEAS-2B human broncho-epithelial cell line (**16**), were grown in bronchial-epithelial growth medium (BEGM; Cambrex, Walkersville, MD), before expansion in T-150 flasks.

Cells (1×10^6) of the MH-S mouse alveolar macrophage cell line (**17**), were grown in RPMI 1640 medium with 10% FBS to 80–90% confluence.

Mouse 3T3-L1 preadipocytes (**18**) were grown to 2 days post-confluence in DMEM with 10% CS and induced to differentiate in DMEM supplemented with 4.5 g/L glucose, 10% FBS, 100 U/mL penicillin, 100 µg/mL streptomycin, 0.5 mM 3-isobutyl-methylxanthine, 1 µM dexamethasone, and 1.7 µM insulin. After 48 hrs, this medium was replaced with DMEM containing 4.5 g/L glucose supplemented with 10% FBS (**19**).

BDS Generation and Collection. The process of BDS generation and collection has been described in detail (**11**).

BDS Exposures. In all cases, the medium was changed immediately before addition of BDS or staining solutions. BDS stock solution was prepared by sonicating 10 mg BDS in 50 mL of medium (3 15-second pulses of a Model 450 Branson Sonifier, Danbury, CT). BEAS-2B cells, MH-S cells, and adipocytes were exposed to 20 µg/mL sonicated BDS for 24 hours prior to fluorescence imaging. For microarray and/or quantitative real time (qRT)-PCR experiments, BEAS-2B and MH-S cells were exposed to BDS for 0, 1, 3, or 24 hours prior to RNA extraction.

Fluorescent Dye Co-Localization. Organelle-specific fluorescent probes (Molecular Probes, Invitrogen; Carlsbad, CA) for lysosomes, autophagosomes, endosomes, mitochondria, endoplasmic reticulum and lipid droplets were used to investigate candidate organelles for the localization of BDS-associated fluorescence. To investigate peroxisomes as candidate organelles we transfected BEAS-2B cells with a plasmid whose protein product is a fusion of the red fluorescent protein DsRed2 with the peroxisomal targeting sequence 1 [PTS1] (**20**).

Fluorescence Microscopy. We used a Zeiss Axiovert 405M inverted fluorescence microscope (40X) with a Microfire Megapixel Digital CCD camera, operated by PictureFrame software (Optronics; Goleta, CA) and a Leica DM RXA2 upright microscope, equipped with differential interference contrast (DIC) optics (40X, 63X, 100X)–with a SensiCam QE 12-bit, cooled CCD camera (Cooke; Romulus, MI), operated by SlideBook software (Intelligent Imaging Innovations; Denver, CO). We used SlideBook to measure lipid droplet dimensions and Adobe Photoshop CS to process all images.

Cytotoxicity Assay. We assessed cell viability with the Cell Counting Kit-8 (Dojindo Molecular Technologies (Gaithersburg, MD). Cells were exposed to sonicated BDS for 1, 4, 8, or 24 hours. Unexposed sets of control cells were used for comparison at each time point.

RNA Isolation. BEAS-2B cells and MH-S cells at or near confluence were scraped into 1 mL of Invitrogen TRIzol Reagent and passed 3X through a 23G needle. RNA was isolated by TRIzol/chloroform extraction, followed by column purification (Qiagen RNeasy Mini Kit). RNA quality and integrity was assessed using the Agilent RNA 6000 Nano Assay Kit and the Agilent 2100 BioAnalyzer (Santa Clara, CA). Total RNA was converted to cDNA using TaqMan Reverse Transcriptase Reagents (Applied Biosystems; Foster City, CA).

pDsRed-Monomer-C-ADFP Plasmid Preparation and Transformation. Human ADFP PCR primers were as designed (**21**) from the ADFP mRNA sequence in GenBank: a) 5'-GGGGCAGGTTTAATGAGTTTTATG-3'; and b) 5'-CCAGGAAGAAAAT GGCATCCGTT-3' (Integrated DNA Technologies; Coralville, IA). PCR (30 cycles) was performed on 50 ng of BEAS-2B cDNA in a PTC-100 thermal cycler (MJ Research, Waltham, MA). To impart red fluorescence to ADFP expressed within cells, the ADFP PCR product was ligated into the pDsRed-Monomer-C In-Fusion Ready Vector (Clontech; Mountain View, CA) as described (**13**). BEAS-2B cells grown on glass coverslips in 60 mm dishes were transformed with 500 ng of the plasmid, in the presence of Lipofectamine LTX and PLUS reagents (Invitrogen). After 48 hours, cells were observed for lipid droplet fluorescence and then exposed to BDS (20 µg/mL) for 24 hours prior to imaging for co-localization.

Gene Microarray Assay. Global gene expression in BEAS-2B cells was assessed from total RNA on Affymetrix Human Genome U133 Plus 2.0 Arrays that were processed at the Research Core Facility of the LSU Health Science Center-Shreveport. All gene chips and instrumentation were from Affymetrix (Santa Clara, CA). All data collected and analyzed here adhere to the guidelines for Minimal Information About a Microarray Experiment (MIAME).

Quantitative Real Time (qRT)-PCR. qRT-PCR was performed on cDNA samples from BEAS-2B and MH-S cells with inventoried TaqMan Gene Expression Assays primer-probe sets (Applied Biosystems) for selected genes. Relative gene expression was determined by the comparative cycle threshold ($\Delta\Delta C_T$) method, with each gene normalized to β-actin (ACTB) for human cells (**22**) or hypoxanthine guanine phosphoribosyl transferase (Hprt1) for mouse cells (**23**), and then compared to the 0 hour control. Results are reported as fold change over control. Standard error of the mean [$(2^{-\Delta\Delta CT}) \pm$ SEM] was used to determine significant at $\alpha = 0.05$.

Statistical Analysis. We used the GLM procedure of the SAS statistical package (version 9.1.3; SAS Institute, Inc., Cary, NC) to compare RT-PCR data, and Dunnett's *t* test to determine statistical differences. Gene chip data, filtered to remove data from non-functional probe sets, were analyzed at each time point by pair-wise comparisons of gene expression in BDS-treated vs. control cells (Expression Analysis, Durham, NC). Gene expression was expressed as log ratio for these comparisons. All transcripts reported here had a fold change (up or down) of at least 1.5X and both an individual p-value and a false discovery rate < 0.05. Gene expression data were analyzed with the network- and pathway-building software, Ingenuity Pathways Analysis 4.0; gene networks and canonical pathways were examined using the Ingenuity Analysis Knowledge Database (Ingenuity Systems, Redwood City, CA).

Results

BDS-Associated Fluorescence in Human Broncho-Epithelial Cells is Time-Dependent

Within 30 minutes after BEAS-2B cells are exposed to BDS, a diffuse blue fluorescence appears in each of the cells (Figure 1a). By 2-4 hours, the diffuse fluorescence is replaced by bright, punctate fluorescence that is visible clearly within cytoplasmic vesicles (Figure 2b). The fluorescence intensifies over the next 20 hours and then plateaus for an additional 48 hours, in the absence further addition of BDS (figure 1c). No nuclear fluorescence is evident.

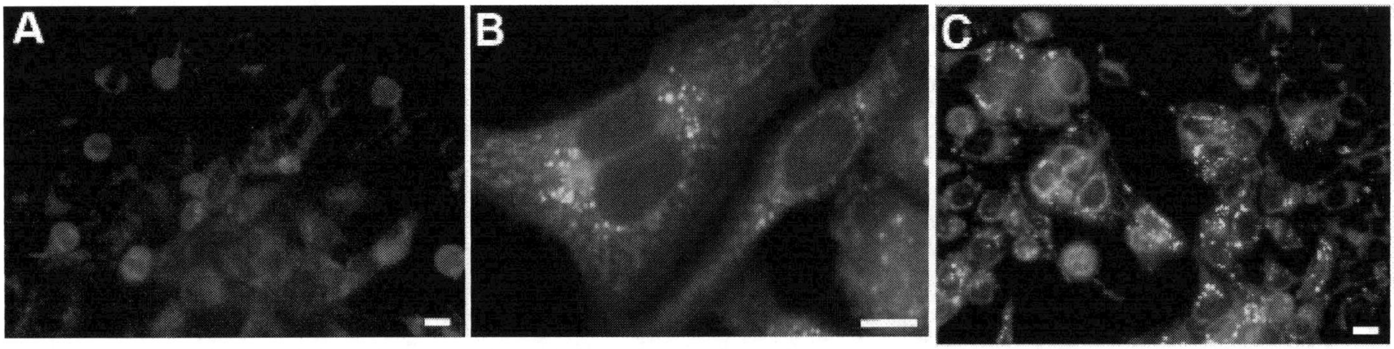

Figure 1: Time-dependent fluorescent responses of BEAS-2B cells exposed to petrochemical combustion-derived ultrafine particles (BDS) sprinkled on the surface of the culture medium. Diffuse fluorescence is visible within 10-30 min (A). There is a transition from diffuse to punctate fluorescence by 2 hours post-exposure (B). Punctate fluorescence increases in intensity through the first 4 hours of exposure to BDS (C). Ultraviolet excitation wavelengths (~360 nm) typically provided the brightest emission spectra for visualization of the BDS-associated fluorescence as shown in A and C; however, the extensive punctate fluorescence shown in B was best demonstrated with a longer excitation wavelength (~480 nm). Bars = 10 μm.

Alveolar Macrophages and Adipocytes also Display BDS-Associated Fluorescence

Mouse alveolar macrophages (Figure 2b) exhibit fluorescence in larger peri-nuclear vacuoles ($1.8 \pm 0.2\mu$) than are present in human bronco-epithelial cells (figure 2a). Fluorescent vacuoles in cultured mouse adipocytes are even larger (2.6±0.3μ) than those appearing in the macrophages (Figure 2c). Development of punctate BDS-associated fluorescence in all three cell types is time-dependent (data not presented).

BDS Exposure Does Not Alter Cell Viability Through 24 Hours of Exposure

Peripheral membrane blebs, often indicative of cell damage, appeared in some BEAS-2B cells after 24 hours of exposure. Therefore, we assessed cell viability of both BEAS-2B and MH-S cells. A formazan-based dye assay indicated that viability was unaffected in each of these cell lines for up to 24 hours after exposure to BDS (data not presented).

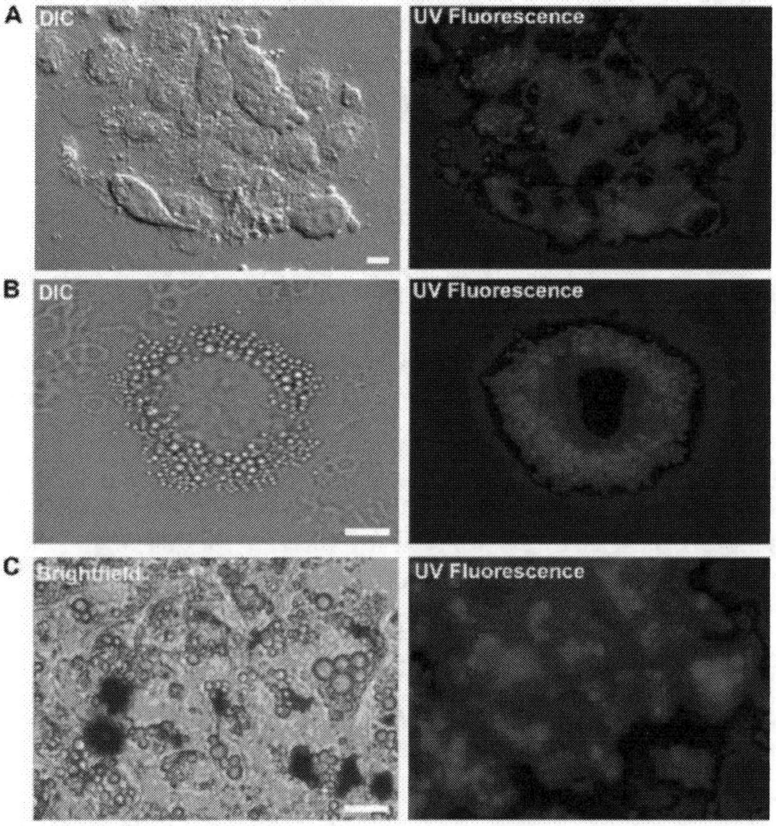

Figure 2. Broncho-epithelial cells, alveolar macrophages and adipocytes display characteristic BDS-associated fluorescence. Differential interference contrast (DIC) (A, B) and bright-field photomicrographs (C) paired with ultraviolet-fluorescence images of BEAS-2B human broncho-epithelial cells (A), MH-S murine alveolar macrophages (B), and 3T3-L1 murine adipocytes (C) after a 24-hour exposure to sonicated BDS. The fluorescent bodies appear as variably sized refractile bodies; peripheral membrane blebs are visible on some cells (A). In contrast to BEAS-2B cells, MH-S cells displayed larger droplets, all of which were fluorescent (B). Large droplets in mouse adipocytes displayed BDS-associated fluorescence (C). Black BDS particles are visible in the bright-field panel of C, and the fluorescent panel demonstrates the lack of fluorescence associated with the particles. Bars 5 10 mm (A), 7.5 mm (B), and 10 mm (C).

BDS-Associated Fluorescence is Localized Within Cytoplasmic Lipid Droplets in Target Cells

Based on light microscope observations of size and cytoplasmic localization, candidate organelles were selected and tested by fluorescence co-localization, to identify those in which the punctate fluorescence was concentrated. The candidate organelles included lysosomes, autophagosomes, endosomes, peroxisomes, mitochondria, endoplasmic reticulum and lipid droplets. Fluorescent markers specific for each organelle were tested for co-localizatjon with the blue BDS-associated fluorescence. The emission maxima of all the markers were in the red region, insuring that the red and blue fluorescence responses would be readily distinguishable from each other, unless they overlapped, in which case the resulting color would be purple.

The BDS-associated blue fluorescence co-localized only with the red cholesterol-BODIPY C_{11} marker, yielding a purple fluorescence (data not presented). Cholesterol-BODIPY C_{11} concentrates within hydrophobic compartments of cells, via the "selective" transport pathway. This pathway directs lipoprotein-derived cholesteryl esters to lipid droplets without their first having to pass through either the Golgi apparatus or lysosomes. Lipid droplet localization of the BDS-associated fluorescence was confirmed by transfecting BEAS-2B cells with a plasmid whose protein product is a fusion of a red dye monomer with a lipid droplet membrane-specific protein, ADRP. The plasmid-transfected cells displayed a large number of fluorescent red cytoplasmic inclusions. One such cell at 100x is shown in Figure 3a. The BDS-associated blue cytoplasmic inclusions in the same cell are shown in Figure 3b. When the two images are merged (Figure 3c) there is complete co-localization of the red and blue fluorescent images, resulting in the purple inclusions (i.e., lipid droplets) seen in Figure 3c.

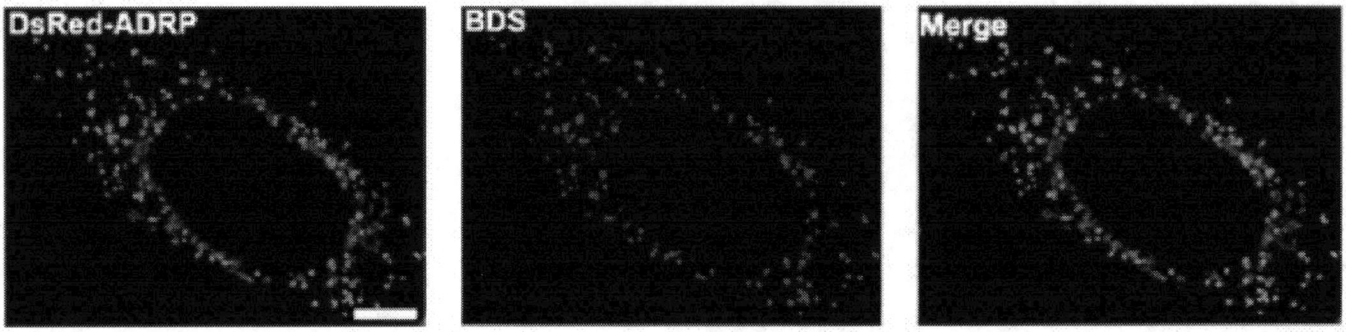

Figure 3: Co-localization of a lipid droplet-specific probe with BDS fluorescence in BEAS-2B cells. Left panel; red fluorescence resulting from lipid droplet localization of the DsRed-ADRP tag; center panel: subcellular localization of the BDS-associated blue fluorescence; right panel: complete co-localization of the red and blue fluorescent images, yielding the purple image shown.

BDS Stimulates Time-Dependent Expression of Aryl Hydrocarbon Receptor (AHR)-Induced and Oxidative Stress Response Genes in BEAS-2B Cells

There were time-dependent changes in gene expression, assessed by analysis of microarray data, in BEAS-2B cells exposed to BDS for 1-24 hr. Of the 448 differentially-regulated genes identified at 24 hours, 70 were down-regulated and 378 were up-regulated. Up-regulated genes (Table 1) included AHR-responsive genes (e.g., AHR repressor, cytochrome P450 1A1 and 1B1) and nuclear erythroid-2 factor 2(NRF2)-mediated oxidative stress response genes (e.g., FOS, NQO1). These results were confirmed in BEAS-2B cells and demonstrated in MH-S cells by qRT-PCR.

TABLE 1. TIME-DEPENDENT MICROARRAY CHANGES IN ARYL HYDROCARBON RECEPTOR– AND NRF2-RELATED GENE EXPRESSION IN BUTADIENE SOOT–EXPOSED BEAS-2B CELLS*

	BEAS-2B Microarray		
	1 h	3 h	24 h
AhR-induced response			
AHRR	1.1	1.9	1.6
ALDH1A3	1.1	2.4	3.2
CYP1A1	2.3	14.7	29.7
CYP1B1	3.3	5.4	4.1
IL1A	3.4	2.5	5.1
IL1B	2.6	2.9	5.3
TIPARP	6.3	5.0	5.7
NRF2-mediated oxidative stress			
FOS	2.1	1.2	2.0
NQO1	1.1	1.2	2.0
TXN	−1.2	1.1	2.1
MAFF	3.6	4.2	5.1
JUN	1.4	1.4	2.3
DNAJB6	−1.1	1.3	2.1

* Results for each gene at each time point are reported as fold change in a single pairwise comparison relative to control sample.

TABLE 2. TIME-DEPENDENT QUANTITATIVE REAL-TIME–PCR EXPRESSION CHANGES IN ARYL HYDROCARBON RECEPTOR–RELATED GENES OF BEAS-2B AND MH-S CELLS EXPOSED TO BUTADIENE SOOT FOR 24 h*

	BEAS-2B		MH-S	
	1 h	24 h	1 h	24 h
AhR-induced response				
AHRR†	n/a	n/a	3.6‡	3.3‡
ALDH1A3	1.5	4.4‡	1.4	1.4
CYP1A1	21.5	108.4‡	10.5‡	6.4‡
CYP1B1	6.1‡	3.5	69.2‡	17.6‡
IL1A	6.8‡	4.9‡	n/a	n/a
IL1B	6.1‡	5.1‡	1.0	1.0
TIPARP	13.1‡	3.5	7.8‡	2.0‡

Definition of abbreviations: n/a = not available.

* Results are reported as fold change relative to control; $n = 4$ for each gene analyzed.

† All gene symbols are expressed as human homologs.

‡ Significantly different from control sample ($\alpha = 0.05$)

Discussion

Previously, we reported physical and chemical characterization of the BDS nanoparticles produced during incomplete combustion of the high volume petrochemical, 1,3-butadiene (**24,11**) and in addition, that human respiratory epithelial cells exhibited punctate cytoplasmic fluorescence following exposure to BDS (**11**). Here, we used fluorescence imaging techniques with organelle-specific probes, combined with analysis of gene expression to demonstrate that a) BDS-associated fluorescence localizes within lipid droplets and b) appearance of this fluorescence is correlated with enhanced expression of Phase I biotransformation and oxidative stress genes. The responses to BDS exposure are not limited to human respiratory epithelial cells, as mouse alveolar macrophages and adipocytes respond in a similar fashion.

The fluorescent responses begin with a diffuse cytoplasmic fluorescence that appears within the first 30 minutes after BDS exposure begins, progresses to fluorescence that localizes within lipid droplets over the next 2-4 hours, and is maintained over the next 72 hours. Lipid droplets were identified by co-localization of BDS fluorescence first with fluorescence of the cholesteryl-BODIPY probe and then with the fluorescent, lipid droplet-specific ADRP probe. The fluorescent responses are due to the PAHs, that are adsorbed to the surface of the BDS nanoparticles and which represent ~16% of the total weight of BDS (**11**). A concomitant time-dependent up-regulation of Ah receptor-associated Phase I biotransformation enzymes, including cytochrome P450s 1A1 and 1B1, revealed both by microarray and qRT-PCR analyses, is entirely consistent with the fluorescent responses. In addition, microarray data from BDS-exposed BEAS-2B cells reveal a time-dependent up-regulation of NRF2-mediated oxidative stress genes. We have reported similar patterns of gene up-regulation in lungs of mice that inhaled BDS daily for 4 days (**14**). The combined results indicate that exposure to BDS, and possibly to combustion products of other petrochemical volatiles, will result in increased cellular oxidative stress.

The localization of environmentally-derived PAHs within lipid droplets is noteworthy for at least two reasons. First, cigarette smoke and diesel exhaust, two ubiquitous sources of readily respirable, PAH-rich particles, would be expected to provide PAHs that could localize within lipid droplets of target cells. A number of the PAHs that are prominent in BDS also are prominent in cigarette smoke (**25**). Second, although it is reasonable to expect that inhaled hydrophobic compounds can be concentrated within lipid droplets of respiratory system cells, our results are the first of which we are aware, in which compounds other than those derived from lipid metabolism are localized within lipid droplets. The long-term availability of PAHs localized within lipid droplets in the lung or other tissues to PAH-metabolizing enzymes has not been investigated. There is a report that diesel exhaust particles on which B(a)P has been adsorbed can be recovered from the lungs of dogs 5 months after exposure (**26**). Similarly, we have found BDS particles in the lungs of mice twenty-eight days after a 2-day BDS exposure ended (unpublished observations). Thus, the combination of persistent lung irritation and the resulting inflammation from inhaled particles + delayed activation of PAH-metabolizing enzymes may promote a precancerous environment in the lung.

Acknowledgement: Figures 2A-C & 3A-C + Tables 2 & 3 originally appeared in:

Ref. **13**. Murphy G, Rouse R, Polk W, Henk W, Barker S, Boudreaux M, Floyd Z, Penn A. Combustion-derived hydrocarbons localize to respiratory cell lipid droplets and up-regulate biotransformation enzymes. *Am J Respir Cell Mol Biol* 2008;38 (5):532-540. The *Am J Respir Cell Mol Biol* is an official journal of the American Thoracic Society.

References

1. Bonvallot V, Baeza-Squiban A, Baulig A, Brulant S, Boland S, Muzeau F, Barouki R, Marano F. Organic compounds from diesel exhaust particles elicit a proinflammatory response in human airway epithelial cells and induce cytochrome p450 1A1 expression. *Am J Respir Cell Mol Biol* 2001;25:515-521.

2. Sayes CM, Reed KL, Warheit DB. Assessing toxicity of fine and nanoparticles: comparing in vitro measurements to in vivo pulmonary toxicity profiles. *Toxicol Sci* 2007;97:163-180.

3. Li N, Sioutas C, Cho A, Schmitz D, Misra C, Sempf J, Wang M, Oberley T, Froines J, Nel A. Ultrafine particulate pollutants induce oxidative stress and mitochondrial damage. *Environ Health Perspect* 2003;111:455-460.

4. Li N, Venkatesan MI, Miguel A, Kaplan R, Gujuluva C, Alam J, Nel A. Induction of heme oxygenase-1 expression in macrophages by diesel exhaust particle chemicals and quinones via the anti-oxidant responsive element. *J Immunol* 2000;165(6):3393-3401.

5. Murphy SA, BeruBe KA, Richards RJ. Bioreactivity of carbon black and diesel exhaust particles to primary clara and type ii epithelial cell cultures. *Occup Environ Med* 1999;56(12):813-819.

6. Ma JY, Ma JK. The dual effect of the particulate and organic components of diesel exhaust particles on the alteration of pulmonary immune/inflammatory responses and metabolic enzymes. *J Environ Sci Health C Environ Carcinog Ecotoxicol Rev* 2002;20(2):117-147.

7. Rao KM, Ma JY, Meighan T, Barger MW, Pack D, Vallyathan V. Time course of gene expression of inflammatory mediators in rat lung after diesel exhaust particle exposure. *Environ Health Perspect* 2005;113(5):612-617.

8. Saber AT, Jacobsen NR, Bornholdt J, Kjaer SL, Dybdahl M, Risom L, Loft S, Vogel U, Wallin, H.Cytokine expression in mice exposed to diesel exhaust particles by inhalation. Role of tumor necrosis factor. *Part Fibre Toxicol* 2006;3:4.

9. Li N, Alam J, Venkatesan MI, Eiguren-Fernandez A, Schmitz D, Di Stefano E, Slaughter N, Killeen E, Wang X, Huang A, Wang M, Miguel AH, Cho A, Sioutas C, Nel AE. Nrf2 is a key transcription factorthat regulates antioxidant defense in macrophages and epithelial cells: protecting against theproinflammatory and oxidizing effects of diesel exhaust chemicals. *J Immunol* 2004, 173: 3467-3481

10. International Agency for Research on Cancer. IARC Monographs on the Evaluation of Carcinogenic Risk to humans: Tobacco Smoking. IARC Monographs 1986;38:168-170.

11. Penn A, Murphy G, Barker S, Henk W, Penn L. Combustion-derived ultrafine particles transport organic toxicants to target respiratory cells. *Environ Health Perspect* 2005;113(8):956-963.

12. Catallo WJ, Kennedy CH, Henk W, Barker SA, Grace SC, Penn A. Combustion products of 1,3-butadiene are cytotoxic and genotoxic to human bronchial epithelial cells. *Environ Health Perspect* 2001;109(9):965-971.

13. Murphy G, Rouse R, Polk W, Henk W, Barker S, Boudreaux M, Floyd Z, Penn A. Combustion-derived hydrocarbons localize to respiratory cell lipid droplets and up-regulate biotransformation enzymes. *Am JRespir Cell Mol Biol* 2008;38 (5):532-540.

14. Rouse R, Murphy G, Boudreaux M, Paulsen D, Penn A. Soot nanoparticles promote biotransformation, oxidative stress, and inflammation in murine lungs. *Am J Respir Cell Mol Biol* 2008;39(2):198-207.

15. Pope CA, 3rd, Burnett RT, Thun MJ, Calle EE, Krewski D, Ito K, Thurston GD. Lung cancer, cardiopulmonary mortality, and long-term exposure to fine particulate air pollution. JAMA 2002; 287(9): 1131-1141.

16. Reddel RR, Ke Y, Gerwin BI, McMenamin MG, Lechner JF, Su RT, Brash DE, Park JB, Rhim JS, Harris CC. Transformation of human bronchial epithelial cells by infection with SV40 or adenovirus-12 SV40 hybrivirus, or transfection via strontium phosphate coprecipitation with a plasmid containing SV40 early region genes. *Cancer Res* 1988;48:1904-1909.

17. Mbawuike IN, Herscowitz HB. MH-S, a murine alveolar macrophage cell line: morphological, cytochemical and functional characteristics. *J Leukoc Biol* 1989;46:119-127.

18. Green H, Kehinde O. An established preadipose cell line and its differentiation in culture. II. Factors affecting the adipose conversion. *Cell* 1975;5:19-27.

19. Floyd ZE, Stephens JM. Interferon-gamma-mediated activation and ubiquitin-proteasome-dependent degradation of PPARgamma in adipocytes. *J Biol Chem* 2002;277:4062-4068.

20. Gould SJ, Keller GA, Hosken N, Wilkinson J, Subramani S. A conserved tripeptide sorts proteins to peroxisomes. *J Cell Biol* 1989;108:1657-1664.

21. Targett-Adams P, Chambers D, Gledhill S, Hope RG, Coy JF, Girod A, McLauchlan J. Live cell analysis and targeting of the lipid droplet-binding adipocyte differentiation-related protein. *J Biol Chem* 2003;278:15998-16007.

22. Fields WR, Desiderio JG, Putnam KP, Bombick DW, Doolittle DJ. Quantification of changes in c-myc mRNA levels in normal human bronchial epithelial (NHBE) and lung adenocarcinoma (A549) cells following chemical treatment. *Toxicol Sci* 2001;63:107-114.

23. Mamo S, Gal AB, Bodo S, Dinnyes A. Quantitative evaluation and selection of reference genes in mouse oocytes and embryos cultured in vivo and in vitro. *BMC Dev Biol* 2007;7:14.

24. Catallo WJ, Kennedy CH, Henk W, Barker SA, Grace SC, Penn A. Combustion products of 1,3-butadiene are cytotoxic and genotoxic to human bronchial epithelial cells. *Environ Health Perspect* 2001;109:965-971.

25. Grimmer G, Naujack KW, Dettbarn G. Gas chromatographic determination of polycyclic aromatic hydrocarbons, aza-arenes, aromatic amines in the particle and vapor phase of mainstream and sidestream smoke of cigarettes. *Toxicol Lett* 1987;35:117-124.

26. Gerde P, Muggenburg BA, Lundborg M, Dahl AR. The rapid alveolar absorption of diesel soot-adsorbed benzo[a]pyrene: bioavailability, metabolism and dosimetry of an inhaled particle-borne carcinogen. *Carcinogenesis* 2001;22:741-749.

Soot Nanoparticles Promote Biotransformation, Oxidative Stress, and Inflammation in Murine Lungs

Rodney L. Rouse[*£], *Gleeson Murphy*[*†], *Marc J. Boudreaux*[‡], *Daniel B. Paulsen*[§], *and Arthur L. Penn*[*]

[*]*Comparative Biomedical Sciences, School of Veterinary Medicine, Louisiana State University, Baton Rouge, LA 70803.* [£] *Present address: U. S. Food and Drug Administratio, CDER, OPS, OTR, Department of Applied Pharmacology Research, White Oak Campus, Silver Spring, MD 20993.* [†] *Present address: U.S. Army Veterinary Medical Corps, Analytical Toxicology Division, USAMRICD, Aberdeen, MD 21010-5400.* [‡] *Division of Biotechnology & Molecular Medicine (BioMMED), School of Veterinary Medicine, Louisiana State University, Baton Rouge, LA 70803.* [§] *Pathobiological Sciences, School of Veterinary Medicine, Louisiana State University, Baton Rouge, LA 70803.*

Urban ambient pollution particles, especially those derived from industrial/combustion processes, are composed of a complex array of organic and inorganic components, often adsorbed to a carbonaceous core. Exposure to ambient particles of a variety of sizes has been associated with increased cardiopulmonary morbidity and mortality and elevation of pro-inflammatory cytokines *in vivo* and *in vitro* (1-7). Ultrafine particles ($PM_{0.1}$; aerodynamic diameter < 0.1 μm) and nanoparticles (aerodynamic diameter < 0.150 μm) have received little attention from regulatory agencies (8), but in numerous experimental settings have been found to elicit a range of toxicological effects (9, 10). The surface characteristics and crystalline structure of ambient particles can be major determinants of pulmonary inflammation and injury (11, 12). Individually or in concert, complex chemicals adsorbed to particles can promote disease processes. Residual oil fly ash (ROFA) containing transition metals induces pulmonary inflammatory infiltrate, edema, and hemorrhage (13). Diesel exhaust particles (DEPs) containing a variety of oxygen radical-generating quinones, polyaromatic hydrocarbons (PAHs), and metal species (14-16) produce pulmonary inflammatory cell infiltration and cytokine production (17-19). A body of *in vitro* and *in vivo* literature has characterized the effects of parent particles and their isolated constituents in humans, as well as in animals (20, 21).

In urban areas, combustion of gasoline, diesel fuels, and industrial organics (simple aliphatics and/or fossil fuels) contributes significantly to the ambient $PM_{2.5}$ fine (22) and ultrafine particulate fractions (22-23). Incomplete combustion of low molecular weight hydrocarbons, e.g. during flaring at refineries, is a real source of complex environmental particulate contamination. Petrochemical-derived particulates resulting, by accident or sabotage, from refinery or pipeline explosions and fires also represent a real hazard both in the United States and abroad. Radicals formed early in combustion interact to form PAHs, including some carcinogens, from less complex structures. PAHs aggregate into nanoparticles, which can extend into branched-chain structures (soots). We have previously characterized the product of incomplete combustion of 1,3-butadiene. This product, butadiene soot (BDS), is both a model mixture and a real-life example of PAH-laden, combustion-derived nanoparticles with potential for environmental contamination and for acute and/or chronic health effects (24). BDS is an organic-rich mixture of 30-50 nm carbonaceous particles to which hundreds of PAH species, including benzo(a)pyrene [B(a)P] and other carcinogens, are adsorbed (24, 25). Both, human bronchoepithelial cells and mouse alveolar macrophages display a distinct punctate blue, PAH-associated cytoplasmic fluorescence following exposure to BDS *in vitro.* The fluorescence localizes to cytoplasmic lipid droplets even as Phase I biotransformation enzymes are induced (26). We hypothesized that inhalation of PAH-containing BDS will result in activation of aryl hydrocarbon receptor (AhR)-associated genes, as is the case with other PAH-rich mixtures (DEPs, cigarette smoke), will cause oxidative stress, and will result in inflammatory changes

METHODS

BDS Exposures: Eighteen 6-week old female Balb/c mice were obtained from Jackson Laboratories (Stock 000651, Bar Harbor, ME). All procedures were IACUC approved and in accord with NIH *Guide for the Care and Use of Laboratory Animals* (27). Ten mice inhaled BDS mixed with HEPA-filtered air (5.0 mg/m^3,4 hr/day, 4 days) and eight inhaled only HEPA-filtered air. Six BDS-exposed mice and four control mice were euthanized and sampled immediately following the fourth day of exposure. The remaining four mice from each group were euthanized and sampled the following day.

Airway Hyperresponsiveness: We assessed airway hyperresponsiveness in unrestrained mice by whole body plethysmography (Buxco, Troy, NY) immediately following the final BDS exposure.

Bronchoalveolar lavage fluid (BALF) Collection and Analysis: We lavaged lungs with phosphate-buffered saline. Leukocytes were categorized by type (macrophage, neutrophil, eosinophil) and macrophages were further graded by particle burden. Group 1 macrophages (MΦ1) had <10% cytoplasmic particle burden. Group 2 macrophages (MΦ2) had 10-50% of the cytoplasm occupied by particles. Group 3 macrophages (MΦ3) had > 50% of their cytoplasm occupied by particles. BALF was analyzed by enzyme-linked immunosorbent assay (ELISA; BD-Pharmingen, San Diego, C) for TNFα, IL-6, and Cxcl2.

Histopathology: The right lung was excised and stored in 0.02 M periodate-0.1 M lysine-0.25% paraformaldehyde (PLP) fixative prior to standard histological sectioning, staining, and evaluation. Histopathology was scored according to 6 parameters; particles found within 1) alveolar macrophages, 2) interstitial macrophages, and 3) bronchial epithelium; numbers of 4) peribronchial neutrophils, and 5) transmigratory neutrophils; and by 6) epithelial damage. Scoring was 0/1 (present or absent) for particulate matter in macrophages and epithelium. All other parameters were scored 0-3 (0, none; 1, mild; 2, moderate; 3, severe).

RNA Isolation: The left lung was stored in RNAlater (Applied Biosystems, Foster City, CA). Total RNA was extracted in TRIzol Reagent (Invitrogen, Carlsbad, CA).

Gene Microarray Assay: Global gene expression was assessed from total RNA on Mouse Genome 430 2.0 Arrays (Affymetrix, Santa Clara, CA). All data collected and analyzed here adhere to the guidelines for Minimal Information About a Microarray Experiment (MIAME).

Gene Expression Analysis: Expression Analysis Systems (Durham, NC) eliminated data from poor performing probe sets (REDI). Principle component analysis was used to determine clustering of experimental units. Treatment group data underwent permutation analysis for differential expression (PADE); including a false discovery rate (FDR) based on a permutation-generated reference curve (technical information on REDI and PADE analyses available at www.expressionanalysis.com). All transcripts included in this study had a fold change of at least 1.5 (up or down), and both an individual p-value and FDR ≤0.05.

Pathway Analysis: We analyzed gene expression data with the network- and pathway-building software, Ingenuity Pathways Analysis 4.0 and examined gene networks and canonical pathways using the Ingenuity Analysis Knowledge Database (Ingenuity Systems, Redwood City, CA).

Quantitative Real Time RT-PCR: We performed quantitative RT-PCR (qRT-PCR) for selected genes on cDNA from lung homogenates with inventoried TaqMan Gene Expression Assays primer-probe sets (Applied Biosystems, Foster City, CA).

Statistical Analysis: We used the UNIVARIATE and TTEST procedures of the SAS statistical package (version 9.1.3; SAS Institute, Inc., Cary, NC) to compare qRT-PCR data from lungs. We used SigmaStat 3.1 software (Systat Software Inc., San Jose, CA) to analyze cytokine data.

RESULTS

Airway Hyperresponsiveness: No significant differences between BDS-exposed mice and controls were noted with unrestrained whole body plethysmography (data not presented).

BALF Cytokines: IL-8 was significantly higher in BDS-exposed mice (data not presented). Following a day of recovery, IL-8 was further elevated in BDS-exposed mice. This finding is consistent with the increased BALF neutrophilia (described below) in these mice. No significant differences were detected in TNFα and IL-6 levels between BDS-exposed and control mice.

BALF Differentials: The differential cytology of BALF from air- and BDS-exposed mice, as well as the grouping of macrophages by particle burden, is presented in Figure 1.

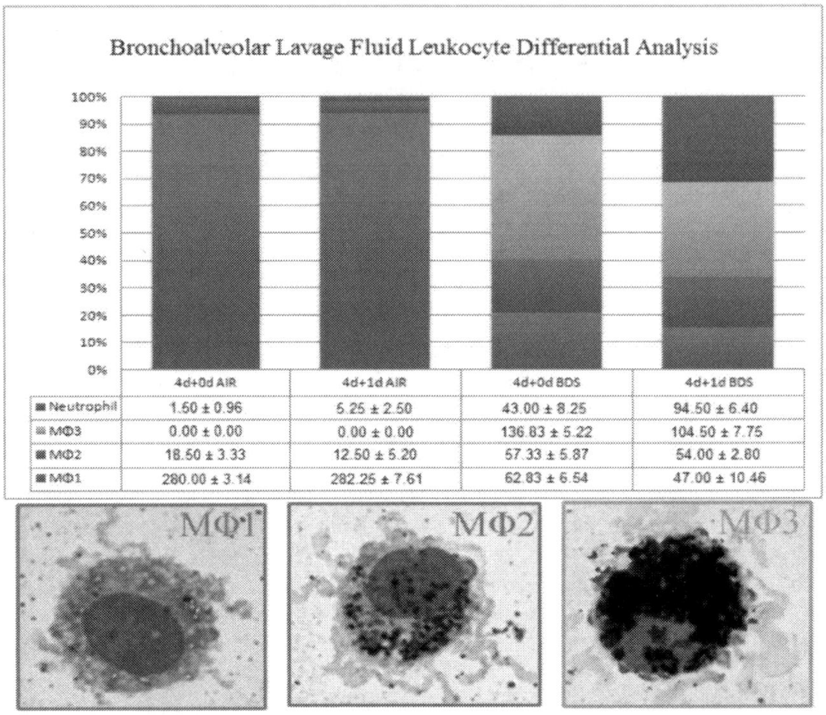

	4d+0d AIR	4d+1d AIR	4d+0d BDS	4d+1d BDS
Neutrophil	1.50 ± 0.96	5.25 ± 2.50	43.00 ± 8.25	94.50 ± 6.40
MΦ3	0.00 ± 0.00	0.00 ± 0.00	136.83 ± 5.22	104.50 ± 7.75
MΦ2	18.50 ± 3.33	12.50 ± 5.20	57.33 ± 5.87	54.00 ± 2.80
MΦ1	280.00 ± 3.14	282.25 ± 7.61	62.83 ± 6.54	47.00 ± 10.46

Figure 1: Differential Leukocyte Counts of BALF from Mice Exposed to Filtered Air or BDS. **In air control mice, essentially all the BALF cells were MΦ1 macrophages with fewer than five particles per cell. The source of those particles is unknown. In the BDS-exposed mice, neutrophil concentration was profoundly increased (>10X); < 1% of neutrophils contained particles. The particle burden of alveolar macrophages also was increased (> 50% of the macrophages present were Mϕ2 or Mϕ3); with one day recovery after BDS exposure, BALF neutrophilia increased in BDS-exposed mice (>2X; p=0.002).**

Lung Histopathology: Figure 2 shows histopathological changes in the lungs of control mice (2A), BDS-exposed mice sacrificed immediately after their final BDS exposure (2B), and BDS-exposed mice sacrificed one day following final exposure (2C). An inflammatory response in BDS-exposed mice sacrificed immediate after final exposure (2B) was evidenced by peribronchial neutrophilia and migration of neutrophils into the bronchoalveolar space. BDS-exposed mice sacrificed one day post-exposure, exhibited increasing neutrophilia, bronchiolar epithelial damage, including basement membrane disruption accompanied by the neutrophilic infiltration (Figure 2C). The BALF and histopathology results indicate that inhalation of these combustion-derived nanoparticles elicits a time-dependent increase in particle accumulation by alveolar macrophages, an associated deep lung neutrophilia, and persistence of particles in the pulmonary interstitium.

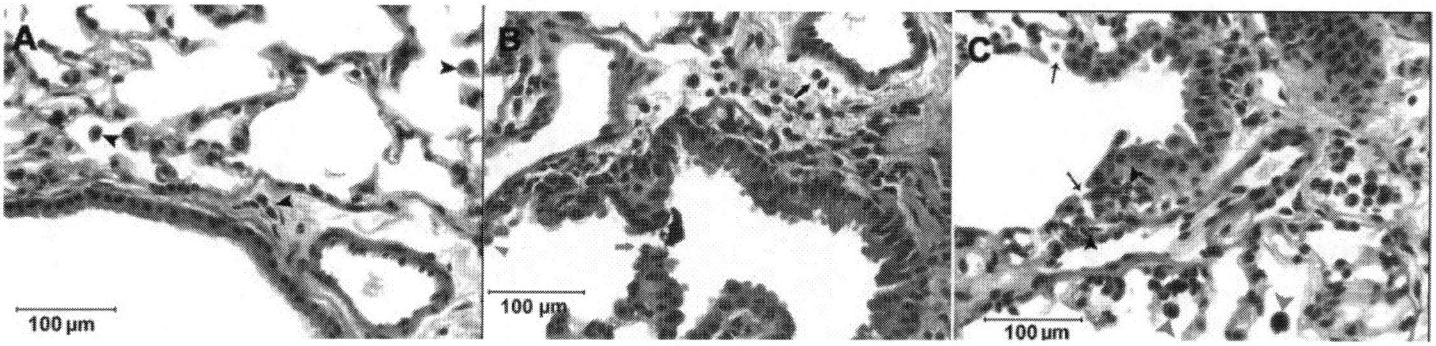

Figure 2: BDS-exposed mice experience a neutrophilic inflammatory pulmonary infiltrate.
A: Photomicrograph of a lung from a mouse exposed to HEPA-filtered air. Notice the scattered resident population of macrophages, in the interstitium and alveolar spaces (black arrowheads). H&E stain.
B. Photomicrograph of a lung from a mouse exposed to 5mg/m3 of BDS for 4 consecutive days and sacrificed immediately. Soot particles are within macrophages (red arrowhead) and bronchiolar epithelial cells (red arrow). Neutrophils are migrating into the bronchiolar interstitium (black arrows) but not through the epithelium. H&E stain.
C: Photomicrograph of a lung from a mouse exposed to 5 mg/m3 of BDS for four consecutive days then allowed rest for one day. Soot particles are within alveolar (red arrowheads) and interstitial macrophages. There is a focus of moderate neutrophilic infiltration and transmucosal exocytosis (black arrowheads) with mild disruption of the continuity of the bronchiolar epithelium (black arrows). An interstitial lymphatic vessel is filled with neutrophils (green arrowhead), most of which contain 1-3 individual soot particles (not apparent at this magnification). H&E stain.

Microarray: A total of 261 transcripts (73 down-regulated and 188 up-regulated) passed all criteria ($\geq$ 1.5 fold-change, $p \leq 0.05$, FDR ≤ 0.05) in mice sacrificed immediately following 4 days of BDS exposure. Ingenuity Systems identified 248 of these transcripts as154 unique genes (33 down-regulated and 121 up-regulated). Aryl hydrocarbon receptor (Ahr)-related genes, oxidative stress response genes, and genes associated with inflammation were up-regulated (Table 1).

Quantitative RT-PCR: We performed qRT-PCR on lung tissue to confirm that BDS inhalation elicited up-regulation of biotransformation enzyme and inflammatory cytokine genes (Table 2). In the lungs of mice sacrificed immediately after BDS exposure, significant increases in expression were confirmed for selected AhR-responsive biotransformation enzymes. Chemokine (C-X-C motif) ligand 2 (Cxcl2; MIP-2; human IL-8 analog) up-regulation was confirmed by qRT-PCR. Il6 up-regulation was demonstrated by qRT-PCR but was not seen on microarray, a phenomenon that we have previously reported for other BALF-derived interleukins (28).

DISCUSSION

We have demonstrated that inhalation of combustion-derived BDS nanoparticles causes acute pulmonary inflammation in the bronchoalveolar space of mice. Metals-rich ROFA, oxygen-rich DEPs, and tobacco smoke particles also produce inflammatory lung infiltrate. Effects of these combustion products have been attributed to metals (13) hydroquinones oxidized to semiquinone radicals that ultimately produce superoxide (29) and other PAH oxidation products. Although oxidation of PAHs in BDS clearly occurs within the cell (Tables 1 & 2), the low oxygen content of BDS indicates that, unlike DEP and cigarette smoke, the initial chemical presentation of BDS to the cell is not primarily in the form of quinones, hydroquinones, and semiquinones. The numerous PAHs, including extensively investigated carcinogens, e.g., B(a)P, found in DEPs and cigarette smoke are metabolized for detoxification and excretion (30, 31). The first step in this process is Phase I biotransformation. Our results reveal up-regulation of Phase I biotransformation genes in response to BDS inhalation.

The oxidized products of select PAHs metabolized by the cytochrome P450 system are capable of forming carcinogenic DNA adducts and/or eliciting oxidative stress responses in cells (32). These effects may be exacerbated by the presence of particles, as in the case of DEPs, where cellular oxidative stress

Table 1: Up-Regulated AhR, Oxidative Stress, and Inflammatory Genes in Lungs of BDS-Exposed Mice

Fold Change (versus Control)	Affymetrix ID	Name	Description
AhR			
5.913	1420796_at	AHRR	aryl-hydrocarbon receptor repressor
8.272	1418752_at	ALDH3A1	aldehyde dehydrogenase 3 family, memberA1
1.917	1417910_at	CCNA2	cyclin A2
15.919	1422217_a_at	CYP1A1	cytochrome P450, family 1, subfamily A, polypeptide 1
64.77	1416613_at	CYP1B1	cytochrome P450, family 1, subfamily B, polypeptide 1
1.506	1423828_at	FASN	fatty acid synthase
2.432	1452160_at	TIPARP	TCDD-inducible poly(ADP-ribose) polymerase
Oxidative Stress			
1.685	1423100_at	FOS	v-fos FBJ murine osteosarcoma viral oncogene homolog
1.823	1455959_s_at	GCLC	glutamate-cysteine ligase, catalytic subunit
1.585	1449279_at	GPX2	glutathione peroxidase 2 (gastrointestinal)
1.603	1448239_at	HMOX1	heme oxygenase (decycling) 1
1.523	1416543_at	NFE2L2	nuclear factor (erythroid-derived 2)-like 2
6.139	1423627_at	NQO1	NAD(P)H dehydrogenase, quinone 1
1.582	1421529_a_at	TXNRD1	thioredoxin reductase 1
Inflammatory/Immune			
2.152	1418133_at	BCL3	B-cell CLL/lymphoma 3
3.407	1422029_at	CCL20	chemokine (C-C motif) ligand 20
2.564	1455577_at	CCL28	chemokine (C-C motif) ligand 28
1.529	1423466_at	CCR7	chemokine (C-C motif) receptor 7
2.294	1417268_at	CD14	CD14 molecule
10.433	1419209_at	CXCL2	chemokine (C-X-C motif) ligand 2
1.722	1449984_at	CXCL3	chemokine (C-X-C motif) ligand 3
2.619	1419728_at	CXCL6	chemokine (C-X-C motif) ligand 6 (granulocyte chemotactic protein 2)
1.71	1421034_a_at	IL4R	interleukin 4 receptor
1.612	1417483_at	NFKBIZ	nuclear factor of kappa light polypeptide gene enhancer in B-cells inhibitor, zeta
5.191	1417262_at	PTGS2	prostaglandin-endoperoxide synthase 2 (prostaglandin G/H synthase and cyclooxygenase)
2.195	1455899_x_at	SOCS3	suppressor of cytokine signaling 3

All gene symbols presented are for the human homolog. These data are from female mice that were sacrificed immediately following final BDS exposure; n = 4 in both groups. Individual microarrays were analyzed for each mouse (non-pooled samples). Gene expression changes were considered significant only if they passed all filtering criteria ($p \leq 0.05$, FDR ≤ 0.05, fold-change ≥ 1.5).

Table 2: Quantitative RT-PCR of lung tissue reveals up-regulation of aryl hydrocarbon receptor responsive biotransformation and inflammatory cytokine genes following inhalation exposure to BDS

Gene Abbreviation	Applied Biosystems Primer-Probe Set ID Number	4d+0d	4d+1d
AhR-Responsive Genes			
Ahr	Mm00478932_m1	-1.03 ± 0.16	1.00 ± 0.20
Ahrr	Mm00477443_m1	59.93 ± 8.95**	5.97 ± 1.22**
Aldh3a1	Mm00839312_m1	7.37 ± 1.05**	1.31 ± 0.28
Cyp1a1	Mm00487218_m1	35.87 ± 4.27**†	11.61 ± 2.44**
Cyp1b1	Mm00487229_m1	67.99 ± 9.83**	9.74 ± 2.11**
Tiparp	Mm00724822_m1	1.90 ± 0.28**	-1.19 ± 0.16
Cytokines			
Cxcl2 (MIP-2; IL-8)	Mm00436450_m1	7.82 ± 1.08**	4.94 ± 0.91**
Ifng	Mm00801778_m1	-1.37 ± 0.13	1.14 ± 0.30
Il-1b	Mm00434228_m1	1.22 ± 0.25	1.91 ± 0.44*
Il-4	Mm00445259_m1	-1.20 ± 0.14	1.07 ± 0.19†
Il-6	Mm00446190_m1	5.51 ± 1.30**	3.17 ± 0.72*
Tgfb1	Mm00441724_m1	1.01 ± 0.11	1.02 ± 0.15
Tnf	Mm00443258_m1	1.50 ± 0.24	1.63 ± 0.45

Results are reported as fold change over control ± standard error of the mean [$(2^{-\Delta\Delta CT})$ ± SEM].
***significance $p<0.05$; **significance $p<0.0001$; †Unequal variance among samples; therefore, Satterthwaite's approximation was used for degrees of freedom in determining significance.**

has been linked directly to inflammatory endpoints (17). We have demonstrated up-regulation both of Nrf2 oxidative stress response genes and of pro-inflammatory genes in the lungs of mice following BDS exposure. NRF2 induces expression of HMOX1, NQO1, GPX2, and TXNRD1 in human tissues (33). Here, expression of Nfe2l2 (mouse homolog of NRF2) induces each of these genes in the lungs BDS-exposed mice. With increasing oxidative stress, inflammatory responses also escalate. In support of this, we demonstrate oxidative stress responses and inflammatory responses in the lungs of mice that have inhaled BDS.

Up-regulation of Il6, Cxcl2 (mouse homolog of IL8), Cxcl3, and Cxcl6 can lead to recruitment of neutrophils and macrophages to the lungs and is consistent with the neutrophilic inflammatory response observed histopathologically in lung sections. These inflammatory effectors may be produced by pulmonary epithelial cells (34) or fibroblasts (35). A significant influx of neutrophils to the bronchoalveolar space also was observed via BALF cytology. Consistent with the neutrophilia, BALF IL-8 levels were elevated in BDS-exposed mice. Increasing neutrophilia was accompanied by the emergence of MΦ2s and MΦ3s in the BALF and pulmonary interstitium (Figure 2C).

Inhalation of particles may elicit not only local effects in the respiratory tract, but systemic effects, as well. An increase in circulating levels of inflammatory cytokines subsequent to PM_{10} exposure has been associated with progression of atherosclerosis (36). Direct translocation of particles from the respiratory tree to the systemic circulation has been described as influencing the subsequent development of cardiovascular disease conditions (37), although this conclusion has been actively challenged by more recent human exposure studies (38, 39). Ultrafine particles deposited in the nasopharyngeal area access the central nervous system via olfactory neuronal transport (40). Studies in rats and dogs have demonstrated that pulmonary epithelial cells have a saturation point with respect to PAHs (41). Combustion-derived ultrafine particles have the potential to translocate to extrapulmonary sites *with* their PAH payload intact, if this PAH saturation point is exceeded. This is unlikely through exposure to ambient environmental particles which usually are present in relatively low concentration and have a low overall PAH burden (42). The saturating concentration might be exceeded however, during an acute exposure to high levels of such particles among firefighters, rescue workers, and local residents following a terrorist attack (as in September 11th), petrochemical industrial accidents, or soldiers and civilian populations exposed to petroleum fires (as in the Gulf Wars).

In summary, we have reported that brief inhalation exposure to a moderate dose of PAH-rich, metals-poor combustion-derived ultrafine BDS particles initiates 1) AhR-responsive biotransformation responses capable of transforming PAHs into more toxic metabolites, as evidenced by increased expression of AhR-responsive genes; 2) cellular oxidative stress, as reflected by up-regulation of Nrf2-mediated oxidative stress response genes and 3) acute pulmonary inflammation, as demonstrated by inflammatory cell infiltration and up-regulation of pro-inflammatory cytokine genes. Finally, our histopathological analyses indicate that freshly-generated inhaled BDS nanoparticles reach the deepest bronchoalveolar spaces in lungs of exposed mice.

References

1. Schwartz J, Dockery DW, Neas LM. Is daily mortality associated specifically with fine particles? *J Air Waste Manag Assoc* 1996; 46(10):927-939.
2. Pope CA, 3rd, Burnett RT, Thun MJ, Calle EE, Krewski D, Ito K, Thurston GD. Lung cancer, cardiopulmonary mortality, and long-term exposure to fine particulate air pollution. *JAMA* 2002; 287(9):1132-1141.
3. Pope CA, 3rd, Burnett RT, Thurston GD, Thun MJ, Calle EE, Krewski D, Godleski JJ. Cardiovascular mortality and long-term exposure to particulate air pollution: Epidemiological evidence of general pathophysiological pathways of disease. *Circulation* 2004; 109(1):71-77.
4. Dominici F, Peng RD, Bell ML, Pham L, McDermott A, Zeger SL, Samet JM. Fine particulate air pollution and hospital admission for cardiovascular and respiratory diseases. *JAMA* 2006; 295(10):1127-1134.
5. Kunzli N, Kaiser R, Medina S, Studnicka M, Chanel O, Filliger P, Herry M, Horak F, Jr., Puybonnieux-Texier V, Quenel P, Schneider J, Seethaler R, Vergnaud J C, Sommer H.
Public-health impact of outdoor and traffic-related air pollution: A European assessment. *Lancet* 2000; 356(9232):795-801.
6. van Eeden SF, Tan WC, Suwa T, Mukae H, Terashima T, Fujii T, Qui D, Vincent R, Hogg JC. Cytokines involved in the systemic inflammatory response induced by exposure to particulate matter air pollutants (pm(10)). *Am J Respir Crit Care Med* 2001; 164(5):826-830.
7. Fujii T, Hayashi S, Hogg JC, Vincent R, Van Eeden SF. Particulate matter induces cytokine expression in human bronchial epithelial cells. *Am J Respir Cell Mol Biol* 2001; 25(3):265-271.
8. United States Environmental Protection Agency. National ambient air quality standards for particulate matter. Standards for particulate matter.: Federal Registry 1997 62(3): 138.

9. Bermudez E, Mangum JB, Wong BA, Asgharian B, Hext PM, Warheit DB, Everitt JI. Pulmonary responses of mice, rats, and hamsters to subchronic inhalation of ultrafine titanium dioxide particles. *Toxicol Sci* 2004; 77(2):347-357.
10. Lundborg M, Dahlen SE, Johard U, Gerde P, Jarstrand C, Camner P, Lastbom L. Aggregates of ultrafine particles impair phagocytosis of microorganisms by human alveolar macrophages. *Environ Res* 2006; 100(2):197-204.
11. Sayes CM, Reed KL, Warheit DB. Assessing toxicity of fine and nanoparticles: Comparing in vitro measurements to in vivo pulmonary toxicity profiles. *Toxicol Sci* 2007; 97(1):163-180.
12. Warheit DB, Webb TR, Colvin VL, Reed KL, Sayes CM. Pulmonary bioassay studies with nanoscale and fine-quartz particles in rats: Toxicity is not dependent upon particle size but on surface characteristics. *Toxicol Sci* 2007; 95(1):270-280.
13. Dreher KL, Jaskot RH, Lehmann JR, Richards JH, McGee JK, Ghio AJ, Costa DL. Soluble transition metals mediate residual oil fly ash induced acute lung injury. *J Toxicol Environ Health* 1997; 50(3):285-305.
14. Kumagai Y, Koide S, Taguchi K, Endo A, Nakai Y, Yoshikawa T, Shimojo N. Oxidation of proximal protein sulfhydryls by phenanthraquinone, a component of diesel exhaust particles. *Chem Res Toxicol* 2002; 15(4):483-489.
15. Li N, Venkatesan MI, Miguel A, Kaplan R, Gujuluva C, Alam J, Nel A. Induction of heme oxygenase-1 expression in macrophages by diesel exhaust particle chemicals and quinones via the antioxidant-responsive element. *J Immunol* 2000; 165(6):3393-3401.
16. Murphy SA, BeruBe KA, Richards RJ. Bioreactivity of carbon black and diesel exhaust particles to primary clara and type ii epithelial cell cultures. *Occup Environ Med* 1999; 56(12):813-819.
17. Ma JY, Ma JK. The dual effect of the particulate and organic components of diesel exhaust particles on the alteration of pulmonary immune/inflammatory responses and metabolic enzymes. *J Environ Sci Health C Environ Carcinog Ecotoxicol Rev* 2002; 20(2):117-147.
18. Rao KM, Ma JY, Meighan T, Barger MW, Pack D, Vallyathan V. Time course of gene expression of inflammatory mediators in rat lung after diesel exhaust particle exposure. *Environ Health Perspect* 2005; 113(5):612-617.
19. Saber AT, Jacobsen NR, Bornholdt J, Kjaer SL, Dybdahl M, Risom L, Loft S, Vogel U, Wallin H. Cytokine expression in mice exposed to diesel exhaust particles by inhalation. Role of tumor necrosis factor. *Part Fibre Toxicol* 2006; 3:4.
20. Diaz-Sanchez D, Tsien A, Fleming J, Saxon A. Combined diesel exhaust particulate and ragweed allergen challenge markedly enhances human in vivo nasal ragweed-specific IgE and skews cytokine production to a T helper cell 2-type pattern. *J Immunol* 1997; 158(5):2406-2413.
21. Salvi S, Blomberg A, Rudell B, Kelly F, Sandstrom T, Holgate ST, Frew A. Acute inflammatory responses in the airways and peripheral blood after short-term exposure to diesel exhaust in healthy human volunteers. *Am J Respir Crit Care Med* 1999; 159(3):702-709.
22. Lighty JS, Veranth JM, Sarofim AF. Combustion aerosols: Factors governing their size and composition and implications to human health. *J Air Waste Manag Assoc* 2000; 50(9):1565-1618; discussion 1619-1522.
23. Oberdorster G, Ferin J, Lehnert BE. Correlation between particle size, *in vivo* particle persistence, and lung injury. *Environ Health Perspect* 1994; 102 Suppl 5:173-179.
24. Penn A, Murphy G, Barker S, Henk W, Penn L. Combustion-derived ultrafine particles transport organic toxicants to target respiratory cells. *Environ Health Perspect* 2005; 113(8):956-963.
25. Catallo WJ, Kennedy CH, Henk W, Barker SA, Grace SC, Penn A. Combustion products of 1,3-butadiene are cytotoxic and genotoxic to human bronchial epithelial cells. *Environ Health Perspect* 2001; 109(9):965-971.
26. Murphy G, Rouse R, Polk W, Henk W, Barker S, Boudreaux M, Floyd Z, Penn A. Combustion-derived hydrocarbons localize to respiratory cell lipid droplets and up-regulate biotransformation enzymes. *Am J Respir Cell Mol Biol* 2008; (in press).
27. Institute of Laboratory Animal Resources. Guide for the care and use of laboratory animals. . Washington, DC: National Academy Press; 1996.
28. Rouse RL, Boudreaux MJ, Penn AL. In utero environmental tobacco smoke exposure alters gene expression in lungs of adult BALB/c mice. *Environ Health Perspect* 2007; 115(12):1757-1766.
29. Pryor WA. Cigarette smoke radicals and the role of free radicals in chemical carcinogenicity. *Environ Health Perspect* 1997; 105 Suppl 4:875-882.
30. Bonvallot V, Baeza-Squiban A, Baulig A, Brulant S, Boland S, Muzeau F, Barouki R, Marano F. Organic compounds from diesel exhaust particles elicit a proinflammatory response in human airway epithelial cells and induce cytochrome p450 1a1 expression. *Am J Respir Cell Mol Biol* 2001; 25(4):515-521.
31. Kim JY, Hecht SS, Mukherjee S, Carmella SG, Rodrigues EG, Christiani DC. A urinary metabolite of phenanthrene as a biomarker of polycyclic aromatic hydrocarbon metabolic activation in workers exposed to residual oil fly ash. *Cancer Epidemiol Biomarkers Prev* 2005; 14(3):687-692.

32. Connery A. Induction of microsomal enzymes by foreign chemicals and carcinogenesis by polycyclic aromatic hydrocarbons. *Cancer Res* 1982; 42: 4875-4917.
33. Yamamoto J, Ihara K, Nakayama H, Hikino S, Satoh K, Kubo N, Iida T, Fujii Y, Hara T. Characteristic expression of aryl hydrocarbon receptor repressor gene in human tissues: Organ-specific distribution and variable induction patterns in mononuclear cells. *Life Sci* 2004; 74(8):1039-1049.
34. Masubuchi T, Koyama S, Sato E, Takamizawa A, Kubo K, Sekiguchi M, Nagai S, Izumi T. Smoke extract stimulates lung epithelial cells to release neutrophil and monocyte chemotactic activity. *Am J Pathol* 1998; 153(6):1903-1912.
35. Sato E, Koyama S, Takamizawa A, Masubuchi T, Kubo K, Robbins RA, Nagai S, Izumi T. Smoke extract stimulates lung fibroblasts to release neutrophil and monocyte chemotactic activities. *Am J Physiol* 1999; 277(6 Pt 1):L1149-1157.
36. Suwa T, Hogg JC, Quinlan KB, Ohgami A, Vincent R, van Eeden SF. Particulate air pollution induces progression of atherosclerosis. *J Am Coll Cardiol* 2002; 39(6):935-942.
37. Nemmar A, Hoylaerts MF, Hoet PH, Nemery B. Possible mechanisms of the cardiovascular effects of inhaled particles: Systemic translocation and prothrombotic effects. *Toxicol Lett* 2004; 149(1-3):243-253.
38. Mills NL, Amin N, Robinson SD, Anand A, Davies J, Patel D, de la Fuente JM, Cassee FR, Boon NA, Macnee W, Millar A M, Donaldson K, Newby D E. Do inhaled carbon nanoparticles translocate directly into the circulation in humans? *Am J Respir Crit Care Med* 2006; 173(4):426-431.
39. Wiebert P, Sanchez-Crespo A, Seitz J, Falk R, Philipson K, Kreyling WG, Moller W, Sommerer K, Larsson S, Svartengren M. Negligible clearance of ultrafine particles retained in healthy and affected human lungs. *Eur Respir J* 2006; 28(2):286-290.
40. Elder A, Gelein R, Silva V, Feikert T, Opanashuk L, Carter J, Potter R, Maynard A, Ito Y, Finkelstein J, Oberdorster G. Translocation of inhaled ultrafine manganese oxide particles to the central nervous system. *Environ Health Perspect* 2006; 14(8):1172-1178.
41. Ewing P, Blomgren B, Ryrfeldt A, Gerde P. Increasing exposure levels cause an abrupt change in the absorption and metabolism of acutely inhaled benzo(a)pyrene in the isolated, ventilated, and perfused lung of the rat. *Toxicol Sci* 2006; 91(2):332-340.
42. Allen J, Dookeran KM, Smith KA, Sarofim AF, Taghizadeh K, Lafleur AL. Measurement of polycyclic aromatic hydrocarbons associated with size-segregated atmospheric aerosols in Massachusetts. *Environ Sci & Tech* 1996; 30(3):1023-1031

167e Speciation of Iron, Carbon, and Sulfur In Diesel Exhaust Particulate from Combustion of Diesel Fuel with and without Addition of 0.1 Wt% Ferrocene

Frank E. Huggins[1], Naresh Shah[1], Artur Braun[1], Gerald P. Huffman[1], Kerry Kelly[2], David Wagner[2], and Adel. F. Sarofim[2]. (1) Chemical and Materials Engineering, University of Kentucky, 533 South Limestone Street, 105 S. J. Whalen Building, Lexington, KY 40506, (2) Institute for Combustion and Energy Studies, University of Utah, 1495 E 100 S Room 107, Kennecott Building, Salt Lake City, UT 84112

The addition of ferrocene, $Fe(C_5H_5)_2$ in small amounts to diesel fuel is known to promote cleaner and more efficient combustion in diesel engines. In addition, it is known to change the particle-size distribution and characteristics of the diesel exhaust particles (DEP) generated during combustion. In this work, bulk spectroscopic methods (^{57}Fe Mössbauer, sulfur XAFS and carbon 1s NEXAFS) have been used to investigate iron, sulfur and carbon forms in DEP derived from combustion of diesel fuel with and without the addition of 1000 ppm of ferrocene. Combustion was conducted in the laboratory with a two-cylinder, direct-injection Kubota model Z482B engine This engine has a displacement of 482 cc and is equipped with a Land and Sea water-break dynamometer for torque load control. The fuel, a 50:50 mixture of the Chevron/Philips reference diesel fuels, T-22 and U-15, had an average cetane number of 46.7 and a sulfur content of 64.1 ppm. Mössbauer spectroscopy showed that the iron in DEP from the ferrocene-doped fuel was largely present in the form of γ-Fe2O3. From the variation of the Mössbauer spectra as a function of temperature, it was deduced that this oxide component in DEP is of nanoparticle size (5 - 10 nm) and not strongly aggregated, presumably due to the separating presence of deposits of carbonaceous material on the particle exteriors. A second iron oxide, α-Fe2O3, which was very minor (~3% of the total iron) in DEP from the ferrocene-doped fuel, but the dominant iron phase in DEP from the undoped reference fuel, was much coarser in size and formed by a completely different mechanism. Carbon 1s NEXAFS spectroscopy showed that the ratio of the unsaturated carbon fraction (associated with "EC", elemental carbon) relative to saturated hydrocarbon and oxygen functionalities (associated with "OC", organic carbon) for DEP derived from the ferrocene-doped diesel fuel was smaller than that for the DEP derived from the diesel fuel without the ferrocene addition. From sulfur XAFS spectroscopy, it was shown that the addition of ferrocene promotes the formation of (i) oxidized inorganic sulfur forms (e.g. sulfate, bisulfate) relative to reduced organic sulfur forms (e.g. thiophenic sulfur), (ii) sulfate relative to bisulfate, and (iii) sulfoxide relative to thiophene forms. These observations on carbon and sulfur speciation are consistent with more efficient combustion of diesel fuel brought about by the addition of ferrocene to diesel fuel. Possible implications of these observations for differences in health effects between different DEP derived from fuels with and without ferrocene will also be briefly discussed.

COAGULATION OF CHARGED NANOPARTICLES: REFINED CLASSICAL APPROACH AND ITS LIMITS

Vladimir Y. Smorodin, University of Maine, Orono, ME, USA

Abstract. Electro-Brownian coagulation of charged nanoparticles has been theoretically investigated, in the range of the moderate Knudsen numbers, $0.1 < Kn < 10$, which correspond to the transition flow regime. New results include: (a) refining a classical theory of the coagulation in the transition regime, (b) criteria of the coagulation of charged fine particles (the "aerosol analogy" of the DLVO theory), (c) the exact charge-size distribution function of charged aerosol ensemble in bipolar ion atmosphere (Boltzmann's equilibrium), (d) criteria of applicability and limits of the refined classical approach. A comparison of theoretical and experimental data demonstrates adequacy of the theory in the range of its validity.

Key Words: boundary sphere method; charge distribution function; coagulation; Knudsen's transition regime; nanoparticles

1. Introduction

Charging of aerosol particles affects the coagulation and evolution of aerosol, in part its size distribution. The efficiency of particulates to penetrate and deposit into the human respiratory tracts and their carcinogenic and mutagenic health effects dramatically depends on their electric charges and sizes, especially, in nano-scale. Since most aerosols carry electrical charge, there are many diverse reasons for studying the coagulation dynamics of charged particles. It presents an interest for both science and technology, e.g., cloud and atmospheric physics, electrostatic propulsion for space vehicles and space thrusters, preventing dust/soot explosion, electrostatic atomization and mass spectrometry, nano-instrumental techniques, nano- and environmental technology, "nano-medicine", etc. (1-6). But still the fundamentals of electrical phenomena in the coagulation of nano-particulates are not clearly understood. Let us review and refine the classical concept of the electro-Brownian coagulation of charged nanoparticles corresponding to the flow transition regime. The main questions we want to answer are: "What are essential peculiarities and criteria of charged nanoparticles coagulation?" and "What is a range of validity of the classical concept of coagulation?"

2. Peculiarities of the coagulation motion of fine particles

Characteristic lengths, times, and scale in coagulation of particles. For analyzing coagulation of particulate matter, first, one has to account for the mean free path of aerosol particulates, or the average distance over which the particle loses all correlation with the initial direction of its motion (the correlation length), and the correlation time[1], τ. The correlation length, l_p is defined as $l_p = \overline{v_p}\tau$, where $\overline{v_p} = \sqrt{8kT/\pi m_p}$ is the thermal velocity of the particle, $\tau = m_p/B$, B is the particle mobility, m_p is the particle mass, k is the Boltzmann factor, and T is the temperature (7). The effective Knudsen number, Kn_p, in aerosol motion is: $Kn_p = l_p\sqrt{2}/2a$; it differs from the molecular "free path" in air. For particulates of unit density the values of the spherical particle radii, a, l_p and Kn_p suspended in air were analyzed in (8). Knudsen numbers for aerosol particles in range $r \approx 1 \div 100$ nm belong to the range: $Kn_p \approx 0.1 \div 46.4$. The transition Knudsen regime between the diffusion and the gas-kinetic (or "free molecular") motion, $Kn_p \in [0.1, 10]$, is an essential part of nanoparticle motion. Let us focus on this range.

The slip flow regime. For extending the continuum diffusion theory results to particle sizes comparable with the correlation length, for moderate Kn_p when Stokes' formula is invalid, Cunningham proposed to modify the Stokes drug force by an empirical factor, $C(Kn_p) = 1 + AKn_p$,

[1] In literature one can find using both couples of terms: the correlation length and the correlation time, and the free path and the relaxation time; their senses are very close, reflecting the same matter from different viewpoints.

where $A \approx 0.70$ for rough spheres (7). Weber introduced another empirical formula which is valid for still smaller particles: $C(Kn_p) = 1 + Kn_p\left[A_1 + A_2 \exp\left(-A_3 / Kn_p\right)\right]$. Then Stokes' drug force and, correspondingly, a diffusion coefficient are corrected: $F_a = 6\pi a\eta V / C(Kn_p)$, $D_a = kTC(Kn_p)/6\pi a\eta V$; here V is the particle velocity, η is the medium viscosity. But even if the correction factor accurately describes the drag change with Knudsen number, unfortunately, it does not lead to a correct form of the coagulation coefficient (9).

The "boundary sphere" approach to the coagulation in the transition regime. As the particle size approaches molecular dimensions, the correction factor, $C(Kn_p)$, becomes very large which is incompatible with kinetic theory. For resolving this anomaly, Fuchs (7) has analyzed the transition regime between continuum and free molecular flow, and developed a famous method of the "boundary sphere" (or the "jump distance"). According to Fuchs, the "probe" particle is included into a concentric sphere of the radius $R_0 + \delta$, where $\delta \sim l$, and $R_0 = a + b$, is the radius of the "influence sphere" presenting a sum of radii of the two colliding particles. Inside of the inner zone, at $R_0 \le r < R_0 + \delta$, a particle movement is regarded as free, and a flux is calculated by an elementary kinetic theory. In the external zone, at $r > R_0 + \delta$, one regards a diffusion. At a boundary, $r = R_0 + \delta$, both fluxes have to be equal. The value of δ is defined as the mean length reached by the reflected particle until the first collision in medium. For coagulation of particles in the Knudsen transition regime, Fuchs derived a formula for the correction factor, $C_F^{a,b} = f(a,b,l,\delta)$ (7). The Fuchs approximation gives correct expressions of the flux onto the probe sphere both in the diffusion and kinetic regimes (at $Kn_p << 1$, and $Kn_p >> 1$, respectively). Therefore Fuchs' correction expresses the transition from diffusive to gas-kinetic coagulation regime just as Cunningham's correction presents the transition from the hydrodynamic to the gas-kinetic regime of particle motion. However Fuchs underestimated and dropped an important effect of viscous forces in the relative particle motion.

Effect of viscous forces. In fact, the relative motion between two colliding particles, in the vicinity of the "sphere of influence" is correlated over the distance of the correlation length order. It is caused by velocity gradients generated with a moving particle in the viscous air. Thus two particles approaching each other have friction coefficients quite different from that of single particle. Last years this problem is thoroughly investigated (10-16). For accounting viscous forces, the effective diffusion (tensor) coefficient is introduced: $D_{ef}(r) = D_0\beta(r)$, where $\beta(r)$ is the "hydrodynamic factor" as a function of the particle separation, r. The effective diffusion coefficient of closing particles depends on the particle separation and loses its additivity: $\overline{D_{12}} \neq \left(D_1 + D_2\right)/2$. A convenient formula to describe the effect of viscous forces on diffusion of two equal spherical particles is derived in (12): $1/\beta(u) = (6u^2 + 4u)/(6u^2 + 13u + 2)$, where $u = (r - 2a)/a$. The effect of viscous forces is to reduce collision efficiency and thereby retard the coagulation rate in the continuum and the transition regime. Especially it is essential in the coagulation of nanoparticles, as we will see later. Thus the Fuchs theory of coagulation has to be corrected by including the hydrodynamic factor (viscous forces).

3. Refining the classical theory of coagulation of charged nanoparticles

Initial model, major forces and effects. For simplicity sake, one can take a case of the electro-Brownian coagulation of identical fine particles of spherical shape. As major forces in coagulation dynamics one regards a contribution of molecular van der Waals', Coulomb's, and "image forces". The hydrodynamic factor is taken into account too. A laminar flow of identical particles at low Peclet's and Reynolds' numbers is considered. The conceptual background of this theoretical approach is based on Fuchs' "flux-matching" classical theory of the transition regime in the laminar aerosol coagulation, a "general" coagulation theory (9, 17), and accounting for the hydrodynamic factor in diffusion.

The equilibrium charge distribution function of aerosols in the bipolar ion atmosphere. For the charging process to be regarded as quasi-stationary, the relaxation time of the ion atmosphere around aerosols should be much less than the mean time between the charge-transfer events. Under these conditions the electric field set up by the non-uniform distribution of the ions around the particulates, i.e., the screening effect, can be ignored. Such conditions are indeed conducted in the laboratory and in the atmosphere. For any initial charge distribution of particulates, a steady-state (or stationary) distribution is eventually established in aerosols (8). A steady-state charge distribution of particles in a bipolar ion atmosphere, when concentrations, mobilitities and other properties of the ions of both signs coincide, conventionally, is described by the Boltzmann distribution (7, 18). But the Fuchs-Sutugin charge distribution (limited with particle sizes $a \geq 10^3$ nm) shows tendencies of deviation from the Boltzmann law (8). However in (19-20) the Boltzmann distribution was confirmed in recent experiments with nano-aerosols in the bipolar ion atmosphere. Suggesting the Boltzmann equilibrium, one can derive a known formula for the average particle charge as a function of its size: $\bar{Z} = f(d) \approx \sqrt{dkT / \pi e^2}$, if $d \geq 50nm$ (18). Here $\bar{Z}$ is the mean number of elementary charge units, e is the electron charge, d is the particle diameter. Let us refine this formula, extending its validity up to zero size. It is easy realized by replacing the method of averaging Z used in (18) with its direct summing (integrating). After simple transformations, a final useful formula is derived:

$$\bar{Z} = \{x\vartheta_3(0, Exp(-x)\}^{-1}, \quad x = \frac{e^2}{dkT}, \quad d \geq 0 \qquad [1]$$

$$\text{where } \vartheta_3(u,q) = \prod_1^{\infty}\left(1 + 2e^{-x(2n-1)} + e^{-2x(2n-1)}\right)\left(1 - e^{-2nx}\right)$$

Here $\vartheta_3(u,q)$ is the elliptic theta function (21). A numerical comparison of difference between this exact and previous formulas shows a large deviation at a<10 nm. Probably it could partially explain the Fuchs-Sutugin results noted above.

The inter-particle potential. In a general form, the total interaction potential of central forces between two particles, $U_\Sigma(r)$, including molecular, $U_M(r)$, electrostatic, $U_C(r)$, and the "image", $U_{im}(r)$, parts, is: $U_\Sigma(r) = U_M(r) + U_C(r) + U_{im}(r)$. The molecular potential of identical particles as a function of the separation distance, within the framework of the pairwise additive concept is found in (22). For particulates in the range $a \leq 10$ nm, the Hamaker constant (22) depends on the particle size due to quantum effects. For nanoparticles the "retardation" effect is vanish (23). For simplicity we can employ a *"point"* approximation of charged equal particles: $a = b \to 0$. If two like charged particles do approach closer, they induce a charge of opposite polarity so that a force of attraction is formed, additionally to initial repel, "the force of image". At smaller distances the image force can change its sign, from repulsion to attraction. In general, this effect is suggested to be marginal for coagulation of large identical like charged particles; but it has to be considered in coagulation of charged particles with uncharged ones. The real value of the image force depends on the particles stage, conductivity, dielectric permittivity, structure, etc. Analyzing contribution of image forces, one has to distinguish two character cases: (A) interaction of charged and neutral particles, and (B) interaction of like charged particles. A third case of interaction of oppositely charged particles is trivial and dropped. (A). *Interaction of the charged particle with the neutral one.* For this case, neglecting the details for brevity, we obtain the condition of dominating the dispersion attractive force, when the image effect can be ignored:

$$Ze^2 / aA << 1 \qquad [2]$$

For estimations, let us take, e.g., soot nanoparticles with a typical value of Hamaker's constant A~10^{-13} erg; then from this criterion, for a particle with Z=1, one can find that the criterion can be satisfied only for particles with d > 250 nm, and larger for Z>1. We find that for nanoparticles image forces must be always taken into account, while for larger particulates they can be dropped when the criterion [2] is satisfied. (B). *Coagulation of two like charged particles.* A general conclusion is the same: analyzing coagulation of charged particles one can drop the image effect if the criterion [2] is satisfied. The image forces of nanoparticles cannot be neglected since they are the same order as molecular forces. Attractive image forces increase coagulation of nanoparticles.

The coagulation coefficient and the enhancement factor. For the case of identical charged nanoparticles in the transition regime a coagulation coefficient is:

$$K(a,a) \approx 16\pi D_a a C_F^{a,a} W_\Sigma(a,a) = \frac{8kT}{3\eta} C_F^{a,a} W_\Sigma(a,a) \qquad [3]$$

Here the "enhancement" factor as a measure of the inter-particle interaction, $W_\Sigma(a, a)$, equals

$$W_\Sigma(a,a) = \left\{ 2a \int_{2a}^{\infty} \frac{\beta(r)}{r^2} \exp\left[\frac{U(r)}{kT}\right] dr \right\}^{-1} \qquad [4]$$

The DLVO-type criterion for the electro-Brownian coagulation of aerosol particles. Let us expand the extremum criteria of the "slow" coagulation of hydrosols formulated in the Derjaguin-Landau-Verway-Overbeek (DLVO) theory, $[dU(r=r_m)/dr]=0$, $U(r=r_m)=0$, for the electro-Brownian coagulation of identical charged aerosol particles. Dropping details, after some transformations, we obtain a simple analytical criterion for coagulation of two like charged dielectric particles driven with molecular, Coulomb's, and image forces:

$$0.544eZ^{1.015}/aA \le 1 \qquad [5]$$

This criterion estimates whether two like charged colliding particles may coagulate.

Criterion of the coagulation of like charged particles in the steady-state bipolar ion atmosphere. Substituting in the formula above the mean equilibrium charge, $\bar{Z} = \alpha\sqrt{dkT/e^2}$, after simplifications we obtain:

$$\alpha^2 kTe^2 / 2aA^2 \le 1 \qquad [6]$$

Here $\alpha = const \sim 1$. In addition to the three driving forces in the particles motion this criterion includes their Brownian energy (the factor kT). Satisfying the condition [6] leads to coagulation (sticking) two colliding like charged particles in the steady-state bipolar ion atmosphere. It can be treated as a condition of the minimal value of Hamaker's constant at a given particle size when coagulation is yet possible, i.e., the attractive dispersion and image forces prevail over repulsive Coulomb's and "distractive" Brownian ones. The case of equality means equilibrium between attractive molecular forces and repulsive electrostatic forces, or, in other words, a "turning point" when a coagulation barrier equals zero.

Criterion of applicability of the "sphere of influence" concept. In the classical theory one does employ a notion of the "sphere of influence" or the "absorbing" sphere": $R_{12} = a_1 + a_2$ (7). Employing a notion of the "sphere of influence" or the "absorbing sphere" in the classical theory, one keeps a suggestion that during (convectional) diffusion of the particle to the sphere, the particle may go to infinity from any vicinity of the sphere, despite the influence of its field. But if the field is strong enough, the moving particle may not have energy for leaving vicinity of the absorbing sphere. In fact, diffusion goes to shorter distance: $R_{ef} > R_{12}$. One has to introduce some "sphere of capture", $R_{ef} \equiv R_c$, to reflect this fact. For example, let us regard two interesting cases: (a) two neutral identical particles with attractive molecular force, and (b) two strongly unlike charged particles with dominating attractive Coulomb force (over molecular one). Elementary calculations show that in the first case (a) always $R_{ef} > R_{12}$ if Hamaker's constant, $A \ge A^* \approx 4.04 \cdot 10^{-14}$ erg. If so, the Fuchs "sphere of influence" concept fails. Since typically A much larger, it means that a notion of the "sphere of influence" in theory of Brownian coagulation (for this case) is not adequate. From other side, if $A \le kT$, coagulation is impossible due to dominating thermal excitations. From our data it follows that even relatively low unlike charging particles leads to dominating the "radius of capture" over the "radius of influence". Influence of attractive molecular forces increases this effect. Therefore, for any unlike charged particles, in general, $R_{ef} > R_{12}$, and the concept of the "sphere of influence" is not adequate. In the case of the interaction of like charged particles, due to compensating the molecular attraction with electrostatic repulsion, the value of critical A^* can be larger. Namely this case was chosen for finding the criterion of coagulation.

4. Comparing and discussing theoretical and experimental data

The hydrodynamic factor. As it follows from our data, a dramatic decrease of the coagulation efficiency as particle size decreases (especially in the range of particle diameters less than 10-20 nm) can be explained with a prevailing of Brownian motion over the molecular attraction in this range, evidently under a strong influence of the viscous forces. On this basis we would interpret experimental data on the soot nano-particles in (24-26) that demonstrate a rapid decrease of the soot coagulation rate in the nano-size area. Additionally, in this case some influence can be also caused by the quantum size dependence of Hamaker's constant.

Concerning the charge distribution of particles in the bipolar ion atmosphere. The conventional approximate formula for the charge distribution overestimates the real (exact) equilibrium charge; the relative deviation between the exact function and the approximation sufficiently grows with decreasing of the particle size in a range 1-15 nm. As shown in experimental measurements of charged soot particles in bipolar ion atmosphere (20), the soot particle fraction in this range may have charges: $Z = \pm1, \pm2$. Therefore, in this region one has to employ our exact formula [1] but not an old approximation. These data give a challenge to experimenters who could realize a thorough verification of the real charge distributions.

Charge effects. Our analysis of the charge effect on the coagulation half-life time demonstrated fitting the experimental in (27-28) on coagulation of the charged organic carbon nanoparticles. It is well fitted at small charges. Further increasing the particle charge leads to a disagreement which can be explained by theoretically overestimating the Coulomb's interaction energy using the "point" approximation. Refining the model may bring us closer to reality.

Theoretical verification. In the range of fine aerosol fraction when the effect of charge has vanished, our theory shows agreement with corresponding results in (29). These authors developed a general comprehensive theory of Brownian coagulation of uncharged particles. Both approaches are able to explain the fast decay of aerosol coagulation in the nano-scale range. Also we have analyzed data on coagulation kernel using the "refined" Fuchs model and other theories. All theories missing the hydrodynamic factor must be revised.

5. Summary

We have reviewed essential aspects of the dynamics of charged aerosol nanoparticles and proposed the refined classical model (RCM) of coagulation of charged nano-sized aerosols. This theory of the laminar electro-Brownian coagulation of charged particles is based on the Fuchs' "boundary sphere" concept which is completed with the hydrodynamic factor; it includes major van der Waals', Coulomb's and "image" forces; for simplicity, coagulation of particles of the equal radii has been only regarded. The RCM and its theoretical tool can be used for describing coagulation of charged and uncharged nano-aerosols in the transition Knudsen range. The limits of the RCM have been discussed. The major new results are: (a) the exact steady-state charge distribution function describing aerosol ensemble in the bipolar ion atmosphere, (b) the DLVO-type criteria of the coagulation of charged particles, (c) the refined electro-Brownian coagulation coefficient and the enhancement factor in the Knudsen transition regime, and (d) criteria of the RCM validity and its limits. A comparison of the theory with other theoretical and experimental data showed a good semi-quantitative agreement in some important cases.

Acknowledgement

I would like to cordially thank Profs. JoAnn Lighty and Adel Sarofim (University of Utah, Salt Lake City, USA) for stimulating my interest to this problem, useful discussions and a partial support, as well as Drs. D'Alessio, D'Anna, and Sagro (Napoli, Italy) for their data.

References

1. Pruppacher, H.R. and Klett, J.D. (1997), Microphysics of Clouds and Precipitation. Kluwer Academic Publ. Dordrecht, Boston, and London.
2. Zhang, H.X., Sorensen, C.M., Ramer, E.R. et al. (1988), In Situ Optical Structure Factor Measurements of an Aggregating Soot Aerosols. *Langmuir,* 4, pp. 867-871.
3. Hospital, A. and Roth, P. (1989), A Molecular Beam Mobility Analyzer Applied to nm-Sized Charged Soot Particles. *J. Aerosol Science*, V. 20, No 8, pp. 1493-1496.
4. Camata, R.P, Atwater, H.A., and Flagan, R.C. (2001), Space-Charge Effects in Nanoparticles Processing using the Differential Mobility Analyzer, *J. Aerosol Sci.,* V. 32, No 5, pp. 583-599.

5. Huang, J., Jayasinghe, S.N., Best, S.M. et al. (2004), Electrospraying of a Nano-Hydroxyapatite Suspension. *J. of Material Sci.,* V.39, pp. 1029-1032.
6. Oberdörster, G., Oberdörster, E., and J. Oberdörster (2005), Nanotoxicology: An Emerging Discipline Evolving from Studies of Ultrafine Particles. EHP, pp. 3-29.
7. Fuchs, N.A. (1964), The Mechanics of Aerosols. Pergamon Press, New York.
8. Fuchs, N.A. and Sutugin, A.G. (1970), Highly Dispersed Aerosols. Ann Arbor Science Publ., Ann Arbor and London.
9. Williams, M.M.R. and Loyalka, S.K. (1991), Aerosol Science. Theory and Practice. Pergamon Press. New York, etc.
10. Spielman, L.A. (1970), Viscous Interactions in Brownian Coagulation. *J. Coll. Interface Sci.,* V. 33, No 4, pp. 562-571.
11. Deutch, J.M. and Oppenheim, I. (1971), Molecular Theory of Brownian Motion for Several Particles. *J. Chem. Phys.* 54, pp. 3547-3561.
12. Honig, E.P., Roebersen, C.J., and Wiersema, P.H. (1971), Effect of Hydrodynamic Interaction on the Coagulation Rate of Hydrophobic Colloids. *J. Coll. Interface Sci.,*36, pp. 97-109.
13. Batchelor, G.K. (1976), Brownian Diffusion of Particles with Hydrodynamic Interaction. *J. Fluid Mechanics*, V. 74, pp.1-31.
14. Jeffry, D.J. and Onishi, Y. (1984), Calculation of the Resistance and Mobility Functions for Two Unequal Rigid Spheres in Low-Reynolds-Number Flow. *J. Fluid Mech.*, 139, pp. 261-272.
15. Wen, S.C. (1996), The Fundamentals of Aerosol Dynamics. World Scientific. Singapore, New Jersey, London, and Hong Kong.
16. Alam, M.K. (1987), The Effect of van der Waals and Viscous Forces on Aerosol Coagulation. *Aerosol Sci. and Techn.*, 6, pp. 41-52.
17. Williams, M.M.R. (1988), Unified Theory of Aerosol Coagulation. *J. Phys. D Appl. Phys.* V 21, pp. 875-898.
18. Reist, P.C. (1984), Introduction to Aerosol Science. McMillan Publ. Co. New York, and London.
19. Karasev, V.V., Ivanova, N.A., Sadykova, A.R., et al. (2004), Formation of Charged Aggregates by Combustion and Pyrolysis: Charge Distribution and Photophoresis. *J. Aerosol Sci.,* V. 35, No 3, pp. 363-381.
20. Maricq, M.M. (2004), Size and Charge of Soot Particles in Rich Premixed Ethylene Flames. *Combustion and Flame.* V.137, No 3, pp. 340-350.
21. Gradshtein, I.S., and I.M. Ryzhik (2000). Tables of Integrals, Series, and Products. Academic Press. San Diego.
22. Hamaker, H.C. (1937), The London van der Waals Attraction between Spherical Particles. *Physica*, 4, pp. 1058-1073.
23. Derjaguin, B.V. and Muller, V.M. (1967), Slow Coagulation of Hydrosols. *DAN SSSR*, V.176, No 5, pp. 1111-1114.
24. D'Alessio, D'Anna, P.Minutolo, L.A.Sagro, and A.Violi (2000), *Proc. Comb. Inst.* 28 (2), pp. 2547-2554.
25. Joutsenoja, T., D'Anna, A., D'Alessio, and M.I. Nazzaro (2001), *Appl. Spectr.*, 55 (2), pp. 130-146.
26. Sagro, L.A., G.Basile, Barone, A.C., D.Anna, A., Minutolo, P., Borghese, A., and D'Alessio (2003), *Chemosphere,* 51 (10), pp. 1079-1090.
27. Kim, S.H., Woo, K.S., Liu, B.Y.H., and M.R. Zachariah (2005), Method of Measuring Charge Distribution of Nanosized Aerosols. *J. Coll. Interface Sci.*, 201, pp. 21-39.
28. Kim, S. H. and Zachariah, M. R. (2004), Enhancing the Rate of Energy Release from NanoEnergetic Materials by Electrostatically Enhanced Assembly, *Advanced Materials*, V.16(20), pp.1821-1825.
29. Narsimhan, G. and E. Ruckenstein (1985), The Brownian Coagulation of Aerosols over the Entire Range of Knudsen Numbers: Connection between Sticking Probability and the Interaction Forces. *J. Coll. Interface Sci.,* V. 104, No 2, pp. 3444-369.

NUMERICAL ANALYSIS OF THE BEHAVIOR OF PHOTOCHEMICAL PRODUCTS IN LOCAL FLOWS OVER CENTRAL JAPAN IN SUMMER SEASON -SENSITIVITY OF OZONE CONCENTRATION TO EMISSION SOURCES OVER COASTAL URBAN AREAS-

Toshihiro Kitada[1)*], and Kiyoshi Okamura[2)]

1) Dept. of Ecological Engineering, Toyohashi University of Technology
Tempaku-cho, Toyohashi, Japan 441-8580
2) Department of Environmental Information, Nagoya Sangyo University
Arai-machi, Owari-asahi, Japan 488-8711
* Corresponding author; E-mail, kitada@earth.eco.tut.ac.jp

Abstract

Air pollution transport over Central Japan was numerically simulated in a typical land and sea breeze situation in summer season. The models used in the calculation were extensions of those described in Kitada et al. (1993) and (1998). Sensitivity of the production of ozone and acidic species to emissions of biogenic non-methane-hydrocarbon (NMHC), and anthropogenic-NMHC and -NOx were focused; the ozone, sulfate, and nitrate are chemically produced in pollutants-rich air mass migrating from coastal big cities to central mountains. Some of the obtained results are as follows: (1) biogenic hydrocarbons could increase daily maximum ozone concentration as much as 20 ppb over the Nohbi and Kanto Plains, and (2) reduction of NOx emission by 20% may result in increase of the ozone concentration by a few percent to almost 20% over the Nohbi Plain, indicating necessity of careful emission control.

Key words : local winds, photochemical smog, ozone, biogenic hydrocarbon, Japan, numerical simulation.

1. INTRODUCTION

The central Japan has characteristic topographical features such as many high mountains and large plains, and is bounded with complex coast lines for Pacific Ocean at the south side and for the Sea of Japan at the north side. Thus various types of local flows develop in this area in summer season. In the daytime flow system pollutants emitted over the huge coastal-urban areas in the plains such as Tokyo, Nagoya, and Osaka likely reach to the mountainous central part in a day, located 150 km from the Pacific Ocean, with producing ozone and others. More than 10 years ago, this phenomenon was extensively studied with observations in the Kanto Plain, i.e. Tokyo area, and the flow system has been named the "extended sea breeze". However, the subsequent behavior and fate of the ozone and other products are not clear. In this study; 3D three-days-simulations, during 26-28 July, 1995, of transport, chemistry and deposition of various chemical species were performed to see the behavior of ozone etc.

produced in the pollutants clouds discharged from the urban area, particularly Nagoya in the Nohbi Plain, for the calculation domain over the Central Japan, with the size of 550 km x 450 km x 6.5 km for east-west, north-south, and vertical directions, respectively The simulation results showed that the high ozone once produced and transported to mountainous area is returned back into the upper layer with its center around 1000 m high over the coastal area on the subsequent day. These were well compared with the vertical profiles of ozone obtained with aircraft observations over the coastal Nohbi Plain and the Ise Bay; the aircraft observations were performed as the part of environmental assessment on a new international airport to be constructed in the Ise Bay Total ozone produced over the Central Japan was evaluated as well as the amount of deposited acidic pollutants.

2. BACKGROUND OF THE PROBLEM

Medium range transport of air pollutants from highly developed urban area to mountainous rural area has attracted concern of many people in applied meteorology and atmospheric environment in Japan, since it occurs in an interesting combination of several types of local winds such as land and sea breezes, mountain and valley winds, and plain-plateau winds, and the transport repeated in summer season probably causes severe damage for eco-systems in the mountainous area. Kurita et al. (1985) pointed out that the pollutants discharged in Tokyo Bay area migrate to the mountains in central Japan, i.e. the travel distance of about 150 km within one day. The distance far exceeded what was expected in simple sea breeze situation in mid-latitudes, and thus the phenomena suggested importance of combined effects of sea breeze, valley wind, and plain to plateau wind for the transport. Later Kondo (1990) numerically verified using meso-scale meteorological model that the presence of the high mountains in the central Japan, which are taller than 2000 m, is a key for the formation of the flow resulting in the medium range transport of pollutants. Kitada et al. (1998) also discussed dominant role of the high mountains of the central Japan in temporal variation of sea breezes over coastal plain area in the Pacific Ocean side, i.e. the Nohbi Plain. It was made clear more than 10 years ago, as described above, that the pollutants released from huge urban areas in the coastal Japan such as Tokyo can be transported more than 150 km within 12 hours mostly during the daytime and reach to the mountains (Kurita and Ueda, 1985). However, the synoptic scale weather with light gradient wind under high pressure system, which is preferable for the development of local winds, usually continues for several days, and the behaviors during those days of the primary and secondary pollutants, once brought in the mountainous area, are not well understood. Furthermore, recently photochemical smog episodes are frequently reported in Japan, and some people stress importance of long range transport from East Asian continent on these episodes. However, ozone observation at a mountainous site in central Japan tells primary importance of local ozone production (Kitada, et al., 2008) In this paper, to make these phenomena clear we have investigated three days behavior of the pollutants emitted in the coastal urban areas using a combined model of mesoscale meteorology (Kitada et al., 1998) and transport/chemistry/deposition of air pollutants (Kitada et al., 1993) together with airborne observation of pollutants over the Nohbi Plain and the Ise Bay area in the central Japan. The period for the investigation was 26 to 28 July, 1995.

3. NUMERICAL MODELS

a. Transport/chemistry/deposition model

The model deals with processes of advection, diffusion, chemical reactions and deposition of trace chemical species, and the basic frame of the model is described in detail in Kitada et al. (1984) , Carmichael et al. (1986) and Kitada et al. (1993). Equations are written for advected species,

$$\rho \frac{\partial C_i}{\partial t} + \rho U_j \frac{\partial C_i}{\partial x_j} = \frac{\partial}{\partial x_j}\left(\rho K_{jj} \frac{\partial C_i}{\partial x_j} \right) + R_i + S_i, \qquad i = 1, 2, \ldots, IL \tag{1}$$

where C_i denotes dimensionless concentration of species i, and R_i and S_i for chemical reaction rate and emission rate due to point source for the species i, respectively, ρ is air density, x_j spatial coordinates (i.e., j=1, 2, and3 for x, y, and z, respectively), $K_{11}(=K_{22}=K_H)$ and $K_{33}(=K_V)$ for eddy diffusivity in horizontal and vertical directions, respectively, and U_j for wind velocity in x_j direction. Meteorological fields such as wind velocity, temperature, and K_{33} were all calculated using mesoscale model described in Kitada et al. (1998). Equation (1) was applied for 44 advected species (i.e., IL=44) such as NO, NO_2, HNO_3, PAN, NH_3, NH_4NO_3, SO_2, SO_4^{2-}, O_3, H_2O_2, various anthropogenic and biogenic hydrocarbons, and others. The reaction mechanism used in the model is based on that of Lurmann et al. (1986) and α-pinene chemistry of Lloyd et al. (1983) with some revision of reaction rate coefficients from Grey et al. (1989) and DeMore et al. (1997). The model in this study consists of 159 chemical reactions. Horizontal boundary conditions for east-west, i.e.x_1, direction, for example, were expressed as follows. For inflow case,

$$- \rho K_H \frac{\partial C_i}{\partial x_1} + \rho U_j C_i = \rho U_1 C_{B,i} \tag{2}$$

where $C_{B,i}$ stands for concentration at outside of the boundary and can be renewed, and for outflow case,

$$\frac{\partial C_i}{\partial x_1} = 0 \tag{3}$$

Vertical boundary conditions were given as follows. At the earth surface boundary,

$$- \rho K_V \frac{\partial C_i}{\partial x_3} = Q_i - \rho V_{g,i} C_i \tag{4}$$

where Q_i denotes the area emission source, and $V_{g,i}$ the dry deposition velocity for species i, which was evaluated by the method calculating resistances to mass transfer from air to surface (Wesely and Hicks, 1977, Sheih et al., 1979, Kitada et al., 1986 etc.). The dry deposition process was applied to the chemical species of NO_2, HNO_3, NH_3, NO_3^-, SO_2, SO_4^{2-}, O_3, PAN, and H_2O_2. At top boundary, basically the same formulations as Eqs. (2) and (3) were applied.

For 36 free radical species such as OH and HO_2, the following algebraic equations were applied by using the steady state assumption:

$$R_1 = 0, \quad i = IL + 1, \cdots, IS \tag{5}$$

For Eqs. (1) through (5), terrain following coordinate transform was applied as Eq. (6):

$$\varepsilon = x_1, \quad \eta = x_2, \quad and\ \sigma = \frac{x_3 - h(x_1, x_2)}{H - h(x_1, x_2)} \tag{6}$$

where h (x_1,x_2) denotes the height of topography above mean sea level(MSL) , and H is the height of top boundary above MSL and was assumed constant at 6.5 km.

b. Mesoscale meteorological model

The input meteorological fields, of flow, temperature, moisture, and eddy diffusivity, to the simulation of air pollution transport were computed with a mesoscale model with $k-\varepsilon$ turbulence module. The basis of the model is described elsewhere (Kitada et al., 1998). Here we explain only what was added into the model in this application. The model is hydrostatic, and the equations of momentum transport for horizontal directions are expressed as follows:

$$\begin{aligned}\frac{DU_i}{Dt} &= -\frac{1}{\rho}\frac{\partial p}{\partial x_{j1}} + \frac{\partial}{\partial x_k}\left(-\overline{u_j u_k}\right) + f_1\left[\delta_{j1}(U_2 - U_{g2}) - \delta_{j2}(U_1 - U_{g1})\right] - f_2\delta_{j1}U_3 \\ &\quad + G_{Uj}\left(U_{obsj} - U_j\right), \qquad j = 1, 2\end{aligned} \tag{7}$$

where U_j and u_j denote the mean wind velocity to be predicted and the turbulent fluctuation, U_{gj} stands for the geostrophic wind component estimated from the observed vertical profiles of the atmospheric pressure, and U_{obsj} the observed winds, where the suffix j is for coordinate direction, i.e. x_j: j=1 for east-west, j=2 for north-south, and j=3 for vertical direction. Furthermore, p denotes the mesoscale pressure perturbation from the synoptic scale pressure field, the Coriolis parameters $f_1 = 2\Omega\cos\varphi$ and $f_2 = 2\Omega\cos\varphi$, the nudging coefficient $Gu_i = (10^{-3}s^{-1}$ which was selected after some numerical experiments), δ_{j1} and δ_{j2} the Kronecker's delta, and $D/Dt = \partial/\partial t + U_j \partial/\partial x_j$ the substantial derivative.
The equation of heat transport:

$$\frac{D\Theta}{Dt} = \frac{\partial}{\partial xj}\left(-\overline{u_j\theta}\right) + \frac{\Theta_0 - \Theta}{\tau_0} \tag{8}$$

where Θ, Θ_0 and θ stand for the mean potential temperature to be predicted, its synoptic scale component to be estimated using aerological data at seven observation sites in central Japan, and the turbulent fluctuation component, respectively, and τ_0 is a relaxation time scale which was set at $10^4 s$, which was adopted after some numerical experiments. The turbulent fluxes in Eqs. (7) and (8) were evaluated using eddy diffusivity and local gradient of corresponding dependent variables, where the eddy diffusivity was calculated applying $k-\varepsilon$ turbulence model. Detail of the model is described elsewhere (Kitada et al., 1998).

4. METEOROLOGICAL INPUTS, EMISSION SOURCES AND SIMULATION CASES

a. Meteorological input data

Numerical simulations were performed for two days from 06 JST (Japan Standard Time) 26 July to 06 JST 28 July, 1995. During these days central Japan had been under the influence of high pressure system with sunny sky and light gradient wind, as seen in weather chart in Fig. 1. Thus these were typical days suitable for the development of local winds such as land and sea breeze, mountain and valley wind, and plain to/from plateau wind. Photochemical smog is also known to likely occur in this situation.

To obtain meteorological fields for the simulations of air pollution transport and chemistry, the meteorological model described above was numerically integrated in the region shown in Fig. 2a. The sizes of the region were 582.9 km $\times$ 480.5 km $\times$ 6.5 km for east-west (x), north-south (y), and vertical (z) directions, respectively. Variable grid sizes were applied with their minimum values of 2.28 km, 1.85 km, and 3 m for x, y, and z directions, respectively, and with the maximum values of 11.43 km, 9.24 km, and 200 m also for x, y, and z directions. The smallest grid size was used for the Nohbi Plain region in Fig. 2b, which is our primarily concerned area. Computed meteorological fields were for air flow, temperature, humidity, pressure, and eddy diffusivity. Observational data were taken into the computation through U_{g1}, U_{g2}, and U_{obsj} in Eq.(7), and Θ_0 in Eq.(8). These observations include: (1) aerological data at 9 sites over Japan with 12 hours interval, (2) data by 6 pilot balloons and 2 rawinsondes up to 2 km in the Nohbi Plain and Ise Bay area with one and a half hour interval during daytime, and (3) wind data at every one hour at about 360 surface meteorological observation sites operated over the whole calculation domain either by Meteorological Agency of Japan or by local governments. To construct synoptic scale pressure field, the kriging method with spherical distribution model was applied for the sparse aerological data.

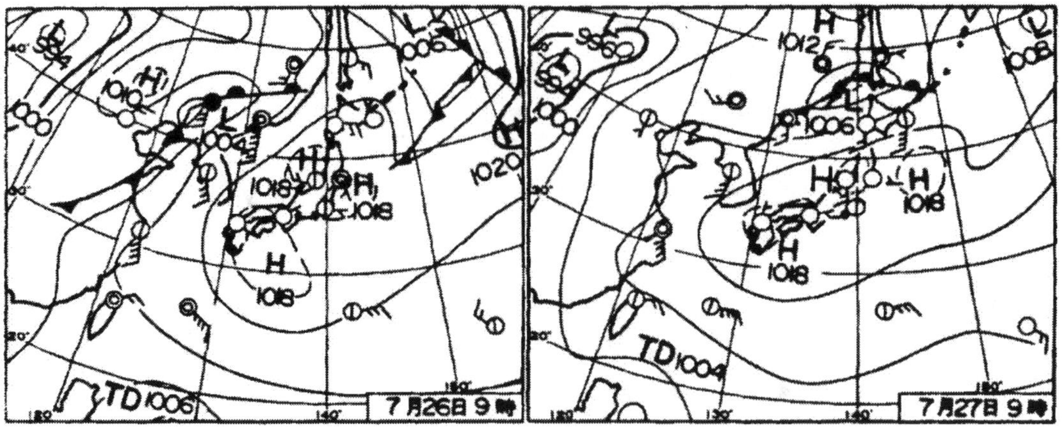

Fig. 1. Weather chart in Japan area at 09 JST (a) 26, (b) 27 July, 1995.

b. Emission sources

Emission sources of NO_x, SO_x, and various anthropogenic non-methane-hydrocarbons (hereafter abbreviated to ant-NMHC) in and near the Nohbi Plain (see calculation domain in Fig. 1b) were estimated from data compiled by the Aichi Prefectural Office (1995). NO_x and SO_x emission inventories were explicitly given with their diurnal variations in the data; they were computed on the basis of consumption of fuels in various sectors such as industries, offices, households, and traffic. The ant-NMHC were grouped into six categories of C_2H_6, C_3H_8, ALKA, C_2H_4, ALKE, and AROM as in the original chemistry model by Lurman et al.(1986). Then their emission values were evaluated as ratios of those to NO_x emission (see Table 1). Two biogenic hydrocarbons of isoprene and α-pinene were considered in the model. Emission sources of isoprene and α-pinene were estimated every one hour using the BEIS (Biogenic Emissions Inventory Systems, Pierce et al., 1990) with vegetation map compiled by Center for Natural Environment, Japan (1993). These emissions depend on solar radiation and temperature, and thus the solar radiation data were obtained at the observation sites operated both local governments and the Meteorological Agency of Japan and the temperatures were output of the meteorological model (Kitada, et al., 1998); actually the temperature at leaf skin is required, and the leaf skin temperatures were estimated by the canopy model (Okamura and Kitada, 1998).

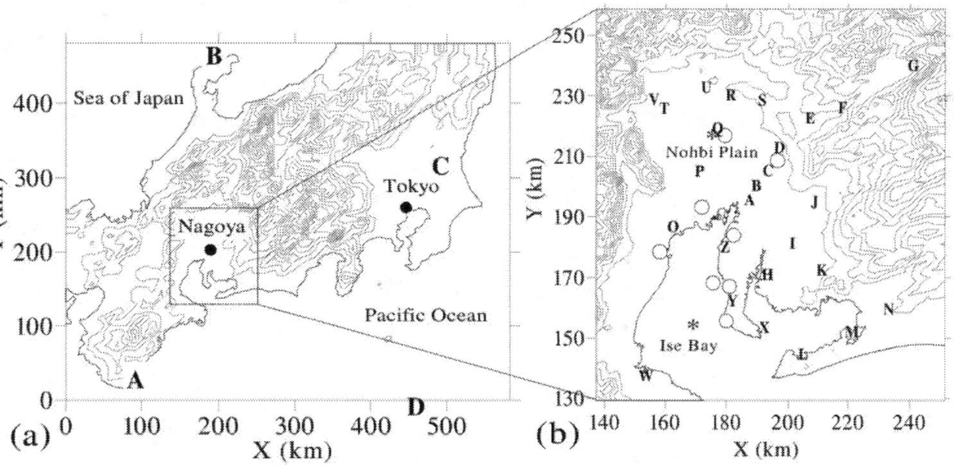

Fig. 2. Calculation domain : (a) whole area, and (b) the area magnified for the Nohbi Plain and Ise Bay. A,B,C, and D in Fig. 2a represent aerological observation sites. A through Z in Fig. 2b are for observation sites at "surface" level, open circles for pilot balloon or rawinsonde sites, and * for the locations where spiral flights were carried out, i.e. one over Ichinomiya ("Q") and another over Ise Bay.

Table 1. Summary of emission sources for the region in Fig. 1b.

Chemical species.	Emission rate (kmol day^{-1})	Ratio[3]
NO_x	7.90×10^3	1
ant−NMHC[1]	7.72×10^3	0.978
C_2H_6	0.47×10^3	(0.059)
$C_3H_8^*$	0.81×10^3	(0.103)
ALKA[#]	2.90×10^3	(0.367)
C_2H_4	0.52×10^3	(0.066)
ALKE[+]	0.70×10^3	(0.089)
AROM[=]	2.32×10^3	(0.294)
bio−NMHC[2]	2.09×10^3	
ISOP	1.6×10^3	
PINENE	0.45×10^3	

1, ant-NMHC = C_2H_6 + C_3H_8 + ALKA + C_2H_4 + ALKE + AROM.
2, bio-NMHC = ISOP + PINENE.
3, Ratio of emissions of various anthropogenic hydrocarbons to that of NO_x.
*, C_3H_8 stands for propane and benzene.
#, ALKA for $\geq C_4$ alkanes.
+, ALKE for $\geq C_3$ alkenes.
=, AROM for alkylbenzenes.

Emission sources in the whole domain (Fig. 2a) were estimated, except for the Nohbi Plain region (Fig. 2b), by using $1^\circ\times1^\circ$ Asian distribution of NO_x and SO_x (Akimoto and Narita, 1994) and $1^\circ\times1^\circ$ global distribution of isoprene (GEIA, 1998). Emissions of ant-NMHCs were computed by multiplying the NO_x emission distribution by the ratios listed in Table 1.

c. Simulation cases

Influence of emission sources of NO_x and anthropogenic and biogenic hydrocarbons on O_3 production/destruction was investigated by numerical simulations. These simulation cases are summarized in Table 2.

Case 701rx, 711rx, and 741rx do not include biogenic NMHC emission, and each of them has different strength of anthropogenic NMHC emissions while NOx emission is fixed. Case 701rxvv, 741rxvv, 751rxvv, and 711rxvv have different ant-NMHCs emission strengthwhile the same bio-NMHCs emission is included into those cases. Case 741rxnoinvv and 741rxnovv represent increased and decreased NOx emission case, respectively, with same conditions as case 741rxvv except for the NOx emission. Case 742rxvv has the same emission condition as case 741rxvv, but it turns off all the emission sources after 00 JST 27 July.

4. RESULTS AND DISCUSSION

a. Computed flow fields

Mesoscale meteorological model was integrated for four days from 00 JST 25 July to 00 JST 29 July 1995. As described above, using these outputs of the meteorological model,

Table 2. Simulation cases.

Case No.	$NO_x^{\#}$	ant-NMHCs#	bio-NMHCs#	Dry Deposition Velocity	Others
701rx	1	1.33	0	a	—
741rx	1	1	0	a	—
711rx	1	0.13	0	a	—
701rxvv	1	1.33	1	a	—
741rxvv (base)	1	1	1	a	—
751rxvv	1	0.67	1	a	—
711rxvv	1	0.13	1	a	—
700rxvv	1	0.0	1	a	—
741rxnoinvv	1.2	1	1	a	—
741rxnovv	0.8	1	1	a	—
741rxno6vv	0.6	1	1	a	—
761rxnovv	0.8	0.8	1	a	—
771rxnovv	0.6	0.6	1	a	—
742rxvv	1	1	1	a	off*
741rxvv2	1	1	2	a	—
741rxvv1.5	1	1	1.5	a	—
741rxvvh	1	1	0.5	a	—
741rxvvnewdepnoleaf	1	1	1	b	—
741rxvvnewdep	1	1	1	c	—

Relative emission strength to those listed in Table 1.
* All emission sources were turned off after 00 JST, i.e. midnight, on 27 July.
(a) Constant dry deposition velocities for land and water surfaces (see text for detail).
(b) Dry deposition velocities formulated using resistances of aerodynamic, surface, and residual (see text for detail).
(c) Dry deposition velocities same as in (b), but for weighted with leaf area index, LAI (see text for detail).

air pollution transport simulations were performed for two days from 06 JST 26 July to 06 JST 28 July.

Although these two days, i.e. 26 and 27 July, were typical summer days suitable for development of various local flows, there were some differences on the development of flow patterns in the Nohbi Plain. The wind fields on 26 July were more typical than those on 27 July. Usually, two types of sea breezes blow in a day in coastal area of the Nohbi Plain (see Fig. 2b); first, rather local sea breeze, the south-westerly, from Ise Bay starts to blow around noon time, and eventually, another strong sea breeze, south-easterly, from Pacific Ocean takes over the local sea breeze and continues nearly until midnight, the formation of which is strongly influenced by the Japanese Alps of the central Japan in Fig. 2a (Kitada et al., 1998). On 27 July, the sea breeze at the first stage was modified from its usual south-westerly to a westerly wind under the influence

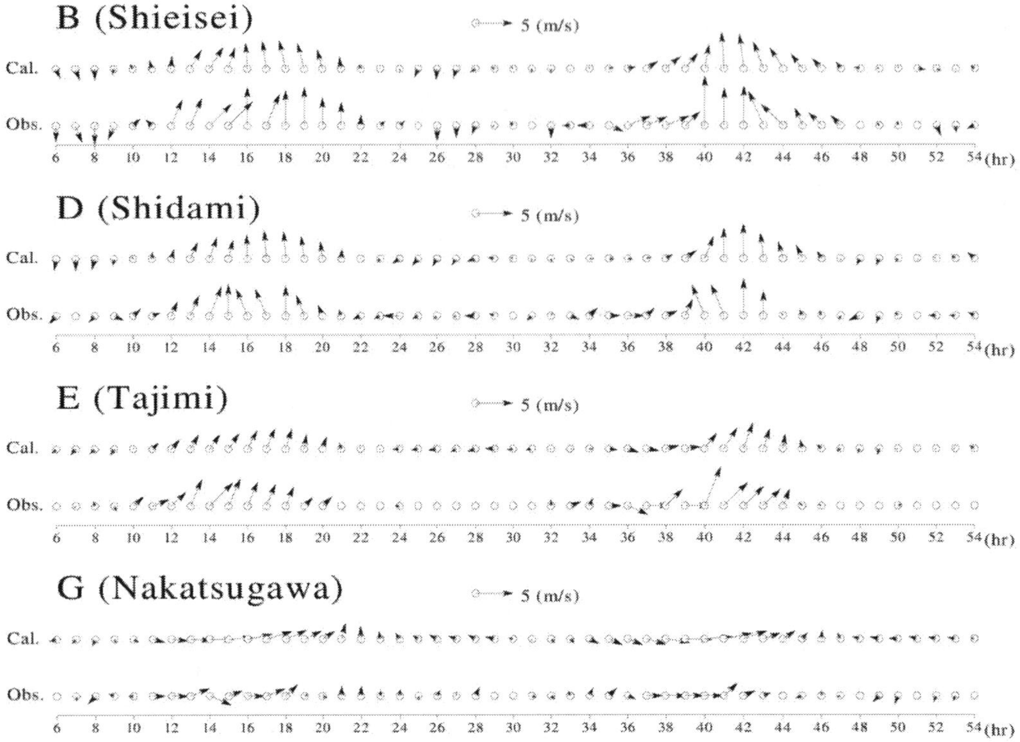

Fig. 3. Calculated vs. observed wind vectors for two days from 06 JST 26 July to 06 JST 27 July, 1995, at sitesB, D, E, and G (see Fig. 2b).

of the westerly or north-westerly winds in upper layer, and the depth of the combined sea breeze and plain-to-plateau-wind layer was suppressed and did not exceed 1 km over the Nohbi Plain in contrast to 26 July. Figure 3 shows hourly variations of both calculated and observed surface wind at 10 m above ground at the several sites, the locations of which are marked with alphabetical characters of B, D, E, and G in Fig. 2b. In Fig. 3 which compares calculated winds with those observed, the winds at B (Shieisei), and D (Shidami) from 36 to 39 hr (i.e., from 12 to 15 JST 27 July) show the "modified westerly" sea breezes compared with winds at the corresponding time on the previous day, i.e. 26 July. Figure 3 demonstrates that the calculation well simulated the observed winds.

In the simulation, winds in upper layer acquired by pilot balloon and rawinsonde were used for 4 dimensional data assimilation. Figure 4 illustrates performance of the assimilation process: Figs. 4a, 4b, and 4c show computed wind fields at 00 JST 27 July, i.e. Fig. 4a for surface wind about 10 m above the ground level(AGL) in whole calculation domain, and Fig. 4b and 4c for those at 500 m and 1000 m above mean sea level(ASL), where the observed winds are also shown with thick dotted arrow, while Figs. 4d, 4e, and 4f are the same as Fig. 4a etc. but for 15 JST 27 July. The results show appropriate modifications of the computed winds, and thus indicate that the nudging coefficient, $G_{Uj}=10^{-3}\ s^{-1}$, the weighted average of observed winds with a weight of the reciprocal of the squared distance between grid point and observation site, and the radius of influence of 100 km below 2 km AGL all worked well. As can be seen in Fig. 4c, however, weak wind situation, i.e. no significant pressure gradient may cause some discrepancy between computed and observed winds. Due to the low wind speed, this discrepancy would not result in serious problem in transport simulation of air pollutants.

Vertical structure of wind shown in Fig. 4 also captured the characteristics found in observation on a different occasion but in the same land-sea breeze situation (Mori et al., 1998).

b. Dynamical behavior of ozone in local winds

As shown in Fig. 2b, the Nohbi Plain-Ise Bay area is bounded by mountains on the north, west, and east. This topographic feature tends to accumulate both anthropogenic and natural trace chemical species in the basin with stable stratification in the upper layer caused by the prevailing high pressure system. Reaction products produced from primary pollutants during their transport from coastal area to inland show the same behavior; ozone can be regarded as a typical species. Figure 5 plots a sequence of the north-south vertical cross sections, passing over Ichinomiya, marked with the symbol

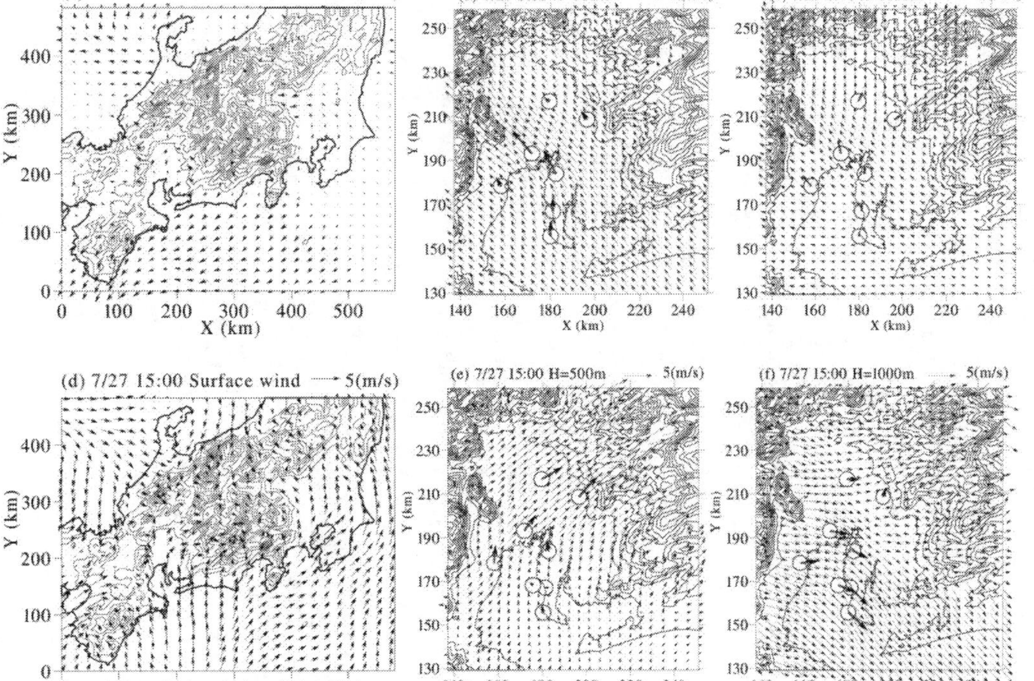

Fig. 4. Computed wind fields : (a) surface wind (at 10m above ground), (b) and (c) upper winds at 500mand 1000 m high, respectively, at 00 JST 27 July. Fig. 4 d, e, f are the same as Figs. 4 a, b,c but for 15 JST 27 July.

"Q" in Fig. 2b, of computed O_3 concentration (case 741rxvv, i.e. BASE case in Table 2)

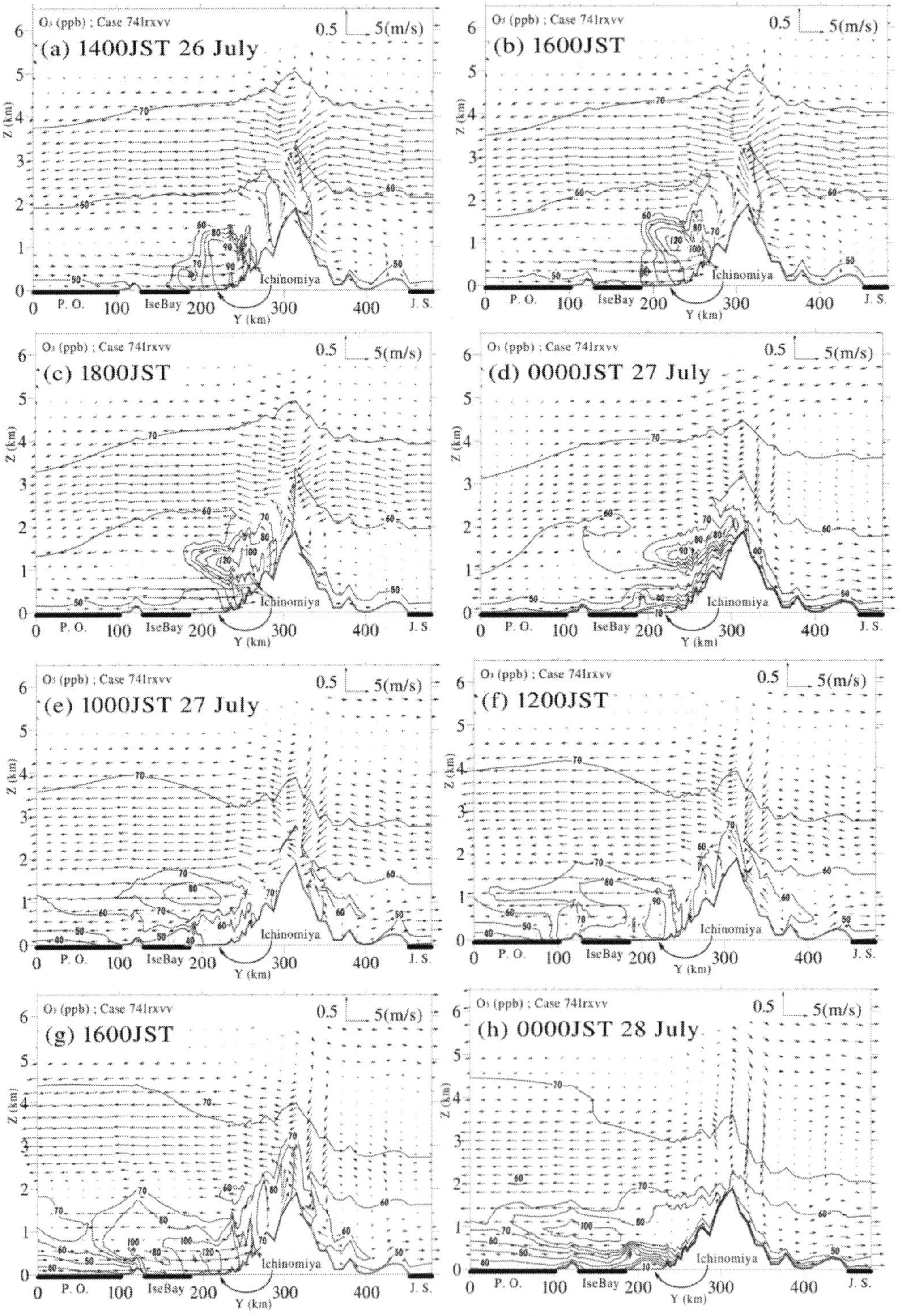

Fig. 5. Vertical cross sections of the computed O_3 (BASE; see Table 2) at x = 175 km (symbol "Q" in Fig. 2b) at (a) 14, (b) 16, (c) 18 JST on 26 July, (d) 00, (e) 10, (f) 12, (g) 16 JST 27 on July, and (h) 00 JST on 28 July.

at 14, 16, 18 JST (26 July), 00, 10, 12, 16 JST(27 July), and 00 JST(28 July), respectively. These figures illustrate the followings: (1) during the daytime O_3 concentration continues to increase near surface level in the pollutant rich air mass, originated from the coastal area and advancing inland with sea breeze and valley wind (Figs. 5a,b,c), (2) high O_3 concentration zone is formed in upper layer at around 1 km high (Fig. 5b,c) over the Nohbi Plain; the high O_3 zone is located near top of the mixed layer developed before arrival of sea breeze and in between the layers of sea breeze/valley wind and its return flow; formation of the zone is attributable to the transport of ozone and its precursors from lower level to the upper layer by both strong thermal convection before the sea breeze arrives and strong updraft when the sea breeze front passes by; these characteristics in the concentration field modified by the passage of sea breeze front agree well with the idealized 2-dimensional sea/land breeze case described in Kitada and Kitagawa (1990), (3) the high O_3 concentration zone in the upper layer at around 1.3 to 1.5km high remains during the nighttime (Fig. 5d), the layer which extends by 100km from the mountainous area to the Nohbi Plain, while O_3 concentration at the surface level is very low over the Nohbi Plain mainly due to NO_x emissions from urban area, (4) the O_3-rich air mass in the upper layer starts to move to the south over the Nohbi Plain-Ise Bay area in prevailing northerly wind in the morning on 27 July (Figs. 5e,f), while the primary pollutants-rich air mass in the lower layer over the plain area also moves to the south over Ise Bay area because of mountain wind and land breeze in the early morning; these pollutants over Ise Bay react to produce O_3 in lower layer in the morning, and (5) at noon (Fig. 5f), the developing mixed layer over the plain area entrains O_3 and probably other chemical species in the upper layer, the pollutants which were formed and emitted in the previous day; thus the background O_3 concentrations in the layer below around 1 km high over the Nohbi Plain show higher values on 27 July (Figs. 5g,h) than those on 26 July (Figs. 5b,d), demonstrating accumulation of the pollutants in the air mass extending over bay, plain, and mountain areas, when high pressure system continues for several days and a diurnal cycle of local winds is repeated.

c. Vertical profiles of O_3 —observation and calculation—

For two days of 26 and 27 July, airborne observation of O_3, NO_x etc. was carried out three times a day. In each flight, vertical profiles of the chemical species were obtained over Ise Bay and over Ichinomiya in the Nohbi Plain; the places are marked with asterisk in Fig. 2b. Figure 6a shows O_3 profiles over Ise Bay at around 09 JST 27 July; both the observed (thin solid line) and the computed (from the BASE case: square, circle, and triangle for 08, 10, and 12 JST, respectively). Observation (thin solid line) in Fig. 6a shows characteristic structure of clear double maxima at 200 m and 1200 m high. Although there is time delay, the computed O_3 profiles also show the same structure, indicating the simulation captures the characteristic behavior of O_3. Origin of the double maxima structure can be made clear by comparing case 741rxvv (BASE) with case 742rxvv (Table 2), where all the emission sources are turned off after 00 JST 27 July. The O_3 profiles in case 742rxvv (Fig. 6b) indicate that the maximum O_3 value in the lay and difference in O_3 level between the observed and the computed profile lower layer is substantially small compared with the BASE case (Fig. 6a), while the maximum in the

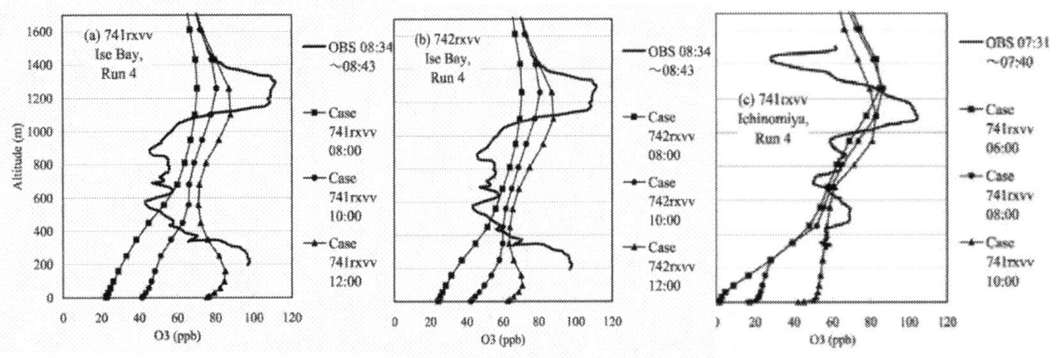

Fig. 6. Vertical profiles of O_3 observed and calculated over (a) and (b) Ise Bay and (c) Ichinomiya. The location of Ichinomiya is marked with "Q" and "*" in Fig. 2b. The observations were carried out from 0730 to 0900 JST 27 July.

upper layer at 1200 m is almost same as in Fig. 6a. This suggests that the maximum in the upper layer consisted of the O_3 produced on the previous day over the Nohbi Plain and mountainous area and transported back by the northerly or northwesterly wind blowing at the 1200 m level in the morning, while the another maximum in the lower layer was mainly caused by the photochemical production from the fresh precursors discharged in the rim area of Ise Bay and carried over Ise Bay by northerly and/or westerly land breeze in the morning of 27 July.

Figure 6c is the same as Fig.6a but for Ichinomiya (Fig. 2b) in the Nohbi Plain at around 08 JST. The computed O_3 profiles (Fig. 6c) also reasonably well simulate the observation. The calculated profiles at 06 and 08 JST show low concentrations near

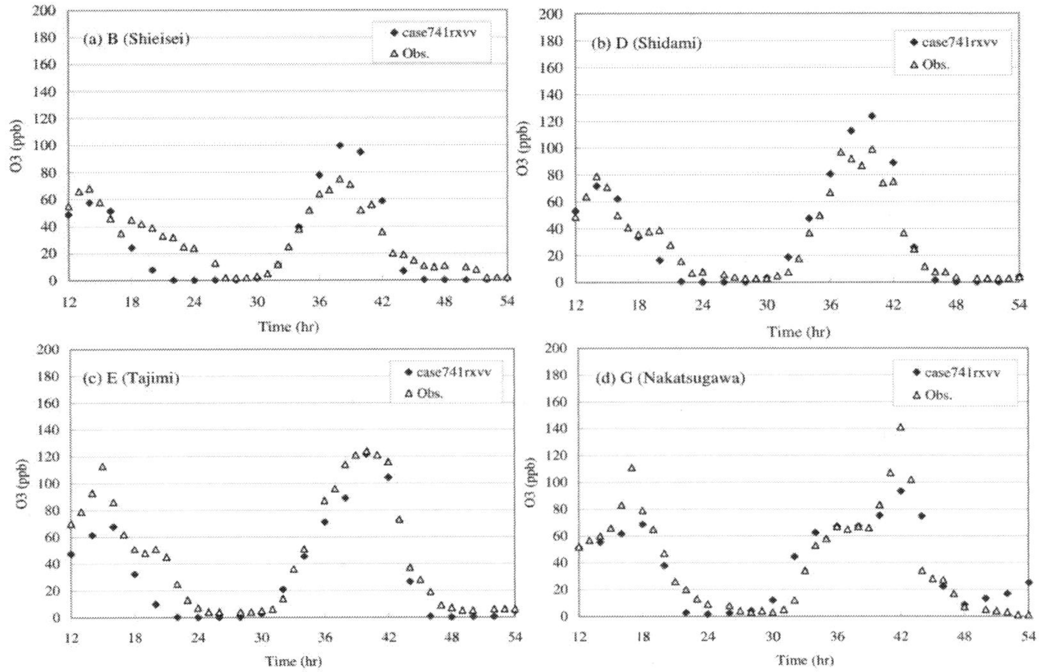

Fig. 7. Comparison between observed and calculated O_3 concentrations at sites B, D, G, and E (case 741rxvv, i.e. BASE). The "12 hr" means 12 JST on 26 July

ground level because NO_x sources are strong and the scavenging of O_3 is faster than its photochemical production during the time.

d. Tracking of O_3 transport along a major river

Local winds tend to develop along river and valley. Thus transport of air pollution also likely occurs along the river. Figure 7 shows temporal variations of both observed and computed O_3 (BASE in Table 2) at the sites of B, D, E, and G (Fig. 2b), which are located along the Shonai River and the Kiso River.

On the 2nd day (27 July) starting from 24 hr in Fig. 7, the observed maximum O_3 concentrations at these sites show clear time delay with the earliest maximum occurring at 14 JST at site "B" near the coast and the latest maximum being at 18 JST at the most inland site "G" (Fig. 2b). This time delay and the sharp peak concentration in the late afternoon at site "G" demonstrate the ozone came along the rivers with polluted air mass from the coastal area. Figure 7 also shows the "BASE" case well simulates these characteristic O_3 pattern at each site.

Temporal variation of the vertical O_3 profile (BASE) at site "G" (Fig. 8) suggests that the O_3 - rich air mass first arrived at the site at 18 JST, then the polluted air mass continued to fill space in the valley until mountain wind and plateau-to-plain wind started to blow around 02 - 04 JST (Figs. 8b, c); the high O_3 air filled the valley up to the ridge height surrounding site G, i.e. around 1200 m in altitude, as indicated with arrow in Fig. 8, and (3) the surface O_3 concentration peaked at 18 JST and decreased

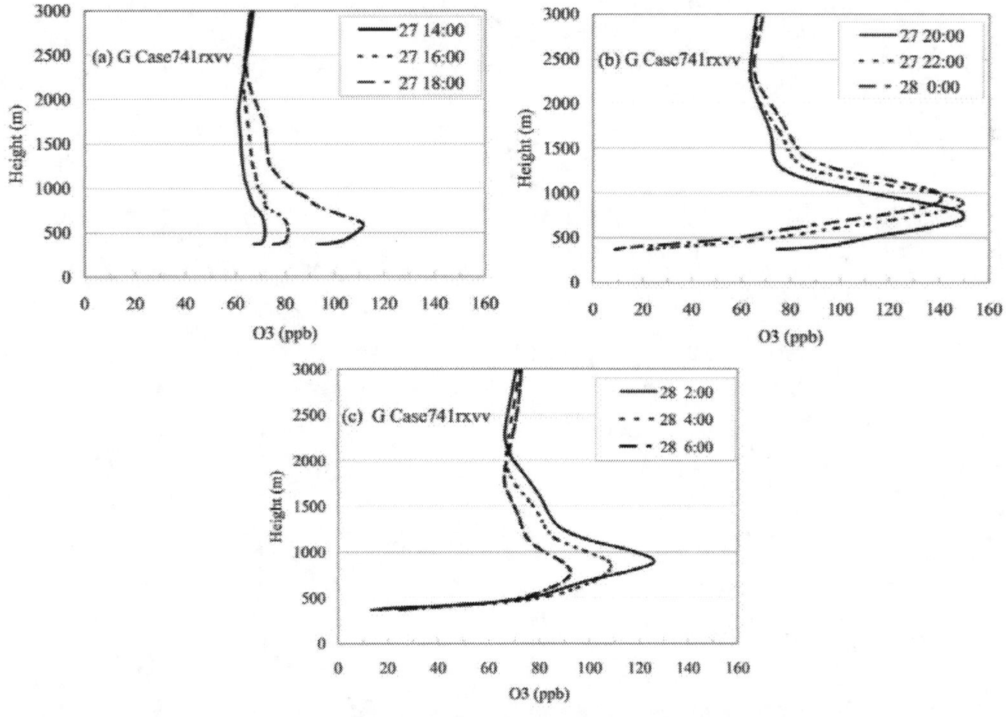

Fig. 8. Computed vertical profile of O_3 at site "G" (BASE case). (a) 14, 16, and 18 JST 27 July, (b) 20, 22 JST 27 July, and 00 JST 28 July, and (c) 2, 4, and 6 JST 28 July. Arrows show the height of ridge surrounding the valley in which site "G" locates.

after that due to destruction within stably stratified surface layer formed by radiational cooling.

e. Effect of biogenic hydrocarbons on O_3 concentration

Quantitative understanding of the effect of biogenic hydrocarbons on O_3 production etc. is important for better prediction of performance of designed emission control on anthropogenic sources. The effect was evaluated by examining cases with and without the natural source (see Table 2).

The "BASE" emission source distributions of isoprene and α-pinene were estimated

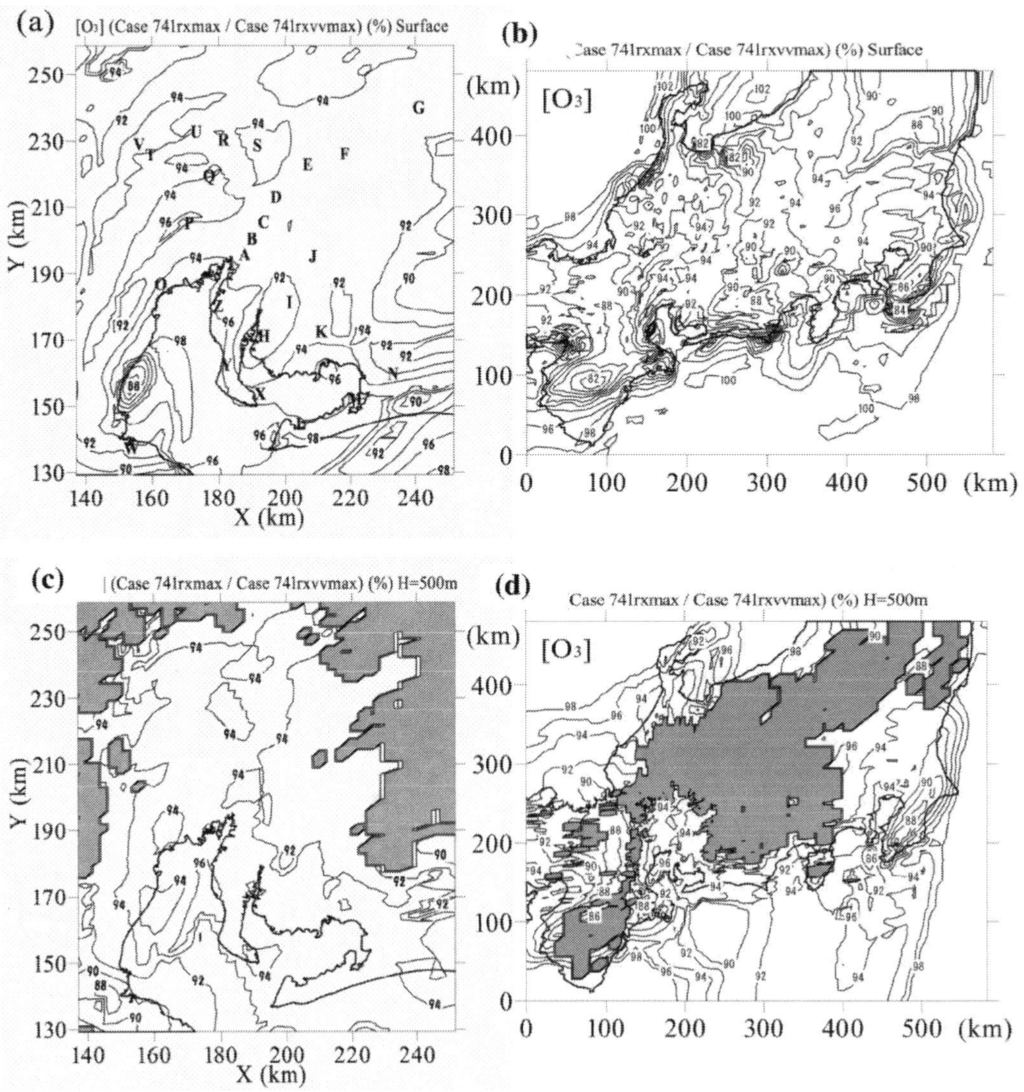

Fig. 10. Horizontal distribution of the ratio of max O_3 in case 741rx to that in BASE (case 741rxvv) on 27 July, 1995: (a) surface level, and (c) 500 m AMSL, and (e) 1200 m ASL in the Nohbi Plain-Ise Bay area, and (b), and (d) are the same but for the central Japan.

using BEIS and a map of vegetation type, as described in the previous subsection (see Table 1). Because of its nature isoprene is emitted only during the daytime, while α-pinene is discharged also during the nighttime.

Figure 9 compares O_3 in the BASE case (case 741rxvv, dashed line) with that in case 741rx (solid line), i.e. without biogenic emissions. Increase of maximum O_3 concentration due to bio-NMHC emissions on the 2nd day, i.e. 27 July, is about 10 ppb at all sites of B, D, E, and G (see Fig. 2b), and thus does not show significant difference among the sites which are aligned along one of the major directions of mass transport in the Nohbi Plain.

Figure 10 illustrates horizontal distribution of the percentage of maximum O_3 in case 741rx (no bio-NMHC sources) to that in case 741rxvv (BASE; with bio-NMHCs sources) at the surface level (Figs. 10a, b), and at the height of 500m AMSL (Figs. 10c, d) on 27 July 1995, showing spatial variation of relative importance of biogenic hydrocarbons on O_3 production; 100% minus contour value in Fig. 10 indicates contribution of bio-NMHC to the maximum O_3 concentration in percentage.

In the Nohbi Plain (Fig. 10a) relatively uniform and small contribution, to the maximum ozone, of bio-NMHC between 6% (=100-94) to 8% (100-92) is predicted, and this means that both NOx and ant-NMHCs are supplied from the coastal emission area enough to produce O_3, and neither an excess-NOx condition nor shortage-of-NMHCs situation does occur during the daytime transport. A similar observation is also possible in the Kanto Plain, Tokyo area at around x=450 km and y=300 km in Fig. 10b. On the other hand, an interesting contrast can be seen in near Osaka area at x=50km and y=150km in Fig. 10b; that is, Fig. 10b shows the maximum ozone concentration in Osaka could be reduced by more than 10 % without biogenic NMHCs emission, and the area widely spreads over the down-sea-breeze region of the anthropogenic emissions in the Osaka Plain, where the sea breeze blows from west to east. The Osaka Plain is smaller than the Kanto and the Nohbi Plains, i.e. distance from coast to mountainous area is much shorter. Thus the polluted air mass from the Osaka Plain reaches the mountains (the Kii Peninsula) in the early afternoon, and biogenic NMHCs in the mountains could more effectively contribute to the ozone production.

Generally speaking, effect of biogenic NMHCs on O_3 production is large when the air mass from urban area is mixed with the biogenic NMHCs earlier in the daytime, and, of course, higher the concentrations of the biogenic NMHCs are, larger their effects on ozone production are. High contributions in the Bohsoh Peninsula at x=450 km and y=200km may be due to pre-loading of bio-NMHCs over coastal mountains into the sea breeze before the breeze passing over anthropogenic emission area. Another high zone at x=250km and y=180km is also due to the pre-loading from high bio-NMHCs emissions in this area where ant-NMHCs emissions are relatively weak. Figure 10b also shows, in most of the central mountainous area, bio-NMHCs contribute to daily maximum O_3 concentration by 6 - 8 %.

The patterns at 500m AMSL (Figs. 10c,d) are almost similar to those at the surface; it is suggested that high contribution of the bio-NMHCs at x=250km over the coast of Pacific Ocean (Fig. 10b) propagated toward the Pacific Ocean (Fig. 10d) due to the effect of synoptic scale flow on the day, blowing southward to south-westward in upper layer.

The bio-NMHCs emitted in mountainous area have to be mixed with the anthropogenic NOx discharged over the coastal urban areas so as to produce O_3, and such a contact of those chemical species at this height is only possible at high mountains relatively close to coastal urban area.

Figures 11a,b,c show the maximum ozone concentration on 27 July in BASE case: (a) surface, (b) 500m ASL, and (c) 1200m ASL. We did not compare the calculated ozone concentration with the observed in the Knato Plain (Tokyo area) and the Osaka Plain since in these areas the anthropogenic emission sources evaluated from $1^\circ \times 1^\circ$

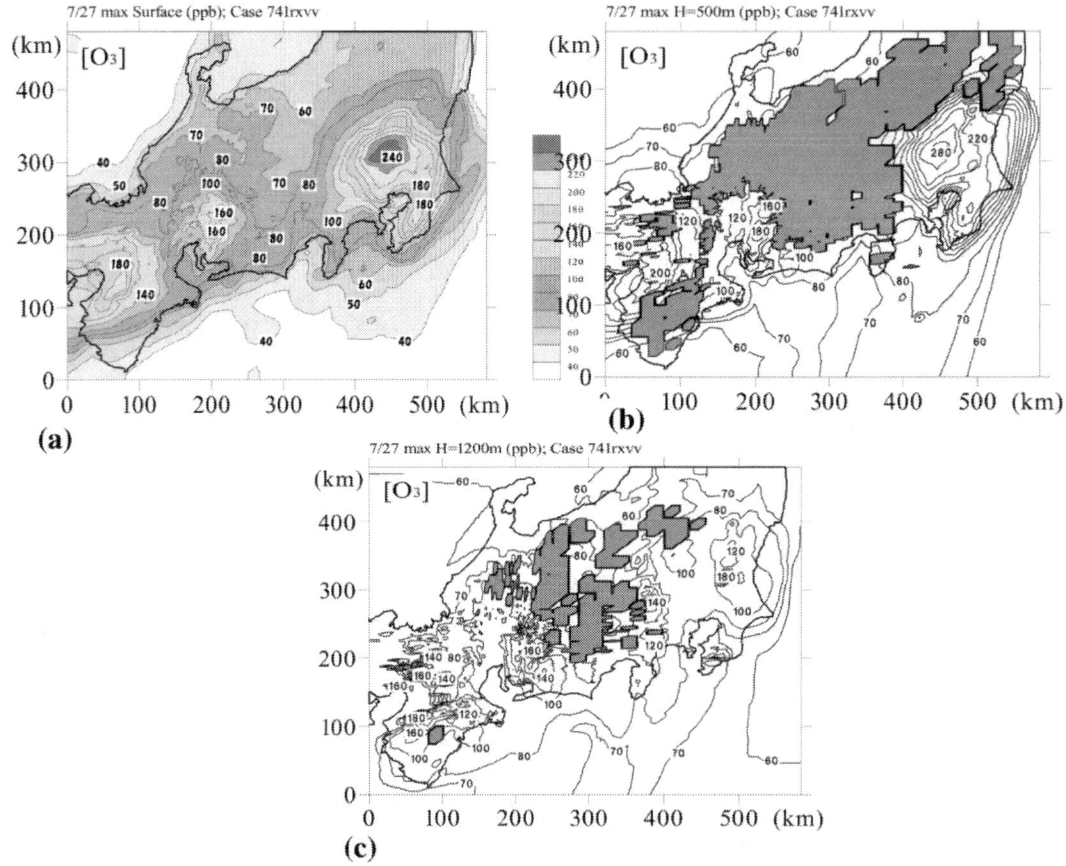

Fig. 11. Distributions of calculated $(O_3)_{max}$ on 27 July at (a) surface, (b) 500m ASL, and (c) 100m ASL in BASE case

global distribution (Akimoto and Narita, 1994) are used and they do not give fine resolution. However from these figures some observations may be possible: first, high O_3 concentration air over 100ppb widely extends in the main coastal urban areas of the Pacific Ocean side, maximum ozone concentration exceeding 100ppb over most of the plains; secondly, the ozone concentrations over plain areas show their maximum values at the middle of mixing layer, as many previous researches point out; thirdly, Fig. 11c suggests that at 1200m ASL the highest ozone concentrations appear over mountainous areas in the Kii Peninsula, east and north of the Nohbi Plain, and west and north of the Kanto Plain.

f. Sensitivity of O_3 concentration to anthropogenic emission sources of NMHCs and NOx

Sensitivity of the maximum O_3 concentration to ant-NMHCs and NOx emission in the Nohbi Plain was investigated using results of cases 701rxvv, 741rxvv, 751rxw, and 711rxvv in Table 2.

Figure 12 plots computed maximum O_3 concentration (on 27 July) at sites B, D, E, and G (Fig. 1b) to ant-NMHCs emissions. These Figs. 12a, b, c show more or less that the maximum O_3 concentration at each site is sensitive to anthropogenic NMHCs emission. For example, at site B in Nagoya a 20% reduction of ant-NMHCs emissions resulted in by 15 ppb cut of the maximum O_3 concentration; similarly, by 14 and 13 ppb at site D and E, respectively. In contrast, at the mountain site G, only small decrease in its maximum O_3, i.e. 2 ppb, was predicted. These again suggest that ant-NMHCs emissions limit O_3 production in the Nohbi Plain, but not in the mountainous area.

Open circle in Fig. 12 shows case 741rxvvh (Table 2); i.e., biogenic NMHCs emissions were cut by 50 % from the BASE case, and it suggests only negligible reduction in max O_3 can occur with halved bio-NMHC emissions.

Both of the comparison of O_3 between cases with and without bio-NMHCs in Fig. 9 and the result in Fig 12 showing almost no change of max O_3 demonstrate highly non-linear nature of the role of bio-NMHC in O_3 chemistry (McKeen et al., 1991, Roselle et al., 1991 etc.). This suggests that (1) at sites B, D, and E, which are located in either central or northeastern part of the Nohbi Plain, NO_x was sufficiently fed for

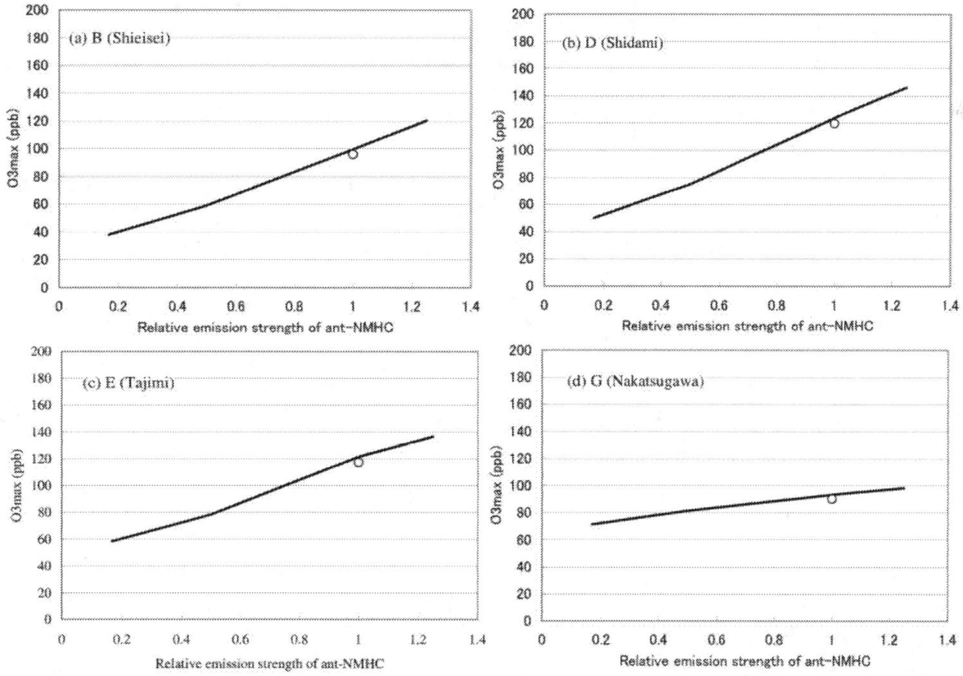

Fig. 12 Sensitivity of maximum O_3 (27 July) to anthropogenic NMHCs emission ; results of cases 701rxvv, 741rxvv, 751rxvv, and 711rxvv are used. ant-NMHCs emission is normalized by that in case 741rxvv. Open circle denotes the case 741rxvvh.

photochemical O_3 production, but ant-NMHC were not, i.e. hydrocarbon-limiting situation, and (2) the site G is in mountainous region (Fig. 2b), amount of ant-NMHC was adequate compared with NO_x, i.e NO_x-limiting situation.

Figure 13 illustrates sensitivity of O_3 concentration at the sites of B, D, E, and G (Fig.

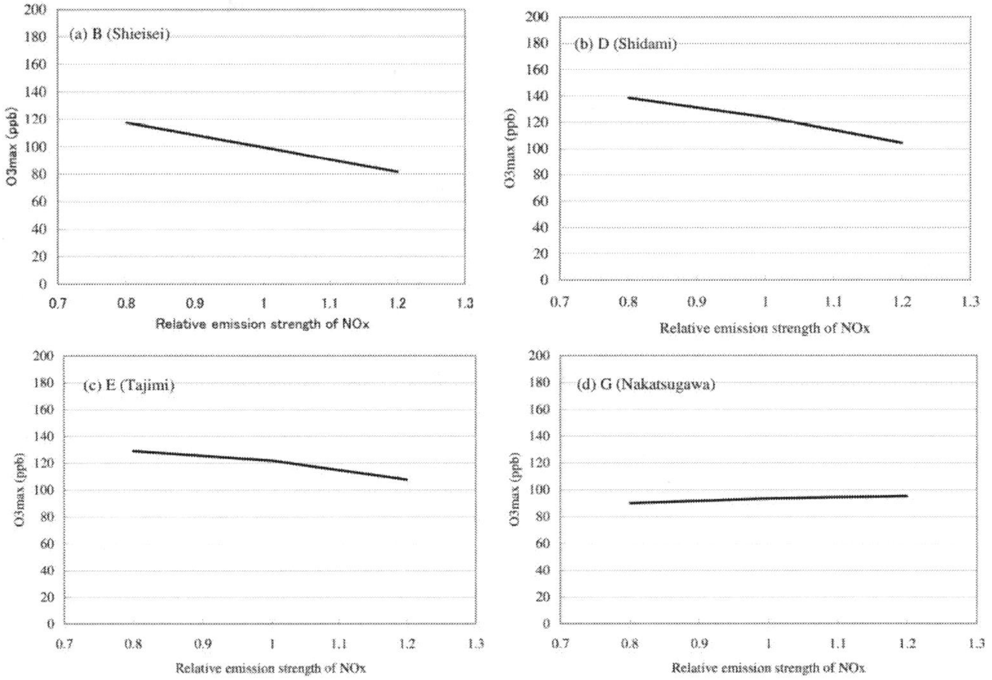

Fig. 13. Sensitivity of maximum O_3 concentration, on 27 July, to NO_x emission : the results of cases 741rxvv, 741rxnoinvv, and 741rxnovv (Table 2) are used. NO_x emission is normalized by that in case 741rxvv (BASE).

2b) to NOx emission in the calculation domain. Figures 13a,b,c suggest ozone production at the sites of B, D, and E in the urban area is in the excess-of-NOx condition; thus decrease in NOx emission is leading to increase in O_3 concentration. In contrast to these urban sites, ozone at the site G in mountainous rural area (see Fig. 2b) shows a slight increase with increased NOx emission, indicating ozone production in the area of "G" is rather in the excess-of-hydrocarbon condition. Abundant biogenic hydrocarbons in the mountain area explain this situation. In summary, the urban area including Nagoya as a central mega-city in the Nohbi Plain is in the excess-of-NOx situation, and extent of the "excess" gradually decreases in the suburbs. Finally in rural mountainous area, the situation changes to the excess-of-hydrocarbon condition.

g Calculated high concentrations of photochemical reaction products in mountainous area during nighttime

As can be expected from the topography in the domain (Fig. 2a), local flows such as sea breeze, valley wind, and plain-to-plateau wind tend to converge to the central mountainous area during daytime. Thus various photochemical products may form high concentration zone at high altitudes, e.g. Fig. 5d.

Figure 14 shows (a) O_3, (b) SO_4^{2-}, and (c) $HNO_3 + NO_3^-$ concentrations at 1500 m high on 00 JST 27 July (BASE case). Two high concentration zones can be seen; one is Chichibu, Tanzawa, and Izu area to the west of Tokyo. The other is Hida and Takayama area to the north of Nagoya. These pollutants are products of photochemical smog reactions and are originated from the huge urban areas of Tokyo and Nagoya. Mountains in Chichibu and Tanzawa have already suffered the damage to vegetation by acidic fog (Igawa et al., 1998).

In Hida area to the north of Nagoya, such an effect on the vegetation has not yet been clarified. However, Fig. 5 strongly suggests the same damage as Chichibu and Tanzawa may be taking place also in the Hida area. Statistical study on the local flows in central Japan indicates that the meteorological condition leading to the pollutants' transport

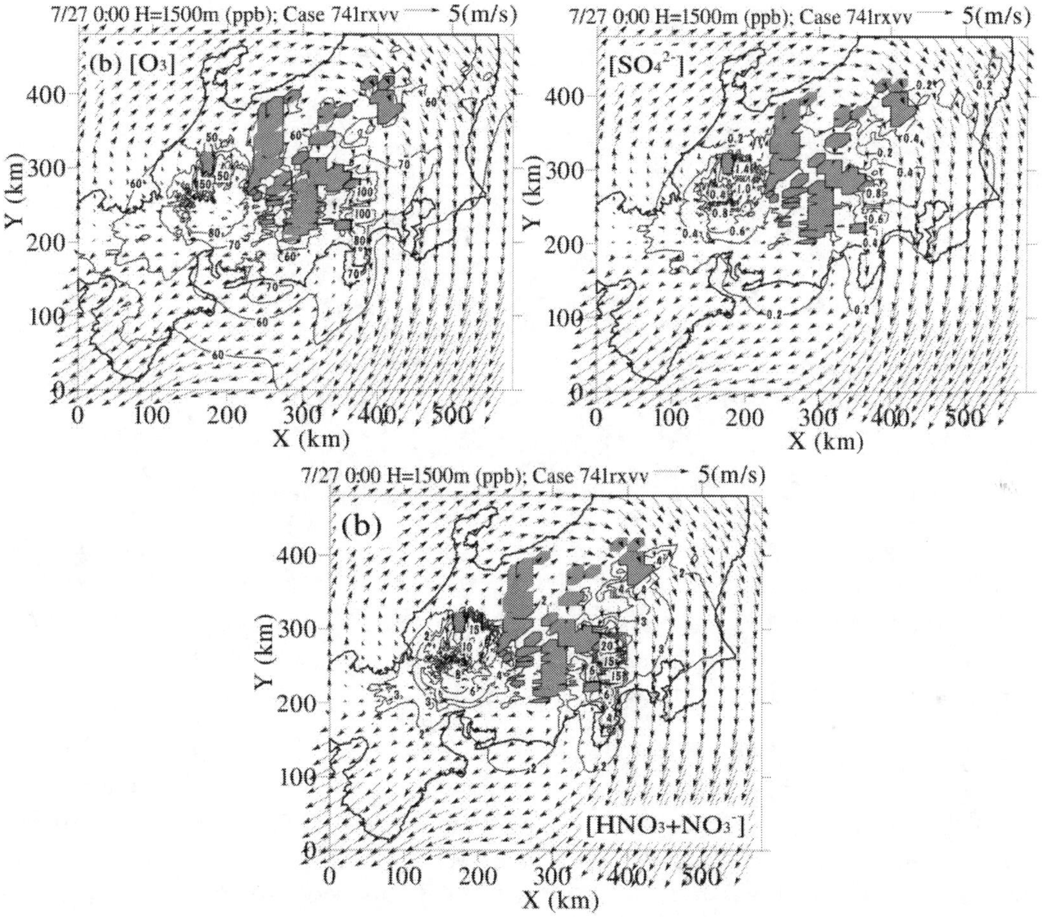

Fig. 14. Computed horizontal cross sections of (a) O_3, (b) SO_4^{2-}, (c) $HNO_3 + NO_3^-$ at 1500 m high on 00 JST 27 July (case 741rxvv, i.e. BASE).

found in Figs. 5 and 14 occurs quite frequently such as for about 80 days during warm season from April to October (Mori et al., 1994).

Figure 15 also shows (a) O_3, (b) SO_4^{2-}, (c) $HNO_3 + NO_3^-$ concentrations for case 741rx, i.e. without biogenic hydrocarbons. Comparison between Figs. 14 and 15 demonstrates effects of biogenic hydrocarbons on photochemical products in remote area : (1) O_3 production can be enhanced by the biogenic hydrocarbons and the concentrations are increased by 20 ppb in Chichibu, Tanzawa, and Izu areas near Tokyo,

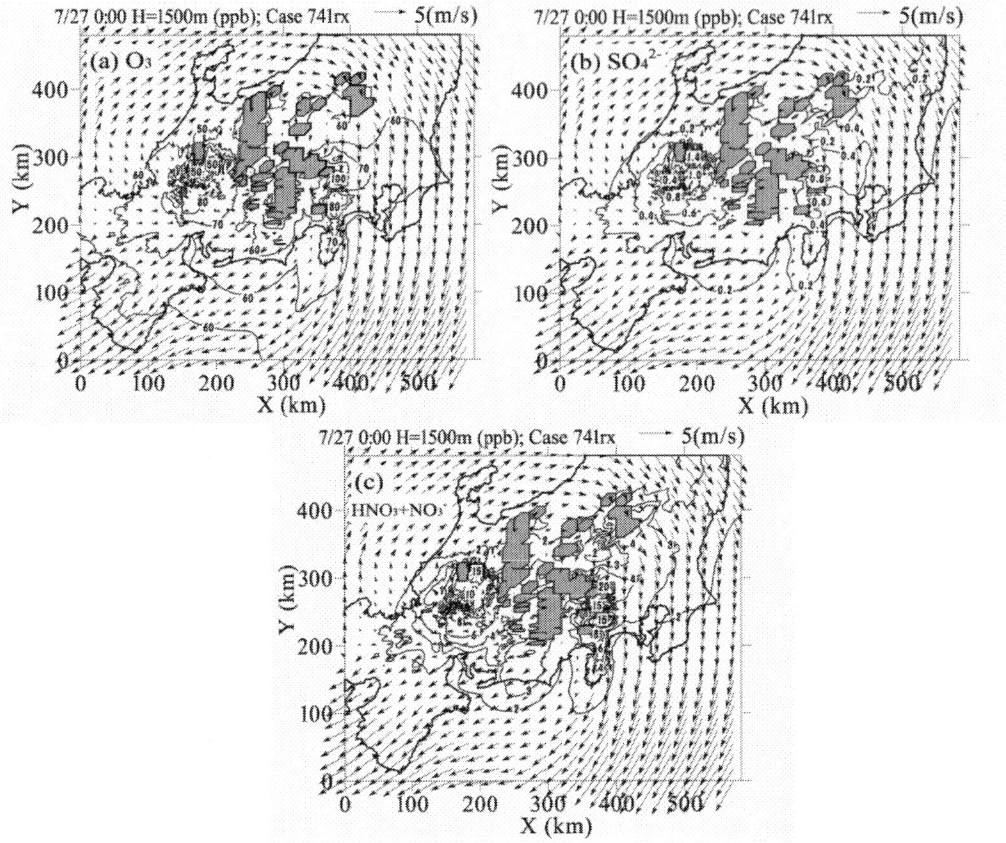

Fig. 15. Same as Fig. 14 but for case 741rx, i.e. without biogenic hydrocarbons.

and by 10 ppb in Hida and Takayama areas near Nagoya ; (2) sulfate concentrations are also increased by about 0.2 ppb; and (3) nitric acid and nitrate are drastically decreased by 5 - 10 ppb, and these lost nitric acid and nitrate are turned to biogenic nitrate.

5. SUMMARY AND CONCLUSIONS

To see the multiple days' behavior of air pollutants released in coastal urban areas in Japan, two days numerical simulations of meso-scale flow and air pollution transport were performed from 06 JST 26 July to 06 JST 28 July 1995 for central Japan. High pressure system, with light gradient wind and sunny sky, was dominant over the region. Thus various types of local flows developed during the simulated time. Furthermore, in the Nohbi Plain 3 dimensional observations using aircraft, pilot balloon; and rawinsonde were performed for the same time. These observed data were utilized in simulations as inputs for nudging computed wind, and for evaluation of computed concentration profiles.

Analysis was focused on (1) clarifying two days' behavior O_3 in developed local wind system. (2) tracking polluted air mass along a river-valley topography. (3)

quantifying effects of biogenic NMHCs on O_3, SO_4^{2-}, and nitrate productions. (4) sensitivity of maximum O_3 concentration to emissions of ant-NMHCs and NO_x.

Part of the obtained results is as follows : (1) Pollutants released in coastal urban area are transported, with being chemically transformed, to mountainous area by midnight, and forms high concentration zone of O_3, SO_4^{2-}, HNO_3 etc. at around 1500 m in altitude. Then, on the following day these pollutants are backed to the upper layer over the coastal area. (2) In summer season over the Nohbi Plain, air mass may be in (non-methane-) hydrocarbon limiting situation on O_3 production, and maximum O_3 concentration was quite insensitive to increase/decrease of NO_x emission by 20 %. (3) Inclusion of biogenic hydrocarbon emission into simulation very much enhanced O_3 and SO_4^{2-} concentrations on mountainside at 1500 m high, but decreased HNO_3 and NO_3^- by providing NO_x pathway to biogenic nitrate.

ACKNOWLEDGMENT

This work was supported in part by the Ministry of Education, Culture, and Science, Japan through Grant-in-Aid for Scientific Research (B), No. 17360256 and Global Environment Research Fund, No. C-051 by Ministry of the Environment, Japan.

REFERENCES

Aichi Prefectural Office, Japan, 1995, Data compiled by Division for Environment.

Akimoto, H., and Narita, H., 1994, Distribution of SO_2, NO_x and CO_2 emissions from fuel combustion and industrial activities in Asia with $1^\circ \times 1^\circ$ resolution. *Atmos. Environ.*, 28:213.

GEIA, 1998, Emission source distribution of biogenic hydrocarbons. Web site http://blueskies.sprl.umich.edu/geia/

Gery, M. W., Whitten, G. Z., Killus, J. P., and Dodge, M. C., 1989, A photochemical kinetics mechanism for urban and regional scale computer modeling. *J. Geophys. Res.*, 94:12925.

Guenther, A., Hewitt, C. N., Erickson, D., Fall, R., Geron, C., Graedel, T., Harley, P., Klinger, L., McKay, W. A., Pierce, T., Scholes, B., Steinbrecher, R., Tallamraju, R., Taylor, J., and Zimmerman, P., 1995, A global model of natural volatile organic compound emissions. *J. Geophys. Res.*, 100:8873.

Igawa, M., Tsutsumi, Y., Mori, T., and Okochi, H., 1998, Fogwater chemistry at a mountainside forest and the estimation of the air pollutant deposition via fog droplets based on th atmospheric quality at the mountain base. *Environ. Sci. Technol.*, 32:1566.

Kitada, T., Carmichael, G.R. and Peters, L.K., 1986, Effects of dry deposition on the concentration-distributions of atmospheric pollutants within land- and sea-breeze circulations. *Atmos. Environ.*, 20:10, 1999-2010.

Kitada, T., Lee, P.C.-S., and Ueda, H., 1993, Numerical modeling of long-range transport of acidic species in association with meso- β -convective clouds across the Japan sea resulting in acid snow over coastal Japan -I. Model description and qualitative verifications. *Atmos. Environ.*, 27A:1061.

Kitada, T. Nagano, M., and Arai, K., 2008, Ozone in mountainous central Japan on the Pacific Ocean side: seasonal variation and characteristic meso-scale transport, *Proc. 15th Seminar of JSPS-MOE Core University Program on Urban Environment, 157-165.*

Kitada, T., Okamura, K., Nakanishi, H., and Mori, H., 2000, Production and transport of ozone in local flows over central Japan –comparison of numerical calculation with airborne observation-. *Air Pollution Modeling and Its Application XIII*, Kluwer Academic/Plenum Pub. Co., 95-105.

Kitada, T., Okamura, K., and Tanaka, S., 1998, Effects of topography and urbanization on local winds and thermal environment in Nohbi Plain, coastal region of central Japan -A numerical analysis by meso-scale meteorological model with k- ε turbulence model. *J. Appl. Meteor.*, 37:1026-1046.

Kurita, H., Sasaki, K., Muroga, H., Ueda, H., and Wakamatsu, S., 1985, Long-range transport of air pollution under light gradient wind conditions. *J. Climate Appl. Meteorol.*, 24:133.

Lloyd, A.C., Atkinson, R., Lurmann, F.W., and Nitta, B., 1983, Modeling potential ozone impacts from natural hydrocarbons - I. Development and testing of a chemical mechanism for the NO_x-air photooxidations of isoprene and α -pinene under ambient conditions. *Atmos. Environ.*, 17:1931-1950.

Lurmann, F.W., Lloyd, A.C., and Atkinson, R., 1986, A chemical mechanism for use in long-range transport/acid deposition computer modeling. *J. Geophys. Res.*, 91:10905.

McKeen, S. A., Hsie, E. -Y., and Liu, S. C., 1991, A study of the dependence of rural ozone on ozone precursors in the Eastern United States. *J. Geophys. Res.*, 96:15,337.

Mori, H., Kitada, T., and Iyoda, K., 1998, Vertical structure of land and sea breezes in the Nohbi plain during a passage of an anticyclone over the central Japan. *Tenki*, Meteorol. Soc. Japan, 45:515 (in Japanese).

Mori, H., Ogawa, H., and Kitada, T., 1994, Characteristics of land and sea breezes in the Nobi plain, and the conditions of occurrence of the "extended sea breeze". *Tenki*, Meteorol. Soc. Japan, 41:379 (in Japanese).

Nakanishi, H., 1996, Emission sources and concentrations of non-methane hydrocarbons over the Nohbi plain in central Japan. *Toyohashi Univ. Technology, B.S. thesis*, 42 (in Japanese).

NASA Panel for Data Evaluation, 1997, Chemical Kinetics and Photochemical Data for Use in Stratospheric Modeling, Evaluation No. 12, NASA JPL.

Shawn, J. R., Thomas, E. P., and Kenneth, L. S., 1991, The sensitivity of regional ozone modeling to biogenic hydrocarbons. *J. Geophys. Res.*, 96:7371.

Sheih, C.M., Wesely, M.L., and Hicks, B.B., 1979, Estimated dry deposition velocities of sulfur over the eastern United States and surrounding regions. *Atmos. Environ.*, 12, 2055-2087.

Slinn, W.G.N., 1977, Some approximations for the wet and dry removal of particles and gases from the atmosphere. *Water, Air, and Soil Poll.*, 7:513.

Wesely, M.L., and Hicks, B.B., 1977, Some factors that affect the deposition rates of sulfur dioxide and similar gases on vegetation. *J. Geophys. Res.*, 100:11,447.

Yokouchi, Y., 1994, Seasonal and diurnal variation of isoprene and its reaction products in a semi-rural area. *Atmos. Environ.*, 28:2651.

180b The Fate and Transport of Ammonia at the Local to Regional Level

Robin L. Dennis[1], Rohit Mathur[1], Jon Pleim[1], and John T. Walker[2]. (1) Atmospheric Modeling Division, U.S. Environmental Protection Agency, Mail Drop E243-04, Research Triangle Park, NC 27711, (2) National Risk Management Research LAboratory, U.S. Environmental Protection Agency, Mail Drop E305-02, Research Triangle Park, NC 27711

Atmospheric deposition of nitrogen (N) is a major contributor to eutrophication of estuarine and coastal waters. It is estimated to contribute 20-35% of the N load to east coast and gulf coast estuaries in the United States. Most of the N loading is contributed indirectly by first depositing on the watershed. Currently, oxidized-N from nitrogen oxide (NOx) emissions from combustion is the major source of N in total (wet + dry) atmospheric deposition. With the anticipated control of NOx emissions, to reduce ozone, model projections suggest that reduced-N from ammonia (NH3) emissions will become the major source of atmospheric N deposition in the future. Agriculture is by far the largest contributor to ammonia emissions. In some disciplinary areas the conventional wisdom is that all of the NH3 emissions deposit locally, an idea that is not consistent with what we know in the atmospheric community. Initial ammonia budget studies of an emissions hot spot and an emissions region with a regional atmospheric model, EPA's Community Multi-scale Air Quality Model (CMAQ) show that while local deposition is significant in the vicinity of NH3 emission regions, a majority of the emissions loft up into the atmosphere away from the surface, are converted to aerosol ammonium and subsequently subject to long-range transport. Close to the source dry deposition dominates and farther from the source wet deposition dominates. Additional budget studies are developed to address how uncertainty in the NH3 dry deposition rate affects the uncertainty in the long-range transport and regional influence of ammonia emissions. A major source of uncertainty related to ammonia deposition is the empirical observation that ammonia flux is really bi-directional. A preliminary bi-directional parameterization has been included in CMAQ. One of the budget studies examines the impact of this bi-directional formulation of ammonia air-surface exchange on long-range transport of ammonia emissions.

180c Development, Testing, and Applications of An Integrated on-Line Meteorology Atmospheric Chemistry Modeling System

Rohit Mathur[1], Jonathan Pleim[1], David Wong[1], Tanya Otte[1], Robert Gilliam[1], Shawn Roselle[1], Jeffrey Young[1], Frank Binkowski[2], and Aijun Xiu[2]. (1) National Exposure Research Laboratory, U.S. EPA, 109 T.W. Alexander Drive, U.S. EPA MD 243-03, Rtp, NC 27711, (2) University of North Carolina, 137 E. Franklin Street, Campus Box 6116, Chapel Hill, NC 27599

Traditionally, atmospheric chemistry-transport and meteorology models have been applied in an off-line paradigm, in which archived output on the dynamical state of the atmosphere simulated using the meteorology model is used to drive transport and chemistry calculations of atmospheric chemistry transport model (CTM). A modeling framework that facilitates coupled on-line calculations is desirable since it (1) provides consistent treatment of dynamical processes and reduces redundant calculations, (2) provides ability to couple dynamical and chemical calculations at finer time-steps and thus facilitates consistent use of data, (3) reduces the disk-storage requirements typically associated with off-line applications, and (4) provides opportunities to represent and assess the potentially important radiative effects of pollutant loading on simulated dynamical features. A coupled on-line atmospheric modeling system is developed based on the Weather Research and Forecasting (WRF) meteorological model and the Community Multiscale Air Quality (CMAQ) air quality modeling system. The flexible design of the system which facilitates configurations for the both on-line and off-line modeling paradigms will be described. Initial testing of the system and the impacts of varying frequency of coupling between dynamical and chemical calculations will be presented. The impacts of including direct radiative forcing of simulated aerosol distributions on modeled dynamical and chemical features will be discussed. Initial evaluation of the model against measurements from routine networks and specialized campaigns will be presented.

180d Carbon Emissions from Tropical Fires

Prasad Kasibhatla, Division of Environmental Sciences and Policy, Duke University, Room A355, LSRC, Durham, NC 27708

Tropical savanna and forest fires are important sources of air pollutants to the global atmosphere. In this talk, I focus on our recent efforts to quantify the magnitude and variability of tropical fire-driven atmospheric emissions using remotely sensed datasets of active fires and burned area in combination with a global biogeochemical model. Satellite based measurements of tropospheric carbon monoxide enable us to characterize the accuracy of the resulting inventory. Finally, I will discuss internannual variability in carbon emissions associated with deforestation in the Amazon and Indonesia in recent years, and explore the causes of these variations.

180e Regional and Global Perspectives of Megacity Air Pollution

Gregory R. Carmichael, Chemical and Biochemical Engineering, University of Iowa, CGRER, 424 IATL, University of Iowa, Iowa City, IA 52242

Megacities are leading drivers of economic and environmental change. Fueled by high population growth and vibrant economies, energy consumption in megacities are large and growing. In Asia megacities are projected to account for ~40% of their country's GNP (gross national product) by 2030. Because fossil fuels will provide much of this energy, emissions of greenhouse gases and air pollutants such as sulfur and nitrogen oxides and particulates could dramatically increase. Without strong intervention the situation will inevitably worsen.

The motorization of urban environments all around the world has produced local smog in hundreds of cities. In aggregate, pollution from megacities and surrounding areas can grow to create regional and global problems. The current interest in transboundary and hemispheric transport of pollutants reflect this. The pressing environmental problems of urban pollution and climate change are closely linked megacity problems sharing common causes and solutions. The fact that air pollution problems and greenhouse gas emissions arise largely from fossil fuel combustion and the important role of aerosols in both air pollution and climate change are illustrative examples. Globally many megacities represent atmospheric brown cloud hotspots; regions with large aerosol radiative forcing of the atmosphere and surface (dimming), with annual mean surface dimming in most tropical mega cities exceeding 20 Wm-2, equivalent to reducing solar irradiance at the top of the atmosphere by more than 10%. The increase in solar energy absorbed by aerosols over the atmosphere of some of these mega cities is comparable to the heat input by energy consumption. Thus in addition to contributing to regional and global climate change, the atmospheric forcing may also contribute to the urban heat island effect.

In this paper the impacts of megacities on regional and global pollution are discussed, drawing upon finding from current international activities including MILAGRO, ABC, and GURME.

180f Closing Comments

Leonard K. Peters, Battelle, xxxxxxx, Xxxxx, KY xxxxx

A brief look back and comments regarding future directions.

203a Surface-Bulk Partitioning of Organic Material In Aqueous Aerosols

V. Faye McNeill, Julia Szprengiel, and Michael Giordano. Chemical Engineering, Columbia University, 801 S. W. Mudd, 500 West 120th St., New York, NY 10027

Internal mixtures of inorganic and organic material are common in tropospheric aerosols, with organic matter typically comprising 10 to 90% of fine aerosol mass. One special class of organic material commonly found in atmospheric aerosols is surface-active species, including fatty acids, proteins, and humic-like substances. It has been suggested that surface-active organic molecules partition to the gas-aerosol interface of aqueous atmospheric aerosols in an "inverted micelle" configuration, thus forming an organic film around the aqueous core. Recent laboratory studies suggest that, if organic films are a common feature of aqueous aerosols, this could significantly affect atmospheric composition and climate. However, despite their potential significance, the existence of surface films on aqueous aerosols has not yet been confirmed unambiguously under a wide range of conditions. We will present the results of pendant drop tensiometry and ellipsometry studies designed to identify the range of aerosol compositions for which the inverted micelle morphology may exist. We will focus on classes of surface active organic compounds which are expected to have distinct phase behavior at atmospherically relevant pH, such as short and long-chain fatty acids, diacids, proteinaceous material, and secondary organic aerosol material, as well as mixtures containing multiple organic species. The effects of surface films on aerosol heterogeneous chemistry will also be discussed.

203b Field and Modeling Studies of Atmospheric Nanoparticles

Alicia Pettibone[1], Sang-Rin Lee[2], Jay Raife[2], Yifang Zhu[3], Chad Bailey[4], and **Charles O. Stanier**[5]. (1) Department of Chemical and Biochemical Engineering, University of Iowa, 4122 Seamans Center, Iowa City, IA 52242, (2) Chemical and Biochemical engineering, University of Iowa, Iowa City, IA 52242, (3) Environmental Engineering, Texas A&M, 700 University Blvd. MSC 361, Kingsville, TX 78363, (4) EPA, (5) Http://www.Cbe.Engineering.Uiowa.Edu/faculty/stanier/, University of Iowa, Iowa City, IA 52242

The two main sources of atmospheric ultrafine particles are combustion (particularly from vehicles) and atmospheric new particle formation. Using models to simulate both sources of size resolved aerosol particles, we will present predictions for the balance of these sources in near-road, urban, and rural continental environments. A key input of these models is the number and surface area distribution from on road vehicles. Progress in a critical review of existing literature on size-resolved activity-based emission factors will be presented. Simulation results will be contrasted with measurements from the rural Midwest U.S., from Mexico City, and from other U.S. locations. During periods where combustion emissions dominate, a carbon dioxide-based emission factors for ultrafine particles can be developed or evaluated from ambient particle measurements. For example, in March 2006, during the MILAGRO field campaign in Mexico City, observations at the T0 location could often be characterized by morning conditions with high particle mass concentrations, low mixing heights, and good correlation between particle number and carbon dioxide, indicative that particle number is controlled by primary emissions. Using this correlation between particle number and carbon dioxide, an average number based-emission factor of 3.4×10^{13}#/vehicle-km has been determined. In the afternoon, the CO_2 level drops during ventilation of the daily polluted layer, and the coupling between CO_2 and particle number breaks down, with particle number sometimes increasing as CO_2 decreases. New particle formation events were observed both simultaneous to and separately from the afternoon ventilation.

203c Understanding the Evolution of Cloud Condensation Nuclei In Urban Plumes: Effects of Aerosol Ageing and Mixing

Akua Asa-Awuku[1], Richard H. Moore[1], Chuck Brock[2], Roya Bahreini[2], Ann Middlebrook[2], Joshua Schwarz[2], Ryan Spackman[2], John Holloway[2], Dave Tanner[3], Greg Huey[3], and Athanasios Nenes[4]. (1) Chemical and Biomolecular Engineering, Georgia Institute of Technology, 311 Ferst Drive NW, Atlanta, GA 30332, (2) NOAA, (3) Earth & Atmospheric Sciences, Georgia Institute of Technology, 311 Ferst Drive NW, Atlanta, GA 30332, (4) Georgia Institute of Technology, Schools of Earth & Atmospheric Sciences and Chemical & Biomolecular Engineering, 311 Ferst Drive, Atlanta, GA 30332

The aerosol indirect effect is a major source of uncertainty for predicting anthropogenic influences on climate; constraining the concentration of particulate matter that can act as cloud condensation nuclei (CCN) is a key component for quantifying this uncertainty. Köhler theory, used to predict CCN activity is well established yet calculating ambient CCN concentrations requires a significant amount of information that characterizes the aerosol size distribution, chemical composition and mixing state. In addition, large uncertainties are associated with predicting these aerosol properties, especially for highly heterogeneous aerosol close to source regions. As a result, it is very important to quantify the CCN prediction uncertainty associated with simplifying assumptions taken in aerosol models. In this study, airborne aerosol and CCN measurements collected aboard the NOAA WP-3D platform during the 2006 Texas Air Quality Study/Gulf of Mexico Atmospheric Composition and Climate Study (TexAQS/GoMACCS) are analyzed. From September 20th through October 10th, 2006 over ten flights characterize the freshly emitted and aged aerosol sampled in the vicinity of Houston, eastern/northeastern Texas and the northwestern Gulf of Mexico. The analysis focuses on the importance of chemical heterogeneity (size-dependant chemistry and mixing state) and aerosol size distribution function for achieving CCN closure, as a function of particle ageing. The relative contribution of particle ageing versus entrainment of background aerosol to the CCN concentration is also examined.

203d Relating Cloud Condensation Nuclei Activity, Volatility, and Droplet Growth Kinetics of Terpene Secondary Organic Aerosol

Gabriella J. Engelhart[1], Akua Asa-Awuku[2], Athanasios Nenes[2], and Spyros N. Pandis[1]. (1) Chemical Engineering, Carnegie Mellon University, 5000 Forbes Ave, Pittsburgh, PA 15213, (2) Chemical and Biomolecular Engineering, Georgia Institute of Technology, 311 Ferst Drive NW, Atlanta, GA 30332

Cloud droplets form on a subset of atmospheric particles (aerosol) called cloud condensation nuclei (CCN). The concentration of these CCN in the atmosphere influences the radiative properties of clouds, thus having an effect on the Earth's climate. Cloud-particle interactions are one of the major challenges in quantifying the indirect climate forcing (Houghton et al., 2001; IPCC, 2007). Understanding CCN activation is an important step towards understanding cloud formation and properties. While inorganic particles are well understood (Köhler, 1936; Pruppacher and Klett, 1997; Svenningsson et al., 2006), many questions still remain about the influence of organic particulate material on CCN activity. A class of compounds called monoterpenes is estimated to make up 11% of the global natural VOC flux of 1150 Tg carbon per year (Guenther, 1995). Sesquiterpenes, while less abundant as a precursor, are important because of potentially large aerosol yields (Griffin, 1999) due to high molecular weight oxidation products.

The ability of secondary organic aerosol (SOA) particles formed during the ozonolysis of beta-caryophyllene, alpha-pinene and other monoterpenes to act as cloud condensation nuclei (CCN) was investigated using a static CCN counter and a cylindrical continuous-flow streamwise thermal gradient CCN counter developed by Droplet Measurement Technologies (DMT). Secondary organic aerosol (SOA) was produced from the dark reaction of either beta-caryophyllene, alpha-pinene or monoterpene mixtures (alpha-pinene, beta-pinene, limonene and 3-carene) with ozone in a 12 m^3, temperature-controlled smog chamber. The initial terpene concentrations were in the 10-30 ppb range and an excess of ozone was used. The CCN concentrations, activation diameter and droplet growth kinetic information were monitored online as a function of supersaturation for several hours and their changes with age were quantified. Additionally, filter samples were collected for analysis of water soluble organic carbon (WSOC). Filter samples allow for the investigation of surfactant characteristics and estimation of thermodynamic properties, such as molecular weight. Volatility of sesquiterpene (beta-caryophyllene) SOA was probed using a thermodenuder.

Both fresh and aged monoterpene SOA are quite active as CCN. The initial concentrations of ozone and monoterpene precursor do not appear to affect the activity of the resulting SOA. However, reactions of the hydroxyl radicals produced during the monoterpene ozonolysis lead to further oxidation of the SOA material and an improvement of their CCN properties with time. A decrease in CCN activation diameter for alpha-pinene SOA of approximately 3 nm hr^{-1} was observed as the aerosol continued to react with oxidants. Hydroxyl radicals further oxidize the SOA particles thereby enhancing the particle CCN activity with time. The initial concentrations of ozone and monoterpene precursor (for concentrations lower than 40 ppb) do not appear to affect the activity of the resulting SOA.

Köhler Theory Analysis (KTA) is used to infer the molar mass of the monoterpene SOA sampled online and offline from atomized filter samples. The estimated average molar mass of online monoterpene SOA was determined to be 180 ± 55 g mol^{-1} (consistent with existing monoterpene SOA speciation studies) assuming complete solubility. KTA suggests that the aged aerosol (both from alpha-pinene and the mixed monoterpene oxidation) is primarily water-soluble (around 65%). CCN activity measurements of the monoterpene SOA mixed with $(NH_4)_2SO_4$ suggest that the organic can depress surface tension up to 10 nM m^{-1} (with respect to pure water). The droplet growth kinetics of monoterpene SOA samples are similar to $(NH_4)_2SO_4$. The CCN activation diameter of alpha-pinene and mixed monoterpene SOA can be modelled to within 10-15% of experiments by a simple implementation of Köhler theory, assuming complete dissolution of the particles, no dissociation into ions, a molecular weight of 180 g mol^{-1}, a density of 1.5 g cm^{-3}, and the surface tension of water.

The model sesquiterpene system, beta-caryophyllene, shows that the CCN contain small amount of hygroscopic material. Although beta-caryophyllene produces higher yields than monoterpene systems the aerosols are less effective as CCN in the atmosphere. A major finding in this work is that the hygroscopic component of beta-caryophyllene SOA is volatile. This leads to discrepancies between CCN instrumentation not observed in monoterpene systems. The finding is potentially quite important for atmospheric measurements impacted by diurnal temperature cycles and thus, potentially, CCN activity diurnal cycles. Evidence of a kinetic barrier to droplet activation was revealed from a comparison of droplet growth kinetics of the CCN and the fraction of WSOC. Filter samples show that the WSOC fraction of the SOA is composed of low molecular weight (<200 g mol^{-1}) compounds that are slightly surface active. This is similar to the findings for monoterpene SOA systems indicating that exact speciation knowledge might not be required for predictive understanding of CCN.

Griffin, R. J., Cocker, D. R., Flagan, R. C., and Seinfeld, J. H.: Organic aerosol formation from the oxidation of biogenic hydrocarbons, J. Geophys. Res., 104, 3555–3567, 1999.

Guenther, A., Hewitt, C. N., Erickson, D., Fall, R., Geron, C., Graedel, T., Harley, P., Klinger, L., Lerdau, M., McKay, W. A., Pierce, T., Scholes, B., Steinbrecher, R., Tallamraju, R., Taylor, J., and Zimmerman, P.: A global model of natural volatile organic compound emissions, J. Geophys. Res., 100, 8873--8892, 1995.

Houghton, J., Ding, Y., Griggs, D., Noguer, M., van der Linden, P., Dai, X., Maskell, K. and Johnson, C.: Climate Change, 2001: The Science 1035 of Climate Change, Cambridge University Press, New York, Cambridge, UK, 2001.

IPCC (Intergovernmental Panel on Climate Change): Changes atmospheric constituents and in radiative forcing in: Climate change 2007: The physical science basis, Cambridge University Press, 2007.

Köhler, H.: The nucleus in and the growth of hygroscopic droplets, Trans. Faraday Soc. 32, 1151--1161,

1936.

Pruppacher, H. R. and Klett, J. D.: Microphysics of Clouds and Precipitation, Kluwer, Dordrecht, 1997.

Svenningsson, B., Rissler, J., Swietlicki, E., Mircea, M., Bilde, M., Facchini, M.C., Decesari, S., Fuzzi, S., Zhou, J., Monster, J., Rosenorn, T.: Hygroscopic growth and critical supersaturations for mixed aerosol particles of inorganic and organic compounds of atmospheric relevance, Atmos. Chem. Phys., 6,1937—1952, 2006.

203e Airborne Size-Resolved Ccn Activity and Droplet Growth Kinetic Measurements In Pristine and Polluted Airmasses

Luz T. Padro[1], Harmony Gates[2], Shane M. Murphy[3], Armin Sorooshian[2], Haflidi Jonsson[4], Richard C. Flagan[5], John H. Seinfeld[2], and Athanasios Nenes[1]. (1) School of Chemical and Biomolecular Engineering, Georgia Institute of Technology, 311 Ferst Dr. NW, Atlanta, GA 30332-0100, (2) Chemical Engineering, California Institute of Technology, 1200 E. California Blvd. - MC210-41, Pasadena, CA 91125, (3) Departments of Environmental Science and Engineering and Chemical Engineering, California Institute of Technology, 1200 E. California Blvd. - MC210-41, Pasadena, CA 91125, (4) Center for Interdisciplinary Remotely Piloted Aircraft Studies, 3200 Imgin Road, Marina, CA 93933, (5) California Institute of Technology, 1200 E. California Blvd. - MC210-41, Pasadena, CA 91125

The effects of aerosol composition (especially from organic species) on the cloud droplet formation process is highly variable and uncertain, but can be largely constrained by size-resolved measurements of Cloud Condensation Nucleus (CCN) activity (SRCM). A fast method for SRCM is Scanning Mobility CCN Analysis (SMCA; Nenes and Medina, in review), where the monodisperse aerosol output from a Differential Mobility Analyzer sampling ambient aerosol (and operated in scanning voltage mode) is concurrently introduced into a Streamwise Thermal Gradient CCN Chamber (STGC) and a Condensation particle counter; inversion of the concentration timeseries from both detectors yield the activation ratio (i.e., fraction of particles that act as CCN) and size of activated droplets, as a function of dry mobility diameter and supersaturation. From such measurements, Köhler Theory Analysis (e.g., Asa-Awuku et al., 2008; Moore et al., 2008; Padró et al., 2007) can be applied to infer the presence of surfactants, the average molecular weight of water-soluble organics and the impact of organics on droplet growth kinetics. In this study, we performed SMCA for aerosols sampled aboard the CIRPAS Twin Otter during the Marine Stratus/Stratocumulus Experiment II (MASE II) in Marina, CA in July 2007. We focus on two research flights, one carried out in marine air and one downwind of a large bovine source in the San Joaquin Valley, California; the differences seen in water-uptake properties, droplet growth kinetics as a function of organic content and character are presented.

203f Multi-Component Organic Aerosols: Relationships among Hygroscopic Growth, Ccn Activity, and Phase State as Measured with Htdma, Ccnc, and AFM

Timothy M. Raymond, Bucknell University, Chemical Engineering Dept., Lewisburg, PA 17837 and Juan Alberto Lopez Ruiz, Chemical Engineering, Bucknell University, Chemical Engineering Dept., Lewisburg, PA 17837.

Previous research on multi-component organic mixtures containing up to 10 different compounds has provided interesting data on aerosol hygroscopic growth. For various mixtures of organics, some experiments have indicated continuous growth while other mixtures show restructuring and distinct deliquescence points. Research on super-micron particles has shown a relationship with phase state of the initial particles and deliquescence which may also extent to sub-micron particles and to cloud activity. This has important implications for the comparison of laboratory and field data on aerosol-water interactions and ultimately for understanding the aerosol indirect effect.

In this work, we will present results of experiments measuring hygroscopic growth (using HTDMA), cloud condensation nuclei activity (using CCNC), and phase state/morphology (using AFM) of various mixtures of up to 10 different common organic aerosol components. Additionally, we have analyzed the aerosol generation process typically used in laboratory studies (solution atomization and subsequent drying) to determine what parameters affect the resulting aerosol and its properties. We have investigated the influence of different types of atomizers, solvents, solution concentrations, drying rates and residence times, and temperatures on the resulting particles.

The results indicate that the particular method used to generate aerosols in the laboratory can have an important affect on aerosol morphology and phase state. The resulting aerosol hygroscopic properties are similarly altered with obvious relationships between hygroscopic growth and CCN activity.

270a Heterogeneous Uptake of Octamethylcyclotetrasiloxane (D4) and Decamethylcyclopentasiloxane (D5) on Reactive Dust Aerosol: Experiments and Atmospheric Implications

Juan G. Navea, Department of Chemistry and Chemical and Biochemical Engineering, University of Iowa, Iowa City, IA 52242, **Charles O. Stanier**, Http://www.Cbe.Engineering.Uiowa.Edu/faculty/stanier/, University of Iowa, Iowa City, IA 52242, Mark A. Young, Chemistry, University of Iowa, Iowa City, IA 52242, Vicki H. Grassian, Chemistry and Chemical and Biomolecular Engineering, University of Iowa, Iowa City, IA 52242, and Shihe Xu, Health and Environmental Sciences, Dow Corning Corporation, Midland, MI 48686-0994.

We have carried out kinetic studies to determine the heterogeneous decay of octamethylcyclotetrasiloxane (D4) and decamethylcyclopentasiloxane (D5) in the presence of atmospheric particles at different relative humidity (RH). The heterogeneous decay of D4 and D5 mediated by atmospheric particles is simulated in an atmospheric chamber and monitored by means of FT-IR spectroscopy. The apparent heterogeneous uptake coefficient is analyzed for different aerosols including kaolinite, hematite, quartz, carbon black, and calcite. A box model is constructed to contrast the heterogeneous processes with the known gas phase degradation of these compounds through reaction with OH, which leads to atmospheric residence times of order days to weeks. Using the model, we quantify the required combination of thermodynamic and kinetic parameters needed for a reduction in the atmospheric lifetime. The model is constructed using three compartments (urban, semi-rural, and rural environments), and includes simulation of gas phase decay, thermodynamic partitioning of D4 and D5, diurnal release patterns, and heterogeneous reactions. The tuning of the box model to match observed concentrations of benzene (which reacts with OH on a similar timescale) is discussed.

270b Evaluation of Long-Range Transport of Atmospheric Particulate Matter Using a Regional Chemical Transport Model

Kristina M. Wagstrom and Spyros N. Pandis. Department of Chemical Engineering, Carnegie Mellon University, 5000 Forbes Ave, Pittsburgh, PA 15213

An important aspect of designing effective control strategies for fine particulate matter (PM) concentrations is the quantification of the contributions of different source types and source regions to PM levels in a given area. Our goal is to quantify how much of the particulate matter in different locations is the result of local sources, short-range transport and long-range transport. We will discuss and compare two approaches to reaching this goal and results found using both approaches. The first approach is focused on source regions and the second is focused on receptor regions.

We have used PMCAMx, a regional three-dimensional chemical transport model, to study these impacts of pollutant sources and transport on particulate matter concentrations in the Eastern United States using PSAT (Particulate Matter Source Apportionment Technology). PSAT is a computationally efficient particulate matter apportionment algorithm that is able to track source contributions to both primary and secondary particulate matter in regional chemical transport models such as PMCAMx.

For the source focused approach, transport of pollutants is studied by tracking the emissions from source regions in order to study how far and in what direction these emissions are being transported. For the receptor region focused approach, transport of pollutants is studied by tracking the pollutants emitted from a series of rings around a receptor region of interest. This approach allows for the fullest understanding of the distance pollutants have been transported before they reach the receptor region we are studying but is slower than the source region focused approach since only one receptor region can be studied at a time. We will compare the results from both of these approaches looking at the impacts to primary organic aerosol, SO_2, particulate sulfate and particulate nitrate. These species are representative of most species of interest in the atmosphere since they include a primary non-reactive species (primary organic aerosol), a primary reactive species (SO_2), a non-volatile secondary PM species (particulate sulfate), a semi-volatile secondary species (particulate nitrate) and a precursor – product pair (SO_2 and particulate sulfate).

The source region focused approach is applied to study the contributions of 11 different source regions in the Eastern United States, ranging in size (both in terms of land area and emissions) from the Ohio River Valley to Little Rock, AR. These impacts are quantified using a variety of measures to account for both the distance that the pollutants are transported and the overall impact the emissions have on pollutant concentrations in the region. The receptor region focused approach is used to evaluate the distance of transport for six receptor regions, ranging from remote regions (such as Great Smoky Mountains National Park) to urban regions (such as Pittsburgh, PA). Influences of several hundred miles have been found for secondary aerosol components from emissions originating in the larger source regions while primary aerosols show impacts predominately on the local scale. Of the source regions studied, the impact to surrounding areas is largest for the Ohio River Valley, the Mid-Atlantic region and Southwestern Pennsylvania and smallest for Little Rock, AR and St. Louis, MO.

270c Decomposition of the Primary Ozonide: Towards a Better Understanding of Ozonolysis Kinetics

Scott A. Epstein, Department of Chemical Engineering, Carnegie Mellon University, 1207 Doherty Hall, 5000 Forbes Ave., Pittsburgh, PA 15213 and Neil M. Donahue, Department of Chemistry / Department of Chemical Engineering, Carnegie Mellon University, 1106 Doherty Hall, 5000 Forbes Ave., Pittsburgh, PA 15213.

Ozonolysis is an important oxidation pathway for gaseous and condensed-phase alkenes in the troposphere. Since ozonolysis products often have lower vapor pressures than the reactant alkenes, these reactions are a significant source of secondary organic aerosol. In addition, ozonolysis reactions are a source of radicals that play a key role in tropospheric chemistry. Also, ozone uptake ages particles on a timescale that is relevant to the typical lifetime of an air mass.

The mechanism of ozonolysis involves many branch points, but in all cases it proceeds through an initial intermediate known as the primary ozonide (POZ). The decomposition of the POZ can proceed in several different ways leading to different subsequent chemistry and product distributions. However, the POZ decomposes too rapidly to be studied under normal atmospheric conditions.

We observe the decomposition of primary ozonides formed from different alkenes on a cold (>90 K) surface using Temperature Programmed Reaction Spectroscopy (TPRS) as the surface is warmed at multiple heating rates. Data from these experiments allows us to estimate the barrier heights for POZ decomposition and, in cases with multiple primary ozonides, the branching ratios of different reaction pathways. In addition, we observe formation of carbonyl products, secondary ozonides, and their subsequent desorption.

Two alkenes with only one unique primary ozonide, (2,3)-Dimethyl-2-butene and Cyclohexene, were previously studied. In addition, a few other linear alkenes with multiple POZ decomposition pathways were examined. Comparison of the POZ decomposition behavior will allow us to make predictions of the mechanistic kinetics for a suite of precursors ranging from fully symmetric straight chain alkenes to alkenes with R or Z isomers to cycloalkenes. These experiments will be extended to include alkenes that are significant secondary organic aerosol precursors in the atmosphere.

270d The Sensitivity of Modeled Ozone to Changes In the Temporal Distribution of Area, Point, Mobile and Non-Road Emissions

Patricia Castellanos[1], Sheryl Ehrman[1], Russell Dickerson[2], Jeffrey Stehr[2], and Dale Allen[2]. (1) Chemical and Biomolecular Engineering, University of Maryland, Building 90, College Park, MD 20010, (2) Atmospheric and Oceanic Science, University of Maryland, College Park, MD 20010

High ozone concentrations that occur via a photochemical mechanism involving nitrogen oxides (NOx) and volatile organic compounds (VOCs) and exceed the 80 ppb 8-hour average National Ambient Air Quality Standard (NAAQS) are a longstanding problem in many Northeast urban/suburban areas (U.S. Environmental Protection Agency 2004). Eulerian grid models or air quality models (AQMs) that simulate atmospheric chemistry, transport, and removal processes of multiple pollutants including ozone, aerosols, and other oxidants that are used as tools in air quality decision making often satisfactorily replicate ozone when compared to measurements (U.S. Environmental Protection Agency 2001; Hogrefe 2007; Godowitch 2008). Regardless, the simulations are subject to uncertainty as a result of parameterizations and approximations embedded in the model algorithms as well as inaccuracies in the meteorological and emissions inputs as a result of measurement uncertainty, missing data, etc. (Placet 2000; Fine 2003; Mallet 2006). By understanding the sensitivities in the model – the models' response to changes in inputs, parameters, or resolution – one can identify areas that contribute to the simulation uncertainty and therefore require additional scientific analysis.

In this work we analyzed the effects of altering emissions temporal distributions on simulations of ozone with a focus on the Northeast. We conducted a summer simulation with the Community Multi-Scale Air Quality Model (CMAQ) v4.5 using a 12 km grid size with the 2002 National Emissions Inventory. We applied three temporal distributions to the four major source categories of emissions and simulated a total of nine cases: 1) four uniform cases, where either point, area, mobile, or non-road emissions remained the same hour-to-hour and day to day, 2) a base case, where all emissions varied according to 2002 NEI estimates, and 3) four high variability cases, where 50% of either point, area, mobile, or non-road nighttime emissions were added to the daytime. We find that the largest sensitivities occurred in urban areas both in the Northeast and Southeast when mobile sources were altered. For a summer simulation, on average, the ozone sensitivity was +/- 1-3 ppb. The largest sensitivities, +/- 2-7 ppb, occurred when an 8-hour average concentration greater than 80 ppb was simulated in the base case. When moderate, 50-80 ppb, 8-hour averages were simulated, the sensitivity was less than 2 ppb. Furthermore, the number of days with 80 ppb 8-hour average exceedances decreased by 2-10 days when mobile sources were altered. The sensitivities from the other source categories, area, point, and non-road sources, were not significant in comparison.

REFERENCE

Fine, J., Vuilleumier,L.,Reynolds,S.,Roth,P.,Brown,N. (2003). "Evaluating Uncertainties in Regional Photochemical Air Quality Modeling." Annual Review of Environmental Resources 28: 59-106.

Godowitch, J. M., Gililand,A.B.,Draxler,R.R.,Rao,S.T. (2008). "Modeling assessment of point source NOx emissions reductions on ozone air quality in the easter United States." Atmospheric Environment 42 (1): 87-100.

Hogrefe, C., Hao,W.,Civerolo,K.,Ku,J.-Y.,Sistla,G.,Gaza,R.S.,Sedefian,L.,Schere,K.,Gilliland,A., Mathur,R. (2007). "Daily Simulation of Ozone and Fine Particulates over New York State: Findings and Challenges." Journal of Applied Meteorology and Climatology 46(7): 961-979.

Mallet, V., Sportisse,B. (2006). "Uncertainty in a chemistry-transport model due to physical parameterization and numerical approximations: An ensemble approach applied to ozone modeling." Journal of geophysical research 111: D01302.

Placet, M., Mann,C.O.,Gilbert,R.O.,Niefer,M.J. (2000). "Emissions of ozone precursors from stationary sources: a critical review." Atmospheric Environment 34: 2183-2204.

U.S. Environmental Protection Agency (2001). CMAQ Model Performance Analysis. Office of Air Quality Planning and Standards.

U.S. Environmental Protection Agency (2004). The Ozone Report: Measuring Progress through 2003. Office of Air Quality Planning and Standards.

270e Use of a Pseudo-Lattice Approach to Model the Mean Activity Coefficient of Water+Organic+Inorganic Aerosols

Elsa Moggia and Bruno Bianco. Department of Biophysical and Electronic Engineering, University of Genoa, Via Opera Pia 11 A, Genoa, Italy

The theoretical modelling of the mean activity coefficient of atmospheric liquid aerosols has been afforded in various ways, but so far only inorganic+water systems have been extensively investigated. Indeed, in most treatments, organic compounds are neglected since their fraction is assumed to be sufficiently small. However, realistic situations can involve large fractions of organic compounds, up to about 65% within particulate matter. To deal with water + organic+ inorganic mixtures of interest in environmental chemistry, some models have been proposed in recent years, which are mainly based on UNIFAC. However, many adjustable parameters are required by such models, these parameters being not directly measurable from experiments but only predictable by means of best-fitting numerical techniques from experimental data. In this context, the pseudo-lattice approach [E. Moggia, B. Bianco, J. Phys. Chem. B, 2007, vol. 111] provides a one-parameter equation available for wide ranges of concentrations and temperatures. Moreover, the unique required parameter is a suitable value of solute concentration, which is experimentally known at least for many simple electrolytes. To use the pseudo-lattice approach for atmospheric aerosols, one starts from considering the organic + water component as a two-component mixture solvent, whose dielectric constant can be either measured or calculated starting from the known permittivities of the components. The application of the model to H2SO4 + water + organic and to HNO3+water+organic solutions will be presented and discussed.

270f Environmental Foresight through Computational Chemistry: Improved Radiative Forcing Predictions for Global Warming Potentials

Paul Blowers, Chemical and Environmental Engineering, The University of Arizona, PO Box 210011, Tucson, AZ 85721-0011 and Maurice Lee, Mathematics, U of Arizona, Tucson, 85721.

Global warming potential values are based on the atmospheric lifetime of chemical species released into the troposphere and on the radiative forcing values of those chemical species. The atmospheric lifetime is evaluated using the chemical degradation rate by hydroxyl radical abstraction of hydrogen atoms from the released parent species for compounds with hydrogen atoms. Radiative forcing, the energy absorbed by the chemical per meter square of troposphere parallel to the Earth's surface per part per billion of concentration, is typically obtained through experimentally measured absorption cross sections followed by atmospheric modeling to obtain results. Pinnock, et al., found a short-cut of estimating radiative forcing through binning the percent transmittance of the Earth's atmosphere into 10 cm-1 bins and then aggregating absorption cross section data. Papasavva, et al., transferred this work for use with computational chemistry based predictions of IR spectra. Our earlier work build upon their approach and extensively showed relatively good agreement with experimentally based modeled values for radiative forcing for a large number of hydrofluoroethers.

One of the criticisms of our and Papasavva's work is that all of the peak intensity for a given frequency was assigned to only one 10 cm-1 bin, ignoring the fact that experimental peaks are distributed about peak heights. In this work, we explore the use of Lorentzian distributions of intensities about peak locations using scaled frequencies from the B3LYP/6-31g* level of theory. Extensive work in our lab has shown that this level of theory reproduces frequency locations extremely well and intensities are accurate compared to the limited data available for comparison.

We compute radiative forcing values for close to 100 chemical species using the Lorentzian distributed peaks in order to improve our earlier predictions and expand on a reproducible methodology for accurate radiative forcing predictions. We find significant reduction of error compared to experimentally modeled results versus our earlier work. This approach provides for a robust method of quickly building databases of radiative forcing results that can be combined with kinetic degradation rates to predict global warming potentials.

270g Data-Driven Statistical Analysis of Global Climate Change

Ian J. Laurenzi, Chemical Engineering, Lehigh University, B330 Iacocca Hall, 111 Research Dr., Bethlehem, PA 18015

In public discourse on the subject of climate change, there is considerable doubt regarding global climate models (Douglas et al. Int. J. Climatol. 2007), particularly with respect to the effects of clouds. Moreover, there have been questions regarding the magnitude of the rate of global temperature change, as well as its error relative to the error in local temperature measurement. In this paper, an "omic" approach to global temperature change is undertaken utilizing publicly-available temperature databases. Robust statistical approaches are employed to analyze temperature timeseries for over 400 cities of various sizes, latitudes and longitudes, and identify positional and seasonal trends in climate change. A detailed analysis of Antarctic temperatures over the past fifty years is also conducted, illustrating the relationship between global temperature and the Antarctic climate.

AN UPDATED SITE SCALE SATURATED ZONE GROUND WATER TRANSPORT MODEL FOR YUCCA MOUNTAIN

Sharad Kelkar[1], Mei Ding[1], Shaoping Chu[1], Bruce Robinson[1], Bill Arnold[2], Arend Meijer[3]

[1] *Los Alamos National Laboratory, Los Alamos, New Mexico, USA*
[2] *Sandia National Laboratories, Albuquerque, New Mexico, USA*
[3] *GCX Inc, 1389 E. Stoney Canyon Cr., Tuscon, AZ 85737, USA*

This paper summarizes the numerical site scale model developed to simulate the transport of radionuclides via ground water in the saturated zone beneath Yucca Mountain.

I. INTRODUCTION

The U.S. DOE has submitted a license application to the Nuclear Regulatory Commission (NRC) to create a geologic repository at the Yucca Mountain, Nevada (Figure 1), for the disposal of spent nuclear fuel and high-level radioactive waste. A number of engineered and natural barriers are expected to retard any potential release of radionuclides from the repository and to reduce their concentrations. In the event of a release, the ground water beneath the Yucca Mountain is a primary medium through which most radionuclides might move from the geologic repository to the accessible environment at the compliance boundary. The role of the saturated zone (SZ) is to delay the transport of radionuclides to the accessible environment and to reduce the concentration of radionuclides before they reach the accessible environment. Radionuclide breakthrough curves are determined by the specific discharge, flow paths and flow distribution of ground water, and the mechanisms of solute transport including advection, dispersion, diffusion, sorption and colloid-facilitated transport. This paper describes the development of a site scale transport model to evaluate these processes.

II. FLOW MODEL

Yucca Mountain is located about 150 km northwest of Las Vegas, Nevada (Figure 1) in the Great Basin section of the Basin and Range province[1]. Yucca Mountain consists of a group of north-trending block-faulted ridges composed of volcanic rocks bounded by basins composed of volcanic tuffs, which pinch out to the south beneath valley-fill alluvium, underlain by a thick carbonate aquifer. In general, the confining units are zeolitic, nonwelded tuffs and the aquifers are welded, fractured tuffs.

The conceptual model of SZ flow in the model area is that ground water flows generally southwards from recharge areas to the north and at higher elevations, through the volcanic rocks towards the Amargosa Desert [2]. Towards the southern portion of the model area, the flow near the water table enters the valley-fill alluvium. Data suggests an upward gradient indicating that, for the immediate Yucca Mountain area and downstream, radionuclide transport would be restricted to the volcanic and alluvial formations. Discharge from the model occurs mainly via the alluvium, at the model boundary to the south and southeast, and through pumping out of wells in the Amargosa Valley.

The overall groundwater system is assumed to be in steady state for the purpose of this modeling. A confined aquifer solution is used for the flow model. The top boundary of the model represents the observed water table levels. The effective continuum approach is used to model flow and transport, which averages fracture and surrounding rock properties in a given grid block, with the modification that several of the important fault zones are included as explicit features[2].

The flow and transport modeling is done with the FEHM computer code[13], a general purpose unsaturated-saturated zone non-isothermal code built around state-of-the-art control volume finite element numerical procedures. Solute transport is simulated using a particle tracking technique with a generalized dispersion tensor that allows for different longitudinal dispersivities in the vertical and the horizontal directions, matrix diffusion, and sorption.

The site scale flow model is a rectangular area about 30km east west and 45 km north south including the Yucca Mountain[2]. The geometry of geologic units was defined in a hydrogeologic framework model (HFM2006) with 27 hydrologic units constructed using data such as geologic maps and sections, borehole lithologic logs, and topographic digital elevation models[3]. Major hydrogeologic features such as fault zones, and areas of hydrogeologic alteration were classified as distinct permeability zones.

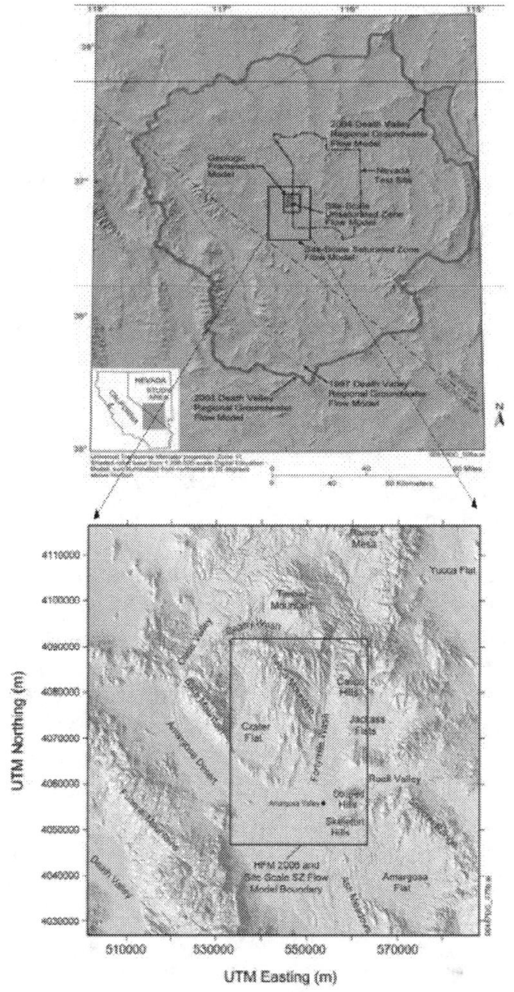

Fig. 1: Location Map Of The Yucca Mountain Project And An Enlargement Showing The Saturated Zone Study Area And Associated Geological Features.

A structured computational grid (Figure 2) was created from the HFM2006 with uniform horizontal 250 m grid spacing. The vertical grid spacing ranges from 10 m at the upper boundary near the water table to 600 m deep within the SZ.The vertical grid extends from an elevation of +2018.5 m to –4000 m (MSL). The permeabilites of various units in the model were optimized using a parameter estimation code (PEST) to achieve a minimum difference between observed water levels and simulated water levels, and also between volumetric/mass flow rates along specific boundary segments simulated by the SZ regional- and site-scale models. Post-development model validation[2] was accomplished by comparing a) predicted hydraulic heads with new water-level data, b) predicted flow paths with those estimated from hydrochemistry and isotope data, c) calibrated parameter values with those derived from well testing, and d) predicted specific discharge with the conclusions of the Expert Elicitation Panel[4].

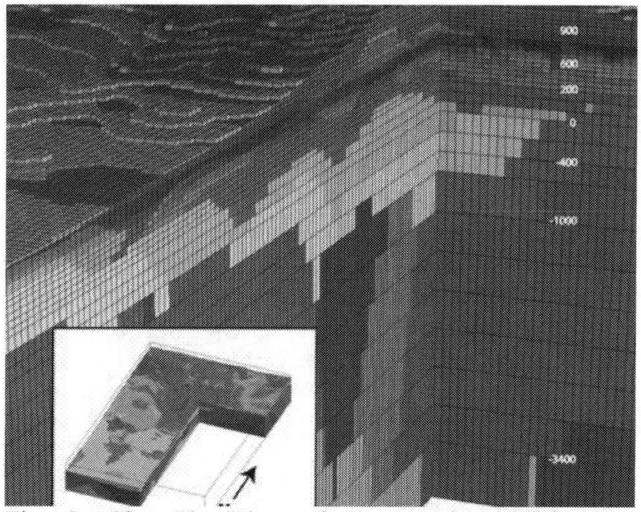

Fig. 2: Close-Up View of Computational Grid (3× Vertical Exaggeration) Showing Cut Away at UTM Easting = 549,000 m and UTM Northing = 4,078,000 m Through the Yucca Mountain Repository.

III. TRANSPORT MODEL

Radionuclides are modeled to enter the SZ via downward percolation of water from the unsaturated zone. Within the SZ model, radionuclides are transported in the upper few hundred meters below the water table with pore water that flows sub-horizontally in a southwardly direction through the highly fractured portions of the tuff, entering the alluvium to the south. Current hydrologic evidence supports the model of fluid flow within fractures in the moderately to densely welded tuffs of the SZ[5]. Fluid flow within fractures occurs preferentially along tortuous channels of largest aperture, with stagnant fluid residing in the low permeability rock matrix. Diffusion of solutes between the stagnant water in the rock matrix and the flowing fracture zones is simulated using a dual-porosity approach for the volcanic formations. In the alluvium, flow and transport is represented using a single porosity continuum model. This is due to the more porous, less fractured nature of the alluvial material. Alluvium is a heterogeneous medium, flow occurs through the relatively more permeable regions within the medium, with the low-permeability regions acting as flow barriers that groundwater flows around rather than through.

Dispersion is caused by heterogeneities at all scales, from the scale of individual pore spaces to the scale of the thickness of individual strata and the length of structural features such as faults. The spreading and dilution of radionuclides that result from these heterogeneities could be important to performance of the potential repository. The largest heterogeneities are represented explicitly in the site-scale SZ flow and transport model[2]. For dispersion at smaller scales, it is assumed that the convective-dispersion model applies with dispersion characterized using an anisotropic dispersion coefficient tensor consisting of longitudinal, horizontal-transverse, and vertical-transverse dispersivities[7].

Field studies of transport and dispersion have been conducted at a variety of length scales (from meters to kilometers) to address the issue of dispersion. The results show a trend toward larger apparent dispersion coefficients for transport over longer distances. Solutes encounter larger-scale heterogeneities at larger scales, and thus, spreading is more pronounced. The dispersivity estimates from C-wells tracer experiments fall within the range of values from other sites[7], suggesting that transport in the fractured tuffs exhibits similar dispersive characteristics. The values used in the simulations of radionuclide transport are somewhat higher than those estimated from the C-wells because of the larger scale that is relevant for radionuclide migration to the accessible environment. Transverse dispersion is conservatively assigned a very low value in the transport model.

Because the site-scale SZ flow and transport model is used to predict transport under a well-withdrawal scenario[8], the potentially complex distribution of radionuclides within a grid block will be averaged in the process of extracting water from the aquifer.

Sorption of radionuclides on rock surfaces is a mechanism that will enhance retardation of the solute movement. Radionuclide-rock interactions potentially can occur on the surfaces of fractures and within the rock matrix. Sorption within the rock matrix is supported by the C-wells tracer experiments[9]. As a conservative approach, sorption on fracture surfaces is not included in the model.

Sorption reactions are represented in the transport model by a constant sorption coefficient (k_d), which is the ratio of the solute adsorbed on the solid to that in solution . The values of these k_d's are sensitive to the radionuclide element being considered and geological variables including rock type, water chemistry, redox condition of the ground water, radionuclide concentrations, and variations in rock surface properties within the major rock types [7]. As a conservative approach, the groundwater was assumed to be in oxidizing conditions. Since there is uncertainty in knowledge regarding the geological variables, probabilistic uncertainty distributions were developed for each radionuclide of interest. The distributions selected were based in part on laboratory data and in part on professional judgment regarding the impact of variables not considered in the experimental program. The experimental data were augmented with the results of modeling calculations using the code PHREEQC[10] to provide a basis for interpolation and extrapolation of the data.

The use of a k_d to represent sorption reactions in a transport calculation requires that the sorption reaction be reversible, linear (isotherm) and instantaneous. On the time scale of radionuclide transport from a repository, these criteria will generally be met[7]. The probability distributions of k_d's are biased towards lower values to compensate for any nonlinear sorption.

Transport of radionuclides via colloids is another mechanism that is of importance at field scale[11]. Solute molecules can absorb onto the colloid particles and be transported along with them. Because the colloidal particles are much larger in size and mass than the radionuclides that sorb onto them, the radionuclides do not affect the transport of the colloid particles. The colloids in the saturated zone can be of several types including natural colloids (typically clay or silica), waste-form colloids resulting from degradation of spent nuclear fuel or glass, and iron-oxyhydroxide colloids resulting from degradation of the waste container[11]. Because of the relatively large size of the colloids, matrix diffusion of these particles is negligible. Several field observations have suggested that a small percentage of colloids has the potential to transport unretarded with groundwater[12], whereas the majority undergo filtration, which can be described by a retardation factor, which is the ratio of the travel time including filtration to that without filtration..The retardation factor is dependent on several factors such as colloid size, colloid type, and geochemical conditions (e.g., pH, Eh, and ionic strength).

A fraction of the total radionuclide inventory (1% to 10%)[11] is expected to reach the SZ as radionuclides that attach reversibly to the colloids, thus partitioning their transit time between the solution, attached to the rock walls, and attached to the colloids. The radionuclides that are reversibly absorbed onto colloids are modeled using the K_c model. K_c (groundwater concentration of colloids times sorption coefficient onto colloids) represents the equilibrium partitioning of radionuclides between the aqueous phase and the colloidal phase with the distribution coefficient[7]. In this approach the diffusion coefficient and the sorption coefficient for the radionuclide are reduced to account for the fraction of time spent by the radionuclide attached to the colloid particles. The distribution coefficient K_c is modeled as a function of radionuclide sorption properties, colloid substrate properties, aqueous chemistry, and colloid concentration, but independent of the properties of the immobile media through which transport occurs. The radionuclides that are irreversibly absorbed onto the colloids are modeled to transport in a manner identical to the colloids onto which they are sorbed.

A site scale transport model was built on the basis of the site scale flow model using particle tracking and incorporating matrix diffusion, dispersion and sorption.

The flow paths and travel times are found to agree with the regional geochemistry data[2], providing additional confidence in the transport model. The ^{14}C ages for water samples from wells within the site scale model domain range from about 11430 years to 16390 years[2], which compare well with the model predictions[7].

VI. MODEL UNCERTAINTIES

Model form uncertainty in regards to the transport model is explicitly addressed by discussing alternative conceptual models. Although the base case model is designed to accurately represent the saturated zone transport, the transport model was selected such that it resulted in transit times faster than those expected using alternative conceptual models. Certain alternate conceptual models were accommodated within the current mathematical model framework by modifying the uncertainty distributions of the effective parameters.

Certain alternative conceptual models lead to transit times greater than the base-case results and were not included in the transport model. These were[7]: 1) the matrix material intervening the fracture zones in volcanics permits significant fluid flow, 2) radionuclides undergo irreversible sorption on the rock surfaces, 3) radionuclides precipitate forming solid rock phases, 4) water table rises due to future climate changes, and 5) solutes diffuse into low permeability zones in alluvium.

The alternative conceptual models implicitly included in the site-scale saturated zone transport model by modifying the uncertainty distributions of the effective parameters were[7]: 1) sorption reactions are not instantaneous, 2) sorption reactions are non-linear, 3) sorption parameters vary locally within the model domain, and 4) in alluvium, high-permeability channels exist that can provide preferential pathways for flow and transport.

V. TRANSPORT PARAMETERS

The input parameters of importance to the transport model, along with the range of values for each are presented in Table 1. Uncertainty in regards to the values of transport model is unavoidable given the sparseness of observed data and the limited amount of information available. Much of the data used for deriving parameter values are from experiments conducted on spatial and temporal scales much smaller than those in the site scale SZ model. Hence uncertainty ranges and distributions for the various transport parameters were developed[7] on the basis of available data from site specific field experiments, laboratory measurements on drill-cuttings and core samples obtained from the site, well logs, relevant data from other sites, and expert elicitation. The details of the basis of development of these uncertainty distributions are presented elsewhere[14]. The base case transport model presents the case of non-sorbing radionuclides. Base-case values for most parameters were chosen to be the median values for the distributions except for sorption related parameters.

TABLE 1. Input Parameters and Range of Values[7] for the Site-Scale Saturated Zone Transport Model

Parameter	**Base-Case (Stochastic Uncertainty Range)**	**Parameter**	**Base-Case (Stochastic Uncertainty Range)**
Specific discharge multiplier	1 (0.112 to 8.933)	**Permeability horizontal anisotropy ratio (N-S/E-W)**	5 (0.05 to 20)
Bulk density in alluvium	1910 (1669 to 2151) kg/m^3	**k_d in alluvium**	0.0 [b] (0 to 10^4) mL/g
Effective porosity, alluvium	0.18 (0.02 to 0.3) fraction	**Colloid retardation factor in alluvium**	1 [b] (8 to 5195)
Flowing interval porosity	0.001 (10^{-5} to 0.1) fraction	**Flowing interval spacing**	25.773 (1.86 to 80) m
Matrix porosity in volcanics	0.15 – 0.25 [a] fraction	**Diffusion coefficient in volcanics**	5.0×10^{-11} $5.0 \times (10^{-12}$ - $10^{-10})$ m^2/s
Matrix sorption coefficient in volcanics	0 [b] (0 to 10,000) mL/g	**Colloid retardation factor in volcanics**	1 [b] (6 to 794)
Groundwater concentration of colloids	0 [b] (10^{-9} to 2×10^4) g/mL	**Sorption coefficient onto colloids**	0 [b] (1 to 1.0×10^7) mL/g
Unretarded fraction of colloids	0 [b] (0.00034 to 0.0017)	**Dispersivity, longitudinal**	10 (0.1 to 2000) m
Dispersivity, transverse, horizontal	0.05 (0.0005 to 10) m	**Dispersivity, transverse, vertical**	0.0005 (5×10^{-6} to 0.1) m

[a] this parameter is not treated stochastically in the TSPA, but assigned fixed values that differ on a unit by unit basis
[b] base case value represents non-sorbing radionuclides.

VI. RESULTS

A plan view of the particle tracks for nine particles with starting locations distributed over the repository footprint are shown in Figure 3, along with flow paths deduced from geochemistry data[2]. A good agreement is noted between the two sets of flow paths. The base case (Table 1) serves as a reference point for exploring the role of processes and features of the system in subsequent simulations. The particle source location was chosen near the middle of the repository footprint at the water table. Calculations were performed for an instantaneous release of particles at the source location.

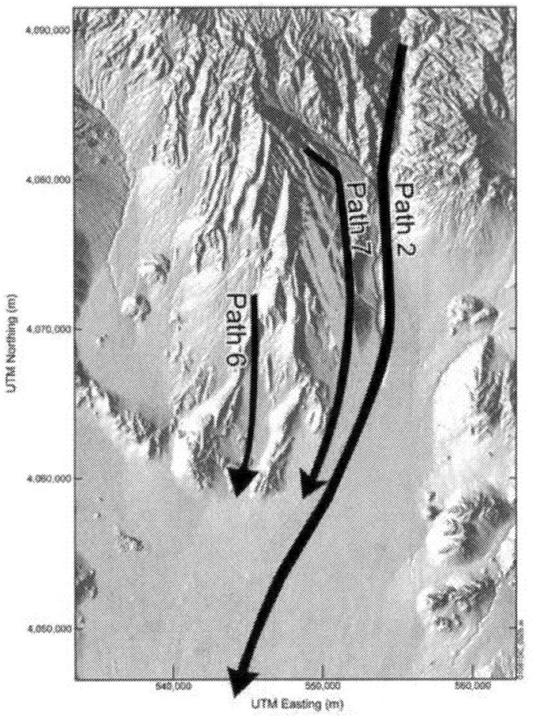

Fig. 3 Particle Tracks Resulting From The Site-Scale Saturated Zone Base-Case Transport Model (In Blue), Compared Against The Transport Pathways Deduced From Hydrochemistry Data (In Black).

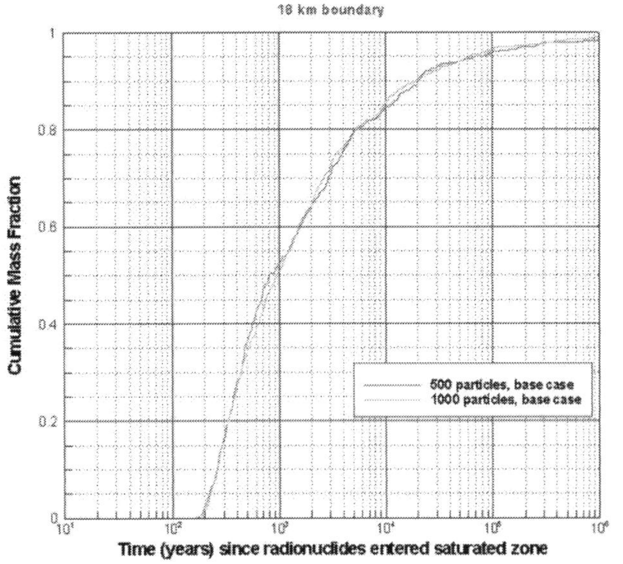

Fig. 4. Breakthrough Curves At The 18-Km Boundary For The Transport Base Case (500 Particles For A Conservative, Nonsorbing Radionuclide) With 1000 Particle Case For Comparison.

The BTC at the boundary of the accessible environment at 18 km was calculated by releasing 500 particles and by outputting the cumulative number of particles crossing an east-west vertical plane across the entire width and depth of the model. This is the BTC for a conservative, nonsorbing radionuclide in the absence of radioactive decay. Breakthrough times on the order of hundreds of years are predicted for the bulk of the mass arriving at the water table, with transport times extending into thousands of years for the slowest moving 20% of the mass. The time for 50% of the initial mass to breakthrough at the 18km compliance boundary (BT) for the curve in Figure 4 is 810 years. For comparison, also shown in Figure 4 by the green curve is the BTC for base case with initial number of particles being 1,000 instead of 500. As expected, although this curve is smoother than the BTC for 500 particles, they compare well with each other.

The results from simulations that evaluate the effect of changes in the specific discharge are shown in Figure 5. In these simulations, the base-case flow model was modified to scale the input recharge fluxes by the same factor as the rock permeabilities to preserve the model calibration. The results show the sensitivity of the output to the level of uncertainty in this parameter. For the case of upper limit of specific discharge multiplier where the base-case fluxes are multiplied by a factor of 8.933, BT is 31 years (with extremely fast fluid flow such as would be expected for the very unlikely case of a high-permeability channel going continuously over the distance of 18 km in a highly faulted region). The lower limit of specific discharge multiplier, a factor of 0.112, leads to a BT of 52,840 years.

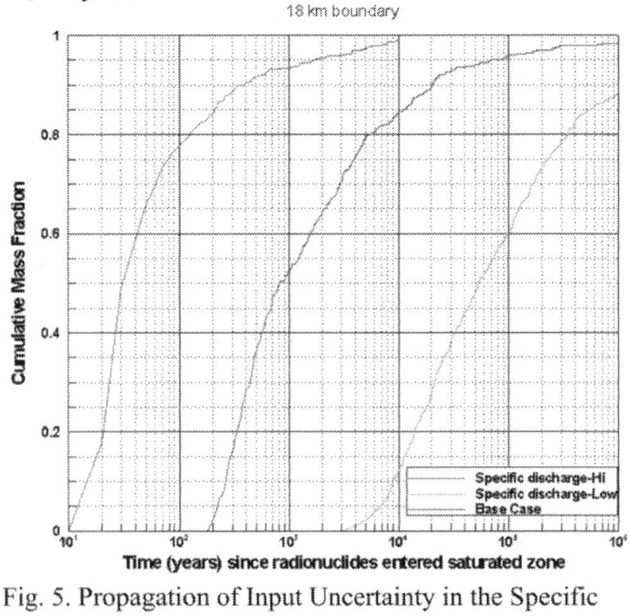

Fig. 5. Propagation of Input Uncertainty in the Specific Discharge to the Output Breakthrough Curves at the 18-km Boundary.

The results from simulations that evaluate the effect of changes in effective diffusion coefficient in volcanics on the BTCs are shown in Figure 6. Also shown in the figure for comparison is the BTC for practically zero diffusion (diffusion coefficient of 10^{-50} m^2/s). The range of 5.0×10^{-12} to 5.0×10^{-10} (m^2/s) in the input value results in the range of 245 to 6,324 years for BT. Note that this parameter affects only transport through the volcanics and not through the alluvium.

Figure 7 shows two simulations of sorbing radionuclides. The curve labeled "Matrix sorption" allows sorption in the matrix continuum of the volcanic tuffs and reflects a small matrix sorption coefficient of 1.3 mL/g. Including only this process yields a BTC similar to the base-case BTC but with significantly delayed transport times. The final curve in the figure shows the influence of sorption in the alluvium along with the fractured volcanics. Travel times largely in excess of 10,000 years are predicted in the saturated zone alone for a sorption coefficient of 6.3 mL/g, a value meant to fall in the range of k_d values for weakly sorbing radionuclides such as neptunium. As seen from these curves sorption in the alluvium can increase the transport time by orders of magnitude for even the weakly sorbing radionuclides such as neptunium.

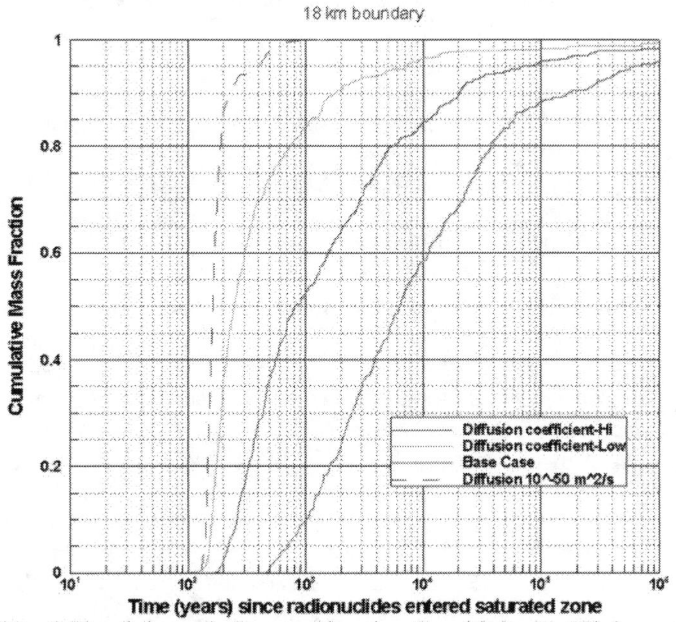

Fig. 6. Breakthrough Curves Showing Sensitivity To High And Low Values Of Diffusion Coefficient.

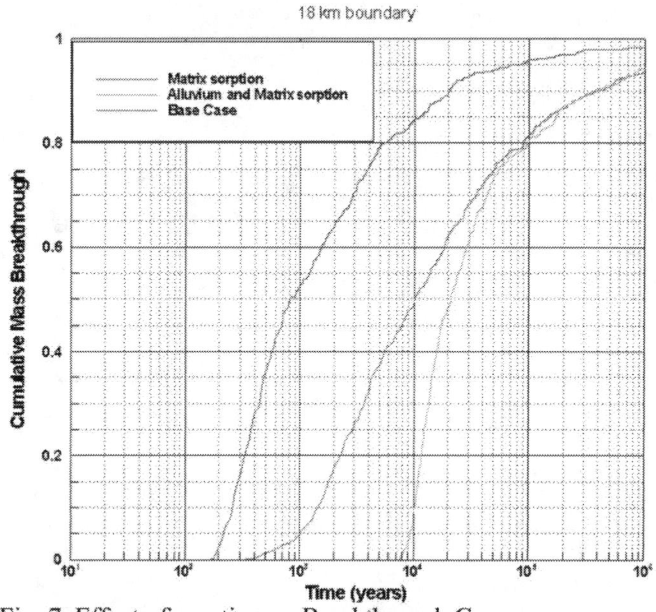

Fig. 7. Effect of sorption on Breakthrough Curves.

A small fraction of colloids with irreversibly attached radionuclides (0.034% to 0.17%, Table 1) has the potential to travel unretarded through the SZ. This fraction has the potential of traveling through within a few tens to a few hundreds of years depending on the value of specific discharge. This fraction is accounted for explicitly in the dose calculations performed for the TSPA. For the largest fraction of the colloids (99.83% or greater), retardation occurs due to the reversible filtration of the colloids themselves in the alluvium and in the fractures within the volcanic units. The radionuclides do not diffuse into the volcanic matrix due to the large size of the colloids. Thus, the matrix diffusion coefficient is set to 0 in this simulation. Figure 8 shows the expected behavior of colloidal-species transport with retardation included in alluvium alone, compared to the base-case aqueous species transport. BT's in the range of 1859 to 688,357 years result for a range of retardation factors of 8 to 5188, indicating significant retardation. Thus, the modeling suggests that for majority of the radionuclide inventory that reaches the saturated zone within the compliance time period, the saturated zone is expected to impart a significant transport time delay.

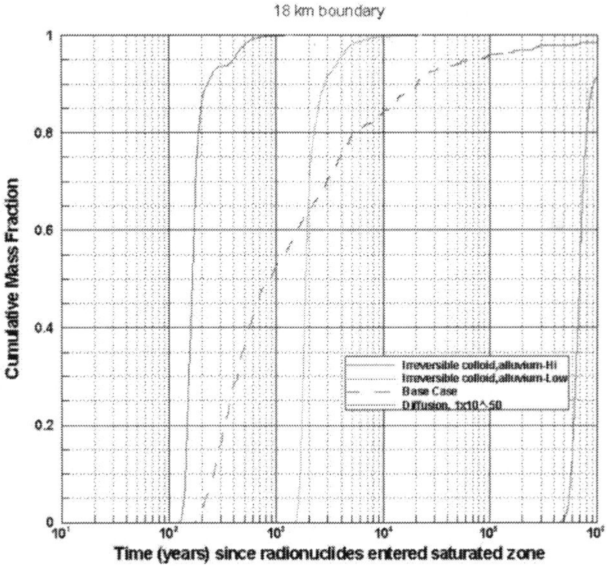

Fig. 8. Comparison of Breakthrough Curves for the Base Case and Radionuclides Irreversibly Attached to Colloids.

The results from simulations that evaluate the effect of changes in distribution parameters for reversible sorption onto colloids in alluvium are shown in Figure 9. For very low values of K_c (10^{-9} m^2/s) the transport is essentially the same as that for base case, and the fastest transport results if the diffusion coefficient in volcanics is also chosen very low, as shown by the solid blue curve giving a BT value around 245 years. The upper limiting value for K_c is of 2×10^3 for americium onto hematite colloids. Using this, the upper limit of the adjusted value of k_d in alluvium becomes 5.0 mL/g. The adjusted minimum effective diffusion coefficient in volcanics of 1.25×10^{-18} m^2/s is used for this case. This is the high curve (labeled "Reversible colloid alluvium-Hi") in Figure 9, giving a BT of 7,023 years, indicating significant delay even for very little matrix diffusion, compared to the base case value of 810 years.

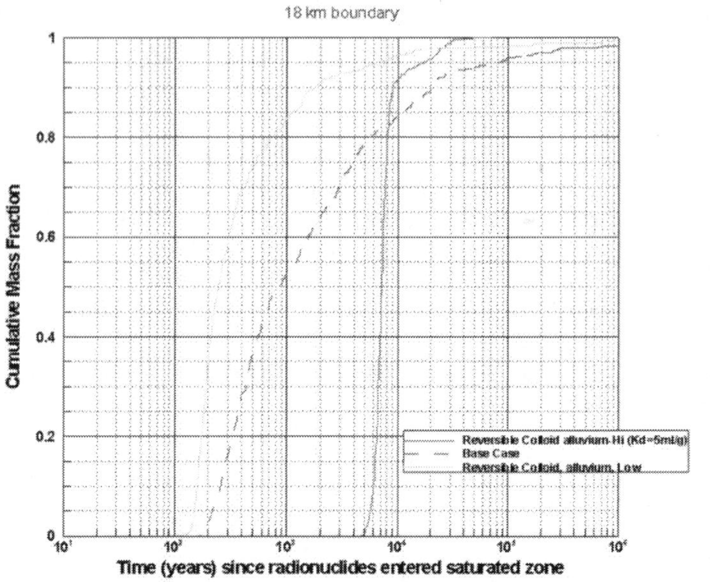

Fig. 9. Propagation of Input Uncertainty in the Colloid Retardation Factor in Alluvium for Reversible Colloids to the Output Breakthrough Curves at the 18-km Boundary.

VI. SUMMARY

A site scale transport model has been developed to simulate the potential transport of radionuclides within the saturated zone at Yucca Mountain for use in the TSPA calculations. It is based on a steady state flow model using the effective continuum approach. The model predicts southerly to south-easterly and sub-horizontal flow near the water table in the vicinity of the Yucca Mountain. The transport model includes the solute transport mechanisms of advection, dispersion, matrix diffusion, retardation due to solute-matrix interactions, and colloid-facilitated transport. The model was validated by comparison with flow paths derived from geochemical data, and by comparing the transit times with those inferred from the ^{14}C data. Uncertainty in the model form is accounted for by considering alternate conceptual models and selecting the base case model that leads to the shortest transit time, i.e. the most conservative model, as well as by modifying the uncertainty distributions of transport parameters.

The base case breakthrough time at the compliance boundary to accessible environment at 18km downstream of the repository for 50% concentration is 810 years. The transport model calculations show that results very with changes in the values of the specific discharge and the matrix diffusion coefficient in volcanic units. Sorption in volcanic rock matrix and in alluvium leads to significant retardation of the sorbing of radionuclides. A small fraction of colloids, explicitly accounted for in TSPA, with irreversibly attached radionuclides (0.17% to 0.034%, Table 1) has the potential to travel unretarded through the SZ within a few hundreds of years depending on the value of specific discharge. However, for the largest fraction of the colloids (99.83% or greater) with radionuclides attached irreversibly, there is significant retardation, with the range of values of BTs between 4600 to 28519 years. For radionuclides attached reversibly to the colloids, the range of values for BTs is 254 to 7023 years for retardation in alluvium alone, and 165 to greater than a million years for retardation in volcanics alone.

ACKNOWLEDGEMENTS

This manuscript has been authored by Los Alamos National Laboratory, and by Sandia National Laboratories. Sandia National Laboratories is a multi-program laboratory operated by Sandia Corporation, a Lockheed Martin Company, for the United States Department of Energy's National Nuclear Security Administration under contract DE-AC04-94AL85000. The United States Government retains and the publisher, by accepting the article for publication, acknowledges that the United States Government retains a non-exclusive, paid-up, irrevocable, world-wide license to publish or reproduce the published form of this manuscript, or allow others to do so, for United States Government purposes. The views expressed in this article are those of the authors and do not necessarily reflect the views or policies of the United States Department of Energy, Sandia National Laboratories, or Los Alamos National Laboratory..

REFERENCES

1. Luckey, R.R.; Tucci, P.; Faunt, C.C.; Ervin, E.M.; Steinkampf, W.C.; D'agnese, F.A.; And Patterson, G.L. 'Status of Understanding of the Saturated-Zone Ground-Water Flow System at Yucca Mountain, Nevada, as of 1995.' Water-Resources Investigations Report 96-4077. Denver, Co. (1996)
2. SNL 2007. *Saturated Zone Site-Scale Flow Model*. MDL-NBS-HS-000011 REV 03. Las Vegas, Nevada: Sandia National Laboratories. ACC: DOC.20070626.0004.
3. SNL (Sandia National Laboratories) 2007. *Hydrogeologic Framework Model for the Saturated Zone Site-Scale Flow and Transport Model.* MDL-NBS-HS-000024 REV 01. Las Vegas, Nevada: Sandia National Laboratories. ACC: DOC.20070411.0003.
4. CRWMS M&O (1998)- '*Saturated Zone Flow and Transport Expert Elicitation Project.*' Deliverable SL5X4AM3. Las Vegas, Nevada.
5. Luckey, R.R.; Tucci, P.; Faunt, C.C.; Ervin, E.M.; Steinkampf, W.C.; D'Agnese, F.A.; and Patterson, G.L. 1996. Status of Understanding of the Saturated-Zone Ground-Water Flow System at Yucca Mountain, Nevada, as of 1995. Water-Resources Investigations Report 96-4077. Denver, Colorado: U.S. Geological Survey. ACC: MOL.19970513.0209.
6. Robinson, B.A.. 'A Strategy for Validating a Conceptual Model for Radionuclide Migration in the Saturated Zone Beneath Yucca Mountain.' Radioactive Waste Management and Environmental Restoration, 19, (1-3), 73-96. Yverdon, Switzerland: Harwood Academic Publishers.
7. SNL (Sandia National Laboratories) 2008. *Site-Scale Saturated Zone Transport*. MDL-NBS-HS-000010 REV 03 AD 01. Las Vegas, Nevada: Sandia National Laboratories. ACC: DOC.20080121.0003.
8. 'Yucca Mountain Science & Engineering Report' Rev 01 2002, DOE/RW-0539-1, U.S.Department of Energy, Yucca Mountain Site Characterization Office, POBox 364629, Las Vegas, Nevada.
9. Reimus, P.W.; Haga, M.J.; Adams, A.I.; Callahan, T.J.; Turin H.J.; Counce D.A. 'Testing and parameterizing a conceptual solute transport model in satirated fractured tuff using sorbing and onosorbing tracers in corss-hole tracer tests.' Journal of Contaminant Hydrology, Elsevier Publishing Co. (2002)
10. Ding, M.; Kelkar, S.; Fabryaka-Martin, J.; Caporuscio, F.; And Meijer, A. 'Surface Complexation Modeling of Radionuclide Sorption in the Saturated Zone of Yucca Mountain Rocks.' International High-Level Radioactive Waste Management Conference, September 7-11, 2008, Las Vegas, NV, USA (2008)
11. Viswanathan, H.S.; Wolfsberg A.V.; And Reimus, P.A. 'Colloid Facilitated Transport in Fractured Rocks: Parameter Estimation and Comparison with Experimental Data.' International High-Level Radioactive Waste Management Conference, March 30-April 3, 2003, Las Vegas, NV, USA. (2003)
12. Kersting, A.B.; Efurd, D.W.; Finnegan, D.L.; Rokop, D.J.; Smith, D.K.; and Thompson, J.L. 1999. "Migration of Plutonium in Ground Water at the Nevada Test Site." *Nature*, 397, (6714), 56-59. London, England: Macmillan Journals.
13. Zyvoloski, G.A.; Robinson, B.A.; Dash, Z.V.; and Trease, L.L. 1997. Summary of the Models and Methods for the FEHM Application—A Finite-Element Heat- and Mass-Transfer Code. LA-13307-MS. Los Alamos, New Mexico: Los Alamos National Laboratory. TIC: 235587.
14. SNL (Sandia National Laboratories) 2008. Saturated Zone Flow and Transport Model Abstraction. MDL-NBS-HS-000021 REV 03 AD 02. Las Vegas, Nevada: Sandia National Laboratories. ACC: DOC.20080107.0006.

Colloid-Facilitated Transport of Plutonium in Saturated Porous Med

Amr Abdel-Fattah[1], S. Doug Ware[1], Paul Reimus[1], and Sean Reilly[2]. (1) Earth and Environmental Sciences Division, Los Alamos National Laboratory, PO Box 1663, Los Alamos, NM 87544, (2) Chemistry Division, Los Alamos National Laboratory, PO Box 1663, Los Alamos, NM 87544

Packed column experiments were conducted to evaluate colloid-facilitated transport of Plutonium, Pu (V), and the suitability of selected synthetic colloids as tracers to natural colloids in saturated alluvium from a well at the southwest corner of the Nevada Test Site. The well is located along a projected flow pathway from the proposed high-level nuclear waste repository at Yucca Mountain, Nevada. Natural colloids and water used in the experiments were collected directly from groundwater pumped from the same well. Pu(V), which was electrochemically prepared and verified by UV-Visible absorbance spectroscopy to be over 99% pure (in the +5 oxidation state), was adsorbed onto the natural colloids in batch sorption experiments conducted in filtered water from the same well prior to the column experiments. After adsorbing about 40% of the total Pu from solution onto natural colloids (60% in solution), the resulting tracer was injected into two separate columns having different porosities, along with tritiated water and two different sizes of fluorescent-dye-tagged carboxylate-modified-latex (CML) microspheres (500- and 190-nm in diameter). Two experiments were conducted in each column, one with a relatively low injection flow rate (~ 1.2 mL/hr, resulting in a residence time of 45-50 hrs) and the other with a higher flow rate (~ 6 mL/hr with a residence time of 8 to 9 hrs). In both columns, results showed that natural colloids traveled through the columns essentially unretarded, with almost 100% recovery at the two flow rates employed. The natural colloids arrived slightly earlier than the tritiated water in all experiments, indicating a slightly smaller effective pore volume for the colloids. No soluble Pu was detected in the column effluents, while Pu adsorbed onto natural colloids was almost completely recovered in all experiments (100% at the higher flow rate and 95% at the longer residence times), indicating that natural colloids are capable of facilitating the transport of otherwise immobile Pu in subsurface alluvium at the NTS. These results are consistent with the very slow desorption of Pu from natural colloids observed in separate batch desorption experiments, which indicated essentially no measurable desorption of Pu over the time scales of the column experiments (~ 3 to 4%). The above results combined suggest that the desorption rates in the columns, while slow, may have been enhanced by the presence of competing mineral surfaces and collisions of the Pu-bearing colloids with these surfaces. Results also showed that the 500-nm CML microspheres were significantly attenuated in the column experiments, with recoveries of about 70% at the higher flow rate and between 45 and 60% at the lower flow rate, depending on the porosity of the column. The 190-nm microspheres, however, exhibited just slightly more attenuation than the natural colloids, suggesting that these microsphers (less than 200-nm in diameter) are good representatives to the transport of natural colloids in saturated alluvium. Colloids smaller than about 200-nm in size particularly play the major role in this enhanced transport phenomenon.

Dynamic modeling of the microbial community prior to and during uranium-bioremediation

Kai Zhuang[1]*, Mounir Izallalen[2], Paula Mouser[2], Hanno Richter[2], Carla Risso[2], Derek R. Lovley[2], Radhakrishnan Mahadevan[1]

[1]Department of Chemical Engineering, University of Toronto, Canada
[2]Department of Microbiology, University of Massachusetts, USA

1: University of Toronto
Department of Chemical Engineering and Applied Chemisty,
Institute of Biomaterials and Biomedical Engineering,
Toronto, ON, Canada M5S3E5

2: University of Massachusetts
Department of Microbiology,
Amherst, MA, USA 01003

*Corresponding author
Email: k.zhuang@utoronto.ca

Additional Contacts
Email: krishna.mahadevan@utoronto.ca

Running title: Community modeling of uranium-bioremediation
Submission to: ISMEJ

ABSTRACT

We have created a dynamic genome-scale metabolic model of *Geobacter sulfurreducens* and *Rhodoferax ferrireducens*, the two primary iron-reducers in uranium-contaminated grounds, in order to understand the community competition prior to and during uranium-bioremediation.

The simulation results agree with experimental measurements, and suggest that the community competition is modulated by two factors: the ability of *G. sulfurreducens* to fix nitrogen under ammonium limitation, and a rate vs. yield trade-off between these two organisms. Prior to acetate amendment, if ammonium is limited, *G. sulfurreducens* dominates due to its ability to fix nitrogen. However, if the system contains abundant ammonium, *R. ferrireducens*, a yield-strategist, is favoured because the acetate flux is very low. During acetate amendment, the high acetate flux strongly favours *G. sulfurreducens*, a rate-strategist. The model also predicts the up-regulation of respiration in *G. sulfurreducens* during nitrogen fixation by sacrificing its biomass yield, leading to an increase in uranium reduction under low ammonium conditions. This model will be an important tool for the designing of effective uranium-bioremediation strategies.

INTRODUCTION

The anoxic ferric microbial community contains abundant numbers of *Geobacteraceae* and *Rhodoferrix* species, dissimilatory iron-reducing bacteria capable of extra-cellular electron transfers (Chaudhuri & Derek R Lovley, 2003; Cummings et al., 2003; Methé et al., 2003; Derek R Lovley et al., 2004; Esteve-Núñez et al., 2005; Finneran et al., 2003). These organisms are capable of using uranium(VI) as a terminal electron acceptor; the process converts soluble uranium(VI) to its precipitate form, uranium(IV), thus prevents uranium from entering the ground water stream. There are significant physiological differences between *Geobacter sulfurreducens* and *Rhodoferax ferrireducens*, the model species of these two genera of iron reducers. Compared to *G. sulfurreducens*, *R. ferrireducens* can oxidize a wider range of substrates, including glucose, using iron(III) as a terminal electron acceptor (Finneran et al., 2003; Chaudhuri & Derek R Lovley, 2003). On the other hand, while *G. sulfurreducens* is dependent on fermenters that supply the typical electron donor for this microorganism, such as acetate and H_2 (Derek R Lovley & Phillips, 1989), it is capable of fixing nitrogen under ammonium deprivation. Furthermore, *G. sulfurreducens* grows faster than *R. ferrireducens* at the cost of energetic efficiency (Chaudhuri & Derek R Lovley, 2003; Finneran et al., 2003; Esteve-Núñez et al., 2005). Acetate amendment has been shown to be an effective bioremediation strategy by stimulate the growth of these iron-reducers; however, after the initial dominance of *Geobacter*, continued acetate injection stimulates the growth of sulfate reducers, causing the reduction of uranium(VI) to stop (Chang et al., 2005; Vrionis et al., 2005; Anderson et al., 2003). In order to devise environmental biotechnology strategies for bioremediation of uranium-contaminated sites (N'Guessan et al., 2008; D R Lovley, 2003), we need a more in-depth understanding of the ecology of this metal reducing community (Figure 1).

Microorganisms in nature exist in complex communities, either in cooperation or in competition. The composition of the community and the metabolic states of its members are highly sensitive to the ever-changing environment. The activities of the community members can modify the environment, which further modifies community composition and behavior. For example, the rumen microbial community composition varies significantly based on the host's dietary input (Tajima et al., 2000), and the community's digestive performance is decided by the community's composition. In addition, given the microbial diversity and abundance (Konopka, 2006), one cannot truly understand nature without understanding microbial communities. However, modeling and physiology studies have mainly focused on pure cultures in the past (McMahon et al., 2007; D R Lovley, 2003).

The advent of systems biology has greatly increased our understanding of the complex processes present at the organism level. Applying a similar systems approach to the study of microbial communities would reveal some insights into the intricate interactions and emergent properties among these communities. (Konopka, 2006) Flux balance analysis is a systems biology modeling methodology that does not require a complete kinetic description, yet is able to incorporate as much genomic and physiological information as possible (Lee et al., 2006). In FBA models, the metabolic pathways are mathematically represented as a stoichiometric matrix. After the applications of thermodynamic and physiologic constraints, linear programming is used to calculate the metabolic fluxes through each of the pathways by assuming a cellular objective function such as the maximization of biomass yield (Jeremy S Edwards et al., 2002; Price et al., 2004; Warren & Jones, 2007). FBA models of pure cultures, such as *Escherichia coli* (Rafael U Ibarra et al., 2002; J S Edwards et al., 2001; Feist & Bernhard Ø Palsson, 2008; Reed & Bernhard Ø Palsson, 2003; Reed et al., 2003; Feist et al., 2007) and *G. sulfurreducens* (R

Mahadevan et al., 2006; Segura et al., 2008), are capable of accurate growth predictions. This modeling approach has been useful for both understanding the behavior of biological systems in complex environment and for engineering purposes (Burgard et al., 2003; Hjersted et al., 2007; Pharkya et al., 2003; Anesiadis et al., 2008; Pharkya & Maranas, 2006; Izallalen et al., 2008).

Previously, Stolyar *et al.* (2007) created the first FBA-based metabolic model of a mutualistic microbial community. In this model, the stoichiometric matrices of *Desulfovibrio vulgaris* and *Methanococcus maripaludis* are connected together directly (Stolyar et al., 2007). The authors assumed that since the two species grew at the same rate, did not explicitly consider the dynamic changes in the biomass concentrations of the individual species. While this approach may be appropriate when the microorganisms are inter-dependent, it is inappropriate in complex ecological settings where the community composition is dynamic.

In this paper, we present a dynamic metabolic model of *G. sulfurreducens* and *R. ferrireducens*, utilizing a dynamic multi-species metabolic modeling framework (DMMM). It is the first model to dynamically integrate multiple genome-scale metabolic models. By modeling the coculture of these two organisms, along with recent physiological and genetic insights, we were able to explain the composition and the dynamic changes in the microbial community prior to and during uranium bioremediation.

METHOD

1. The Dynamic Multi-species Metabolic Modeling framework

A community metabolic model must account for the metabolic exchanges between species and with the environment, as well as the changes in biomass of the relevant species. The

DMMM framework is based on the dynamic flux balance analysis developed by Mahadevan *et al.* (Radhakrishnan Mahadevan et al., 2002; R. Luo et al., 2006). In the DMMM framework, each species is treated as an object containing a metabolic model and a set of operations. The metabolic model can be described either by Monod-kinetic equations or by FBA, so long as the model can predict a set of metabolic exchange fluxes and a biomass flux, given a set of input parameters. The species are embedded inside an environment object. The environment object also contains the metabolite and biomass concentrations inside the environment. The simulation procedure is summarized as the following: [1] the metabolite production/consumption rate is calculated by multiplying the flux with the respective biomass. [2] The sum of production/consumption rates for each metabolite is integrated over the simulation time.

2. Modeling the anoxic ferric iron-reducing community

The well-studied *G. sulfurreducens* and *R. ferrireducens* are used to represent the *Geobacteraceae* and *Rhodoferax* genera in our study of the anoxic ferric iron-reducing community. The previously published *G. sulfurreducens* genome-scale FBA model (R Mahadevan et al., 2006) was used to represent the *Geobacteraceae* species, and a newly developed *R. ferrireducens* genome-scale FBA model was used to represent the *Rhodoferax* species. The *Geobacter* model is expanded and updated with the current physiological data - specifically, the cost of nitrogen fixation is increased to 16ATP from 8ATP, based on experimental observations. Maximization of biomass yield is assumed to be the cellular objective for both organisms. The acetate, glucose, Fe(III), Fe(II), CO_2, NH_4 exchange fluxes of both organisms represent the sources and sinks of metabolites from and into environment for each species.

The DMMM framework is applied to the modeling of this iron-reducing community (Figure 2). As in the dFBA formulation (Radhakrishnan Mahadevan et al., 2002), the flux of primary carbon and the energy source are restricted by their concentrations in the environment using a Michaelis-Menten equation. The acetate transport kinetics of *G. sulfurreducens* has been determined by measuring the C-14 labeled acetate uptake (Richter et al., Personal Communication). Three different acetate uptake mechanisms, each with different K_s and V_{max}, were observed in that study. The *G. sulfurreducens* acetate uptake flux in the DMMM framework is constrained by the summation of three experimentally identified Michaelis-Mentin expressions. The acetate uptake rate of *R. ferrireducens* is calculated from previously published batch growth data (Finneran et al., 2003), and its K_s is assumed to be the same as the lowest of the *G. sulfurreducens* K_s. The ATP maintenance requirement for *G. sulfurreducens* is 0.45 mol ATP/gDw, and the ATP maintenance requirement for *R. ferrireducens* is calculated to 0.7 mol ATP/gDw (Risso et al., Personal Communication).

$$v_{ac}^{Gs} = \frac{14.3S}{S+0.405} + \frac{2.37S}{S+0.023} + \frac{1.60S}{S+0.008}$$

$$v_{ac}^{Rf} = \frac{1.85S}{S+0.008}$$

3. Death phase modeling

Microorganisms enter death phase due to a failure to meet the maintenance energy requirements — neither *G. sulfurreducens* nor *R. ferrireducens* can grow once Fe(III) is depleted. The death rate of an organism has been calculated by multiply the yield (gDw/mol substrate) with the substrate maintenance flux (mol substrate/gDw)(Rittmann & McCarty, 2001; VanBriesen, 2002). We can find the death rate of the organism using the constraint-based metabolic model in a similar fashion. The substrate maintenance flux is the acetate flux needed to meet the ATP maintenance energy requirements, and can be found by minimizing for acetate while fixing the growth flux to zero. The death rate is calculated as the negative of the growth flux when the acetate uptake flux is set to the maintenance flux and the ATP maintenance requirement is set to zero.

4. Pre-injection simulations

Prior to acetate injection, acetate is generated in the anoxic sediments as a fermentation product. The uptake rates and concentration of organic chemicals, including acetate, has been determined experimentally in various sediment environments. Although no such experiment was done specifically for the anoxic ferric environment of interest, we can assume the rates and concentration to be very low (King & Klug, 1982; de Graaf et al., 1996). Acetate fluxes of 3 μM/day, 5 μM/day, 10 μM/day, 50 μM/day are used in the simulations, which is comparable to the acetate that would be generated from the environmental glucose turnover rate. From field experiments (Mousser *et al.*, Personal Communication), we know that the ammonium concentration are typically low, although high ammonium locations do exists occasionally. The steady state ammonium concentration is assumed to be either 0.4 mM or 0.005 mM in order to simulate either ammonium limitation or excess conditions. Fe(III) is assumed to be not limiting prior to injection, and is set to 30 mM. The initial biomass concentration for both *G. sulfurreducens* and *R. ferrireducens* are assumed to be 0.00001g/L, which corresponds to 10^7 cells/L. This is a reasonable value for an unamended aquifer with low organic carbon. (cite JCH) Environmental dilution rate is assumed to be 0.01 hr-1 (Yabusaki et al., 2007).

5. Post-injection simulations

During the acetate injection field experiment, a high acetate flux is artificially created in order to induce a non-substrate-limiting condition. In our simulation, a very high acetate flux of

1mM/day is used. The ammonium concentration is assumed to be similar to that of the pre-injection situation. While ammonium concentration is mostly very low, there is one significant outlier, well D08, with a maximum concentration of 0.4 mM and an average concentration of 0.3 mM. As with the pre-injection simulations, both ammonium excess and ammonium limiting conditions are simulated. Fe(III) limitation is hypothesized to occur after a prolonged period of acetate injection. Previous sediment sampling at bioremediation sites reports Fe(III) concentration to be in the range of 5 - 40 μmol/g (Anderson et al., 2003; Vrionis et al., 2005; Yabusaki et al., 2007), which includes both bioavailable and non-bioavailable Fe(III). If we assume sediment density to be 2 g/ml and 50% of Fe(III) are bioavailable, 2.5 - 20 mM of Fe(III) is bioavailable. For the post-injection simulations, the initial Fe(III) concentration is assumed to be 10mM (why not use 15 for both post and preinjection simulations rather than 30 there and 10 here…). As with the pre-injection simulations, initial biomass for both organisms are set to 0.00001g/L.

6. Validation and sensitivity analysis

Conditions of test-wells D08 and D05 during 2007 field experiment are simulated. The initial acetate concentration is set to zero, and the initial ammonium concentration is set to the *in situ* measurement values at time zero. The fluxes of ammonium and acetate is determined the solution of the constraint-based metabolic model. The *in silico* acetate injection is stopped at day 10, in accordance with the field-scale experiment. The *in situ* 16S rRNA measurement of *G. sulfurreducens* and *R. ferrireducens* abundance is used to calculate the *G. sulfurreducens and R. ferrireducens* fractions at day 0, 9, 18. Simulations are initialized with the *in situ G. sulfurreducens* to *R. ferrireducens* ratios (1:4 for D08 and 7:4 for D05). Since we do not know the organisms' biomass without acetate amendment and the bioavailable ferric iron concentration, we assume that the total biomass prior to injection is in the range of 0.000001 g/L to 0.0001 g/L, and the bioavailable Fe(III) is in the range of 2.5 - 20 mM. Nine simulations are performed at different initial biomass concentrations and Fe(III) concentrations for each well. *Geobacter* fraction, defined as [Gs]/([Gs]+[Rf]), is used as the metric to compare *in situ* and *in silico* organism abundance.

RESULTS

1. Pre-injection

The microbial competition prior to acetate injection was simulated for 800 hours in eight simulations. The simulations were differentiated by different acetate fluxes and whether ammonium is limiting. The acetate concentration in all eight simulations reached a steady state below 10 nM (Figure3), which is comparable in order of magnitude to previously reported acetate concentrations in relevant environment (de Graaf et al., 1996). In the ammonium excess simulations, ammonium is never exhausted; in the ammonium limiting simulations, ammonium is rapidly exhausted (not shown).

The fraction of *G. sulfurreducens* in the community is a measure of the relative success of the two organisms. A fraction value above 0.5 indicates *Geobacter* success, and a value below 0.5 indicates *Rhodoferax* success. Since the acetate concentrations reach a steady state by the 800th hour in all simulations, we can assume that the *Geobacter* fraction at this time is indicative of the overall success of the two organisms. This value is dependent on both acetate flux and whether ammonium limitation exists.

In the ammonium excess cases, the *Geobacter* fraction increases as the acetate flux increase, indicating that *G. sulfurreducens* is more successful in environments of higher acetate fluxes. In the ammonium limiting cases, the initial trend is similar until the point at which ammonium is exhausted. After that point, the *Geobacter* fraction increases at a rapid rate (Figure 3). These simulations suggest that *R. ferrireducens* has an advantage over *G. sulfurreducens* in a low substrate flux environment, however, due to its inability to fix atmospheric nitrogen, it cannot compete against *G. sulfurreducens* when ammonium is limited. A similar trend is observed in the *in situ* rRNA samples of *Geobacteraceae* and *Rhodoferrix* species. In well D08, which has a high ammonium concentration, the fraction of *Geobacteraceae* measured prior to acetate injection is 20%. In wells of low ammonium concentrations, including D05 and D02, the measured fractions are 63% and 58% respectively.

2. Post-injection

Microbial competition during acetate-injection bioremediation is simulated for 400 hours in two simulations, differentiated by whether ammonium is in excess or limiting. Both *G. sulfurreducens* and *R. ferrireducens* biomass increase initially; however, *G. sulfurreducens* outcompetes *R. ferrireducens* rapidly after the initiation of acetate injection, indicated by the rapid increase in *Geobacter* fraction value (Figure 4 f). In the simulation with high ammonium concentration and flux, ammonium is never exhausted. In the low ammonium simulation, ammonium is exhausted at around 50 hours. *G. sulfurreducens* enters the nitrogen fixation metabolic state - additional energy is spent on fixing nitrogen, leading to a decrease in biomass production. While *R. ferrireducens* cannot fix nitrogen, it is capable of maintaining its cell mass as maintenance does not require a nitrogen source (Figure 4 b). In both cases, once Fe(III) is exhausted, both organisms enter decay phase. Fe(III) is exhausted slightly later in the nitrogen limiting case (Figure 4 c). Acetate is never exhausted during the simulation; however, if more Fe(III) is available, acetate can reach exhaustion temporarily before Fe(III) exhaustion. (Fe(III) sensitivity simulations are not shown).

3. Validation and sensitivity analysis

We chose to simulate the microbial activities in test-wells D05 and D08 during the 2007 field experiment in order to validate our model, because they represent the two extreme cases with respect to ammonium availability. The ammonium concentration in D05 is much lower than the ammonium concentration in D08. The competition dynamics are similar to the simulations in the previous section. The *Geobacter* fraction appears to be sensitive to the initial biomass concentration, especially when ammonium is in excess. The simulations initiated with 0.00001 g/L and 0.000001 g/L of biomass agreed well with the experimentally measured microbial fraction in the respective wells (Figure 5), while the simulations initiated with 0.0001 g/L did not. This suggests the cellular density is in the range of 10^6 and 10^7 cells/L prior to acetate injection, a value that is consistent with previous publications (Yabusaki et al., 2007) and our simulation assumptions in the previous sections. The *Geobacter* fraction is insensitive to the bioavailable Fe(III) concentration. We have also performed sensitivity analysis on the kinetic parameter of *R. ferrireducens* by increasing and decreasing its Vmax and Ks on acetate by 20%. We found no significant changes in the simulation results. (Not shown).

4. Changes in the metabolic states

One advantage of the genome scale dFBA models is their capability to predict changes in metabolic states. We were able to detect the several different growth phases of *G. sulfurreducens* and *R. ferrireducens*. Beyond the donor-limited growth state and acceptor-limited decay state, ammonium-limited states were predicted for both organisms. Although we do not have ammonium uptake parameters, we can still predict the behavior of the organisms in the extreme cases. For *R. ferrireducens,* a pure maintenance phase is predicted where acetate is oxidized for maintenance purpose only and the biomass flux is zero. *G. sulfurreducens* has an ammonium-limited nitrogen fixation phase. When ammonium becomes limited, *G. sulfurreducens* switches to the nitrogen fixation state at an energetic cost. The energetic cost is reflected in the 30% reduction in growth yield during the nitrogen-fixation phase (Figure 6).

We compared the normal donor-limited growth phase of *G. sulfurreducens* with the nitrogen fixation state (Figure). An increased flux during the nitrogen fixation state is shown in red, whereas a decreased flux is shown in blue. More acetate enters the TCA cycle during the nitrogen fixation state and less acetate enters through acetate kinase, leading to less pyruvate production through pyruvate oxidoreductase. There is also an increase flux through ATP synthase (energy production), as well as through NADH dehydrogenase and cytochrome C reductase (electron transport chain). This indicates that during the nitrogen fixation state, respiration is elevated to provide additional ATP used for nitrogen fixation, at the cost of biomass production.

DISCUSSION

1. Growth Competition

It was previously unknown whether the dominance of *Geobacter* species during bioremediation reflected the pre-existing dominance of *Geobacter* species in uranium-contaminated grounds. The 16S rRNA analysis of sediment samples and nitrogen concentration measurements show that prior to acetate injection, the community distribution seems to be modulated by ammonium concentration — *Geobacter* species have an advantage under low ammonium conditions, while *Rhodoferax* species have an advantage under high ammonium conditions. As soon as acetate is injected into the ground, *Geobacter* species appear to be dominant regardless of ammonium concentrations, until other factors such as Fe(III) availability limits growth. This shows that the dominance of *Geobacter* species during bioremediation is induced by the acetate injection and does not reflect the community composition prior to injection.

It seems that *Geobacteraceae* and *Rhodoferrix* species have adopted very different evolutionary strategies despite being in the same environment. Thermodynamics dictates that organisms either wastefully utilize resources to maximize specific growth rate, or economically utilize resources to maximize growth yield (Thomas Pfeiffer & Sebastian Bonhoeffer, 2003; T Pfeiffer et al., 2001; Thomas Pfeiffer & Sebastian Bonhoeffer, 2004; Stefan Schuster et al., 2008). From growth kinetics of *G. sulfurreducens* and *R. ferrireducens*, we can see that *G. sulfurreducens* appears to be optimized for faster rate while *R. ferrireducens* is optimized for higher yield, thus we can consider *G. sulfurreducens* as a "rate-strategist" and *R. ferirreducens* as a "yield-strategist". Pfeiffer *et al.* (2001) showed that a high substrate flux condition favors rate-strategists, while yield-strategists thrive under lower substrate flux. (Pfeiffer:2001p581,

Schuster:2007). During uranium bioremediation, the acetate flux to the contaminated ground is very high, thus favoring the rate-optimizing *G. sulfurreducens.* Prior to the acetate injection, the survival of the microorganisms depends on the minute flow of organic chemicals such as glucose and acetate. Since *R. ferrireducens* has high yield on both substrates, it has significant advantage over *Geobacter* species. When ammonium is unlimited, *R. ferrireducens* significantly outcompetes *G. sulfurreducens*. However, since *Rhodoferax* species cannot fix nitrogen, their growth is limited under low ammonium conditions, allowing *Geobacter* species to succeed. Recent data (unpublished) suggests that it is possible that the subsurface clades of *Geobacteraceae* can use glucose. Therefore, the benefit of *Rhodoferrix*'s capability to use both substrates relative to Geobacteraceae is debatable.

2. Nitrogen fixation state

Since nitrogen fixation is associated with an energetic cost, environments with higher ammonium availability will have higher biomass concentrations. During the 2007 field experiment, well D08 had the highest ammonium concentration measurements. Despite having a significantly lower acetate concentration than the other wells due to water flow path, *Geobacteraceae* is significantly more abundant in this well. This suggests that ammonium is a major limiting substrate in anoxic ferric sediments. (Mousser *et al.*, Personal Communication).

The ammonium limitation in the natural environment has a significant impact on the effectiveness of uranium-bioremediation. The rate of uranium reduction can be calculated as the product of specific uranium reduction rate and biomass concentration. During nitrogen fixation, the specific uranium reduction rate can increase due to the up-regulation of respiration, at the expense of biomass production. This hypothesis is supported by experimental data. Well D08 has a significantly lower uranium reduction rate compared to the wells with lower ammonium concentrations, despite having the highest biomass concentration, possibly due to the increased respiration in the wells with limited ammonium. However, although D05 has a lower ammonium concentration than D02, D02 has a higher uranium reduction rate. The predicted respiratory up-regulation during the nitrogen fixation of *G. sulfurreducens* as well as the prediction of a maintenance state for *R. ferrireducens* illustrate a significant advantage of using dFBA models — their ability to detect metabolic changes in response to environmental cues. Ultimately, such models can be integrated with dynamic optimization techniques to predict the optimal ammonium concentrations at which the total uranium reduction over time is maximized. This insight also suggests bioremediation strategies that artificially reduce the ammonium concentration to induce nitrogen fixation at an energetic cost, thus achieving a higher uranium reduction rate. This strategy is reminescent of a recent study where the introduction of a futile cycle in *G. sulfurreducens* increased the energy demand and resulted in increased respiration, but lowered biomass yields (Izallalen et al., 2008).

3. Understanding bioremediation

Since the community composition shifts in favor of *G. sulfurreducens* immediately after acetate injection begins, *R. ferrireducens* plays a minor role in the overall consumption of acetate, ammonium and ferric iron. We can assume that uranium bioremediation is dominated by the interactions between *Geobacteraceae* and the environment, so we focus our attention on the behavior of *Geobacter* species during acetate injection. Our simulations suggest that the rapid utilization and exhaustion of bioavailable ferric iron lead to the decay of *Geobacter* species; this agrees with the previous hypothesis that the decrease in uranium removal upon prolonged acetate injection is caused by iron limitation. Continued acetate injection after the exhaustion of ferric iron leads to an excess of acetate, which stimulates the growth of other acetate oxidizing organisms. While exhaustion of the electron acceptor limits the growth of *Geobacter* species, other organisms can outcompete them for acetate, thereby hindering uranium bioremediation.

Sensitivity analyses suggest that both the acetate injection rate and the bio-available ferric iron concentration modulate the length of time *Geobacter* species can maintain their dominance. We cannot manipulate Fe(III) availability in nature easily; however, we can adjust the rate of acetate injection. The 2007 field experiment shows termination of acetate at day 10 does not lead to an immediate termination of uranium reduction, since acetate remains in the system until washed out by ground water flow. One strategy might be to decrease the acetate injection rate prior to Fe(III) depletion. Since *Geobacter* species will still dominate the community, the lowered acetate flux will be used for uranium reduction. There will not be enough acetate to stimulate the growth of other organisms. Although such a strategy is conceptually viable, the predictive design of such a strategy requires the aid of a computational model that integrates microbial physiology with geochemistry as well as the expansion of the current modeling effort to incorporate the growth physiology of sulfate reducers such as *Desulfobacter* species (cite a reference).

REFERENCES

Anderson, R. T., Vrionis, H. A., Ortiz-Bernad, I., Resch, C. T., Long, P. E., Dayvault, R., Karp, K., Marutzky, S., Metzler, D. R., Peacock, A., White, D. C., Lowe, M., & Lovley, D. R. 2003. Stimulating the in situ activity of Geobacter species to remove uranium from the groundwater of a uranium-contaminated aquifer. *Applied and Environmental Microbiology* **69**: 5884-91.

Anesiadis, N., Cluett, W. R., & Mahadevan, R. 2008. Dynamic metabolic engineering for increasing bioprocess productivity. *Metabolic Engineering*. Retrieved August 16, 2008, from http://www.ncbi.nlm.nih.gov/pubmed/18606241

Burgard, A. P., Pharkya, P., & Maranas, C. D. 2003. Optknock: a bilevel programming framework for identifying gene knockout strategies for microbial strain optimization. *Biotechnology and Bioengineering* **84**: 647-57.

Chang, Y., Long, P. E., Geyer, R., Peacock, A. D., Resch, C. T., Sublette, K., Pfiffner, S., Smithgall, A., Anderson, R. T., Vrionis, H. A., Stephen, J. R., Dayvault, R., Ortiz-Bernad, I., Lovley, D. R., & White, D. C. 2005. Microbial incorporation of 13C-labeled acetate at the field scale: detection of microbes responsible for reduction of U(VI). *Environmental Science & Technology* **39**: 9039-48.

Chaudhuri, S. K., & Lovley, D. R. 2003. Electricity generation by direct oxidation of glucose in mediatorless microbial fuel cells. *Nature Biotechnology* **21**: 1229-32.

Cummings, D. E., Snoeyenbos-West, O. L., Newby, D. T., Niggemyer, A. M., Lovley, D. R., Achenbach, L. A., & Rosenzweig, R. F. 2003. Diversity of Geobacteraceae species inhabiting metal-polluted freshwater lake sediments ascertained by 16S rDNA analyses. *Microbial Ecology* **46**: 257-69.

Edwards, J. S., Ibarra, R. U., & Palsson, B. O. 2001. In silico predictions of Escherichia coli metabolic capabilities are consistent with experimental data. *Nature Biotechnology* **19**: 125-30.

Edwards, J. S., Covert, M., & Palsson, B. 2002. Metabolic modelling of microbes: the flux-balance approach. *Environmental Microbiology* **4**: 133-40.

Esteve-Núñez, A., Rothermich, M., Sharma, M., & Lovley, D. 2005. Growth of Geobacter sulfurreducens under nutrient-limiting conditions in continuous culture. *Environmental Microbiology* **7**: 641-8.

Feist, A. M., Henry, C. S., Reed, J. L., Krummenacker, M., Joyce, A. R., Karp, P. D., Broadbelt, L. J., Hatzimanikatis, V., & Palsson, B. Ø. 2007. A genome-scale metabolic reconstruction for Escherichia coli K-12 MG1655 that accounts for 1260 ORFs and thermodynamic information. *Molecular Systems Biology* **3**: 121.

Feist, A. M., & Palsson, B. Ø. 2008. The growing scope of applications of genome-scale metabolic reconstructions using Escherichia coli. *Nature Biotechnology* **26**: 659-67.

Finneran, K. T., Johnsen, C. V., & Lovley, D. R. 2003. Rhodoferax ferrireducens sp. nov., a psychrotolerant, facultatively anaerobic bacterium that oxidizes acetate with the reduction of Fe(III). *International Journal of Systematic and Evolutionary Microbiology* **53**: 669-73.

de Graaf, W., Wellsbury, P., Parkes, R. J., & Cappenberg, T. E. 1996. Comparison of Acetate Turnover in Methanogenic and Sulfate-Reducing Sediments by Radiolabeling and Stable Isotope Labeling and by Use of Specific Inhibitors: Evidence for Isotopic Exchange. *Applied and Environmental Microbiology* **62**: 772-777.

Hjersted, J. L., Henson, M. A., & Mahadevan, R. 2007. Genome-scale analysis of Saccharomyces cerevisiae metabolism and ethanol production in fed-batch culture. *Biotechnology and Bioengineering* **97**: 1190-204.

Ibarra, R. U., Edwards, J. S., & Palsson, B. O. 2002. Escherichia coli K-12 undergoes adaptive evolution to achieve in silico predicted optimal growth. *Nature* **420**: 186-9.

Izallalen, M., Mahadevan, R., Burgard, A., Postier, B., Didonato, R., Sun, J., Schilling, C. H., & Lovley, D. R. 2008. Geobacter sulfurreducens strain engineered for increased rates of respiration. *Metabolic Engineering*. Retrieved August 16, 2008, from http://www.ncbi.nlm.nih.gov/pubmed/18644460

King, G. M., & Klug, M. J. 1982. Glucose Metabolism in Sediments of a Eutrophic Lake: Tracer Analysis of Uptake and Product Formation. *Applied and Environmental Microbiology* **44**: 1308-1317.

Konopka, A. 2006. Microbial ecology: Searching for principles. *Microbe* **1**: 175-179.

Lee, J. M., Gianchandani, E. P., & Papin, J. A. 2006. Flux balance analysis in the era of metabolomics. *Brief Bioinform*: bbl007.

Lovley, D. R. 2003. Cleaning up with genomics: applying molecular biology to bioremediation. *Nature Reviews. Microbiology* **1**: 35-44.

Lovley, D. R., Holmes, D. E., & Nevin, K. P. 2004. Dissimilatory Fe(III) and Mn(IV) reduction. *Advances in Microbial Physiology* **49**: 219-86.

Lovley, D. R., & Phillips, E. J. P. 1989. Requirement for a Microbial Consortium To Completely Oxidize Glucose in Fe(III)-Reducing Sediments. *Applied and Environmental Microbiology* **55**: 3234-3236.

Luo, R., Liao, S., Tao, G., Li, Y., Zeng, S., Li, Y., & Luo, Q. 2006. Dynamic analysis of optimality in myocardial energy metabolism under normal and ischemic conditions. *Molecular Systems Biology* **2**: 2006.0031.

Mahadevan, R., Bond, D. R., Butler, J. E., Esteve-Nuñez, A., Coppi, M. V., Palsson, B. O., Schilling, C. H., & Lovley, D. R. 2006. Characterization of metabolism in the Fe(III)-reducing organism Geobacter sulfurreducens by constraint-based modeling. *Applied and Environmental Microbiology* **72**: 1558-68.

Mahadevan, R., Edwards, J. S., & Doyle, F. J. 2002. Dynamic flux balance analysis of diauxic growth in Escherichia coli. *Biophysical Journal* **83**: 1331-40.

McMahon, K. D., Martin, H. G., & Hugenholtz, P. 2007. Integrating ecology into biotechnology. *Current Opinion in Biotechnology* **18**: 287-92.

Methé, B. A., Nelson, K. E., Eisen, J. A., Paulsen, I. T., Nelson, W., Heidelberg, J. F., Wu, D., Wu, M., Ward, N., Beanan, M. J., Dodson, R. J., Madupu, R., Brinkac, L. M., Daugherty, S. C., DeBoy, R. T., et al. 2003. Genome of Geobacter sulfurreducens: metal reduction in subsurface environments. *Science (New York, N.Y.)* **302**: 1967-9.

N'Guessan, A. L., Vrionis, H. A., Resch, C. T., Long, P. E., & Lovley, D. R. 2008. Sustained removal of uranium from contaminated groundwater following stimulation of dissimilatory metal reduction. *Environmental Science & Technology* **42**: 2999-3004.

Pfeiffer, T., Schuster, S., & Bonhoeffer, S. 2001. Cooperation and competition in the evolution of ATP-producing pathways. *Science (New York, N.Y.)* **292**: 504-7.

Pfeiffer, T., & Bonhoeffer, S. 2003. An evolutionary scenario for the transition to undifferentiated multicellularity. *Proceedings of the National Academy of Sciences of the United States of America* **100**: 1095-8.

Pfeiffer, T., & Bonhoeffer, S. 2004. Evolution of cross-feeding in microbial populations. *The American Naturalist* **163**: E126-35.

Pharkya, P., Burgard, A. P., & Maranas, C. D. 2003. Exploring the overproduction of amino acids using the bilevel optimization framework OptKnock. *Biotechnology and Bioengineering* **84**: 887-99.

Pharkya, P., & Maranas, C. D. 2006. An optimization framework for identifying reaction activation/inhibition or elimination candidates for overproduction in microbial systems. *Metabolic Engineering* **8**: 1-13.

Price, N. D., Reed, J. L., & Palsson, B. Ø. 2004. Genome-scale models of microbial cells: evaluating the consequences of constraints. *Nature Reviews. Microbiology* **2**: 886-97.

Reed, J. L., & Palsson, B. Ø. 2003. Thirteen years of building constraint-based in silico models of Escherichia coli. *Journal of Bacteriology* **185**: 2692-9.

Reed, J. L., Vo, T. D., Schilling, C. H., & Palsson, B. O. 2003. An expanded genome-scale model of Escherichia coli K-12 (iJR904 GSM/GPR). *Genome Biology* **4**: R54.

Rittmann, B. E., & McCarty, P. L. 2001. *Environmental Biotechnology*.

Schuster, S., Pfeiffer, T., & Fell, D. A. 2008. Is maximization of molar yield in metabolic networks favoured by evolution? *Journal of Theoretical Biology* **252**: 497-504.

Segura, D., Mahadevan, R., Juárez, K., & Lovley, D. R. 2008. Computational and experimental analysis of redundancy in the central metabolism of Geobacter sulfurreducens. *PLoS Computational Biology* **4**: e36.

Stolyar, S., Van Dien, S., Hillesland, K. L., Pinel, N., Lie, T. J., Leigh, J. A., & Stahl, D. A. 2007. Metabolic modeling of a mutualistic microbial community. *Molecular Systems Biology* **3**: 92.

Tajima, K., Arai, S., Ogata, K., Nagamine, T., Matsui, H., Nakamura, M., Aminov, R. I., & Benno, Y. 2000. Rumen Bacterial Community Transition During Adaptation to High-grain Diet. *Anaerobe* **6**: 273-284.

VanBriesen, J. M. 2002. Evaluation of methods to predict bacterial yield using thermodynamics. *Biodegradation* **13**: 171-90.

Vrionis, H. A., Anderson, R. T., Ortiz-Bernad, I., O'Neill, K. R., Resch, C. T., Peacock, A. D., Dayvault, R., White, D. C., Long, P. E., & Lovley, D. R. 2005. Microbiological and geochemical heterogeneity in an in situ uranium bioremediation field site. *Applied and Environmental Microbiology* **71**: 6308-18.

Warren, P. B., & Jones, J. L. 2007. Duality, thermodynamics, and the linear programming problem in constraint-based models of metabolism. *Physical Review Letters* **99**: 108101.

Yabusaki, S. B., Fang, Y., Long, P. E., Resch, C. T., Peacock, A. D., Komlos, J., Jaffe, P. R., Morrison, S. J., Dayvault, R. D., White, D. C., & Anderson, R. T. 2007. Uranium removal from groundwater via in situ biostimulation: Field-scale modeling of transport and biological processes. *Journal of Contaminant Hydrology* **93**: 216-35.

ACKNOWLEDGEMENT

This research was supported by the Office of Science (BER), U.S. Department of Energy, Cooperative Agreement No. DE-FC02-02ER63446 and Grant No. DE-FG02-07ER64367, as well Canada Foundation for Innovation and University of Toronto Open Fellowship. Special thanks: Laurence Yang for insightful discussions and Emma Janssen for grammatical corrections.

Figure 1. Conceptual model of uranium-bioremediation.

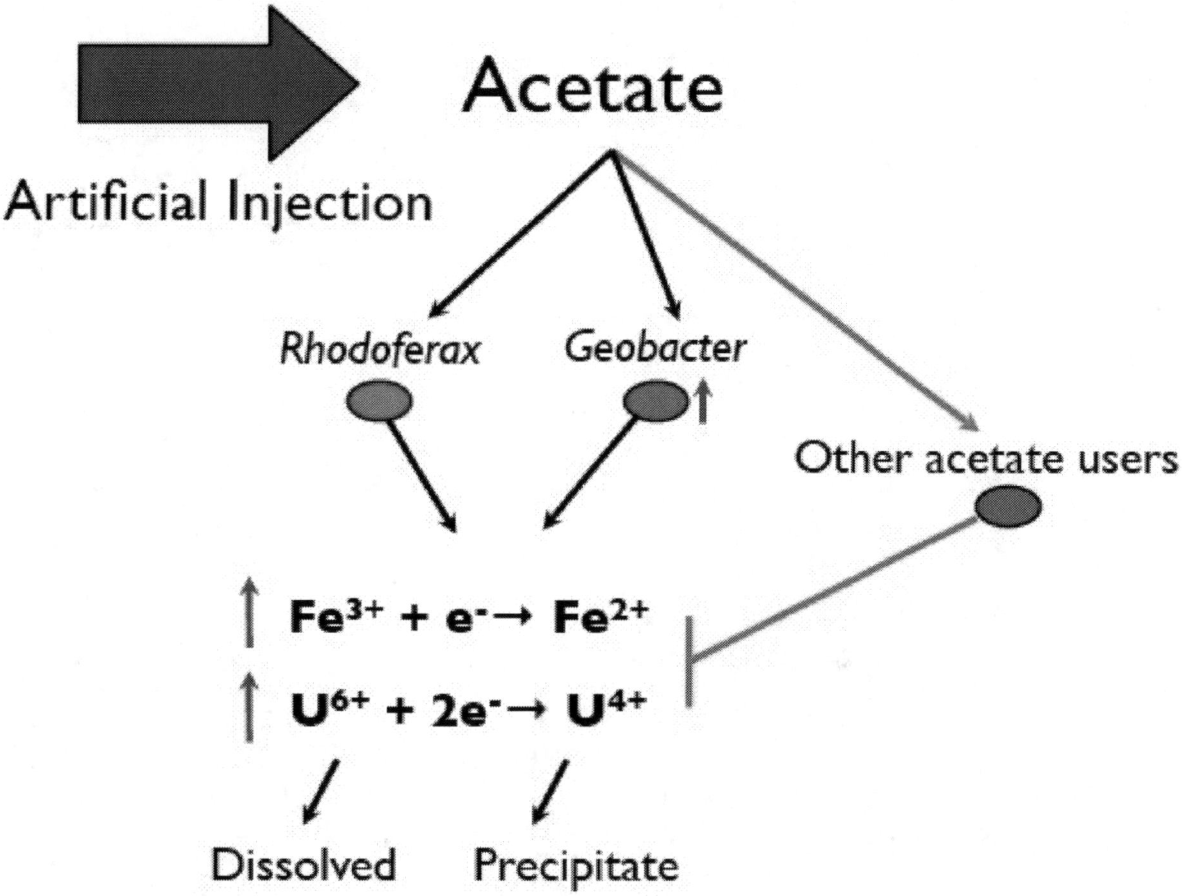

Figure 2. The DMMM framework applied to the modeling of iron reducers.

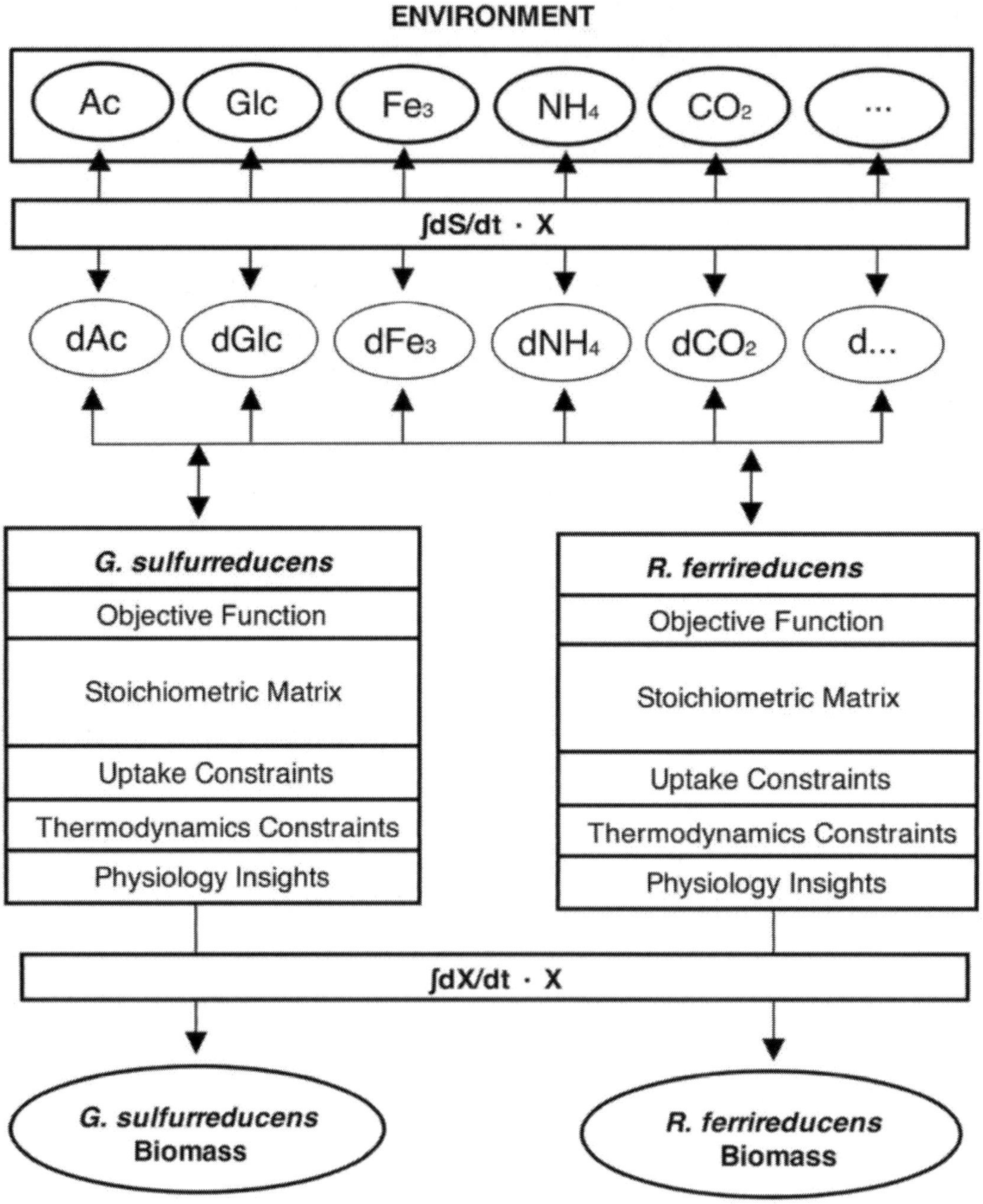

Figure 3. Iron-reducer competition prior to acetate amendment.

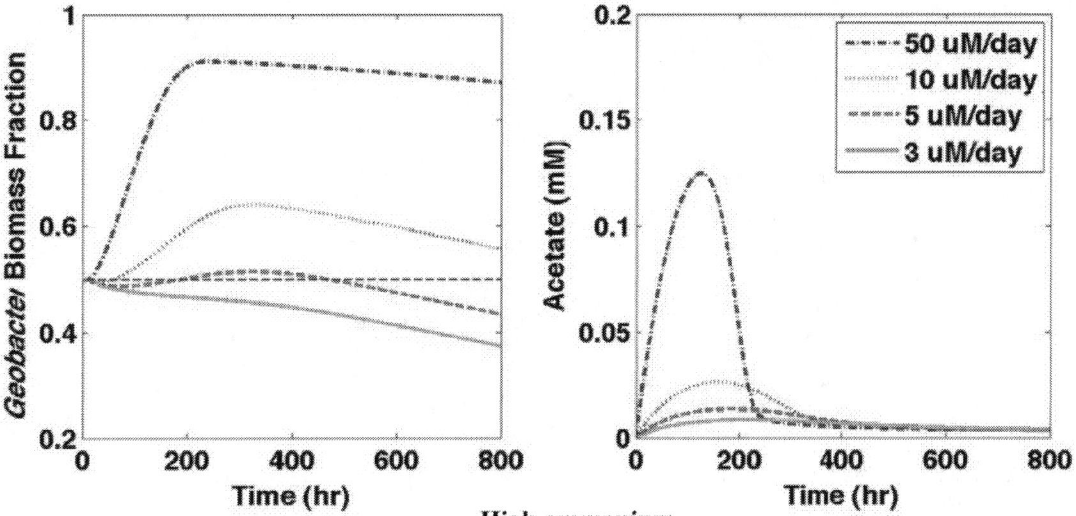

High ammonium

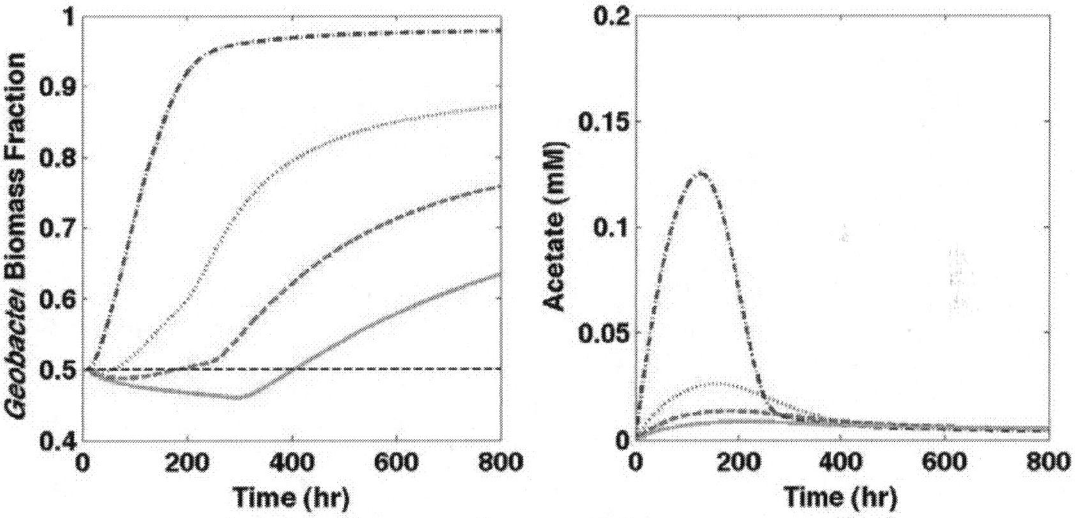

Low ammonium

Figure 4. Iron-reducer competition prior during acetate amendment.

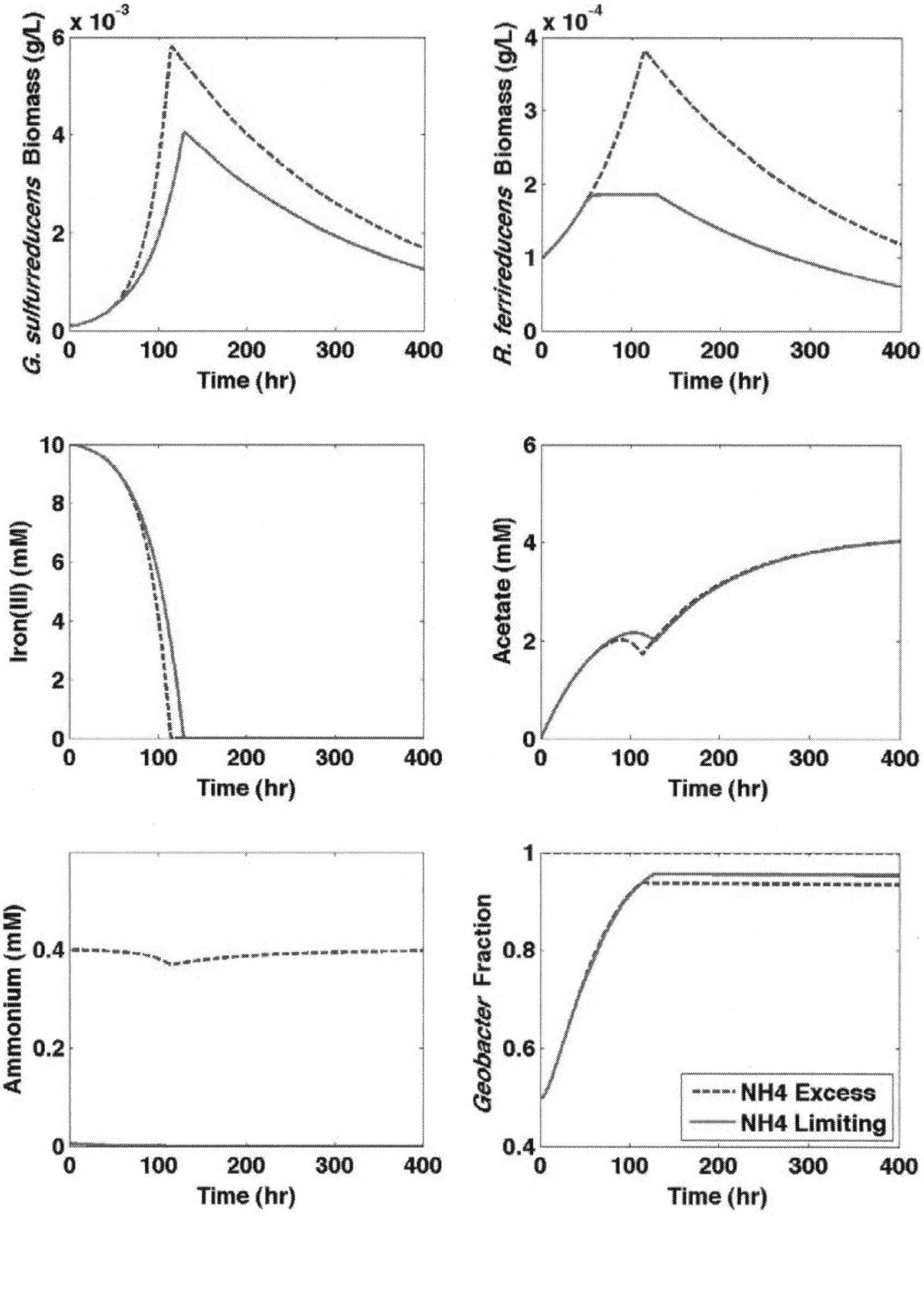

Figure 5. Model validation. ▲ is the experimentally measured *Geobacter* fraction.

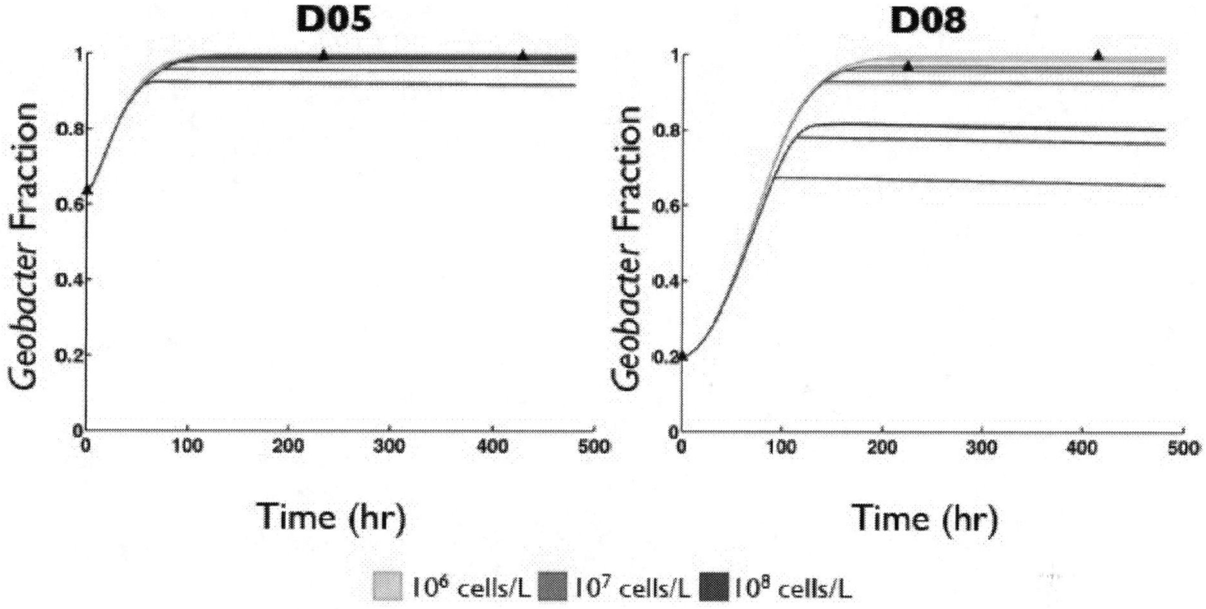

Figure 6. Normal growth vs. nitrogen-fixation dependent growth of *G. sulfurreducens*

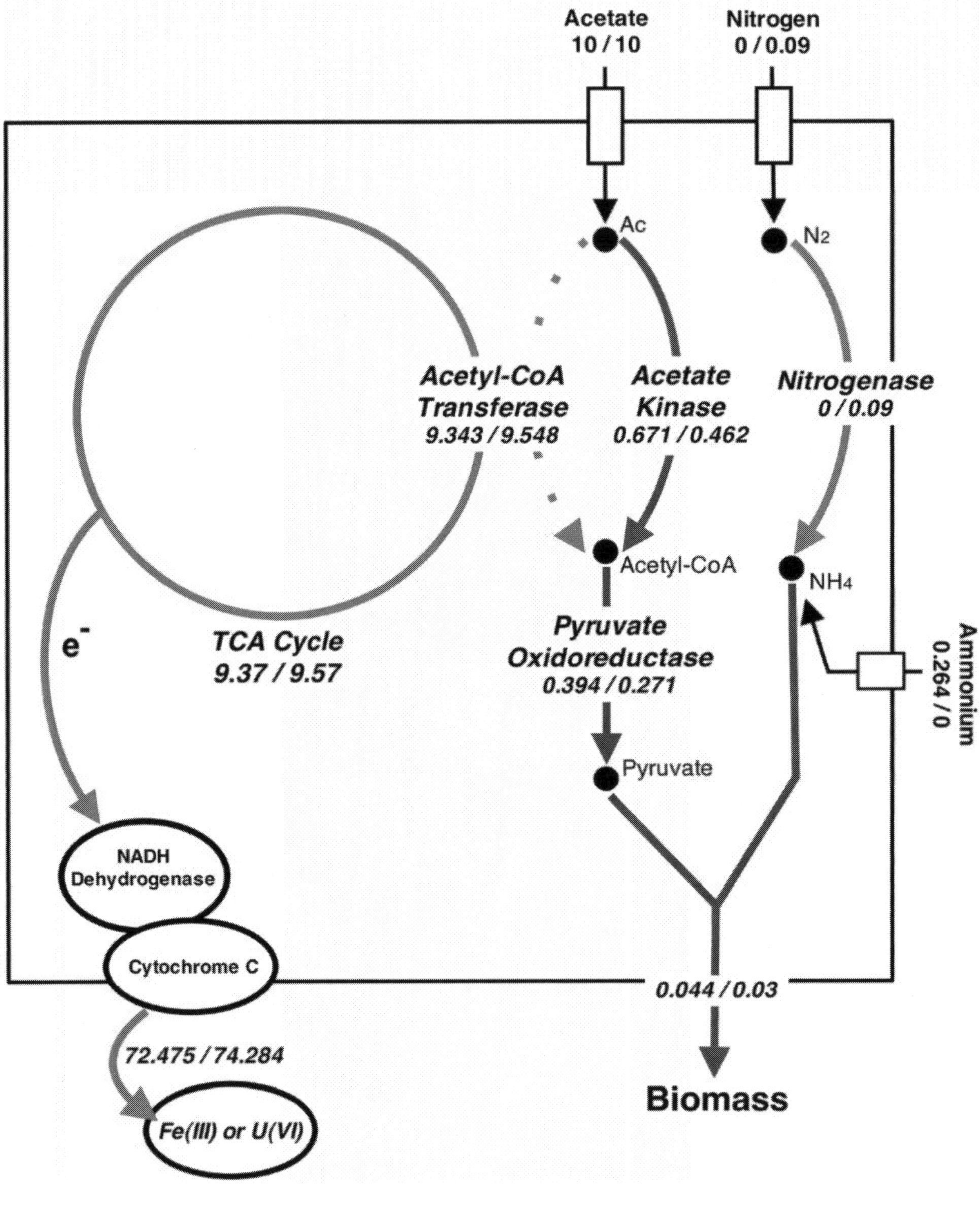

281c Migration of Chemotactic Bacteria toward Residual Contamination in a Sand Column with Structured Heterogeneity

Meng Wang and Roseanne M. Ford. Chemical Engineering, University of Virginia, 102 Engineers' Way, PO Box 400741, Charlottesville, VA 22904-4741

Understanding the migration of bacterial populations in natural groundwater environments is critical for the successful implementation of bioremediation to contaminated sites. One factor often limiting the effectiveness of bioremediation is the transport of bacteria which are able to degrade chemical contaminants to low permeable regions where contaminants are difficult to remove. Chemotaxis, the directed migration of a bacterial population in response to a chemical concentration gradient, attracts bacteria toward chemically favorable regions to enhance their growth and survival. Many groundwater contaminants also elicit chemotactic responses in bacteria.

A laboratory-scale packed column comprised of a coarse-grained sand core with an embedded chemoattractant source surrounded by a fine-grained annulus was used to simulate a structured heterogeneity in saturated granular media. Chemoattractant diffused outward from the core creating a chemical gradient in the radial direction. A pulse injection of chemotactic bacteria was introduced which responded to the chemoattractant by migrating from the fine-grained annulus toward the coarse-grained core. Bacteria that migrated toward the chemoattractant in the core were retained for a shorter time in the column than bacteria in control experiments without chemotactic effects. Breakthrough curves were monitored to assess chemotaxis on bacterial migration in the packed columns. A range of chemoattractant concentrations, flow rates and columns structures were investigated. Mathematical models were developed to quantify the effect of chemotaxis relative to other transport processes in laboratory column. Results from this study will be used in mathematical models to predict the fate and transport of chemotactic bacteria in contaminated groundwater systems with structured heterogeneity.

Enhanced Anaerobic Bioremediation Shuts Down a TCE Plume Groundwater Treatment System

William K. Glynn (glynnwk@cdm.com), Michael E. Miller, and Kevin P. Molloy (CDM, Cambridge, Massachusetts, USA)

ABSTRACT: At a manufacturing facility, rows of injection wells were used to gravity feed 60% sodium lactate solution into the shallow sand aquifer in order to promote enhanced anaerobic bioremediation (EAB) of trichloroethylene (TCE). Concentrations of TCE were as high as 20,000 µg/L, and bioremediation successfully treated them to less than 50 µg/L during four years of active treatment. Intermediate daughter products cis-1,2-dichloroethene (cDCE) and vinyl chloride (VC) did temporarily accumulate and sub slab depressurization was required to mitigate indoor air risks. The complete dechlorination daughter product, ethene, was also generated steadily throughout the process. The site groundwater cleanup goals established by the State of Connecticut were drinking water criteria for offsite groundwater. Monitored natural attenuation is expected to reduce residual contamination levels to meet drinking water criteria within the next several years. A predictive plume model was developed from historical data. The model demonstrated that the chloroethenes attenuated naturally, although slowly, and their concentrations decreased with distance downgradient from the source. The model results were also used to negotiate alternative on-site cleanup goals that were calculated to be protective of human health at downgradient off-site receptors. These higher cleanup standards were acceptable to the regulatory agency as the endpoint for active remediation and the start of monitored natural attenuation.

INTRODUCTION

At a manufacturing operation in Connecticut, chlorinated ethene solvents were found in the shallow aquifer mostly beneath the facility building. The contaminants were predominantly TCE with historical groundwater concentrations as high as 50,000 µg/L in the source area. Tetrachloroethene (PCE) was also a source chemical, although present at lower concentrations up to 600 µg/L. Historically total 1,2-DCE concentrations were as high as 23,000 µg/L, and although usually below detection limits, VC concentrations have been as high as 600 µg/L when observed. Redox conditions in the affected groundwater ranged from moderately reducing in the source area to aerobic farther downgradient.

Hydrogeology. The shallow aquifer starts at approximately 4 ft (1.2 m) below ground surface (bgs) and is composed of sand that is underlain by varved clay at approximately 14 ft (4.3 m) bgs as shown in the cross-section in Figure 1. The chloroethene plume is approximately 10 ft (3.0 m) thick, 500 ft (150 m) long in the direction of groundwater flow, and 140 ft (43 m) wide. Pumping test data indicated a horizontal hydraulic conductivity of 50 ft/day (15 m/day), from which we estimated a natural groundwater velocity of 0.8 ft/day (0.2 m/day) to the west.

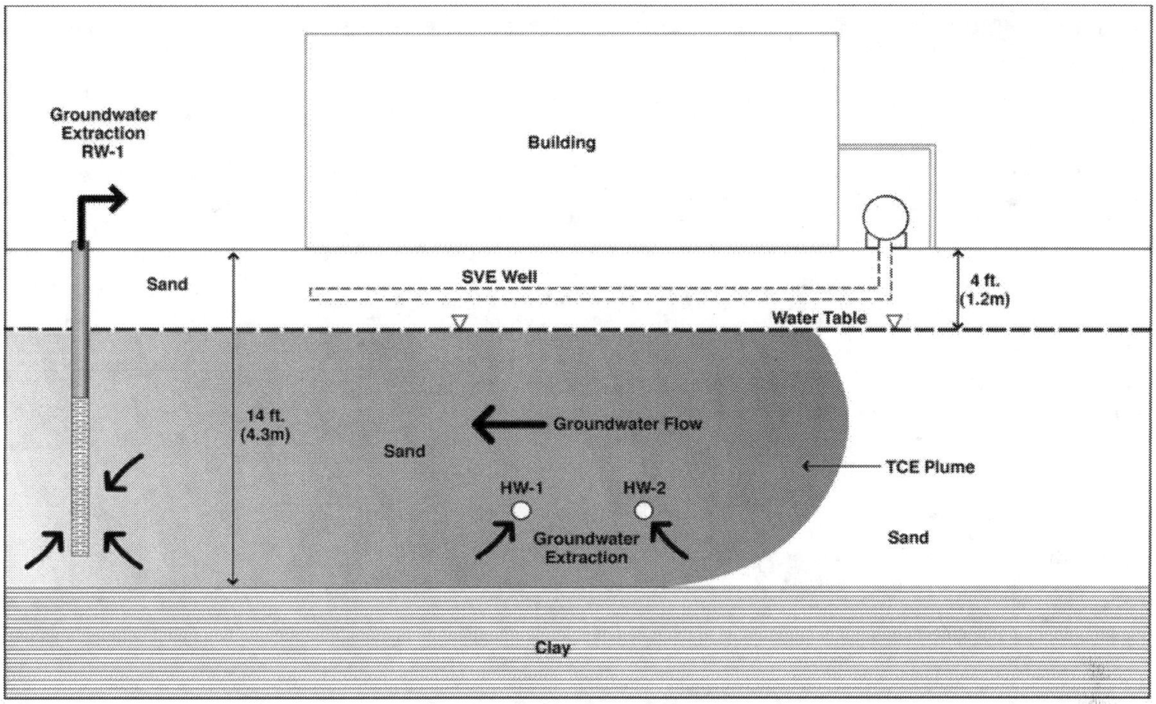

FIGURE 1. Schematic Site Cross Section Showing the Shallow Aquifer Stratigraphy.

Remediation History. Chloroethenes in the vadose zone, between the water table below and the building slab above, were removed through horizontal soil vapor extraction (SVE) wells that were run from July 1997 through September 1999. Groundwater contamination was pumped from below the building and treated using air stripping and activated carbon. The groundwater was extracted through two horizontal wells oriented perpendicularly to the direction of groundwater flow, one in the farthest upgradient part of the source area and the second approximately 120 ft (35 m) downgradient. A third vertical extraction well is located farther downgradient, beyond the building footprint at the property boundary (see Figure 1). Under the influence of groundwater pumping, the horizontal hydraulic gradient approximately doubled, resulting in an estimated enhanced groundwater velocity of approximately 1.6 ft/day (0.5 m/day).

Operation of the groundwater pump and treat system began in May 1997, and by 2001 annual chloroethene mass removal had leveled off to approximately 3 lb/yr (1.4 kg/yr). Starting in July 2004, anaerobic bioremediation was gradually phased in, while simultaneously phasing out groundwater pump and treat in the source zone. Recently, the downgradient vertical extraction well and treatment system was shutdown and active treatment suspended. The bioremediation system is described in detail below.

IN SITU BIOREMEDIATION SYSTEM DETAILS

Anaerobic bioremediation was initiated through the addition of 60% (w/w) aqueous sodium lactate solution, directly into the chloroethene plume. Sufficient lactate was supplied to provide a dissolved organic carbon (DOC) groundwater concentration of approximately 2,000 mg/L in the biotreatment zone. The lactate solution was gravity fed directly into 2-inch (5.1 cm) diameter injection wells, arranged in rows oriented perpendicularly to the direction of groundwater flow and starting upgradient in the most contaminated portion of the source area (Figure 2). The injection wells were screened from 1 ft (0.3 m) below the water table to the top of the underlying clay layer.

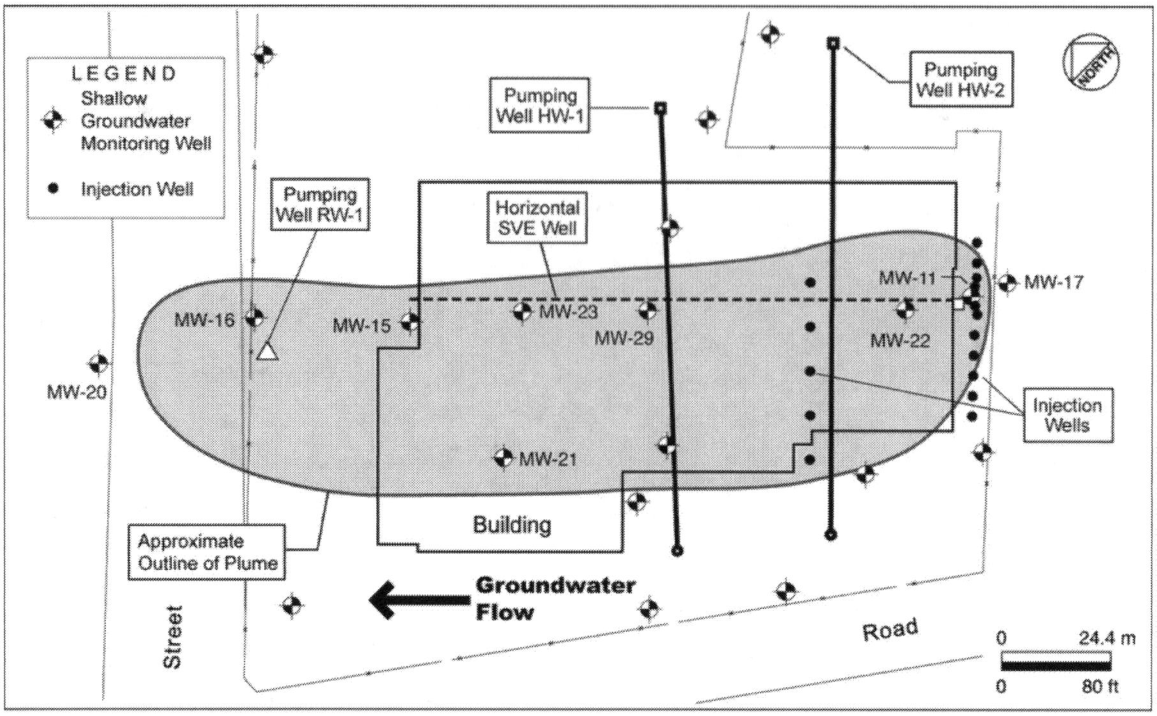

FIGURE 2. Site Plan Showing the Locations of Groundwater Extraction Wells, Organic Carbon Injection Wells, and Monitoring Wells.

In July 2004, the first lactate addition was performed utilizing the farthest upgradient row of injection wells (Row 1). Groundwater pumping from the nearest downgradient extraction well HW-2 helped to distribute this first addition of DOC throughout the treatment zone. Within two months, DOC was detected in groundwater sampled from HW-2, so its pumping was stopped to prevent further removal of the added lactate. Subsequent lactate injections involved both Row 1 and Row 2, located approximately 120 ft (35 m) down gradient from Row 1. The second injection was begun in April 2005, and two months later groundwater extraction from HW-1, farther downgradient, was stopped as well. Thus, as noted above, only the vertical extraction well RW-1 continued to operate to mitigate potential off-site chloroethene migration until recently. Additional lactate injections were conducted in June 2006 and October 2006. The injection system was expanded by one additional, farther downgradient row of injection wells just west of HW-1. Two additional lactate injections were conducted in August 2007 and February 2008.

Periodically, groundwater samples are collected from monitoring wells within the biotreatment zone and tested for pH, redox potential, dissolved oxygen (D.O.), nitrate, ferrous iron, sulfate, methane, ethane, ethene, DOC, metabolic acids, and halogenated volatile organic compounds (HVOCs) to follow the progress of bioremediation and the relevant groundwater geochemistry. Approximately five months after each lactate injection, most of the added carbon has been depleted as observed by DOC concentrations falling below 100 mg/L, and further addition of lactate is required to continue the anaerobic dechlorination process. Lactate metabolism appears to follow a pseudo-first order decay model, with a half life of approximately one month.

BIOREMEDIATION RESULTS

Within one month of the second lactate injection in May 2005, methanogenic redox conditions were established and maintained thereafter in the biotreatment zone, with methane groundwater concentrations ranging from 3,000 to 23,000 µg/L, and sulfate groundwater concentrations staying below 10 mg/L. At the same time, the groundwater redox potential was in the -100 to -220 mV range, and pH held steady between 6.5 and 7.

Almost one year into the biotreatment, after the second lactate injection, TCE concentrations had increased from approximately 5,000 to 20,000 µg/L. This was likely due to increased solubilization of the high concentration organic carbon source (Bioavailability Enhancement Technology™, U.S. patent no. 6,783,678). Dechlorination daughter product cDCE, VC, and ethene concentrations increased even more sharply at the same time. Thus, even as more TCE dissolved into the groundwater, complete dechlorination of the parent compounds all the way to ethene was occurring.

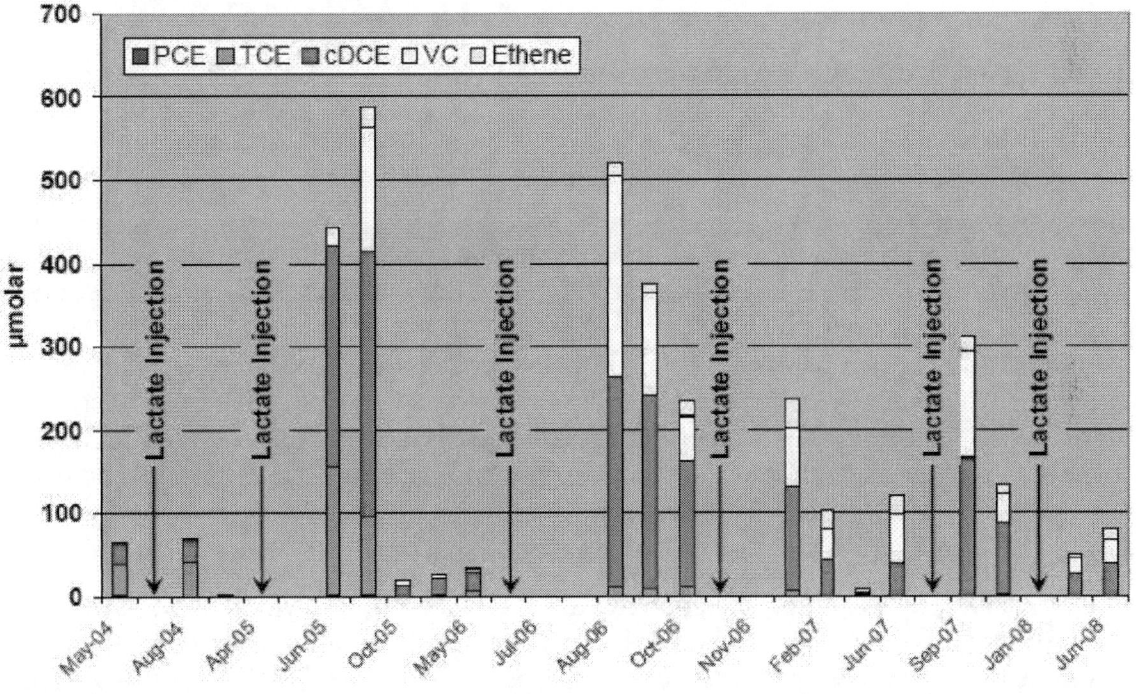

FIGURE 3. Chloroethene Transformations in Molar Units Observed at MW-22 during Biotreatment, where the Arrows Indicate Lactate Injection Events.

Starting in June 2005, TCE groundwater concentrations began to decrease steadily, while cDCE, VC, and ethene groundwater concentrations rose. Still later, around August 2006, the concentrations of cDCE and VC peaked and then began to fall, while ethene concentrations continued to increase slowly. Due to the increased VC concentrations, a portion of the SVE system was temporarily reactivated under the building. By February 2007, TCE groundwater concentrations were less than 50 µg/L in the source area and ethene was as high as 990 µg/L. PCE, originally present at much lower levels than TCE, decreased from approximately 200 µg/L to less than 5 µg/L during the same treatment period. The chloroethene transformations at the most contaminated source area well, MW-22, are shown in molar units in the bar chart in Figure 3. The concentration increases attributed to Bioavailability Enhancement Technology™ decreased in magnitude at MW-22 with each lactate injections. This suggests that the source mass is being successfully depleted using this technology. Similar results were observed at the other treatment zone wells.

In August 2006, groundwater samples from the source area biotreatment zone were collected for analysis by quantitative polymerase chain reaction (QPCR). The results indicated 5.5×10^5 copies per mL of *Dehalococcoides spp.* (DHC) 16S rRNA gene and 2.1×10^5 copies per mL of the vinyl chloride reductase functional gene, vcrA, at monitoring well MW-22. Thus, abundant dechlorinating bacteria are present in the biotreatment zone with the appropriate gene functionality to dechlorinate PCE and TCE completely. Bioaugmentation with a commercially available DHC-containing culture was not applied to the site, as it was not expected to further increase chloroethene biodegradation activity.

In June 2008, TCE groundwater concentrations continued to decrease steadily, while cDCE, VC, and ethene groundwater concentrations leveled off. Still later, around August 2008, the concentrations of cDCE and VC peaked and then began to fall, while ethene concentrations continued to increase slowly.

DEVELOPMENT OF ALTERNATIVE CLEANUP LEVELS

The site is located within the Connecticut River watershed and the state-mandated groundwater cleanup goals were set to drinking water criteria (known in Connecticut as the Groundwater Protection Criteria – GWPC) as follows: PCE = 5 μg/L, TCE = 5 μg/L, cDCE = 70 μg/L, and VC = 2 μg/L. As described above, the anaerobic bioremediation program has initiated complete dechlorination of the parent compounds all the way to ethene. However, site groundwater chloroethene concentrations are not expected to reach GWPC within a remediation timeframe within four years of active remediation. Therefore, we collaborated with the Connecticut Department of Environmental Protection (CT DEP) to develop alternative cleanup levels (ACLs), protective of public health but higher than drinking water standards. Active remediation of the plume would end once ACLs are reached, and a long term monitoring program would begin, as long as natural attenuation would be capable of eventually attaining GWPC.

The nearest residences are located approximately 150 ft (45 m) down gradient of the western boundary of the site. CT DEP required that intermediate cleanup standards protect against health risks posed by HVOC volatilization into the basements or first floors of these houses. It would not be necessary to meet the stricter drinking water standards in the shallow aquifer there, since potential drinking water wells at these homes would be screened in a deeper aquifer beneath the clay layer and therefore not in communication with the chloroethene plume. The relevant standards would be the Connecticut Residential Volatilization Criteria (RVC): PCE = 340 μg/L, TCE = 27 μg/L, cDCE = 830 μg/L, and VC = 1.6 μg/L. GWPC remain the ultimate remediation goals to allow final site close-out. The strategy of ending active remediation after higher RVC-based ACLs are attained would be acceptable to CT DEP only if we could demonstrate that natural attenuation processes are viable and would continue to shrink the HVOC plume going forward.

Natural Attenuation Model The historical groundwater monitoring data illustrated the behavior of the undisturbed plume under natural conditions that would again be in effect after active remediation was ended. The data reviewed was composed of chloroethene concentrations in groundwater samples from seven wells located along the centerline of the plume, starting in the upgradient source area and continuing downgradient beyond the site boundary: MW-11, MW-22, MW-29, MW-23, MW-15, MW-16, and MW-20 (see Figure 2). They were collected over the seven-year period from 1990 through 1996, during quarterly monitoring rounds, producing up to 24 data points per well depending on the wells original installation. During this time PCE, TCE, and DCE were observed at each well. During the seven-year period, each chloroethene concentration at each well fluctuated, but showed no temporally increasing or decreasing trends. The plume appeared to be in a steady state, and concentrations decreased with increasing distance downgradient from the source area.

Natural background conditions for the shallow aquifer appeared to be aerobic prior to lactate injections, as indicated by recent monitoring data at background wells, where D.O. levels ranged between 5 and 8 mg/L. It is likely that the processes responsible for plume attenuation with distance are aerobic biodegradation for TCE, DCE, and VC, as well as dispersion for all of the chloroethenes.

Because the pre-remediation plume was stable, we calculated average concentrations over the seven-year period for the chloroethenes at each of the centerline wells. For each compound, separate plots of the natural logarithm of the time-averaged concentration at each monitoring well versus distance downgradient showed linear behavior, indicating first order decay with distance. The slopes and correlation coefficients determined from each log plot are shown in Table 1. These slopes are distance decay constants that allow the calculation of chloroethene concentrations at any point along the plume centerline, based on concentrations observed at a monitoring well by the following formula:

$$C_1 = C_2 \times \exp[m(x_1 - x_2)]$$

where C_1 and C_2 are specific chloroethene compound concentrations at locations x_1 and x_2, and m is the compound-specific distance decay constant. Multiplying the calculated distance decay slopes by the estimated groundwater velocity (0.77 ft/day) converts distance to travel time and yields approximate natural attenuation decay rates. These rates are also listed in Table 1 along with calculated natural attenuation half lives.

TABLE 1. Distance Decay Constants for the Chloroethenes Calculated from Pre-Remediation Groundwater Monitoring Data.

Compound	Slope (1/ft)	Corr. Coeff. (r^2)	Decay Rate (1/day)	Half Life (yr)
PCE	-0.0025	0.7384	-0.0019	2.0
TCE	-0.0040	0.8189	-0.0031	0.97
cDCE	-0.0039	0.9012	-0.0030	0.99
VC	-0.0043	0.5016	-0.0033	0.90

Regulatory Agreement. The nearest house is approximately 670 ft (205 m) downgradient from the source area. By regulatory agreement, the RVC must be met in the shallow groundwater beneath this residence and beneath all locations farther downgradient. We used the above distance decay equation to back-calculate groundwater concentrations at site monitoring wells that would result in compliance with the RVC at this nearest residence. Furthermore, CT DEP directed us to add an extra level of caution around VC since it poses the greatest volatilization hazard. During anaerobic bioremediation, DCE acts as a source of VC via the dechlorination pathway. During the initial stages of biotreatment, the average VC to DCE concentration ratio was approximately 0.15. Thus, we conservatively set the DCE intermediate limits equal to the VC values calculated as above divided by the 0.15 VC/DCE concentration ratio. Finally the intermediate concentration goal for PCE was set at the RVC at all locations rather than values calculated by the distance attenuation model, since the RVC (340 μg/L) is relatively high and already attained throughout the plume. See Table 2 for a summary of the final mandated cleanup goals (GWPC), the volatilization criteria (RVC), and the negotiated ACLs to be met at the downgradient property boundary and within the source area in order to end active groundwater remediation.

TABLE 2. Summary of Final Cleanup Goals and Negotiated Alternative Cleanup Levels to End Active Groundwater Remediation.

Criterion	Point of Compliance	PCE (μg/L)	TCE (μg/L)	cDCE (μg/L)	VC (μg/L)
GWPC	Everywhere	5	5	70	2
RVC	Off-site residential	340	27	830	1.6
ACL	Downgradient property boundary	340	50	19	2.9
ACL	Source area	340	309	119	18

Based on the calculations described above, CT DEP agreed to the following groundwater quality dependent timeline for phasing out remediation and eventually closing the site. When the ACLs are met at the downgradient property boundary (locations MW-16 and RW-1), plume migration control will no longer be necessary and the groundwater pump and treat system will be shut down. When the ACLs are met in the source area (locations MW-11, MW-22, and MW-29), volatilization risks at downgradient residences are no longer a concern and active bioremediation will be ended. At this point the remedy becomes natural attenuation with semi-annual groundwater monitoring. After two years of semi-annual results that meet the ACLs, the monitoring frequency will be decreased to biannual until the GWPC are reached and site closeout can be arranged.

CONCLUSIONS

In situ biostimulation with 60% sodium lactate solution effectively promoted anaerobic dechlorination of PCE and TCE within one year of initial lactate addition. At first, intermediate daughter product DCE and VC groundwater concentrations increased significantly, and only began to decrease after TCE parent compound concentrations were depleted. Throughout the bioremediation process, ethene concentrations gradually increased, indicating that complete dechlorination was occurring. Bioaugmentation with a DHC-containing bacterial culture was not necessary to achieve complete dechlorination at this site.

Anaerobic bioremediation is successful and is expected to reduce chloroethene concentrations by several orders of magnitude within approximately two to four years of treatment initiation, which is an improvement over the pump and treat system whose performance would likely continue for more than ten years. However, the state-mandated cleanup goals are very low and still not easily attainable by bioremediation. The model developed from pre-remediation groundwater data showed that under normal aerobic conditions, natural attenuation was occurring, with chloroethene concentrations decreasing exponentially with distance downgradient. The model results were used to develop intermediate cleanup goals for the on-site plume that would result in protection of human health at downgradient off-site residential receptors. These ACLs formed a rational and attainable goal to end active groundwater treatment that was economically acceptable to the responsible party and environmentally acceptable to the regulator. This approach can serve as a model to advance a remediation project while balancing reasonable expectations of treatment performance with practical cleanup goals that protect human health and the environment.

281e Btex Treatment in a Solid-Liquid Two-Phase Partitioning Bioscrubber: Comparison of Stirred Tank and Airlift Configurations

Jennifer V. Littlejohns, Department of Chemical Engineering, Queen's University, 19 Division St, Dupuis Hall, Kingston, ON K7L 3N6, Canada and Andrew J. Daugulis, Chemical Engineering, Queen's University, 19 Division St, Dupuis Hall, Kingston, ON K7L 3N6, Canada.

Benzene, toluene, ethylbenzene and o-xylene, collectively known as BTEX, are toxic VOCs that are present in many gaseous waste streams produced during site remediation, particularly soil vapor extraction (SVE). Traditionally, physical or chemical treatment methods have been used in order to stay within regulatory emission guidelines. However, biological treatment methods, such as biofilters, which provide a low-cost, energy efficient and effective alternative to the destruction of toxic volatile chemicals (Datta and Allen, 2005), have been shown to be successful in treating low concentrations of BTEX components in gaseous waste streams (Oh and Bartha, 1997; Torkian et al., 2003). Although biofilters are effective for the treatment of low concentrations of VOCs, there has been recent interest in biological BTEX treatment beyond the current abilities of biofilters; that is, the biological treatment of high and fluctuating loadings of BTEX (Stewart et al., 2001; Mason et al., 2000).

A two-phase partitioning bioscrubber (TPPB) has been used successfully to treat waste gases containing high and fluctuating concentrations of individual BTEX compounds (Nielsen et al., 2005a; Boudreau and Daugulis, 2006). TPPBs consist of a cell containing aqueous phase and a nontoxic, non-bioavailable second phase that can sequester high and fluctuating concentrations of toxic substrates and release them to the aqueous phase for subsequent degradation based on metabolic demand. This provides the ability to treat high concentrations of toxic compounds with low water solubility by alleviating toxic levels in the aqueous phase. Organic solvents have traditionally served as the immiscible phase in a TPPB, however, these systems are typically limited to combinations of single VOCs and pure strains of microorganisms due to the potential bioavailability of the organic solvent second phase. Significantly, a treatment system for waste gases containing BTEX must allow for the use of a bacterial consortium, as there is no known pure strain of bacteria with the ability to effectively degrade all BTEX components (Bielefeldt and Stensel, 1999). Recently, solid polymer beads have been used in place of the organic solvent in TPPBs which provide the ability to use a bacterial consortium as polymers are commonly nonbioavailable to a wide range of bacteria (Prpich and Daugulis, 2004).

This presentation will report performance of a bench scale 3L stirred tank TPPB consisting of an aqueous phase containing a bacterial consortium and a polymeric phase of small silicone rubber pellets (solid volume fraction 0.1) used to treat gaseous waste streams containing a mixture of high and fluctuating concentrations of BTEX. Such a mixture of VOCs has not been treated in a TPPB to date. Silicone rubber was systematically selected as the sequestering phase due to its superior affinity for BTEX as determined by its favourable diffusion and partition coefficients and its ability to enhance the oxygen transfer rate into a TPPB (Littlejohns and Daugulis 2007). Fluctuations imposed on the TPPB system in the form of inlet loading step changes reveal that the system can achieve overall removal efficiencies of greater than 95% while obtaining overall elimination capacities of up to 281.5 mg/Lh. TPPB operation, however, succumbs to toxic substrate levels between step changes of 6 and 10 times the nominal loading value (360 to 600 mg/Lh). Aqueous phase and polymer phase concentrations that were monitored during these step changes show that the polymer phase sequesters large amounts of BTEX (70% to 93% of total BTEX in the working volume), thus buffering the aqueous phase from toxic levels.

Scale up of mechanically agitated TPPBs may not be feasible due to their excessive energy inputs. Airlift bioreactors require less energy and ease design scale up, and therefore may be a potential practical replacement for traditional stirred tank configurations. Performance of a bench scale 10L airlift TPPB containing a bacterial consortium and a polymeric phase of silicone rubber pellets (solid volume fraction 0.1) will be discussed and performance comparisons will be made between both stirred tank and airlift configurations. In addition, several critical parameters affecting successful airlift TPPB operation will be discussed including solid phase suspension and VOC mass transfer.

References

Bielefeldt, A., Stensel, H., 1999. Modeling competitive inhibition effects during biodegradation of BTEX mixtures. Water. Res. 33, 707-714.

Boudreau, N., Daugulis, A., 2006. Transient Performance Of Two-Phase Partitioning Bioreactors Treating A Toluene Contaminated Gas Stream. Biotechnol. Bioeng. 94, 448-457.

Datta, I., Allen, D., 2005. Chapter 6 Biofilter Technology. In: Shareefdeen, Z., Singh, A. (Eds.). Biotechnology for Odor and Air Pollution Control. Springer, Heidelberg, Berlin. pp. 125-145

Littlejohns, J., Daugulis, A., 2007. Oxygen transfer in a gas-liquid system containing solids of varying oxygen affinity. Chem. Eng. J. 129, 67-74.

Mason, C., Ward, G., Abu-Salah, K., Keren, O., Dosoretz, C., 2000. Biodegradation of BTEX by bacteria on powdered activated carbon. Bioprocess. Eng. 23, 331-336.

Nielsen, D., Daugulis, A., McLellan, P., 2005a. Transient Performance of a Two-Phase Partitioning Bioscrubber Treating a Benzene-Contaminated Gas Stream. Environ. Sci. Technol. 39, 8971-8977.

Oh, Y., Bartha, R., 1997. Construction of a bacterial consortium for the biofiltration of benzene, toluene and xylene emissions. World J. Microbiol. Biotechnol. 13, 627-632.

Prpich, G., Daugulis, A., 2004. Polymer development for enhanced delivery of phenol in a solid-liquid Two-Phase Partitioning Bioreactor. Biotechnol. Progr. 20, 1725-1732.

Stewart, W., Barton, T., Thom, R., 2001. High VOC loadings in multiple bed biofilters: Petroleum and industrial applications. Environ. Prog. 20, 207-211.

Torkian, A., Dehghanzadeh, R., Hakimjavadi, M., 2003. Biodegradation of aromatic hydrocarbons in a compost biofilter. J. Chem. Technol. Biotechnol. 78, 795-801.

339a Leaching of Brominated Flame Retardants (BFRs) and Metals from Electronic Wastes

R. Mark Bricka, Chemical Engineering, Mississippi State University, Box 9595, Mississippi State, MS 39762 and Otho Barnes, Civil Engineering, Mississippi State University, PO Box 9595, Mississippi State, MS 39762.

Electronic waste has become a serious problem in landfill leachates. Electronic waste is defined as waste that has become contaminated with toxic metals and brominated flame-retardants that are commonly found in electronic parts. Toxic metals such as aluminum, cadmium, and lead as well as brominated flame-retardants are believed to cause serious health problems.

The key brominated flame-retardants that are targeted for study include: Tetrabromobisphenol A (TBBPA), 2,2', 4', 4'5 pentabrominated diphenyl ether, 2,2'4'4'6' pentabrominated diphenyl ether, and decabrominated diphenyl ether. These isomers have been identified as being the most common flame-retardants used in electronic products. In addition toxic metals also are present in E-waste and can be mobile. This study is focused on the mobility of BFR and metals from E-wastes.

The electronic devices that will be investigated in the experiments are electronic circuit, LCD screen, and the outer casing of the laptops. This paper will present the results for the quantity of flame-retardants and metals that are mobile and leached from the electronic waste material. Dymamic leach tests as well as the toxicity characteristic- leaching procedure (TCLP) were conducted as part of this study.. The analysis that will be used to evaluate the characterization of these brominated flame-retardants is the Gas Chromatography with Electron Capture Detector, as well as Inductively Coupled Plasma analysis. ICP was used to measure the metals leached from the e-waste components.

A summary for the results from this study as well be presented. In addition details of the impact of these components on the environment will also be discussed.

339b Photocatalytic Degradation of 1,3 Dinitrobenzene In Aqueous Suspension for Site Remediation

Randy D. Weinstein[1], **Dorothy W. Skaf**[1], and Amanda M. Grannas[2]. (1) Department of Chemical Engineering, Villanova University, 800 Lancaster Ave., Villanova, PA 19085, (2) Department of Chemistry, Villanova University, 800 Lancaster Ave., Villanova, PA 19085

The decomposition of 1,3 dinitrobenzene as a surrogate for chemical explosives in the presence of titanium dioxide and zinc oxide dispersions under simulated solar light has been investigated. Gas chromatographic analysis for the remaining 1,3 dinitrobenzene was used to monitor reaction progress. A wide range of experimental parameters were varied to determine conditions that would provide practical conditions for spill remediation. Variables studied include solution additives, suspended solids, dissolved oxygen and mixing. Significant increases in the 1,3 dinitrobenzene degradation rate were obtained through the impact of solution additives. Reaction rate expressions were developed to fit the experimental results.

339c Delivery and Targeting of Functional Nanoparticles In DNAPL Remediation

Vijay T. John, Chemical and Biomolecular Engineering, Tulane University, 300 Lindy Boggs Bldg., New Orleans, LA 70118, Jingjing Zhan, Department of Chemical and Biomolecular Engineering, Tulane University, New Orleans, LA 70118, and Gerhard Piringer, Department of Earth and Environmental Science, Tulane University, New Orleans, LA 70118.

Effective in situ remediation of contaminated groundwater plume requires the successful delivery of reactive iron particles through soil. This study reports the transport characteristics of nanoscale zero-valent iron particles that are encapsulated in porous silica sub-micron particles through a novel aerosol-assisted technology. These particles can be transported through model soils more efficiently than commercially available reactive nanoscale iron particles (RNIPs). Column tests show the composite particles elute readily while RNIP is trapped at the inlet of the column. Capillary experiments further prove that RNIP clogs the pores due to rapid aggregation, but pore plugging does not occur for the composite particles. The partitioning characteristic of the particles was investigated by a novel capillary video microscopy technique. Our results again indicate that the iron/silica composite particles preferentially accumulate and localize at the TCE/water interface, making dechlorination more efficient. Such particles with enhanced mobility hold promise for in situ ground water remediation.

339d Electrokinetic Enhancement of Air Sparging for Groundwater and Soil Remediation

Pedro E. Arce[1], Thorbjorn Holm[2], and **Mario Oyanader**[1]. (1) Chemical Engineering, Tennessee Technological University, Prescott Hall, Room 214 , 1020 Stadium Drive, Box 5013, Cookeville, TN 37083, (2) Chemical Engineering, Royal Institute of Technology, Stockholm, Sweden

Air Sparging technology, which is more frequently used for organic compound clean ups in soil remediation, has been studied in conjunction with EK to analyze the potential of achieving higher contaminant removal. In other words, the main focus of this contribution is the research and identification of potential enhancements of these two technologies working together. In the first part of this effort, the modeling of a hypothetical dynamic system has been developed for the air injection only. In the second part, the system has been modified to account for electroosmotic forces working on the liquid phase. The non linear nature as well as the coupling of the transient state mass balance model obligates a rigorous numerical solution of the system. The differential model equation yields the pressure profile vertical directions of the target soil. The effects of different induced pressure at the soil surface on the pressure profile underground are illustrated. The change in the dynamic response of the system under the stress of the pressure indicates the progressive movement of air mass thru the porous media. Results of the simulations show the healthy effect of the EK enhancement, on the cleaning process, by reducing the time required to reach a cleaning threshold of the operation, significantly. Details about the model formulation and simulations results will be discussed in the contribution as well as potential ideas for experimental validation.

Science and Technology for Sustainable Water Supply

Menachem Elimelech

Department of Chemical Engineering
Yale University
New Haven, CT 06520-8286

Increasing global water scarcity and degradation in the quality of available water supplies present a grave challenge to the present and future health, welfare, and prosperity of the world's people. Desalination of seawater and saline aquifers and reuse of industrial or municipal wastewaters are viable avenues to increase water availability beyond that attainable from the hydrogeological cycle. Traditional methods for water purification, however, are chemical and energy intensive. Highly effective, low-cost, robust technologies for augmenting water supplies are needed, with minimal impact on the environment.

Recent advances in the science and technology of water purification are presented, focusing on desalination and water reuse technologies. Major developments in these technologies are made possible due to recent advances in materials science, nanotechnology, and the fundamental understanding of the physics/chemistry of the solid-water interface. Among the topics discussed in this presentation are the development of fouling-resistant membranes for water treatment, use of interfacial force measurements to elucidate the antifouling mechanism of water treatment membranes, highly permeable reverse osmosis membranes incorporating nanomaterials, bio-inspired high flux desalination membranes, robust systems for wastewater reuse, and novel desalination technologies that may dramatically alter the energy/water nexus. These technologies are discussed in the context of the global challenges in water supply and energy.

390a Photochemical and Electrochemical Advanced Oxidation Processes

Christoph Hoislbauer, University of Technology Graz, Department of Chemical Engineering and Environmental Technology, Inffeldgasse 25/C/II, Graz, 8010, Austria and Matthäus Siebenhofer, Department of Chemical Engineering and Environmental Technology, Graz University of Technology, Inffeldgasse 25/C/II, Graz, A-8010, Austria.

Recent studies have shown that many pharmaceutical substances can be detected in wastewater effluents. An increasing body of evidence indicates that antibiotics, hormones and antiepileptics are responsible for microbial resistance against active substances and feminisation of higher organisms, respectively [1]. Low concentration of pharmaceuticals ranging from nanograms to micrograms per litre may cause problems in municipal sewage treatment plants. Advanced Oxidation Processes (AOP) may offer appropriate measures and have therefore been investigated.

Photochemical and electrochemical Advanced Oxidation Processes (AOP/EAOP) are efficient oxidation operations for degradation of persistent pollutants such as pharmaceuticals, pesticides, substituted aromatic hydrocarbons, dyes and complexing agents. The UV/H2O2 -, UV/O3- processes and the Photo-Fenton process are representative photochemical AOPs. Anodic oxidation is an electrochemical AOP. Both technological groups generate reactive hydroxyl radicals ($E°=2,80$ V), which mineralize contaminants or at least convert pollutants into less toxic or biodegradable products.

The aim of the project is the comparison and documentation of approved photochemical and electrochemical processes for different substance classes. This catalogue should help to evaluate a proper technology for a specific application. Criterion for comparison is the potential of degradation and mineralization, as well as the specific energy consumption.

Several substances, the selection was mainly based on the properties of substituents, were investigated. All experiments were carried out with a start concentration of 500 ppm of the substance. Photochemical degradation was carried out in a Heareus Noblelight UV-reactor system (UV-RS-2), equipped with a 15 W mercury low-pressure lamp. Electrochemical degradation was performed in open cell electrolysis with diamond coated electrodes and 0.05 M Na2SO4 for conductivity. The results demonstrate different rates of mineralization and degradation related to the substituent of the model substance and the absorption properties in the UV Range. m-Cresol for example shows a higher rate of degradation than the complexing agent ethylenediaminetetraacetic acid because of the different absorbability at 254 nm. In general electrochemical experiments show a simultaneous degradation and mineralization whereas photochemical experiments show first degradation of the substance and then mineralization.

The paper summarizes the present status of research and experience.

[1] Marc-Olivier Buffle, Jochen Schumacher, Elisabeth Salhi, Martin Jekel, Urs von Gunten. Water Research 40 (2006) 1884 – 1894

390b Advanced Oxidation Processes for Destruction of Endocrine Disrupting Chemicals In Water Treatment: Kinetic Modeling and Free-Radical Reaction Mechanisms

Badri N. Badriyha[1], Wonho Song[2], **Varadarajan Ravindran**[2], and Massoud Pirbazari[2]. (1) San Diego State University, Department of Civil and Environmental Engineering, San Diego, CA 92182, (2) Civil and Environmental Engineering, University of Southern California, Los Angeles, CA 90089-2531

Advanced oxidation process (AOPs) have emerged as powerful technologies and are widely used in drinking water treatment for the destruction of endocrine disrupting chemicals (EDCs) identified by the U.S. Environmental Protection Agency and the World Health Organization. These EDCs could be organic or inorganic contaminants including pesticides, industrial solvents and chemicals, pharmaceutical and personal care products, and petroleum organics present in drinking water supplies. This study focues on the use of different AOPs for the destruction of organic EDCs, and these processes include ozone, hydrogen peroxide and ultraviolet radiations, and/or combinations thereof, including direct photolysis, ozonation, hydrogen peroxide-ultraviolet radiation (HP-UV) oxidation, and hydrogen peroxide-ozone (HP-ozone) oxidation. These AOPs were studied for the destruction of the chlorinated pesticides alachlor and heptachlor to concentrations below their regulated maximum contaminant levels (MCLs) of 0.4 mg/L and 2 mg/L, respectively. The study investigated the reaction mechanisms and pathways as well as formation of intermediates. A new aspect of the study addressed in this presentation involves the kinetic modeling of free-radical reactions, and examining the predictive capability of the modeling approach in various operating and process condtions.

Reaction mechanisms and pathways were proposed for some of the AOPs and target contaminants, and the formation of intermediates and products were studied. More importantly, the adverse effects of free-radical scavengers including natural organic matter (NOM) such as humic acid, and the carbonate species on the reaction kinetics were investigated. The efficient and cost-effective application of these processes required a good understanding of the associated complex kinetics, reaction mechanisms, and free-radical reactions. Modeling the process kinetics of these AOPs in the presence of NOM had far-reaching implications in their decomposition into smaller molecules and generation of disinfection byproducts (DBPs) such as aledehydes and ketones.

Ozonation and HP-ozone oxidation of heptachlor were compared under different conditions such as pH, initial pesticide concentration, and influence of free-radical scavengers such as carbonate species and NOM. The results showed that direct ozonation facilitated decomposition of the pesticide, while the hydroxyl radical was not effective as primary oxidant. The influence of hydrogen peroxide and NOM on the free radical reaction kinetics were studied. The HP-ozone oxidation processes was found to be more effective than ozonation only under certain process conditions. In both cases, oxidation products for heptachlor were formaldehyde, acetaldehyde, glyoxal, glyoxylic acid and heptachlor epoxide, albeit subtle variations in reaction mechanisms.

The parameters influencing the process destruction efficiencies of these AOPs were investigated through a systematic approach of combining experimental data acquisition with a kinetic modeling technique. A free-radical reaction kinetic model was developed for the HP-UV process to predict the concentrations of all principal species, including alachlor, NOM, hydrogen peroxide, carbonate species, and intermediate radicals. The proposed model, that addresses all the important aspects of previous models, incorporates additional features such as gradual pH decrease during the oxidation period (attributed to NOM mineralization), and used a lumped parameter approach for quantification of NOM. The alachlor decomposition efficiency was experimentally evaluated as a function of pH, NOM concentrations, total carbonate concentration, hydrogen peroxide concentrations, and the UV intensity. The modeling approach provided insight into the complex kinetics and reaction mechanisms involved in the HP-UV system under the influence of NOM, and the oxidation products of alachlor. This approach could be extrapolated to other AOPs for optimization of process parameters for a given water quality matrix to achieve efficient and cost-effective .

Furthermore, the study evaluated the adverse effects of NOM and carbonate species as free-radical scavengers on the decomposition kinetics. It was shown that larger hydrogen peroxide dosage achieved very marginal improvements in alachlor decomposition efficiency in the presence of NOM and carbonate species. The investigation further demonstrated that the HP-UV process was very effective in the destruction of alachlor in the presence of NOM, and that it could be applied for the destruction of other ECDs as well. Model sensitivity studies also provided a qualitative evaluation of the influence of different process variables on decomposition efficiency of target contaminant(s).

The AOPs under investigation were very effective in destruction of alachlor and heptachlor, meeting their MCL requirements at short reaction times and low oxidant dosages. The proposed AOP reaction mechanisms could adequately explain the product formations. The addition of hydrogen peroxide to UV radiation accelerated the decomposition of alachlor due to faster generation of hydroxyl radicals. The free-radical scavenging of carbonate species and NOM adversely affected the oxidation efficiencies. In UV-based AOPs, the radiation shielding due to NOM also affected the oxidation kinetics of target contaminants.

Key words: Alachlor, heptachlor, advanced oxidation processes, direct ultraviolet photolysis, ultraviolet radiation-hydrogen peroxide oxidation, ozonation, hydrogen peroxide-ozone oxidation, hydroxyl radicals, oxidation products, kinetic modeling.

PHOTOCATALYTIC DEGRADATION OF BISPHENOL A IN AN AQUEOUS SUSPENSION OF TITANIUM DIOXIDE.

Luisa F. Gómez[1], Victor M. Sarria[2], Gabriel Camargo[3] Joaquin Tirano

[1]*Chemical Engineering department, Los Andes University. Bogotá, Colombia.*
[2] *Chemical department, Los Andes University. Bogotá, Colombia.*
[3] *Chemical Engineering department, Los Andes University. Bogotá, Colombia.*
[4] *Chemical Engineering department, Los Andes University.* Bogotá, Colombia.

Introduction:

Bisphenol A (BPA) is a monomer used in the epoxy resin and polycarbonate manufacture. This compound is catalogued like an endocrine disruptor that causes several diseases and syndromes. This is distributed as a result of its multiple use, to its low degradability, to that it can be transported by air, ground and water, to its bioaccumulation in the trophic chain because is accumulated in the fat of the organisms and, to that are transmitted to the descendants through the mother during the gestation. Human exposition to this non biodegrable substance is increasing in the time, particularly through contaminated water by industrial remainders. Considering that the water is one of main means where it is deposited as remainder of industrial processes is tried to create systems for water treatments and purification of the same. Nevertheless, the traditional technologies that are used for the separation of organic substances of the treated water are based on processes of adsorption with activated coal or by drag with air. This implies that these processes only transfer the polluting agents of their watery phase to other that is also contaminated, therefore the problem persists. This problem has generated the necessity to implement new processes to treat non biodegradable compounds [1-4].

At the moment, exists a group of technologies based on processes of destruction of the polluting agents by means of hidroxyl radicals (•OH) which have a high potential of oxidation and attribute the capacity to oxidize or to degrade almost the totality of the polluting agent transforming it into inoffensive compounds to the environment and even can arrive until the mineralization or CO_2 formation. These technologies are called "advanced processes of oxidation" (APO´s), which are beginning to be implemented in North America, Europe and Japan [5-7].

In this work, oxidation process is studied; particularly the photo heterogeneous degradation with titanium dioxide in water is evaluated like alternative for the BPA degradation. Use a photo catalytic reactor with illumination UV and the quantification of the removal of the polluting agent was carried out using liquid chromatography (HPLC). The influence of titanium dioxide amount, BPA concentration, reaction temperature and the catalyst state like suspension and immobilized were also determinate.

Materials and methods:

Bisphenol A Aldrich in solid state; Titanium dioxide (TiO_2) Degussa P25 with a density of 3.8 gr/cm^3 were used. Use a photo catalytic reactor made up of a UV radiation lamp (Lexmana) of 40W of power and emission of 360nm (length 124cm. and diameter 2cm) and a Pirex tube that surrounds the radiation source (diameter 4 cm and length 110cm). Drained space between the lamp and the pirex tube corresponds to the area by where the solution flows. The recirculation of this solution is carried out by means of hoses of silicone rubber (0.5 cm of diameter), which connected to a pump to generate a flow of 400 mL/min.

Experimental procedure of the TiO_2 in suspension.

Four different concentrations from Bisphenol A were prepared: 0; 0.01; 0.05; 0.1 mM in a liter of water. Four different amounts from titanium dioxide were used 0; 100; 250; 500 mg/L of TiO_2. The solution of water, dioxide titanium and Bisphenol A was shaken during approximately 10 minutes with a magnetic agitator and an agitation plate, before being recirculated by the photo catalytic reactor. Each test was of 4 hours and samples were taken hourly. After, the solution was filtered with plastic filter with Pall ring of 0.2 µm nylon acrodisc, and later to be analyzed in HPLC equipment. In the figures 1 and 2 are shown a schema and a picture of the photo reactor

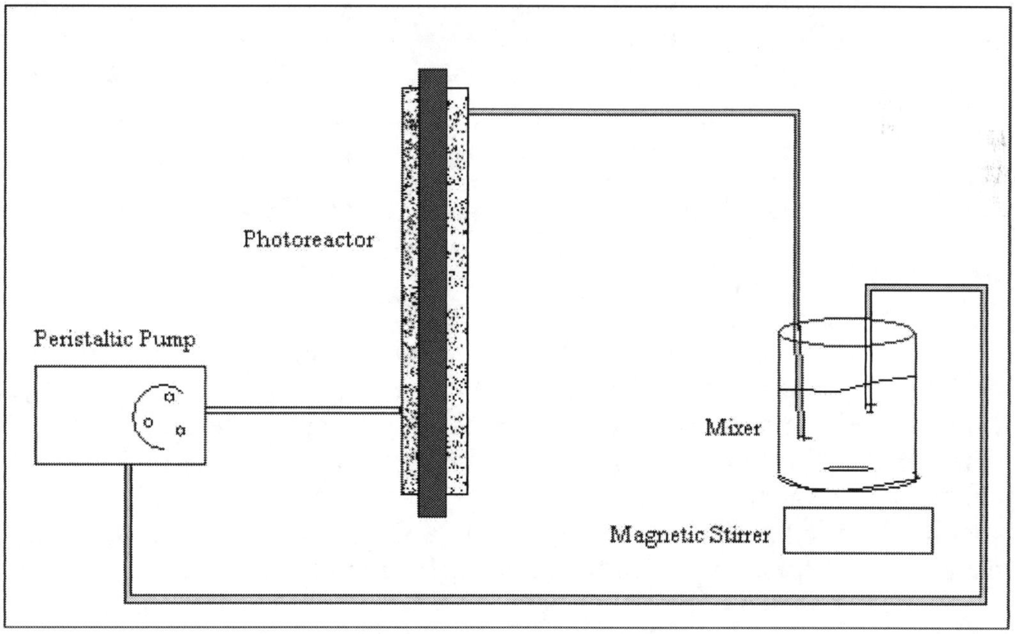

Fig. 1 Schema of photoreactor for photocatalytic decomposition of Bisphenol A

Experimental procedure of the TiO_2 immobilized

Films of three materials were developed; acrylic, acetate and glass, with the purpose of to find the best alternative to fix TiO_2.

A solution concentrated of 5g/L of TiO_2 in water was prepared, which was deposited on each surface. As far as the acrylic and to acetate, it was let dry the solution to room temperature, with the purpose of to evaporate the water of the solution. Whereas the glass was activated with increases of 50°C of temperature until arriving at 400°C during one hour, using a muffle 47900 Barnstead [8]. These films were located inside the photo reactor. The quantification of the BPA concentration was realized with HPLC equipment as with the catalyst in suspension. The TiO_2, was proved by XRD analysis as shown in fig 2.

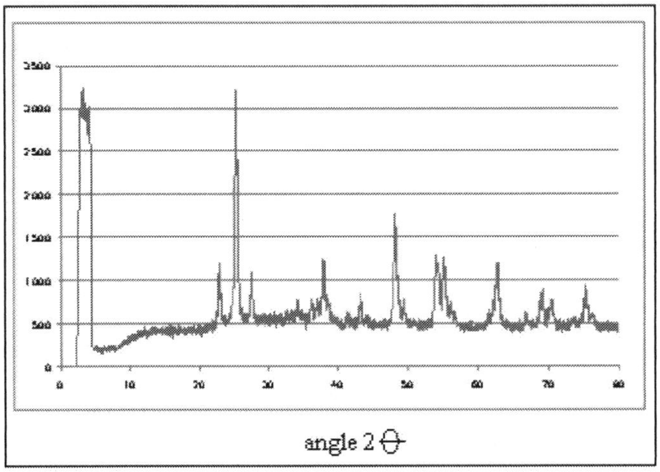

Fig. 2. XRD Analysis for the catalyst TiO2

Analysis HPLC

High Performance Liquid Chromatography (HPLC) was used to quantify the concentration of Bisphenol A. The conditions used in this quantitative analysis of BPA were: A movable phase of 20mM of sodium phosphate like solution buffer (pH 7) and acetonytrile in proportion 65%:35% and a volumetric flow of 1 mL/min. An analytical column of adjustment of diodes with a length of 250mm with +0.35/0.55 V and temperature of 20 ± 3°C was used (Allsphere ODS 2-5µ),. The used method was isocratic, injecting 10µL and the time of analysis by each sample was of 30 minutes.

Results and Discussion:

The concentration of BPA with respect to the time was correlated by means of the equation of reaction rate of first order, since report a good adjustment of the data obtaining coefficients of correlation between 0.8-0.9. This linearization allowed to find the reaction constant apparent (K_{ap}).

Catalytic Photodegradation of BPA using TiO_2 in suspension:

The different experiments show to the behavior of the apparent constant of reaction (K_{ap}) with respect to the concentration of catalyst in suspension 22.54 ± 2.61°C. It is observed that for a low concentration of Bisphenol A owns high values of K_{ap}. This indicates that the photo degradation reaction happens faster for low concentrations because exists less competition by the actives sites assets in a certain amount of catalyst [9]. It is observed, in the figure 2, that the tendency of K_{ap} with respect to the amount of TiO_2 is similar for the three concentrations of BPA. Initially the apparent constant owns an ascending behavior when amounts of titanium dioxide are smaller or equal to 100 mg/L and descendant for higher amounts. The highest K_{ap} (0.0154 min^{-1}) is obtained when the initial concentration of BPA is 0.01mM and the amount of TiO_2 is 100 mg/L.

The diminution of the apparent constant of reaction when increasing the amount of TiO_2 to an amount superior to 100 mg/L, can be explained from the following three phenomena, which can limit or inhibit the photo reaction catalysis falling K_{ap}

• Intermediary formation in the photo catalytic process

The photo catalysis heterogeneous with TiO_2 can disturb to the BPA molecule generating intermediary products of degradation which can generate competition to occupy the active sites of the catalyst [10-11].

• Effect Screening

A high amount of TiO_2 particles in suspension contributes to the formation of a screen that it prevents that the light photons activate to the catalyst [12].

• Adsorption of BPA in the surface of the catalyst (TiO_2)

Considering the expression of Langmuir- Hinshelwood and that the reaction is of first order, one correlates the constants of surface k, adsorption K_c and the apparent constant reaction K_{ap}. This analysis showed that the rank of K_c is between 0.05-0.41 $min.^{-1}$, whereas the rank of k is between 0.0003-0.0012 min^{-1}. These results indicate that the adsorption reaction is faster than the reaction in the surface, allowing to predict that the limit stage of the degradation of BPA is the surface reaction. For 100 mg/L of TiO_2, the constant of the reaction in surface k tends to increase and the equilibrium constant of K_c adsorption tends to fall down, whereas for high concentrations happens inverse, this indicates that the reaction in the surface becomes slower when increasing the catalyst concentration [13-15].

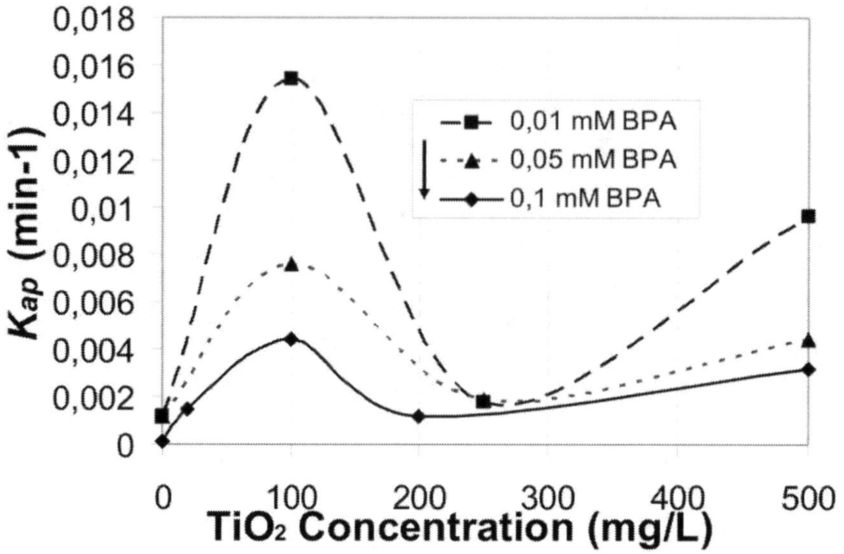

Fig 3. . Apparent constant Rate vs. TiO_2 amount

The photo catalysis reaction was evaluated with 4 different temperatures (22,54 ± 2.61°C, 35,17 ± 2,71, 46,82 ± 3,45 and 58,51 ± 4,06 °C). These values allowed to predict the experimental activation energy for a rank of temperature between 20°C-60°C from the Arrhenius equation, the value found was 14,54 is of is of KJ/mol and the form factor (A) was equal to 1.47 [8, 12, 13, 16].

Photo catalytic degradation of BPA using immobilized TiO_2

The support of titanium dioxide used were; acetate, acrylic and glass. The TiO_2 supported in acetate generated a loosening of catalyst of 95% at the end of the test. The acrylic used like TiO_2 support generated adsorption of the solution, reflected in the increase in the film weight. Whereas catalyst support in glass obtained better results removing 47.65% of BPA and coming off 52% of TiO_2 at the end of four hours of process. In order to improve the fixation efficiency the use of a binder must consider or to realize a pre treatment to the film surface using a diluted solution of HF and a washing with NaOH to increase the number of radicals *OH and to improve the contact with titanium dioxide and the glass supports [17].

References:

[1] Andrade AL, Pacheco A, Cynara, Nóbrega da Cunh AS (2006). *Disruptores endocrinos: potencial problema para la salud pública y medio ambiente.* Rev. Biomed; 17:146-150.

[2] Olea N, Avivar C. Disruptores endocrinos en la biología de la reproducción. http://www.asebir.com/congreso/granada/disruptores.pdf Citado Abril 2007

[3] Montezchua E, Zamarripa H, Leyva E (2003): Degradación *fotocatalítica de soluciones de alta concentración de paraguat.* Revista de contaminación ambiental, vol. 19. pg 117-125.

[4] Santamarta J: Nuestro futuro robado. La amenaza de los disruptores endocrinos. http://www.caps.pangea.org/quadern/29/5santamarta.pdf . Citado Abril 2007.

[5] Valladares. J. E Fotocatalizadores y Energía Solar en la desintoxicación de aguas contaminadas. Aplicaciones Potenciales. Àvailable in http:// transferenciamty.itesm.mx/transferencia44/eli-05htm. (Cited 8/82007)

[6] Sarria V. M Procesos de Oxidación Avanzada. http://ciencia.uniandes.edu.copdf/oxida06.pdf. Cited Abril 2007

[7] Sarria V. M (2005) Procesos de Oxidación Avanzada. http://ciencia.uniandes.edu.copdf/oxida06.pdf. Cited Abril 2007

[8] Bartolomew C.H, Farrauto R.J (2005). *Fundamentals of Indusrial Catalytic processes.* Wiley Second edition.

[9] Watabe Y, Kondo T, Morita Mb, Tanaka N, Haginaka J, Hosoya K (2004): *Determination of bisphenol A in environmental water at ultra-lowlevel by high-performance liquid chromatography with an effective on-line pretreatment devicea.* Journal of Chromatography A, 1032 :45–49

[10] Watanabe N, Horikoshi S, Kawabe H, Sugie Y, Zhao J, Hidaka H (2002). *Photodegradation mechanism for bisphenol A at the TiO2/H2O interfaces.* Chemosphere 52 : 851–859.

[11]Legado químico. Contaminación de la infancia. http://www.greenpeace.org/raw/content/espana/reports/legado-qu-mico-contaminaci-n.pdf.. Citado Abril 2007.

[12] Gálvez J. B., Rodríguez S. M., Peral J., Sánchez B. y Cardona A. I. *Diseño de reactores para fotocatálisis: evaluación comparativa de las distintas opciones.* http://www.cnea.gov.ar/xxi/ambiental/CYTED/17cap11.pdf

[13] Chen D., Aya K Ray . Photodegradation kinetic of 4 nitrophenol in TiO_2 Suspension. Water Research. Vol 32 No 11 3223 -3234 (1998).

[14] Youll L., Xiaodong L., Junwen L Jing Y Photocatalytic degradation of methyl orange by TiO_2 coated activated carbon and kinetic Study. Water Research Vol 40 1119 – 1126 (2006)

[15] Cheheb, Z, Albouchi F., Nasrallah S. B.(2007) *Measurement of thermal radiative and conductive properties of semitransparent materials using a photothermal crenel method.* Journal of Quantitative Spectroscopy & Radiative Transfer 109 (2008) 620–635.

[16] Behnajady M.A, Modirshahla N., Daneshvar N., Rabbani M., *Photocatalytic degradation of an azo dye in a tubular continuous-flow photoreactor with immobilized TiO2 on glass plates.* Chemical Engineering Journal 127 (2007) 167–176.

390d Applicability of the Fe-Assisted Photocatalytic Reactions to Dissimilar Compounds: Positive Effect of Fe Ions on Mineralization Rates of Maleic Acid, Oxalic Acid, Formic Acid, Paraquat, Phenol and 4-Chlorophenol

Aaron Ortiz Gomez, Chemical and Biochemical Department, University of Western Ontario, 1151 Richmond Street, London, ON N6A 5B8, Canada, Benito Serrano, Chemical Engineering, Universidad Autonoma de Zacatecas., Campus Siglo XXI, Edificio 6, Ejido La Escondida,, Carretera Zacatecas Guadalajara, Km. 6, Zacatecas Zacatecas., 98000, Mexico, Jesus Moreira del Rio, Chemical and Biochemical Engineering, University of Western Ontario, 1151 Richmond Street, London, ON N6A 5B8, Canada, and Hugo de Lasa, Chemical and Biochemical Engineering, The University of Western Ontario, London, ON N6A 5B9, Canada.

Photocatalysis, despite its advantages, presents slow reaction rates. Thus, efforts have focused on the improvement of photocatalytic reactors designs and on the enhancement of photocatalytic activity through various techniques, which include the addition of metallic ions to reduce the recombination of charges (h+/e-).

The activity of a photocatalyst is due to its ability to promote the formation of non-selective hydroxyl radicals upon being irradiated with UV light (Fujishima et al., 2000). In previous studies (Ortiz-Gomez et al, 2007) it was demonstrated that the addition of Fe ions accelerates the rate of mineralization of various hydroxylated aromatics (phenolic compounds). As a result, a reaction mechanism was proposed in which the effect of Fe ions was due to the decrease in the h+/e- recombination, which in turn leads to higher catalyst activity.

Given that hydroxyl radicals are non-selective, the same behavior of the Fe-assisted PC reactions is expected in the oxidation of other organic compounds with dissimilar molecular structure. To provide supporting evidence that the proposed mechanism applies to all photocatalytic reactions, experiments with maleic acid, oxalic acid and formic acid, as well as with a toxic herbicide, paraquat, were performed. These runs included both photocatalytic reactions with no Fe (unpromoted PC reactions) and photocatalystic reactions assisted with Fe ions (Fe assisted PC reactions). These experiments were carried out in a 6-L photoreactor, the Photo-CREC photocatalytic reactor. In order to determine that the enhancement was due to the addition of Fe and not a function of the reactor size or design, a 16-L reactor, the Photo-CREC Solar Simulator, was also employed. In this case, phenol and 4-chlorophenol were selected as model pollutants for both unpromoted and Fe-assisted PC reactions.

It is demonstrated that in all cases Fe-assisted PC oxidations using 10 ppm Fe ions were faster than the unpromoted PC reactions. This is shown using model pollutants with very different chemical structures (short-chain carboxylic acids containing 1, 2 and 4 carbons, and a pesticide with a two aromatic ring structure). It is also demonstrated that similar enhancements in the mineralization rates were obtained in reactions performed in the second reactor setup, whose dimensions and characteristics are markedly different from the Photo-CREC unit with a 6-liter capacity.

Overall quantum yields (OQY) were calculated to measure the extent of influence of Fe in these photocatalytic reactions. In all cases the OQY were calculated taking the formation of CO_2 molecules at 99 % conversion of the model pollutant as a basis for comparison. For the carboxilyc acids the increase in OQY was above 50% while for the paraquat a smaller percentage is observed (between 23 and 32%). Similar results were obtained for experiments performed with phenol and 4-chlorophenol in the Photo-CREC Solar Simulator.

These results, along with colorimetric and spectroscopic analyses (EDX, XPS), confirm that Fe enhances the reaction rate by increasing the photo-catalyst activity regardless of the model pollutant. The enhancement is seen on simple molecules (carboxylic acids) and complex molecules (paraquat), and observed also on a different reactor setup up with different model pollutants (4 phenol and 4-chlorophenol). This enhancement is explained due to lower recombination rates of h^+/e^-, which lead to higher catalyst activities.

References.

Fujishima et al (2000), Titanium dioxide photocatalysis. Journal of Photochemistry and Photobiology C: Photochemistry Reviews 1, 1-21.

Ortiz-Gomez et al (2007), Enhanced Effect of Fe^{3+} and Fe^{2+} ions on the photocatalytic oxidation of organic compounds: Reaction Mechanism over the TiO_2 surface., AIChE Annual Meeting, Salt Lake City, Nov. 4-9, 2007.

390e Chelate-Modified Hydroxyl Radical Reactions for Detoxification of Chlorinated Organics: Experimental Results and Model Development

Scott R. Lewis[1], a. Montague[1], Y. Li[1], S. Daunert[2], L. Bachas[2], and D. Bhattacharyya[3]. (1) Chemical & Materials Engineering, University of Kentucky, 177 F. Paul Anderson Tower, Lexington, KY 40506, (2) Chemistry, University of Kentucky, 177 F. Paul Anderson Tower, Lexington, KY 40506, (3) Chemical and Materials Engineering, University of Kentucky, 177 F. Paul Anderson Tower, Lexington, KY 40506

Oxidative techniques utilizing free radicals have proven effective for the destruction of toxic organic compounds such as trichloroethylene (TCE) and polychlorinated biphenyls (PCBs). TCE is a groundwater contaminant commonly found in large pools known as dense non-aqueous phase liquids (DNAPLs). The most common oxidative destruction techniques used for groundwater remediation applications are modified forms of the Standard Fenton reaction, in which Fe(II) reacts with hydrogen peroxide to form Fe(III), a hydroxide ion, and a hydroxyl radical. This reaction takes place at low pH and cannot be performed effectively at near-neutral pH due to ferric hydroxide precipitation. This problem can be alleviated through the use of a non-toxic chelate (L), such as citrate or polyacrylic acid (PAA). The addition of a chelate allows the reaction to take place at near neutral pH and control hydrogen peroxide consumption. This chelate can be incorporated either into the aqueous phase in soluble form or in a polymer matrix.

In addition to experimental data, we have modeled the effects of the chelate:Fe ratio on contaminant degradation and hydrogen peroxide consumption. The chelate reduces the amount of Fe(II) available in the solution phase for reaction with hydrogen peroxide, affecting the production of hydroxyl radicals. The generation of these, as well as other, radicals determines the rate of degradation of TCE in the aqueous phase and the rate of disappearance of TCE DNAPLs (present in droplet form). Many remediation strategies have successfully degraded TCE present in the aqueous phase, only to experience a rebound from the dissolution of the DNAPL. In order to degrade the contaminants present in an efficient manner, it is necessary to develop a model to predict the degradation of contaminants dissolved in the aqueous and non-aqueous phases. This model is comprised of free radical concentrations and iron species distributions obtained by simultaneously solving equilibrium equations for Fe-chelate interactions and rate laws associated with pertinent hydroxyl radical reactions.

Support of this research has been provided by NSF-IGERT, NIEHS/SBRP, and DOE-KRCEE.

Pretreatment of Primary Sludge Prior to Anaerobic Digestion

Saad Aldin, Sayed Elbeshbishy, Fuzhou Tu, George Nakhla, Madhumita Ray[1*]
[*]Department of Chemical and Biochemical Engendering, University of Western Ontario, London, Canada.
London, Ontario, Canada, N6A 5B9

ABSTRACT

The efficacy of ultrasonication as a pretreatment process for wastewater sludge prior to anaerobic digestion is evaluated with respect to the extent of sludge solubilization, particle size distribution, biogas production, and protein degradation. Primary sludge from municipal wastewater was treated with different ultrasonic intensities varying sonication time. Results show that SCOD/TCOD ratio, volatile fatty acids, ammonia, soluble protein, and gas production increased with sonication time at a constant frequency, while lipid, total protein and particle size of the sludge decrease with sonication time.

Key Words: Anaerobic digestion, ultrasonication, particle size distribution, sludge pretreatment, gas production.

INTRODUCTION

Anaerobic digestion, which is the most widely used sludge stabilization process worldwide, utilizes anaerobic bacteria to break down about 40-50% of the organic matter of the sludge to form valuable biogas (methane) and to reduce pathogen content of the sludge prior to disposal. Anaerobic degradation of particulate material and macromolecules is considered to follow a sequence of four steps: hydrolysis, acidogenesis, acetogenesis, and methanogenesis. In the case of sewage sludge digestion, the biological hydrolysis has been identified as the rate-limiting step (Eastman and Ferguson, 1981; Shimizu et al., 1993), causing large residence times in the fermenters. Pretreatment of sewage sludge by mechanical, chemical, or thermal disintegration methods can improve the subsequent anaerobic digestion (Chiu et al., 1997; Doha´nyos et al., 1997; Hiraoka et al., 1984; Mueller et al., 1998). However, there is a lack of information as to

[1]*Tel. 1-519-661-2111 ext. 81273; Fax. 519-661-3498; email: mray@eng.uwo.ca

how different types of sludge disintegration can impact the sludge parameters and subsequent changes in the digestion process, since most work has only focused on the solubilization aspect of pretreatment techniques.

Ultrasonication, an advanced oxidation process is a well-known method for the break-up of microbial cells to extract intracellular material (Harrison, 1991). When the ultrasound wave (>20 kHz) propagates in a medium such as sludge, it generates a repeating pattern of compressions and rarefactions in the medium. The rarefactions are regions of low pressure (excessively large negative pressure) in which liquid or slurry is torn apart. Micro-bubbles or cavitation bubbles are formed in the rarefaction regions. As the wave fronts propagate, micro-bubbles oscillate under the influence of positive pressure, thereby growing to an unstable size before they violently collapse. The collapsing of the bubbles often results in localized temperatures up to 5000 K and pressures up to 180 MPa (Suslick, 1990; Flint and Suslick, 1991). The sudden and violent collapse of huge numbers of micro-bubbles generates powerful hydro-mechanical shear forces in the bulk liquid surrounding the bubbles (Kuttruff, 1991). The collapsing bubbles disrupt adjacent bacterial cells by extreme shear forces, rupturing the cell wall and membranes. The localized high temperature and pressure could also assist in sludge disintegration. At high temperatures, lipids in the cytoplasmic membrane are decomposed, resulting in ruptures within the cell membrane, through which intracellular materials are released in to the aqueous phase (Wang et al., 2005). The objective of this study was to determine the effect of ultrasonication on various parameters of the sludge, examine the digestion efficiency and to the effect of different types of sludge disintegration on the anaerobic sludge stabilization process, mainly the extent of gas production.

MATERIALS AND METHOD

Experimental set-up

A lab scale ultrasonic probe was used to treat primary sludge from a municipal wastewater plant. The ultrasonic probe was supplied by Sonic and Materials (model VC-500, 500 W, and 20 kHz). The primary sludge was obtained from Adelaide municipal WWTP in London, Ontario. Figure 1 illustrates the ultrasonic process. Primary sludge was sonicated for 1, 3, 6, 15, 30, 45, 60 minutes, and for all the experiments reported in this work, the amplitude was set to 100% and the sonication pulse was set to 2 seconds on and 2 seconds off to control the temperature rise of the sludge.

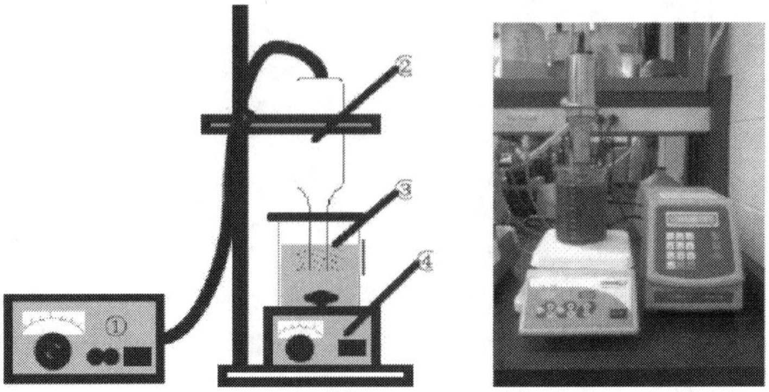

Figure 1 Schematic representation of the lab-scale ultrasonic set-up.

(1) Ultrasonic generator, (2) Ultrasonic probe, (3) 500mL Beaker with primary sludge, (4) Magnetic stirrer

Sludge parameters such as COD, TSS, VSS, volatile fatty acids (VFA), particle size distribution (PSD), lipid, ammonia and total, soluble and bound protein were monitored. The gas production was measured using batch specific methanogenic activity (SMA) to determine the efficacy of the pretreatment process. Except for soluble, bound and total protein, all other parameters were analyzed according to the Standard Methods (, Lowry's method however was used to analyze protein.

RESULTS AND DISCUSSION

Sonication Energy/Power Input:

The economy of an ultrasound system is primarily governed by the power (W or kW) or energy (J or kJ) input needed to achieve effective sludge disintegration. Thus, quantification of energy/power input to obtain a desired degree of disintegration is critical to evaluate the relative efficiency of ultrasound systems. In this study, all results are reported based on sonication time. Figure 2 displays the linear relationship between sonication energy and sonication time.

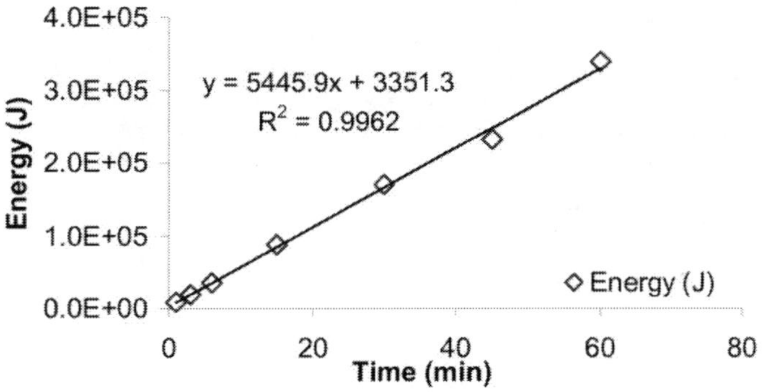

Figure 2: Sonication energy (J) as a function of time.

Chemical Oxygen Demand (COD):

The soluble chemical oxygen demand (SCOD) and total chemical oxygen demand (TCOD) as a function of sonication time are presented below. Results show that TCOD increased initially with sonication time and remained stable thereafter. SCOD, however, drastically increases due to the sonication of the organic matters. Figures 3 a, b, c illustrate the effects of sonication on TCOD, SCOD and SCOD/TCOD ratio, respectively.

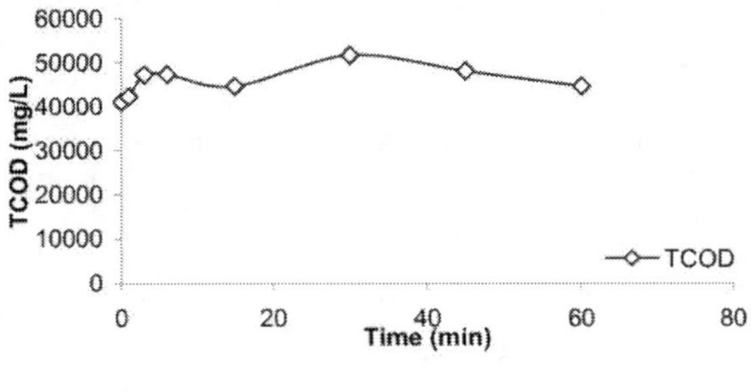

Figure 3a. TCOD as a function of sonication time

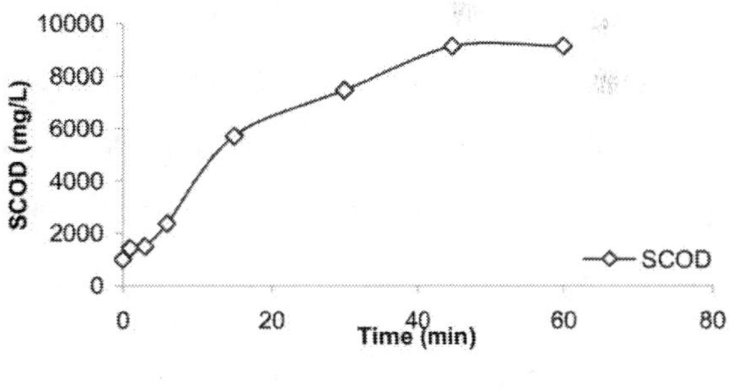

Figure 3b. SCOD as a function of sonication time

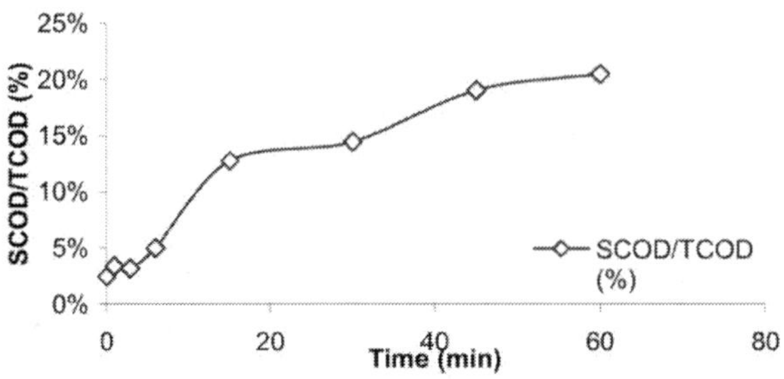

Figure 3c. SCOD/TCOD ration as a function of sonication time

The increase in SCOD/TCOD ratio is due to the release of extracellular polymeric substances (EPS), which consists of polysaccharides, proteins etc. which are embedded in the floc matrix (Drewsa et al., 2006) and disintegrated due to sonication. However, the maximum SCOD/TCOD ratio for treated sludge reached only about 20% after 45 minutes treatment, which indicated that majority of the particle matter did not solubilize with sonication.

Particle Size Distribution:

The sonicated sludge samples were analyzed using a Malvern Mastersizer laser beam diffraction granulometer. The particle size distribution by volume fraction as a function of sonication time is shown in Figure 4. The precision of measurements below 0.5 mm is low and a small number of particles exceeding 1000 mm in diameter would be present in samples, and these were ignored.

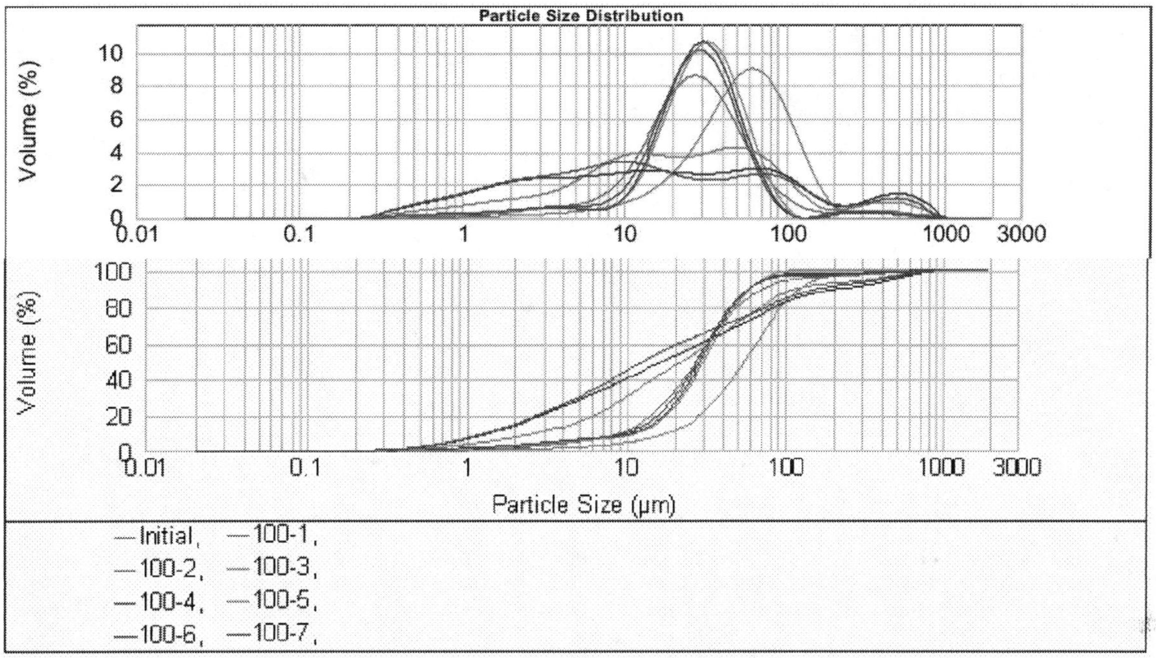

Figure 4. Top panel; particle size distribution for treated and untreated sludge. Bottom panel; the cumulative particle size distribution for treated and untreated sludge.

The cumulative distribution (bottom panel, Figure 4) shows clearly the change in particle size as the first curve from the right represents the initial untreated sludge. As sonication time increases the particle size decreases shown by the curves on the left. It is also interesting to note that sonicated time between 1 to 15 minutes seems to have no significant change in particle size; however the effect starts to show after 30 minutes.

Volatile Fatty Acids (VFA)

Volatile fatty acids are important organic substances, which are easily biodegraded by the microorganisms, and are important for the methane production in sludge digestion and helpful to phosphorus release in nutrient removal. VFA assimilation is more efficient than fermentation. Because of that, nitrates and orthophosphates removal is more effective if wastewater contained higher amount of volatile fatty acids.

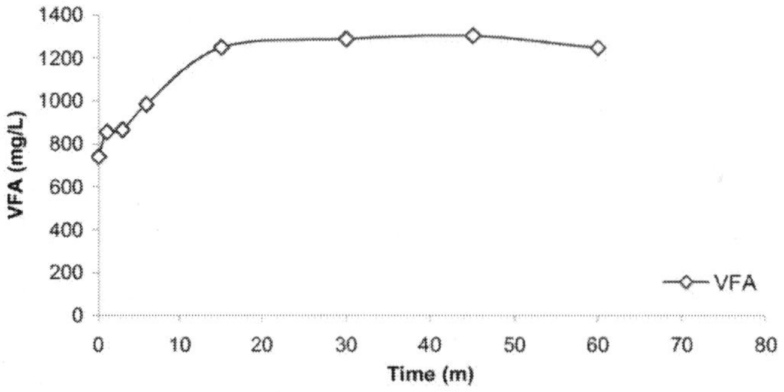

Figure 5, VFA as a function of sonication time

Figure 5 depicts VFA concentration for different sonication time; VFA distinctly increases when the treatment time was less than 20 minutes and reaches plateau after 20 minutes. However, although absolute concentration of VFA increased with sonication time, the ratio of VFA/SCOD decreased from about 0.46 to 0.15 at 20 minutes. Higher VFA/SCOD ratio is generally better for the overall performance of anaerobic digestion.

Lipids:

Figure 6 depicts lipid concentration at different sonication time. It clearly reveals that lipid concentration decreases exponentially with sonication time, and reaches a plateau only about 30 minutes of sonication. This is possible if lipid reacts with the hydroxyl radicals generated in the system due to ultrasonication, and could improve the performance of anaerobic digestion as lipids are difficult to break down in an aerobic digestion.

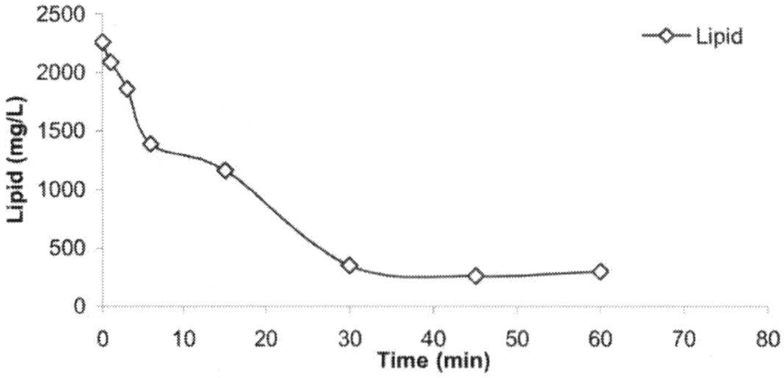

Figure 6. Lipid concentration with sonication time

Total Suspended Solids (TSS) and Volatile Suspended Solids (VSS):

Figure 7 a, b illustrates the TSS and VSS amount as a function of sonication time. A slight decrease in the values of the TSS and VSS can be noticed, this decrease is also possible due to the reaction of the organics in VSS and TSS with the hydroxyl radical reaction and also due to the dissolution of VSS to SCOD.

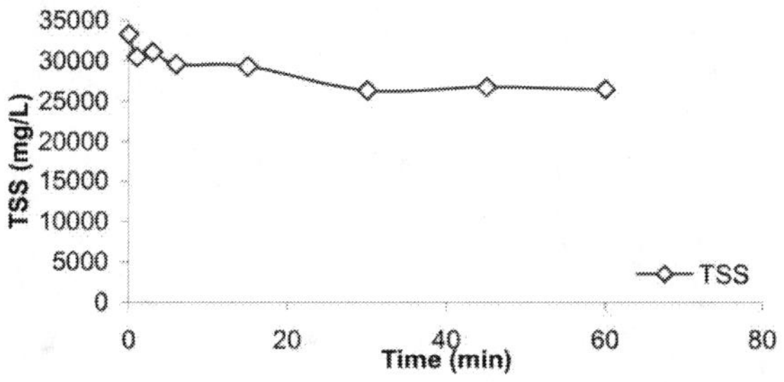

Figure 7a, TSS as a function of sonication time

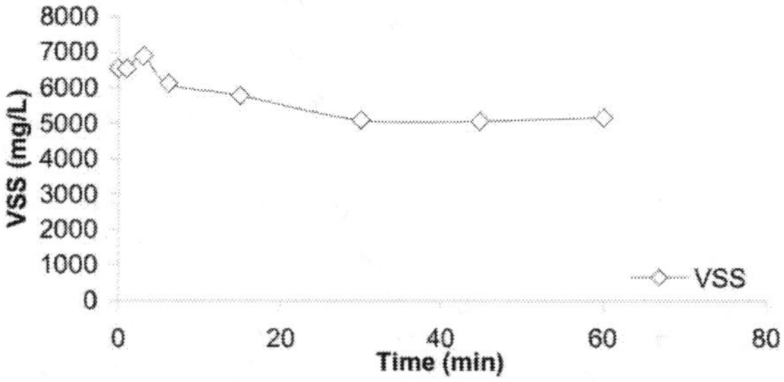

Figure 7b, VSS as a function of sonication time

Ammonia:

Ammonia concentration increases when the treatment time was less than 20 minutes and reaches a plateau after that. Figure 8 shows the ammonia concentration as a function of sonication time.

This might be due to the break down of the biomass, the limit of ammonia was not critical to cause inhibition in the anaerobic digestion.

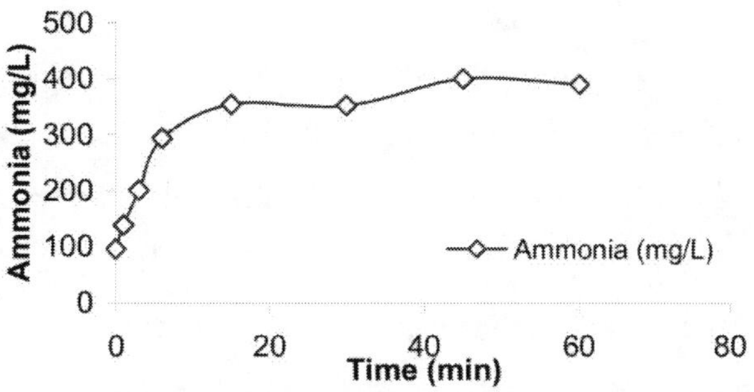

Figure 8. Ammonia as a function of sonication time

Protein:

Proteins in wastewater and sludge are generally divided into there fractions: soluble, bound/labile (loosely attached with the cells) and tightly bound fractions (within the bacterial cells). Labile proteins are thought to become readily bioavailable giving rise to higher odor potential. Total (cell + bound), soluble and bound protein were monitored in this study. The total protein was unlike the soluble protein distinctly decreased with time, while the soluble protein increased with sonication time. Due to cell rupture, total protein, which is the cellular and extracellular protein loosely bound to the cell disintegrated, some were changed into bound protein and loosely attached with the cell, and mostly were dissolved into soluble protein. It is important to say that increased soluble protein concentration increases the performance of anaerobic digestion.

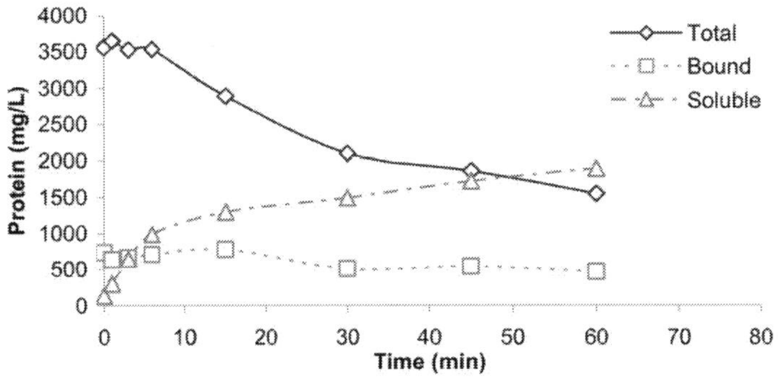

Figure 9. Total, Bound and Soluble protein as a function of sonication time

Specific Methanogenic Activity (SMA):

Anaerobic biodegradability study was conducted to determine the effect of sonication on the anaerobic gas production, specifically methane production. The sonicated sludges for different sonication time were used for SMA at 37°C, using 250 mL bottle capped with Teflon septum. 50 mL anaerobic seed and 50 mL treated primary sludge were added together into the bottles. All the bottles were sealed after purging the headspace with nitrogen to eliminate the present of oxygen/air. The experiment was continued until the bottles stopped producing biogas. Daily biogas was measured by inserting needle attached to a syringe (100 mL and 20 mL). Methane composition was measured by Gas Chromatography (GC) SRI 310C with a packed column.

Figure 10, displays the percent gas produced from the untreated batch and the seven treated samples, and Figure 11 exhibits the volume of methane produced from the untreated batch and the seven treated samples. The experiment was stopped after 17 days due to the lack of gas production.

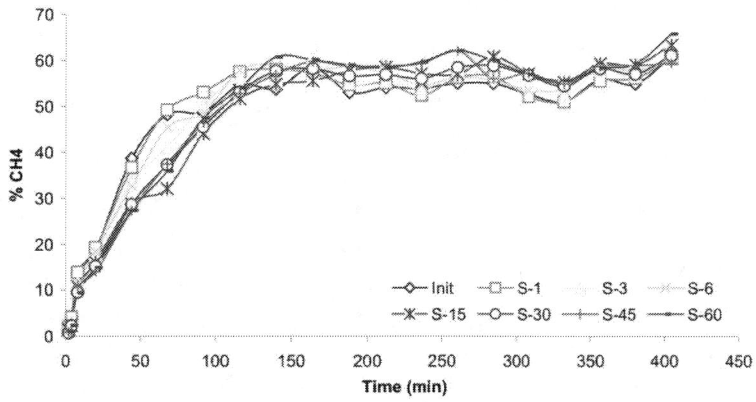

Figure 10. Percentage of methane production for initial and sonicated samples for 17 days

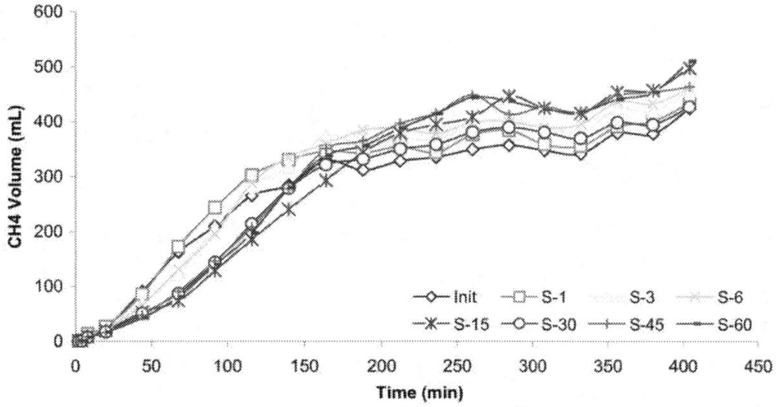

Figure 11. Volume of methane production for initial and sonicated samples for 17 days

Assessing the figures above, percentage methane produced seems to be fluctuating between 54% and 65% for treated and non treated samples after 100 minutes, and there is no clear advantage in sonication in the sludge. However, the volume of gas production seems to have an interesting pattern, the initial rate of gas seems to be high at low sonication time and decreases with sonication time, however, final gas produced seems to higher as sonication time increases, which needs to be further investigated.

CONCLUSIONS

The ultrasonic disintegration was used as pre-treatment for primary sludge obtained from municipal wastewater plant, London Ontario. The important parameters for better performance in anaerobic digestion such as SCOD, VFA and soluble protein increased due to ultrasonication, whereas particle size and lipid concentration decreased with sonication indicating good solubilization of particulates and larger molecules due to sonication. The final gas production also increased with sonication time, although initial rate of gas production was lower for the sonicated sludge.

ACKNOWLEDGEMENT

This research was financially supported by Trojan Technologies and NSERC. The help of Reddy Kandlakuti, Meysaa Hussein, and Nesma Hashem in many analyses are well appreciated.

REFERENCES

A. Tiehm, 2003, Wat. Res. Vol. 35, No 8, pp.,"Ultrasonic Waste Activated Sludge Disintegration for Imporving Anaerobic Stabilization"

Andrega Zgnajnar Gotvajn, 2004, Faculty of Chemistry and Chemical Technology, University of Lijubjana, Askerceva 5, Ljubljana, Slovenia, "Combination of Fenton and Biological Oxidation for Treatment of Heavily Polluted Fermentation Waste Broth"

C.P. Chu, 2001, Wat. Res. Vol. 35, No. 4, pp 1038-1046, "Observations on Changes in Ultrasonically treated waste-activated sludge"

Chiu Y.-C., Chang C.-N., Lin J.-G. and Huang S.-J., 1997, Water Sci. Technol. 36(11), 155 – 162. "Alkaline and ultrasonic pre-treatment of sludge before anaerobic digestion"

D.C. Seo, 2007, Water Science & Technology Vol 55 No1-2 pp 251-259, "Treatment of Non-Biodegradable cutting oil wastewater by Ultrasonication-Fenton oxidation process"

Doha´ nyos M., Zabra´ nska´ J. and Jen´cek P., 1997, Water Sci. Tech. 36(11), 145 – 153,

"Enhancement of sludge anaerobic digestion by using of a special thickening centrifuge"

Dr. M. Ray, 2008, Department of Chemical and Biochemical Engineering, Anaerobic Digestion of Solid Waste and Maximization of Biomethanation"

Drewsa, A., Vocksa, M., Iversena, V., Lesjeanb, B. and Kraumea, M., 2006, Desalination, 192, 1-9, "Influence of unsteady membrane bioreactor operation on EPS formation and filtration resistance"

E. Gonze, 2002, Chemical Engineering and Processing 42 (2003) 965-975, "Ultrasonic Treatment of an aerobic activated sludge in a batch reactor"

E. Riera-Franco de Sarabia, 2000, Elsevier Science B.V. All Rights Reserved, "Application of high-power ultrasound to enhance fluid/solid particle separation process"

Eastman J. A. and Ferguson J. F., 1981, JWPCF 53(3), 352 - 366, "Solubilization of particulate organic carbon during the acid phase of anaerobic digestion"

Harrison, S.T.L. 1991, Biotechnol. Adv. 9, 217–240, "Bacterial cell disruption: a key unit operation in the recovery of intracellular products"

Hiraoka M., Takeda N., Sakai S. and Yasuda A., 1984, Water Sci. Technol. 17, 529 - 539, "Highly efficient anaerobic digestion with thermal pre-treatment.".

January Bien, 1997, Wat Sci.Tech. Vol. 36, pp. 101-106, "Changes of Some Sewage Sludge Parameters Prepared with an Ultrasonic Field"

Kuttruff, H, 1991, Essex, England, "Ultrasonics Fundamentals and Applications. Elsevier Scienc"

Mueller J., Lehne G., Schwedes J., Battenberg S., Na¨ veke R., Kopp J., Dichtl N., Scheminski A., Krull R. and Hempel D. C. 1998 "Disintegration of sewage sludges"

Mukesh Goel, 2004, Department of Chemical and Bimolecular Engineering, National University of Singapore, 4 Engineering Drive 4, Singapore 117567, Singapore, "Sonochemical Decomposition of Volatile and non-Volatile Organic Compounds – A comparative Study"

O. Schlafer, 2000, Ultrasonics 38, pp. 711-716, "Improvement of Biological Activity by Low Energy Ultrasound Assisted Bioreactors"

O. Schlafer, 2002, Ultra Sonic 40 pp. 25-29, "Ultrasound Stimulation of Micro-Organisms for enhanced biodegradation"

Panyue Zhang, 2005, Department of Enviromental Science and Engineering, Tsinghua University, Beijing 100084, China, "Ultrasonic Treatment of Biological Sludge: Floc Disintegration, Cell Lysis and Inactivation"

Raf Dewil, 2006, Department of Bioengineering, University of Antwerp, B-2020 Antwerp, Belgium, "Ultrasonic Treatment of Waste Activated Sludge"

S Lafitte-Trouqué, 2002, Biorsource Technology 84 pp. 113-118, "The Use of Ultrasound and y-irradiation as pre-treatments for the anaerobin digestion of waste activated sludge at mesophilic and thermophilic temperatures"

Samir Kumar Khanal, 2007, Critical Reviews in Environmental Science and Technology, 37:277-313, "Ultrasound Applications in Wastewater Sludge Pretreatment: A Review"

Shimizu T., Kudo K. and Nasu Y. 1993, Bioengng. 41, 1082 – 1091, "Anaerobic waste activated sludge digestion – a bioconversion and kinetic model. Biotechnol"

Suslick, K. Sonochemistry, 1991, Science 247, 1439, 1990. Flint, E.B. and Suslick, K.S., Science 253, 1397–1399, "The temperature of cavitations"

T.Mao, 2004, Water Science and Technology Vol 50 No 9 pp 91-97, "A Comparison of Ultrasound Treatment on Primary and Secondary Sludges"

Wang, F., Wang, Y., and Ji, M., 2005, J. Hazard. Mater. B123, 145–150, "Mechanisms and kinetic models for ultrasonic waste activated sludge disintegration"

X.Q. Cao, 2004, The R & D Centre for Sustainable Environmental Biotechnology, Beijing Inst. of Civil, "Enhanced Sludge Decomposition by Ultrasound"

Ying Shih Ma, 1998, Wat. Sci.Tech. Vol 38, No. 6, pp. 253-260, "Effect of Pre-Sonication on Removal of Organic Matters Resulting from Chlorinated Humic Acids"

Ying-Chih Chiu, 1997, Wat. Sci Tech Vol. 36, No. 11, pp. 155-162, "Alkaline And Ultrasonic Pretreatment of Sludge Before Anaerobic Digestion"

Kinetic Studies on Enhanced Mercury Adsorption in Zeolite Nay and Waste Fcc Catalyst

J. Juyo[I], J.Tirano[I], G. Camargo[I] , L. Giraldo[III] & J. Moreno-Piraján[II].

[I] Faculty of Engineering. Department of Chemical Engineering. Los Andes University. Colombia.

[II] Faculty of Sciences. Department of Chemistry. Los Andes University. Colombia.

[III] Faculty of Sciences. Department of Chemistry. Universidad nacional de Colombia.

ABSTRACT: The mercury (II) removal was studied using two adsorbents: waste FCC Catalyst and NaY zeolite, those materials were characterized under test such as TGA, XRD, IR, immersion enthalpy and N_2 adsorption isotherms. The zeolite superficial area is 773 m^2/g and that of the FCC catalyst is 135 m^2/g. The maximum mercury adsorption capacity was 1.8561mg/g for the FCC catalyst and 2.6845 mg/g for the zeolite. The equilibrium isotherms were described with the Langmuir and Freundlich linear models. Besides, the kinetics was described with the Lagergren and Pseudo-second order, the latter fit better in the experiments. After the use of the adsorbents, they were regenerated, acquiring their superficial areas as 675 and 87 m^2/g for the NaY zeolite and the FCC catalyst, respectively.

Key Words: Mercury removal, Adsorption Kinetics, NaY Zeolite, waste FCC Catalyst.

1 INTRODUCTION

Without earth`s water, no living organism can grow or reproduce if there is scarcity of this recourse, besides the water availability is necessary not only for the Industry but also for the human life. Heavy metals, as mercury, are contaminant pollutants that have to be treated to prevent environmental and community disasters.

The present investigation compares the mercury (II) adsorption kinetics using two materials, the NaY zeolite and the waste FCC catalyst, the latter provided by petroleum Colombian Company (ECOPETROL). These materials were characterized with nitrogen adsorption isotherm at 77K, XRD (X-ray diffraction), FTIR (Infrared spectra), TGA (Thermo gravimetric analysis) and immersion calorimetric test.

The adsorption process was studied from two kinds of solutions ($HgCl_2$+H_2O and $HgCl_2$+H_2O +$CaCl_2$) [1], with different conditions such as temperature (298.55K and 308.15K), residence time and pH. The linear models of Langmuir and Freundlich were used to describe the mechanism, and both fit differently based on each experiment. Adsorption kinetics was analyzed with the Lagergren and Pseudo-second order, the latter represents better the adsorption system.

2 MATERIALS Y METHODS

2.1. Materials

The adsorbents were (A) Waste FCC catalyst which is a complex material composed by Y zeolite, Kaolin and other additives. After the process in ECOPETROL this adsorbent has from 0.2 %wt to 0.4%wt of Coke[2], and (B) the NaY zeolite, its structure has low density, catalytic and adsorbent properties and a high hydration level. All the chemicals used in this investigation were not purified, and all the solutions were diluted with distillate water.

2.2. Methods

2.2.1. Material Characterization

Both adsorbents were subject to a different characterization tests presented next:

1. *BET model:* It starts with the development of the polarization theory of DeBoer y Zwicker, which is based on the dipolar moments happening when the adsorption layers are forming, it permits know the surface area of the material within other characteristics[3].
2. *X-ray diffraction:* It shows the crystalline structure of the solids. When the ray is directed towards the solid, it will generate a diffraction pattern based on Bragg's law.
3. *Infrared spectra:* when the electrons absorb radiation, they pass from a low energy level to a higher one, increasing their energy. This process results in vibrations of different intensity showed in specific regions of the spectra [4].
4. *TGA and DTA:* TGA test determine the weight changed and the structure degradation in the solid being subject to temperature ramps under an inert atmosphere. While DTA compare between two materials temperature variations and behaviors.
5. *Immersion Enthalpy:* A solid immersed into an inert liquid generates heat; this immersion heat is related to a layer formation over the solid surface. The immersion enthalpy is an enthalpy change at constant temperature that develops when the solid is submerged into the liquid and it does not react or dissolve [5].

2.3. Quantification of mercury

This method is based on the radiation absorption at 253.7 nm that mercury atomic vapour suffers; in order to do this is necessary to use cold vapour without flame. With this process the mercury is reduced to its elementary state following these reactions:

$$2MnO_4^- + 16H^+ + 5Hg^0 \rightarrow 2Mn^{2+} + 5Hg^{2+} + 8H_2O \quad (1)$$

$$MnO_4^- + 8H^+ + 5Hg^+ \rightarrow Mn^{2+} + 5Hg^{2+} + 4H_2O \quad (2)$$

$$Sn^{2+} + Hg^{2+} \xrightarrow{H^+} Hg^0 + Sn^{4+} \quad (3)$$

The mercury vapour is drag by an inert gas along an absorption cell made of quartz: this cell is aligned with the spectrophotometer light beam. This experiment produces a signal proportional to the number of free atoms in the cell [6].

2.4. *Adsorption process*

2.4.1. *Equilibrium studies*

The aqueous isotherms were made at two different temperatures (298.55K and 308.15K), controlling pH at 9- 10.5 to be sure that the Ion 2+ is, actually, present in the solutions. The mercury concentrations used were varied from 50 to 700 ppm and the adsorbent doses were between 40 and 200 mg.

1. *Langmuir model:* this model was formulated in 1918, and was the first model trying to interpret the adsorption phenomena, it has the following assumptions [7]: First, The whole surface has the same adsorption activity, and then is a flat surface and energetically homogeneous. Each particle is going to be adsorbed just for one active site. Second, There is not interaction between the adsorbed particles. Finally, The process is directed by just one mechanism and it happens in the monolayer.
The Langmuir Isotherm is presented next:

$$\theta = \frac{KC_e}{1-KC_e} \tag{4}$$

2. *Freundlich model:* Generally this model is used in aqueous adsorption processes and has the following assumptions: The adsorption process is not happening only in the monolayer, The active sites energy is not the same and the adsorption and desorption velocities varies with the force or the energy of each site. The Freundlich model is represented by the following equation:

$$v_{ad} = K_f C_e^{1/n} \tag{5}$$

K_f is the Freundlich adsorption coefficient and n is an empirical coefficient. Suggestions about better process fit have been made: with some systems is best the Freundlich model [8].

2.4.2. *Kinetics studies*

With an initial concentration of 400 ppm was carried out the experimentation, each sample was taken every 30 minutes during 4 hours. The results were approached to the equation of Lagergren first order:

$$Log(q_e - q) = Log(q_e) - \frac{K_{ad}t}{2.303} \tag{6}$$

Where q and q_e are the amount of mercury adsorbed (mg/g) at some time t (min) and at the equilibrium time, respectively. In a logarithmic scale should be prepared a figure with (q_e-q) on the y-axis and t on the x-axis, this, in order to calculate the adsorption constant, K_{ad} [9].
Second, the pseudo-second order was determined along with its constant, K_2 (7th equation). As well as the first order, q (mg/g) is the amount adsorbed at some time t (min), and q_e is the amount adsorbed at the equilibrium time [10].

$$\frac{t}{q} = \frac{1}{K_2 q_e^2} - \frac{t}{q_e} \tag{7}$$

3 RESULTS AND DISCUSSION

3.1. Materials characterization

1. *N_2 adsorption isotherm:* The analysis conditions are shown on Table 1. The adsorbate was nitrogen and the bath temperature was 77.4 K for both adsorbents. The Isotherms for the materials are shown in Figure 1 and 2, for the waste FCC catalyst and NaY zeolite, respectively. The FCC isotherm is described by the isotherm type II because of its small area, and its micropores volume, besides the isotherm shows a fast saturation on the monolayer. On the other hand, the NaY zeolite is correlated with the isotherm adsorption type I, and the reason is that the monolayer saturation occurs at a higher relation between the pressures, P/Po.

Table 1. Analysis Conditions for the nitrogen adsorption.

Conditions	Waste FCC	NaY Zeolite
Weight sample	0.1242 g	0.0895 g
Degasification temperature	250 °C	250 °C
Degasification time	23.0 hrs	12.0 hrs
Analysis time	622.2 min	191.2 min

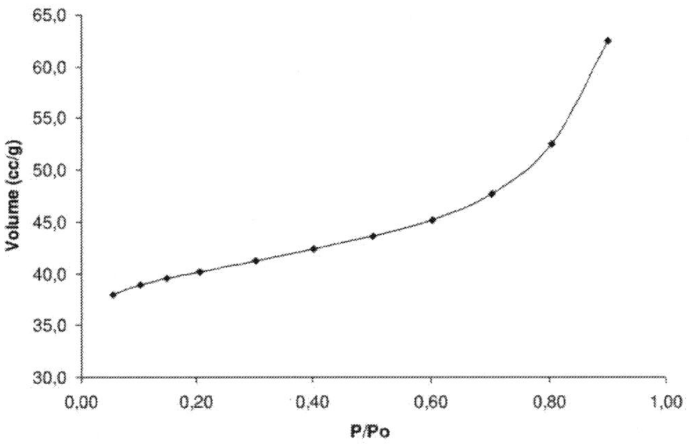

Figure 1. N_2 adsorption isotherm, waste FCC catalyst.

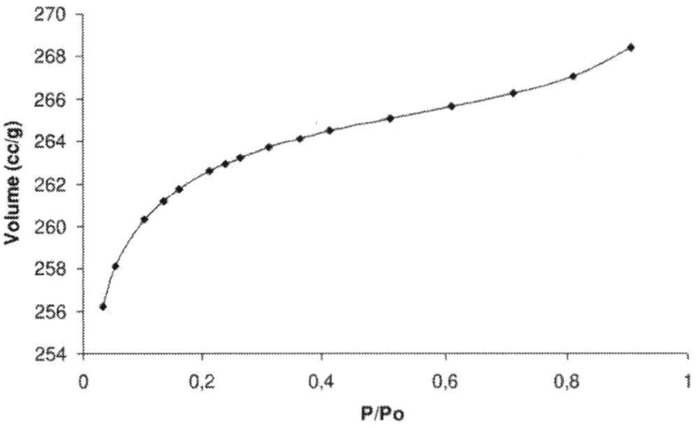

Figure 2. N_2 adsorption isotherm, NaY zeolite.

The surface area was calculated. Is 135 m^2/g for the waste FCC catalyst, and 773 m^2/g for the NaY zeolite, this let conclude that the mercury adsorption should be more in the zeolite structure than in the structure of the catalyst. Moreover the pore diameter of the two materials are 12.0 Å (NaY) and 14.2 Å (FCC) and this allows the mercury Ion (diameter: 2.2 Å) to enter in the structure of the two adsorbents. The hysteresis for the two adsorbents shows the possible presence of mesopores. But in both structures the micropores proportion is more than the mesopores volume. It is possible that the micropores were not filled in completely at the relative partial pressure used, and then the mesopores could be filled before the complete filled of the micropores.

2. *X-ray diffraction:*

In XRD pattern of the waste FCC catalyst are some aluminum oxides peaks at 2-theta angles of 6, 11, 12, 16, 21, 25, 28 and 39 and comparing between the standard pattern is possible to find the Faujasita structure [11]. The standard peaks of the NaY zeolite are: 2-theta= 6, 10, 12, 15, 17, 21, 24, 31 and 35 [12] present on the pattern obtained in XRD pattern of the NaY zeolite. The highest peak in the FCC pattern is at 26.85 Bragg angle, while the NaY zeolite is 6.22. Those angles are typical for an amorphous structure (FCC) and a crystalline structure (NaY). Besides, the crystalline percentage is 28 for the waste FCC, and 95.7 for the zeolite [12, 13].

3. *Infrared spectra:* The infrared spectra of the waste FCC has the following characteristic bounds: at 3443 cm^{-1} wavelength are O-H stretching vibrations, at 1634 cm^{-1} C=N stretching, at 1402 cm^{-1} NOx vibrations, at 1012 cm^{-1} flexion of the bound C-OH, at 789 cm^{-1} C-Cl stretching, and at 460 cm^{-1} can be presence of sodium. The presence of the sulfur compounds in the waste FCC can be an advantaged to the mercury adsorption due to the affinity between those compounds [14]. For the NaY zeolite, the broad bands observed at 3420, 1631, 1082.76 cm^{-1} are due to stretching of O-H, C=S and C-O bounds, respectively. The other visible bands in the up corner of the figure are at 835.76, showing characteristic bounds of C=C-R flexion, and at 612.93 cm^{-1} C-S stretching.

4. *Thermal analysis:* the waste FCC catalyst´s TGA shows the high thermal stability of the structure and permits recognize the low amount of water molecules present on the FCC, moreover the DTA shows that the waste material suffers an exothermic process while is heated between 100 to 1000°C The TGA for the NaY zeolite shows the high degree of hydration of the compound, because the weight lost that suffers is about 25% of its initial weight, before the 400ºC ramp. There is a lose of water molecules, and let know that the structure will not lose another kind of molecules different from water ones, at the same time is developed an endothermic process, that turns into an exothermic process when the most water molecules had been lost. After the flat region occurs an exothermic process; this allows a degradation of the zeolitic structure after the 800ºC ramp. In this investigation the highest temperature, except the recuperation one, was 35ºC and then there was not structural degradation for neither of the two materials.

5. *Immersion Entalphy:* this test was made to probe that the process was going to be physical. And effectively, the immersion process was physical for both materials, these can be observed with the low heats produced when the solid was wet with both solvents. With this experiment is possible to appreciate that both adsorbents have the capacity to accept the mercury Ions because its diameter is 2.2 Å, and the diameter of the dichloromethane is 3.3 Å and 6.0 Å the one of the tetrachloride, both bigger than the diameter of the mercury. Furthermore, the immersion enthalpy developed let know that the surface area

of the NaY zeolite is bigger than the one of the waste FCC catalyst due to the proportionality between the heat of immersion and the surface area. It also shows the difference in adsorption capacity of the two adsorbents.

3.2. Adsorption Process

The adsorption isotherms and the kinetics experiments were performed according to Table 2:

Table 2. Experimental conditions.

Condition	Meaning	1	2	3	4	5	6	7	8
Tempera-	25.4 ºC	X	X	X	X				
	35ºC					X	X	X	X
pH	between 9 and 10.5	X	X	X	X	X	X	X	X
Solution	H_2O + $HgCl_2$(S1)	X	X			X	X		
	H_2O+ $CaCl_2$+$HgCl_2$(S2)			X	X			X	X
Adsorbent	FCC	X		X		X		X	
	NaY		X		X		X		X

3.2.1. Adsorption Isotherms

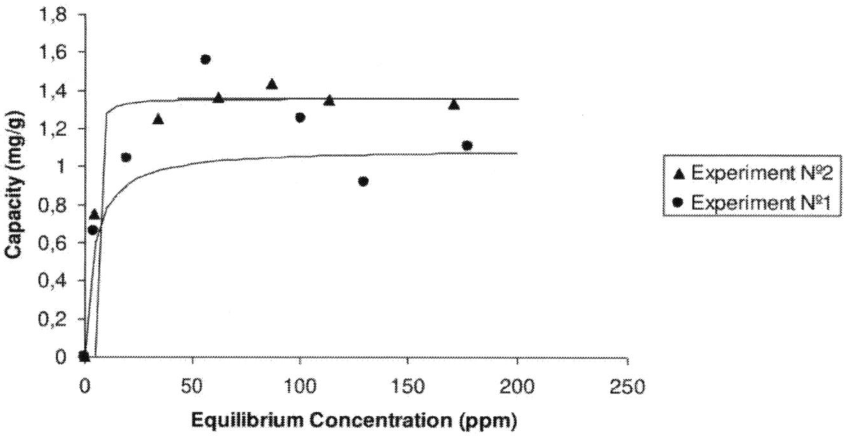

Figure 3. Two of the Aqueous Isotherm, T=25ºC, pH = 9-10.5.

Table 3. Adsorbent´s capacity for each experiment.

Experiment	Q(mg/g)
1	1.4671
2	1.0946
3	1.8561
4	1.4550
5	1.3301
6	2.4162
7	2.0629
8	2.6845

To verify the adsorbent behavior for each experiment was made a duplicated and for the analysis was taken into account the pH, solution, temperature and doses of the adsorbent effects.

Effect of pH: In the experimentation two pH´s, below 2 and above 9, were going to be used [15]; this to guarantee the mercury Ion form (Hg^{2+}) in the solutions. But below pH=3 the structure of the adsorbents collapsed and no profitable results were obtained, reason for which there are not results of that kind presented in this paper and all the experimentation process were carried out with pH values between 9 and 10.5. Th pH value was controlled with HCl and NaOH solutions.

Effect of temperature: At 25.4ºC the capacity for the waste FCC catalyst was 1.4671 mg/g and 1.0946 mg/g for the NaY (S1), while at 35ºC the capacity of the FCC decrease 9.33% and the zeolite capacity increased 54.7%. So, the capacity of the NaY increases with temperature but is important to not carry out the process with high temperatures in order to not accelerate the desorption process.

Effect of the solution type: The adsorption was higher using the $CaCl_2$ solution. At 25ºC the capacity of both adsorbents increased 43% with the waste FCC and 24.8% for the NaY. While at 35ºC the difference between solutions were increased 8.7% for the zeolite and 35.4% For the catalyst this can be advantageous for the process due to no contaminant pollutant is alone in the wastewater, and is necessary to involve other species in the removal of this toxic compounds.

Effect of doses and type of adsorbent: How it was mentioned in the effect of the temperature the waste FCC is best at lower temperatures (1.4671 – 1.8561 mg/g), on the contrary the zeolite mercury removal was higher at higher temperatures. The removal could be higher if were used more adsorbent quantity, although for investigation experimentation is not necessary use to much adsorbent. These conditions can be improved for the process scale up.

3.2.2. Adsorption mechanism

A good linear approach is given by a correlation coefficient $R^2 > 0.98$, however in this process was not possible to obtain $R^2>0.9$ for Freundlich, except for the first experiment that has a $R^2= 0.7853$. There were experiments where Freundlich adjust better than Langmuir and vice versa. Reason for which the adjustments were individually for each experiment, in other words, the Isotherms were adjusted with the model that has higher correlation coefficient. Some approximations are showing next:

Table 4. Langmuir and Freundlich parameters for the experiments.

	Langmuir			Freundlic		
Experi-	**K(g/l)**	**V_m(mg/g**	**R^2**	**K_f**	**n**	**R^2**
1	1,4	1,3564	0,997	0,9	12,936	0,7
2	0,2	1,0866	0,987	0,5	7.8452	0,9
3	0,0	1.5873	0,797	0,0	0,9576	0,9
4	0,0	2,2872	0,897	0,0	1,3493	0,9
5	0.7	1,6326	0,943	0,2	2,5322	0,9
6	1.9	3,0959	0,873	0,0	1,0333	0,9
7	1,7	0,0325	0,281	0,0	0,9401	0,9
8	0,0	5,8445	0,925	0,1	1,6725	0,9

3.2.3. *Adsorption Kinetics*

As well as 25.4ºC experiments the 35ºC experiments have a strong adsorbate-adsorbent interactions, besides when the zeolite is used in the batch procedure the Lagergren graphics (Y-Axis) has larger differences values, while the pseudo-second order has not that differences (less than 0.95%). For that fact and due to the coefficient correlation obtained in the calculations has been chosen the pseudo-second order kinetics to model the system. The Kinetics approaches are showed in Figure 4 and 5:

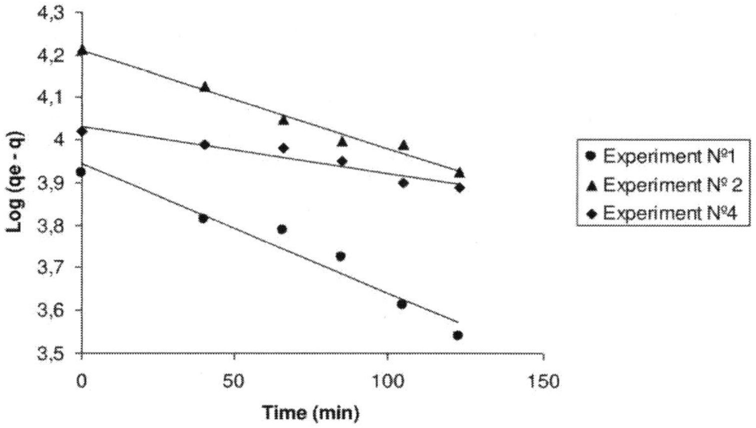

Figure 4. Lagergren approaches for some experiments.

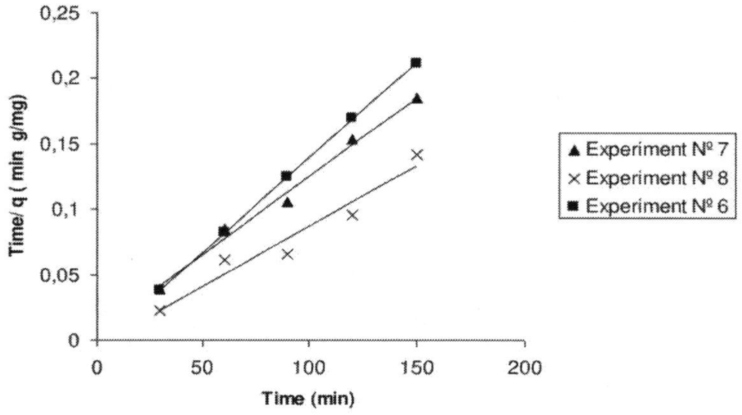

Figure 5. Pseudo-second order approaches for some experiments.

Table 5. Lagergren y Pseudo-second order parameters for the Hg(II) adsorption.

Order	Lagergren			Pseudo second		
Experiment	**K_1(min^{-1})**	**q_e (mg/g)**	**R^2**	**K_2(min^{-1})**	**q_e (mg/g)**	**R^2**
1	0.069	2.209	0.9838	0.058	1.523	0.9456
2	0.052	1.625	0.9482	0.042	1.00	0.8834
3	0.043	1.604	0.9104	0.038	1.565	0.8830
4	0.025	2.107	0.9235	0.019	1.004	0.9376
5	0.009	1.262	0.8196	0.008	2.431	0.9896
6	0.006	2.176	0.8309	0.010	3.768	0.9457
7	0.011	1.21	0.9552	0.009	1.111	0.9464
8	0.006	2.288	0.8846	0.011	3.633	0.9898

4 CONCLUSIONS

The mercury removal capacity for the waste FCC does exceed neither the zeolite capacity, nor other materials in other studies; this is because of its low surface area. Although, the sulphur compounds in the waste FCC catalyst, could helped in the adsorption process.

The Pseudo-second order can model the mechanism well to scale up the process.

Finally, is important for this type of process to know the ionic charge of the surface and give an accurate treatment to the solution involved in the adsorption.

5 REFERENCES

[1] P. HUTTENLOCH, K. ROEHL, & K. CZURDA, Use of copper shavings to remove mercury from contaminated groundwater or wastewater by amalgamation. Environ. Sci. Technol. (2003). Science Direct.

[2] R. SADEGHBEIGI, Fluid Catalytic Cracking Handbook, 2nd Edn., 2000, GULF PROFESSIONAL.

[3] BRUNEAUR, EMMETT & TELLER. Adsorption of gases in multimolecular layers (1938), George Washington University, United States.

[4] T. SOLOMONS, Organic Chemistry, 2002, 2nd Edn., Limusa.

[5] J. SILVESTRE, C. GOMÉZ DE SALAZAR, A. SEPÚLVEDA, F. RODRÍGUEZ, Characterization of microporous solids by inmersion calorimetry (2001). ScienceDirect.

[6] R.A. SALINAS, (2002) Validation of the analytical methodology to find mercury on water with the cold vapour method. No published work. Universidad Nacional de Colombia, Colombia.

[7] J.M. SMITH, Chemical Engineering Kinetics, 1981, 3rd Edn., Boston, United States.: McGraw-Hill, c1981.

[8] COONEY. Adsorption Design for Waste water Treatment (1998). London, England.: Lewis Publisher.

[9] C. NAMASIVAYAM, & S.BENTHILKUMAR, (1996). Recycling of industrial solid waste for the removal of mercury (II) by adsorption process. Elsiever science Ltda. ScienceDirect.

[10] B. BUENO, M. MESQUITA & L. TOREM, (s.f.). Remoción de Iones metálicos tóxicos de efluentes líquidos por biosorción.

[11] C. HUANG, Y. TSENG & K. HSU, (2005), The pozzolanic activity of a waste FCC catalyst and its effect on the compressive strenght of cemtitious materials. ScienceDirect.

[12] B. LAPIERRE, R. PARTRIDGE, P. REISCHMAN & T. KERR, (1989). Method for preparing acid stable zeolites and high silica zeolites prepared by it. Patent Nº4840930 of United States.

[13] N. MARTINEZ, J. LUJANO, N. ALVAREZ, F. MACHADO & C. LOPEZ, (1994). Method for recharging spend fluid catalytic cracking process catalyst. Patent Nº5360774 of United States.

[14] S. EVANGELISTA, E. DEOLIVEIRA, G. CASTRO, L. ZARA & A. PRADO, (2007). Hexagonal mesoporous silica modified with 2-mercaptothiazoline for removing mercury from water solution. ScienceDirect.

[15] P. CARROT & M. RIBEIRO, (1997). Influence of surface ionization on the adsorption of aqueous Mercury chloroclompexes by activated carbons, Pergamon, ScienceDirect.

390h Simultaneous Electrochemical Oxidation and Coagulation for the Treatment of Dye-Containing Wastewater by Using RuO2 Coated Titanium Electrode

Rita Farida Yunus, Environmental Science and Engineering, National University of Singapore, Blk E2, #05-05, Division of Environmental Science & Engineering, 5 Engineering Drive 2, NUS, Singapore, 117576, Singapore, Yuming Zheng, Division of Environmental Science and Engineering, National University of Singapore, 10 kent ridge, Singapore, Singapore, K.G.Nadeeshani Nanayakkara, Division of Environmental Science & Engineering, National University of Singapore, 10 Kent Ridge Cresent, Singapore, 119260, Singapore, and **J. Paul Chen**, Environmental Science and Engineering, Natioanl University of Singapore, Singapore, Singapore.

Dye-containing wastewater from colorants manufacturing industries presents serious problems to the water environments due to its greater toxicity, carcinogenicity, mutagenicity, and poor biodegradability. Electrochemical technologies appear to be promising, since these technologies are low cost and able to provide full decolourization of dyes. In this study, an efficient electrochemical technology for the removal of Rhodamine 6G, a fluorescent-rigid dye, by using RuO_2 coated titanium mesh as dimensionally stable anode was developed and studied. Our results show that, the decolourisation process is favorable under acidic conditions (pH 2 – 4) in the presence of 0.1 – 0.2 M of NaCl as electrolyte salt and the applied current of 1.0 – 1.9 A, and full decolourisation of dye-containing wastewater can be achieved within 5 min under the optimum condition. The simultaneous electrochemical oxidation and coagulation are involved in the removal of dye. Total organic carbon (TOC) analysis reveals, under the optimum conditions (pH 2, 1.9 A, and in the presence of 0.2 M NaCl), the electrochemical process significantly reduces the organic content (~ 95 % of TOC removal) in the wastewater within 5 min. Above 55% of the TOC is removed by forming solid foam products via an electrocoagulation process, and another ~ 40 % of TOC is mineralized by electrochemical oxidation. FTIR and GC/MS spectroscopy were used to analyze the solid foam products and intermediate products to probe the removal mechanism of the electrochemical process.

390i Photocatalytic Activity Assessment of TiO2 for Water Treatment: Substrate-Specificity and Its Implications

Wonyong Choi and Jungho Ryu. School of Environmental Science and Engineering, Pohang University of Science and Technology (POSTECH), Pohang, 790-784, South Korea

As TiO2 photocatalysis has been widely investigated for its environmental application, many researchers aim at development of new and/or modified photocatalysts showing higher activities. The assessment of the activity of photocatalysts is an integral part in the development and characterization of various photocatalytic materials. However, how the photocatalytic activities can be represented and quantified and how the activities of different photocatalysts should be compared are not straightforward. In this study, we tried to reveal the multi-dimensional nature of the photocatalytic activity in a systematic way. The photocatalytic reactivities of 8 commercial TiO2 samples (both anatase and rutile) were quantified by employing 19 test substrates (phenols, organic acids, amines, chlorohydrocarbons, dyes, inorganic ions, etc.) in terms of their degradation rates. The measured activities of photocatalysts exhibited complex behaviors that depend on what compound was used as a test substrate. Even the relative activity order among the tested photocatalysts was markedly influenced by the kind of test substrates. This highly substrate-specific activity of TiO2 photocatalysts hinders the straightforward comparison of which sample is better than others. No single substrate for the activity test can represent the activity of a specific photocatalyst in a comprehensive sense. Although there seems to be no simple correlation between the activity and the property of photocatalysts, the substrate-specific activity will be analyzed and discussed in terms of various parameters such as surface area, crystallinity, surface charge, and substrate adsorption. The reaction rates for DCA and iodide oxidation are proportional to the surface area of TiO2 while the rates for phenol, RhB, and MB degradation are rather inversely proportional to the surface area. There was also rough correlation in photocatalytic activities among test compounds which have similar adsorption behavior. Meanwhile, the effect of surface charge and the particle size of TiO2 turned out to be less significant at given conditions resulting in no correlations with photocatalytic activities. This study clearly indicates that the photocatalytic activity assessment and standardization should be cautiously adjusted to the specific purpose and the condition of photocatalyst applications.

454a Renewable Energy Technologies: Corona Discharges in Water Contaminated with Oil Spillage

Chinyere P., Mbachu, Chemical Engineering, Tennessee Technological University, 1020 Stadium Drive Prescott Hall, P.O.Box 5013, Cookeville, TN 38505

Poverty is generally considered as a major cause of food insecurity and poverty alleviation is essential in improving access to food. The Niger Delta of Nigeria, one of the world's top oil producers, has been the center of violent despair, neglect, poverty & poor environment for years.

86 percent of the population, majority of women and children, live in rural communities and work at subsistence farming. They cannot afford a good education for their children due to their low income levels. Depletion in natural water and lack of development in the oil rich region has led to disintegration of these traditional communities. Renewable energy technologies as a profitable alternative venture is a veritable tool that can help provide food and jobs for teeming youth and women. The coastal communities has remarkable potentials for development, but the government as well as the multinationals have not understood and tapped the potential role of renewable energy technologies in managing the crisis in the Nigeria.

This focuses on the understanding and further developing of pulsed corona discharges in water decontaminants (such as oil spillage decontamination).

454b Optimization of the Ultrasonic Cavitation-Induced Removal of Nitric Oxide from Flue Gases Using Taguchi Statistical Experimental Design Methodology

Yusuf G. Adewuyi, Nymul E. Khan, and Naresh N. Mahamuni. Chemical Engineering, North Carolina Agricultural & Technical State University, 1601 East Market Street, Greensboro, NC 27411

The combustion of fossil fuel releases a large amount of NOx and SO2 into the atmosphere which is a major threat to human health and environment. The efficient, cost effective and environmentally friendlier removal of NOx and SO2 from flue gases poses a considerable industrial problem. Ultrasonic oxidation of NOx and SO2 can be a solution. In our previous work we have demonstrated the feasibility removing NOx in the absence or presence SO2 (with complete and simultaneous removal of SO2) by ultrasonic cavitation and looked at the effects of parameters like US intensity, flow rate, chemical oxidant, additives and composition of flue gas on NOx removal using a sonochemical scrubber. To aid the cost-effectiveness and commercialization of the sonochemical removal process, an understanding of the effects of multiple factors as well as the influence of the individual factors on the overall sonication effectiveness is essential to establish the optimal conditions for the process. In this presentation we will discuss our current effort in the use Taguchi design of experiment (DOE) method to optimize the ultrasonic oxidation process, looking at a lot more process and experimental variables. The variables we have considered include pH, temperature, intensity of ultrasound, position of the horn, type of reactor, flow rate of gas, flow rate of liquid, gas phase composition. In addition to obtaining the optimum set of conditions, the percentage contribution of each experimental variable to the overall efficiency sonochemical removal of NOx will be determined.

Ultrasonic or acoustic cavitation involves the use of high-frequency sound waves in water to generate highly reactive species, particularly the hydroxyl (•OH), hydrogen (H•), hydroperoxyl (HO2•), and superoxide anion (•O2-) radicals and hydrogen peroxide (H2O2). We have previously demonstrated the utility of the Taguchi statistical method to design experiments and optimize the sonochemical removal of carbon disulfide in aqueous solutions. The effects of temperature, ultrasonic intensity, irradiation gas and frequency on the sonochemical oxidation process were studied. In the current work we have, for the first time, applied the method to a gas-liquid sonochemical system under ultrasonic irradiation. This technique will help us gain an insight into the relative significance of the various experimental factors necessary for the effective removal of NOx in a sonochemical scrubber under actual operating levels.

454c Modeling the Ultrasonic Cavitation-Enhanced Removal of Nitric Oxide from Flue Gases In a Bubble Column Reactor

Nymul E. Khan, Naresh N. Mahamuni, and Yusuf G. Adewuyi. Chemical Engineering, North Carolina Agricultural & Technical State University, 1601 East Market Street, Greensboro, NC 27411

The removal of NOx by ultrasonic cavitation is a fairly new area. In our previous works we have looked at the effects of different variables on the absorption of NOx from flue gases in presence of ultrasound. The fractional conversions of NO were found to range from 60% to 85%, while complete removal of SO2 was observed for all the inlet gas concentrations studied. Aimed at a more fundamental understanding of absorption of NOx using ultrasound, a model for such a system incorporating the mass transfer and kinetics approach based upon the rate constants for the reactions of NOx with •OH radicals was developed. The model is based on the simultaneous absorption and reaction of NOx in the liquid phase with •OH radicals generated due to ultrasonic waves. It is a simplified model since the cavitation bubble dynamics has not been included. The ultrasonic portion was incorporated as an •OH radical generation term using previously available correlation of number of •OH radicals with ultrasonic intensity and frequency. The results of the conversion of NOx predicted and validated by experimental results will be discussed. The quantification of the all-important mass transfer coefficient for this cavitation-gas absorption system achieved by the interpretation of experimental results using model will also be discussed. Some models discussing liquid-phase reactions in the presence of ultrasound abound in the open literature. However, reactive mass transfer models coupling gas absorption with liquid-phase reactions are rare. This paper and results that will be discussed are unique in that respect.

454d Falling Film UV-Reactor for off Gas Purification

Christoph Gruber and Matthäus Siebenhofer. Department of Chemical Engineering and Environmental Technology, Graz University of Technology, Inffeldgasse 25/C/II, Graz, A-8010, Austria

Nitrogen oxides and sulphur dioxide have an important role in acid rain production. Most flue-gas desulphurisation methods base on limestone scrubbing and concerning NOx abatement SCR technology is state of the art. There is still demand of new processes for low temperature applications or when catalyst poisoning is expected. Especially simultaneous SOx and NOx control is still of high interest.

Photochemical reactions and photochemical treatment of hazardous substituents became increasingly popular in recent years. Particularly in waste water treatment as well as in off-gas purification Advanced Oxidation Processes are subject of investigation. Dependent on the overall kinetics of photochemical reactions different reactor systems are used. To enable treatment and investigation of a broad range of constituents a flexible reactor design like a falling film reactor is chosen in this project where absorption with photochemical reaction of constituents in the gas phase and the liquid phase was investigated.

Investigation of absorption processes in falling film columns in laboratory scale has several advantages. It is possible to investigate mass transfer because of well defined mass transfer area. The set up allows investigation of gas phase reactions, the combination of photochemical reactions in both the gas phase and the liquid phase, and photochemical liquid phase reactions.

This paper discusses different basic reaction engineering measurements for the analysis of the reactor performance and application for SOx and NOx control is covered. Absorption of SO2 with photochemical oxidation was successfully tested. Dual step liquid phase nitrate reduction with urea was successfully tested too. While irradiation of NO2 in the gas phase will retard absorption efficiency due to cleavage, simultaneous irradiation of SO2 and NOx in the gas phase and absorption leads to different result which will be presented.

Studies on the Decontamination of Surfaces Exposed to Cytotoxic Drugs

Marek Kuzma, Laboratory of Molecular Structure Characterization, Institute of Microbiology, Prague, Czech Republic, Jaroslav Červený, Laboratory of Molecular Structure Characterization, Institute of Microbiology, Prague, Czech Republic, Jiří Švrček, Department of Organic Technology, ICT - Prague, Prague, Czech Republic, Kamila Syslová, Department of Organic Technology, ICT - Prague, Prague, Czech Republic, Petr Kačer, Department of Organic Technology, ICT - Prague, Prague, Czech Republic, Libor Červený, Department of Organic Technology, ICT - Prague, Prague, Czech Republic, Libor Pánek, BLOCK a.s., Prague, Czech Republic

Introduction

It is known a broad spectrum of decontamination techniques for biological and chemical contaminants from different surfaces and new physical and chemical processes continue to be developed[1]. The main reasons for development of new methods for decontamination are both dangerous properties of some decontamination agents, and possible hazardous residues formed after their application. An optimal chemical decontamination agent should demonstrate high-level disinfection properties, ability to remove and/or breakdown of biochemical and chemical contamination, high activity, material compatibility, lack of toxicity to humans and the environment, odorless, non-staining, unrestricted disposal, prolonged reuse life and shelf life, easy to use, ability to be monitored for concentration, cost-effective, and may be validated[2]. **V**apor **P**hase **H**ydrogen **P**eroxide (VPHP) seems to be one of the favorable methods that fit the above-mentioned remarks.[3]

VPHP represents a new technology, which was widely proved as an appropriate method for decontamination of rooms. The attacks on many cities revealed the difficulty of decontaminating buildings that had been exposed to biological weapons. The well-known effects of mold on allergies and asthma are also creating demand for improved building air quality. VPHP has become the method of choice for many bio-decontamination requirements in the pharmaceutical, biomedical and healthcare sectors. VPHP is used to treat pharmaceutical manufacturing clean rooms and laboratory toxicology rooms.

There is rising interest in application of VPHP for decontamination of chemical warfare and biologically active compounds. But there is only limited information about influence of VPHP on chemical substances.[4] The aim of this study was to examine the removal and deactivation of chemical substances from surfaces. Because VPHP is still „frontier" area of research, one of the goal was to explain basics of this decontamination process and study parameters that play important role in this process (temperature, relative humidity, concentration of gaseous hydrogen peroxide, synergistic effect of UV radiation[5]).

Using mass spectrometry (GC/MS, LC/MS) and NMR were studied decomposition products of chemical substances. Furthermore, using HPLC with ESI-MS/MS or GC/MS were monitoring kinetics of decontamination.

Experimental

Experimental equipment

Appropriate unique, sophisticated equipment, called "peroxybox", was developed for laboratory tests of VPHP for chemical substances decontamination (Fig. 1). It allows to do controlled decontamination of chemical substances by VPHP under various parameters (temperature, hydrogen peroxide concentration, relative humidity) and monitor the kinetics of decontamination of organic substances. Furthermore, it is possible to study the synergistic effect of VPHP with UV (wavelenght 254 nm) radiation too.

Decontamination procedure

The typical VPHP decontamination cycle consists of four phases in this peroxybox, namely: dehumidification, conditioning, decontamination and aeration. The tested compound is deposited on a microscopic slide and placed in the holder. The holder is inserted into the peroxybox, where is left for a defined period in the atmosphere of VPHP. The holder can be pushed out segment-by-segment without disrupting the atmosphere in the peroxybox to follow the kinetics of decomposition. The microscopic slide is afterwards washed with appropriate solvent and studied by GC/MS, and LC/MS or is evaporated to dryness and reconstituted in deuterated solvent for NMR analysis.

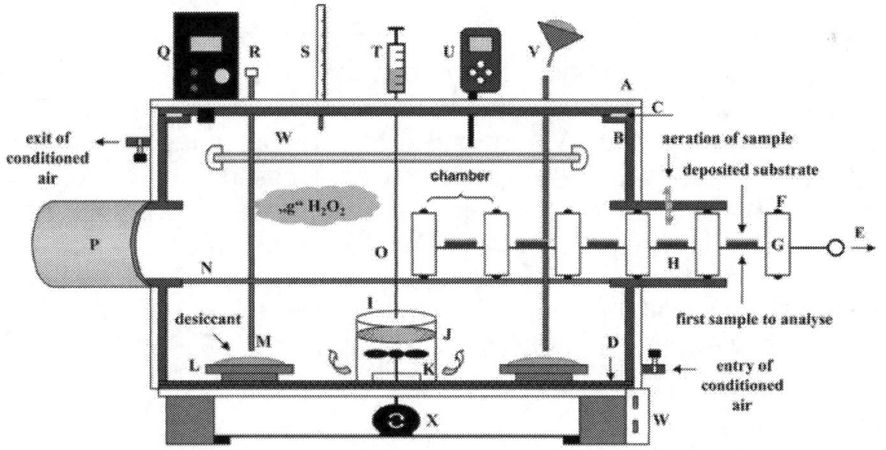

A thermal isolation
B body of peroxybox (glass)
C seal of cover (silicon)
D heating surface
E holder (stainless steel)
F o-ring (viton)
G seal of chamber (teflon)
H microscopic slide (glass)
I evaporation stand (glass)
J evaporation surface (stainless steel net with cotton or glass fibres)
K fan (PVC)
L grid at dessicant (stainless steel)
M transporter of desiccant (glass)
N rail guide (glass)
O cannula (teflon)
P compensatory bag (PVC)
Q sonda of VPHP concentration
R plug (polypropylen)
S thermometer
T injektor of aqueouse H_2O_2 (polypropylen syringe or membrane pump)
U hygrometer
V loader of desiccant
W UV lamp
X motor of fan
Y electric source

Figure 1: The model of the peroxybox.

LC/MS

HPLC system consisted of dual pump Varian ProStar 210, degasser and autosampler Varian 410 (Varian Inc, USA) connected to a triple quadrupole mass spectrometer Varian 1200L (Varian Inc., USA).

GC/MS

It was used Varian Saturn 2000 (Varian Inc, USA) with ion trap allowing tandem mass spectrometry.

NMR

NMR spectra were measured on a Varian ^{UNITY}*Inova*-400 spectrometer (399.89 MHz and 100.55 MHz, respectively) and Varian Mercury 300 (300.07 MHz, and 75.47 MHz, respectively). Residual signal of solvent was used as an internal standard. ^{1}H NMR, ^{13}C NMR, COSY, TOCSY, ROESY, gHSQC, and gHMBC spectra were measured using standard manufacturers' software (Varian Inc., Palo Alto, U.S.A.). 1D NMR spectra were zero filled to fourfold data points and multiplied by window function (for ^{1}H NMR with two-parameter double-exponential Lorentz-Gauss function, for ^{13}C NMR was applied line broadening 1 Hz) before Fourier transformation to improve resolution.

Results and discussion

The initial tests of the peroxybox proceeded on a series of substituted benzaldehydes to get basic information about decontamination process. The aldehydic group is very sensitive to VPHP but decontamination process is sensitive to substitutents of the benzene ring. The introduction of hydroxyl group led to change in the degradation pathway. The aldehydic group was not only oxidized to a carboxylic acid but also replaced by hydroxyl group. Substitution by halogens like chlorine and bromine prevented any transformation. Vanillin (4-hydroxy-3-methoxybenzaldehyde), which was chosen as the model substrate for next experiments, for it's stability on air, was decomposed in relatively short time and could be easily monitored by gas chromatography. The major primary degradation products of degradation of vanillin by VPHP are shown in Fig. 2.

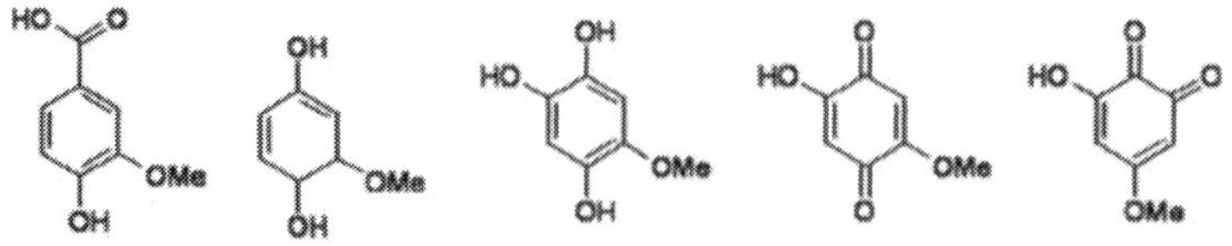

Figure 2: The major primary products of degradation of vanillin by VPHP.

The Fig. 3 demostrates the degradation of vanillin under different conditions. The curves (a)-(c) represent the attempts that were provided in the same operating mode (constant - temperature, and initial relative humidity). It was approved the synergistic effect UV with VPHP, which is related to radical formation which are stronger oxidation agents than hydrogen peroxide. Application of desiccant during the degradation process has negative influence on the degradation rate. It seems that the microcondensation has a positive influence on the course of degradation process. The „dry" VPHP degradation process (Fig. 3c - with desiccant) had a very long inductive period (2 hours) after which relatively fast degradation followed. This behavior can be associated with the formation of an optimal relative humidity or definite value of microcondensation.

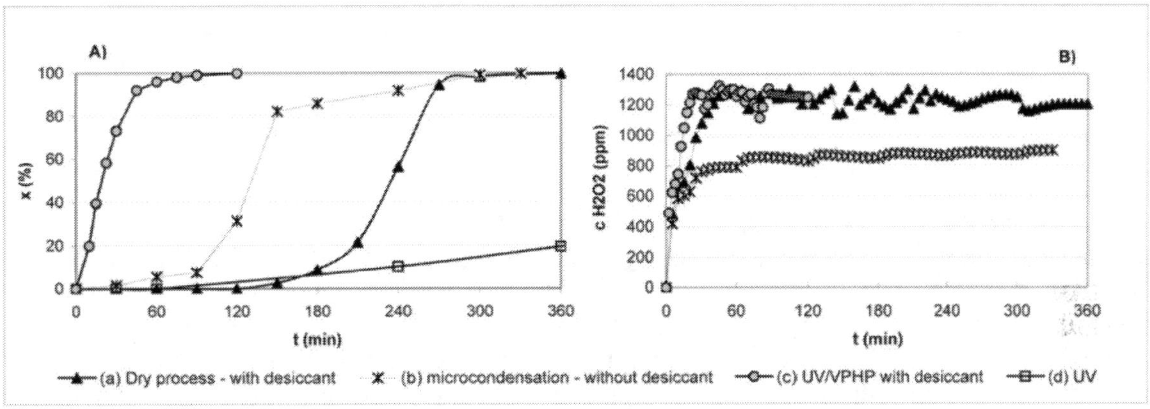

Figure 3: Degradation of vanillin by VPHP under the different conditions: A) the dependence of conversion of vanillin on time, B) time path of VPHP concentration.

Fig. 4 demonstrates degradations of vanillin (dry process – with desiccant) that were provided by different values of initial relative humidity. The higher the initial humidity is the higher is the initial reaction rate.

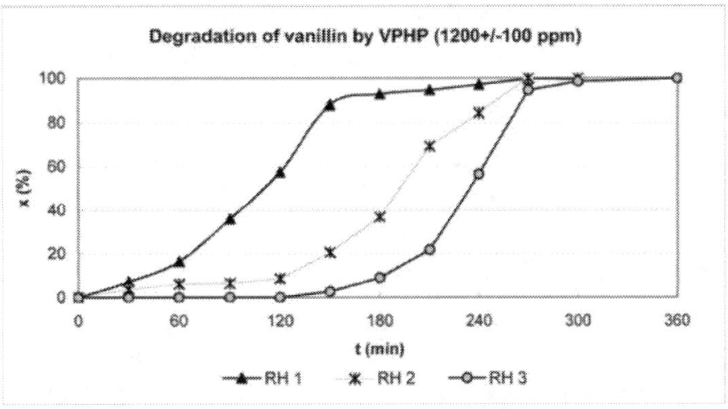

Figure 4: Degradation of vanillin by VPHP under different initial relative humidity

One of the pharmacologically most important groups of indole alkaloids is the ergoline, or ergot alkaloids. These alkaloids are isolated from the dried sclerotium of the fungus Claviceps purpurea (Hypocreaceae) (ergot). VPHP degradation was tested on lisuride, pergolide, nicergoline, ergotamine and bromocryptine. Ergot alkaloids or their derivatives were possible to decompose by VPHP. Degradation of ergopeptines leads to a cyclol moiety (peptide part of ergopeptines) like only decontamination product. It can be explained by initial N-oxide formation at nitrogen N-6 of ergoline or ergolene moiety. An N-oxide can not be formed in cyclol part of ergopeptines, which further remains unchanged.

It is in good agreement with our experiments on tertiary amines. All compounds contain tertiary aminy were sensitive on VPHP. It was established Cope elimination[6] as an initial mechanism of their decontamination (creation an amine oxide to form an alkene and a hydroxyl amine).

Conclusion

The work was focused on the decontamination by VPHP. It was build the appropriate laboratory equipment for tests of VPHP method for organic compounds. It was tested on a series of benzaldehydes derivatives and ergot alkaloids. The decontamination process was followed by GC/MS, LC/MS, and NMR to get deeper insight into the process. It was possible to determine sensitive functional groups and study the kinetic of degradation under the different conditions. It was revealed substantial structure sensitivity on the course of the decontamination. N-Oxide formation is the most probably the initial step in the ergot-alkaloid decontamination. Ergopeptine decomposition product is the intact cyclol moiety. VPHP seems to be a method of choice for degradation of organic molecule like pharmaceutical substances.

Acknowledgements

The authors thank the Ministry of Industry and Trade of the Czech Republic (Grant No FT-TA 4/071), IMIC institutional research concept (AVOZ50200510), and Ministry of Education of Czech Republic (Grant CEZ: MSM 223 100001) for financial support.

References

1. J. V. Rogers, C. L. K. Sabourin, Y. W. Choi, W. R. Richter, D. C. Rudnicki, K. B. Riggs, M. L. Taylor, J. Chang: *J. Appl. Microbiol.* **99** (2005) 739-748.
2. W.A.Rutala, D. J. Weber: *Infect. Control. Hosp. Epidemoil.* **20** (1999) 69-76.
3. N. A. Klapes, D. Vesley: *Appl. Environ. Microbiol.* **56** (1990) 503-506.
4. V. J. Pereira, H.S. Weinberg, K. G. Linden, P. C. Winter: *Environ. Sci. Technic.* **41** (2007) 1682-1688.
5. G. W. Wagner, D. C. Sorrick, L. R. Procell, M. D. Brickhouse, I. F. Mcvez, L. I. Swartz: *Lagmuir* **23** (2007) 1178-1186.
6. A. C: Cope, E. R. Trumbull: *Org. Reactions* **11** (1960) 317.

454f Hydrolysis Enhancement of Swine Wastewater by Ultrasonication

Seungmin Na, Sun-Mee Kim, and **Young Haeng Lee**. Center for Environment Technology Research, Korea Institute of Science and Technology, 39-1 Howelgok-Dong, Sungbuk-Gu, Seoul, 130-650, South Korea

In recent decades, intensive large-scale swine production has increased rapidly in South Korea. Accordingly, the production of swine wastewater has been drastically increased and currently amounts to 150,000 ton per day in South Korea. Swine wastewater is widely known to be a high-strength wastewater with a pollutant organic load which is much higher than domestic sewage. Therefore, minimizing the impact of swine wastewater on the environment is the single most pressing challenge facing the agriculture industry in South Korea.

Anaerobic digestion has been extensively used in the treatment of swine wastewater to stabilize and convert organics to methane gas. However, slow or limited degradation due to very slowly degradable organic matters in swine wastewater complicates the management of swine wastewater and leads to a very long retention time in anaerobic digesters.

Therefore, to accelerate hydrolysis, many chemical, microbiological, mechanical, thermal pre-treatment processes have been applied. Particularly, ultrasonic process is emerging as a popular method for wastewater sludge disintegration. Ultrasonic process leads to the formation of cavities in the liquid phase. Cavitational collapse produces intense local high pressure and temperature on liquid–gas interface, turbulence and high shearing phenomena in the liquid phase

The objective of this work was to investigate the effects of ultrasonic treatment on the solubilization of swine wastewater in order to improve the potential of anaerobic digestibility. The feature of this work was to carry out the comparison of the rate and extent of ultrasonic solubilization of swine wastewater under various operational conditions.

Solubilization of the swine wastewater was achieved by ultrasonic treatment process under various operational conditions. Ultrasonic treatment was shown to have a significant effect on solubilization of the swine wastewater. Ultrasonic treatment resulted in the high degree of solubilization of particulate organics and the degree of solubilization increased with increasing supplied energy. The highest extent of an increment of SCOD concentration and SCOD/TCOD ratio at the end of operation time of 60 min was 109.7 and 117.5%, respectively, under 120 W power output and 20°C operating temperature conditions. The observed highest hydrolysis rate constant described by pseudo-first order rate constant was 2.94 h^{-1} under the same conditions. Based on the estimated activation energy from modeling using the Arrhenius equation, ultrasonic solubilization of the swine wastewater under higher supplied energy conditions was more dependent on the operating temperature, which was consistent with the experimentally obtained results. Investigation on the effect of gas type and gas delivery methods for ultrasonic solubilization under 60 W power output and 20°C operating temperature conditions, oxygen gas bubbling through the liquid showed the highest degree of 67% increment of soluble organics in the swine wastewater which was possibly attributed to the influent of oxygen in an increase of radicals during the sonolysis. In view of the observed successful solubilization of the swine wastewater, the hybrid anaerobic digestion reactor configurations leading to both ultrasonic hydrolysis and microbial methane gas production are under investigation for the development of an efficient anaerobic digestion system.

454g Reductive Dehalogenation of Trichloroethylene Using Aerosol-Assisted Fe/silica Nanoparticles

Gerhard Piringer, Department of Earth and Environmental Science, Tulane University, New Orleans, LA 70118, **Jingjing Zhan**, Department of Chemical and Biomolecular Engineering, Tulane University, New Orleans, LA 70118, Yunfeng Lu, Department of Chemical & Biomolecular Engineering, Tulane University, New Orleans, LA 70118, Vijay T. John, Chemical and Biomolecular Engineering, Tulane University, 300 Lindy Boggs Bldg., New Orleans, LA 70118, and Gary L. McPherson, Tulane University, Department of Chemistry, New Orleans, LA 70118.

Nanoscale zero-valent iron (ZVI) particles are a preferred option for the reductive dehalogenation of trichloroethylene (TCE). However, it is difficult to transport these particles to the source of contamination due to aggregation. This study describes a novel approach to the preparation of ZVI nanoparticles that are efficient and effectively transported to contaminant sites. We describe the synthesis of silica-supported iron nanoparticles using aerosol technology. While the encapsulated iron nanoparticles are reactive for trichloroethylene remediation, environmentally benign silica particles serve as effective carriers for nanoiron transport. Aerosol silica particles with functional groups such as ethyl tails on the surface preferentially adsorb hydrophobic TCE during environmental remediation. This increases the local concentration of TCE in the vicinity of iron nanoparticles, thus promoting the degradation of TCE by iron. These iron/silica aerosol particles with controlled surface properties have the potential to be efficiently applied for in situ remediation and permeable reactive barriers construction.

REMOVAL OF ORANGE II DYE FROM WASTEWATER BY CATALYTIC WET AIR OXIDATION-CATALYST SELECTION AND REACTION KINETICS

Pinar Ozdural[1], Jale F. Akyurtlu[2], Mujde Erten-Unal[1], Vidyasagar Guggilla[2] , Ates Akyurtlu[2]

[1]*Department of Civil and Environmental Engineering, Old Dominion University, Norfolk, VA, USA*
[2]*Chemical Engineering Department, Hampton University, Hampton, VA,USA*

ABSTRACT

The catalytic wet air oxidation (CWAO) of Orange-II, a mono-azo dye, was investigated with different catalysts. Seven catalysts were screened for their TOC and Orange-II removal capacities with a 3-parameter 2-level half factorial experimental design with replicates. The two most efficient and economic catalysts for TOC and Orange-II abatement were determined. A lumped parameter kinetic model was proposed to describe the kinetics of oxidation of Orange-II with chosen catalysts. Within 50 minutes of reaction, 80% TOC removal and 87% Orange-II removal were observed. There was very good agreement between the developed model and the experimental data.

INTRODUCTION

Although the art of dyeing textile materials has been practiced for nearly 5,000 years [1] , until 1856 dyes could only be extracted from natural sources like plants and minerals which made them rare and highly valuable. Colored fabrics were only reserved for royalty and noble people who could afford them. If it weren't for 18-year-old William Henry Perkin who synthesized the first organic dye "mauveine" in a laboratory in 1856 while trying to synthesize quinine as an anti-malaria medication [1], this might have still been the case today. Today, there are numerous synthetic organic dyes and pigments used for coloring everything from clothes to plastics to drugs to food. The dye manufacturing industry has certainly come a long way since the accidental discovery of the first synthetic organic dye. Whereas once it was a sign of wealth and status, today anyone can enjoy wearing colorful t-shirts, or eating cake with colorful decorations on it, thanks to the variety of dyes that are available. Unfortunately, most of the time wastewaters generated as byproducts of dyeing processes are not treatable with conventional treatment methods, are toxic to aquatic life[1], and are not as aesthetically pleasing in lakes and rivers as they are on clothing. Cooper [2] states that while the public is more ready to accept blue, green, or brown rivers, "unnatural" colors such as red and purple usually cause more concern. Most synthetic dyes cannot be removed from wastewaters by conventional methods, e.g. activated sludge systems, and end up in receiving water bodies if not treated. Even 1 mg/l of dye can lead to visible color in rivers and lakes. As discharge standards become more stringent, the need for advanced treatment systems becomes necessary for the treatment of dye containing wastewaters.

Orange II (also known as Acid Orange 7) is one of the dyes that are produced in large amounts in the United States. It is commonly used in pharmaceutical, food, and cosmetics industries as well as in the textile industry, and it is not amenable to conventional biological treatment. This research focuses on catalytic wet air oxidation (CWAO) as a treatment alternative for removing of Orange-II dye from wastewaters.

MATERIALS AND METHODS

The experimental phase of this study focuses on the treatment of synthetically prepared Orange II containing waters by catalytic wet air oxidation. All chemicals were used as received from the manufacturer.

All experiments were carried out in an autoclave reactor equipped with gas introduction lines, vent line, sample line, a propeller for mixing, pressure control unit, and a thermocouple to monitor reaction temperature. Initially, 7 catalysts were screened for their TOC and Orange-II removal capacities. For this purpose, a two-level, 3 parameter half-factorial experimental design was constructed with replicates for each catalyst. The 3 parameters that were incorporated into the screening process were reaction temperature, total pressure, and catalyst loading. This meant that 8 experiments (including replicates) were carried out for each catalyst. The high and low level values for screening experiments can be seen in Table 1 below.

Table 1: Half-factorial design parameter levels

Parameter	Low value	High value
Reaction Temperature, °C	80	120
Total Pressure, psi	300	400
Catalyst Loading, g	2	4

During the catalyst screening experiments, the concentration of Orange-II dye was held constant at 100 mg/l. In a typical run, 600 ml of distilled water and 60 mg Orange II (to yield 100 mg/l) were placed in the reactor with required amount of catalyst according to the design matrix. The reactor was closed and purged with nitrogen gas for a few minutes. The furnace was turned on and the reactor was heated up to the desired temperature under nitrogen atmosphere. This ensured that there was no oxygen to initiate the oxidation reaction in the reactor until desired reaction temperature was reached. When the desired reaction temperature was reached, oxygen was introduced into the reactor, and the stirrer was turned on. This time was marked as the starting time for the experiment. Experiments were run for 120 minutes and a final sample was collected at this time. Control experiments were also run without catalyst under nitrogen atmosphere to confirm that thermal degradation of the dye was not taking place during reactor heat-up.

All collected samples were analyzed for Total Organic Carbon (TOC) and Orange-II concentrations. For measuring Orange-II concentrations, a dye solution of 50 mg/l was scanned with a UV-VIS spectrophotometer. Orange-II displayed maximum absorbance at 486 nm, which agrees with values mentioned in literature. UV absorbance of all collected samples was measured and the corresponding dye concentrations were calculated with a calibration curve constructed with dye standards up to 50 mg/l. The efficiencies of the different catalysts were determined by comparing TOC removal capacities per gram of active metal in the catalyst. Table 2 lists catalysts that were screened during this study.

Table 2: List of catalysts screened for Orange-II oxidation efficiency

Catalyst #	Composition	% Active Metal in Catalyst
1	CuO/ZnO	100
2	CuO/Al_2O_3	10
3	$CuO/(Fe, Mg)Cr_2O_4$	100
4	$Pt/SnO_2/CeO_2$	0.5
5	CuO/CeO_2	2.5
6	Mn/CeO_2	70
7	Mn/CuO	100

In order to choose the best catalyst, a number of factors were considered including the price of metals that make up the catalyst and the efficiencies for Orange-II and TOC removal. The best two catalysts were chosen to be modeled. These two catalysts were further investigated for reaction kinetics and rate expression.

A theoretical triangular lumped parameter kinetic model was developed to explain the reaction kinetics. This model assumes that Orange-II, which accounts for all of the initial TOC concentration, breaks down via two pathways; one to complete oxidation to carbon dioxide and water, and a second path into intermediates because of partial oxidation. The intermediates formed are then oxidized into final products, carbon dioxide and water. Figure 1 depicts this triangular lumped model. According to this model, at any given time, the total TOC concentration in solution would be equal to concentrations A and B added together, $[TOC]_t=[A]+[B]$

Assuming that each step is first-order, we get:

$$\frac{d[A]}{dt} = -(k_1 + k_2)[A] \quad \text{and} \quad \frac{dB}{dt} = k_2[A] - k_3[B].$$

Solving these equations and rearranging, the model becomes:

$$\frac{[TOC]}{[TOC]_0} = \frac{[A]+[B]}{[A]_0} = e^{-(k_1+k_2)t} + \frac{k_2\left(e^{-(k_1+k_2)t} - e^{-k_3 t}\right)}{-k_1 - k_2 + k_3}$$

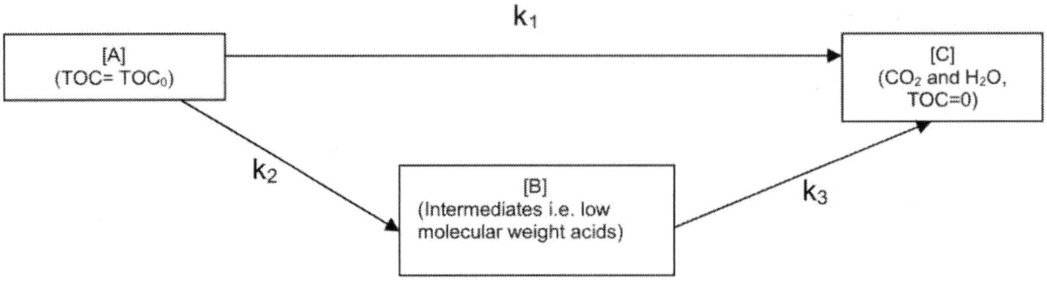

Figure 1: Triangular lumped kinetic model

For the kinetics experiments, the catalyst concentration was fixed at 3 g/l, the initial Orange-II concentration was fixed at 100 mg/l and the total pressure was fixed at 400 psi. Samples were taken at different intervals, and analyzed for TOC and Orange-II concentrations. The results were analyzed with non-linear regression to determine the reaction rate constants k_1, k_2, and k_3 at different temperatures, and to determine whether the theoretical model explained the actual reaction kinetics. The results are discussed in the next section.

RESULTS AND DISCUSSION

Figures 2 and 3 summarize the results of the catalyst screening experiments. It is evident without a doubt that the 0.5 % Pt/SnO_2/CeO_2 is the best catalyst in terms of oxidation capacity. Its TOC removal per gram metal was about 20 times more than the SOR 10-IV catalyst, which was the second best. Unfortunately, platinum is a very expensive metal and costs about 6,000 times more per gram than copper as of September 2008. For this reason alone, a large scale industrial application of platinum catalyst would not be feasible in this case.

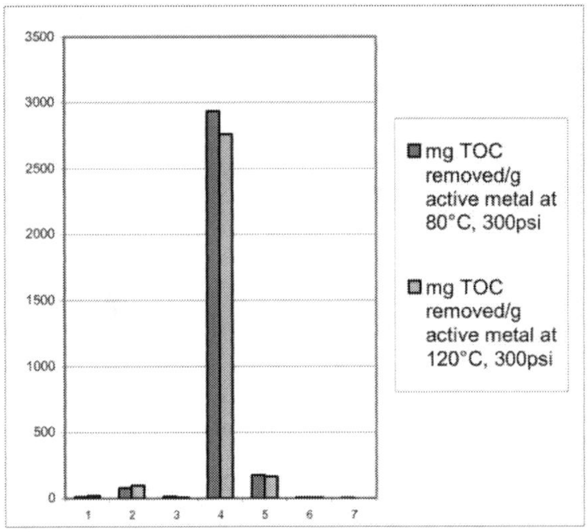

Figure 2: mg TOC removed/g active metal at 300 psi

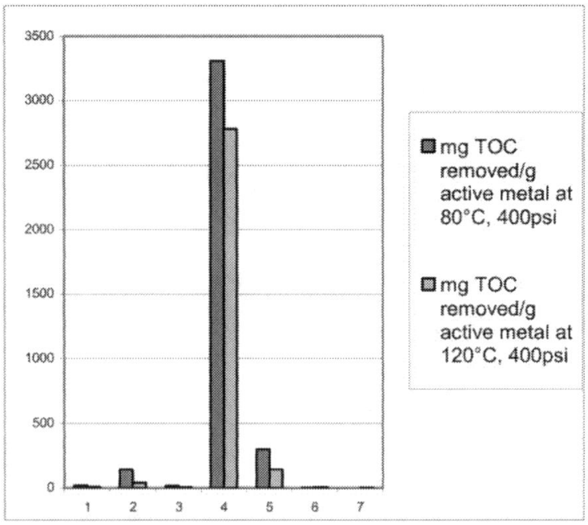

Figure 3: mg TOC removed/g active metal at 400 psi

In the following figures 4 and 5, Pt/SnO_2-CeO_2 catalyst was removed from the graphs so that the other catalyst results can be viewed better. From these graphs, CuO/CeO_2 has the second highest TOC removal per gram active metal.

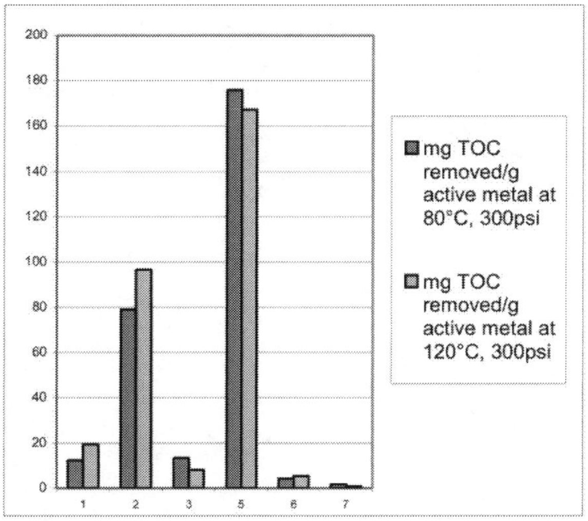

Figure 4: **mg TOC removed/g active metal at 300 psi**

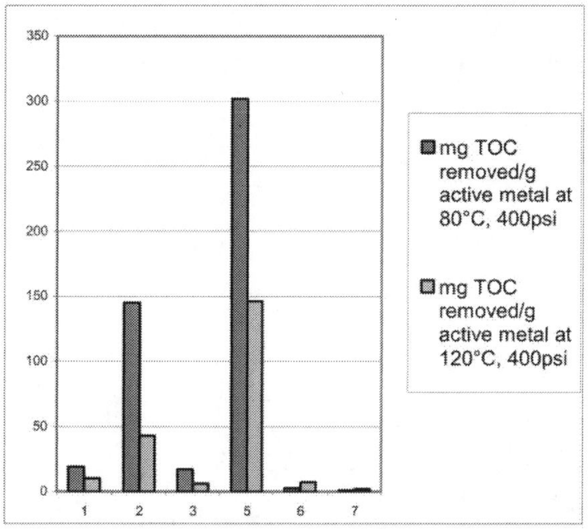

Figure 5 **: mg TOC removed/g active metal at 400 psi**

Table 3: %TOC and Orange-II removals at 120 minutes

Catalyst	% TOC removal at 120 min.	% Orange-II removal at 120 min.
CuO+ZnO	77	85
$CuO+Al_2O_3$	80	87
CuO+ (Fe, Mg)Cr_2O_4	47	99
$Pt/SnO_2/CeO_2$	91	100
$CuO+CeO_2$	46	55
Mn/CeO_2	22	75
Mn/CuO	5	35

Table 3 shows the TOC and Orange-II removal percentages for all the catalysts with 2 grams of catalyst, at 80°C and 400 psi. One of the most important observations from the results is that TOC removal and dye removal are not correlated. For example, the Cu-Chromite catalyst removed only 47% of the total TOC, but Orange-II removal was 99%. This is due to the fact that there are intermediates formed during the reaction, which account for the TOC. However, the double nitrogen bond of Orange-II is broken, so the color is not there anymore. When all factors are considered including TOC removed per gram active metal and the prices of the metals, along with the percent of TOC and dye removed per gram catalyst, the catalysts that perform the best are the CuO+ Al_2O_3 catalyst and the $CuO+CeO_2$ catalyst.

A graph of TOC concentrations of samples taken at different time intervals for CuO/ Al_2O_3 at different temperatures can be seen in Figure 6. From this graph, it is clear that there are two distinct reaction steps taking place, one fast and the other slow. Once oxygen is introduced into the reactor, the oxidation reaction proceeds very rapidly and within the first 10 minutes TOC is reduced by almost 75%. Then the reaction proceeds relatively slower.

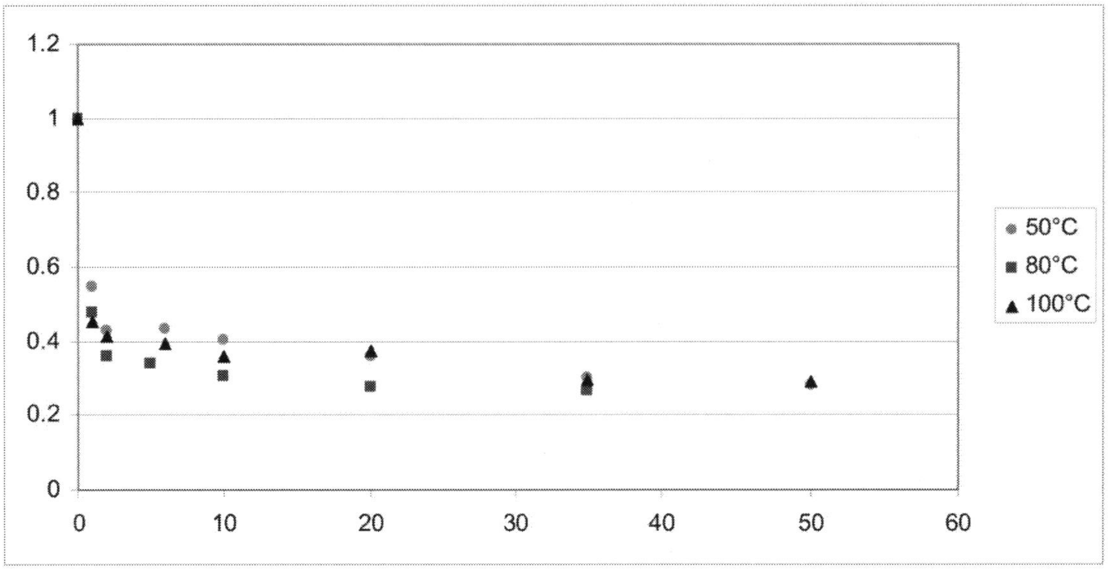

Figure 6: TOC removal with $CuO/\ Al_2O_3$ at different temperatures

The reaction rate constants computed with non-linear regression for different temperatures are given in Table 4. The theoretical model that was constructed fits the experimental data very closely. A graph of actual versus model predicted TOC/TOC_0 values with the catalyst $CuO/\ Al_2O_3$ at 80°C can be seen in Figure 7.

Table 4: Reaction rate constants at different temperatures

Temperature, °C	k_1 (min^{-1})	k_2 (min^{-1})	k_3 (min^{-1})
50	0.948285	0.232279	0.009507
80	1.021212	0.518596	0.007806
100	1.473231	1.006182	0.007306

The formation of intermediates is slow compared to the complete oxidation of Orange-II, but the slowest step of the reaction is the oxidation of formed intermediates into carbon dioxide and water. This is expected, because the formed intermediates are most probably low molecular weight acids which are very hard to completely oxidize at such low temperatures. This also explains why the k_3 values are almost the same and very small at these temperatures.

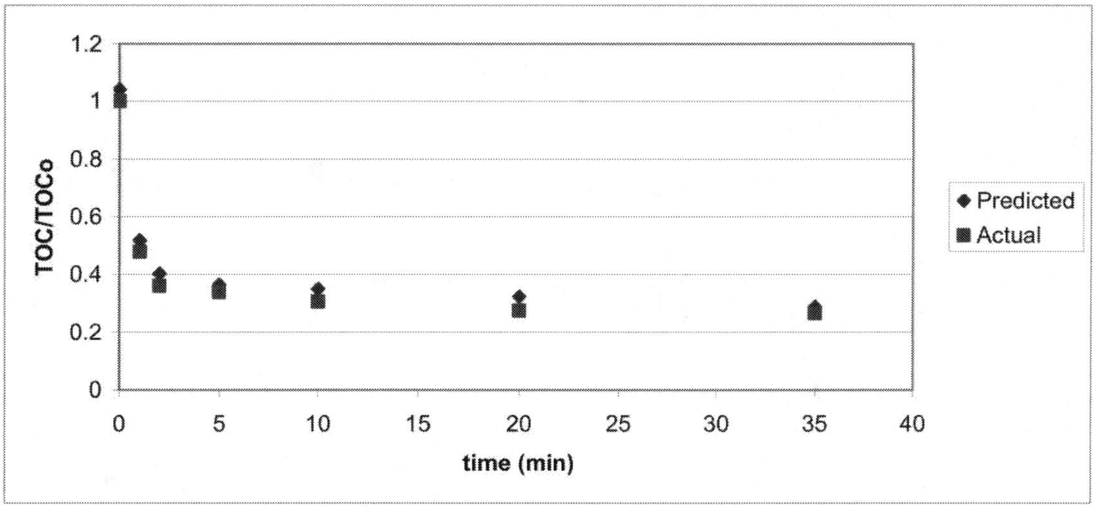

Figure 7: Actual vs. model predicted TOC/TOC_0 at 80°C with CuO/Al_2O_3

CONCLUSIONS

The compound studied, Orange-II, is oxidized by the CWAO process. Among the different catalysts that were tested, CuO/CeO_2 and CuO/ Al_2O_3 were found to be the most economic and efficient catalysts. The degradation of Orange-II was found to be partial with the formation of some byproducts; however, during the first 50 minutes of reaction, almost 80% of the total TOC was completely oxidized. A lumped parameter kinetic model was constructed to explain the reaction kinetics. There was very good agreement between the experimental data and the model predicted values.

REFERENCES

1. Zollinger, H., *Color Chemistry-Synthesis, Properties, and Applications of Organic Dyes and Pigments*. 2003, Zurich: Wiley-VCH.
2. Cooper, P., ed. *Colour in Dyehouse Effluent*. 1995, Oxford-The Alden Press.

Degradation of Acetaldehyde with Doped TiO_2 Photocatalyst Under Visible Light Irradiation

Xiangxin Yang[1], Chundi Cao[1], Keith Hohn[1], Larry Erickson[1], Ronaldo Maghirang[2], Kenneth Klabunde[3]*
[1]Department of Chemical Engineering, Kansas State University, Manhattan, Kansas 66506
[2]Department of Biological & Agricultural Engineering, Kansas State University, Manhattan, Kansas 66506
[3]Department of Chemistry, Kansas State University, Manhattan, Kansas 66506

Abstract

C-, V-doped TiO_2 photocatalysts are prepared by a sol-gel process. These catalysts are highly active for the degradation of acetaldehyde both under visible irradiation (> 420 nm) and in the dark, especially for 2.0% V-containing co-doped TiO_2. Characterization results suggest that vanadium ions are introduced both on the surface and into the bulk of TiO_2. A free electron, induced by the formation of V^{5+} in the sublayers of TiO_2, is delocalized and promoted into the conduction band by thermal energy and further transferred to O_2, generating superoxide radical anion ($O_2^{\bullet -}$) that is responsible for degradation of acetaldehyde in the dark. The doped elemental carbon increases the visible light sensitivity, the surface area, and the dispersion of vanadium.

Keywords: Visible Light; Carbon; Vanadium; TiO_2; Photocatalyst;

* To whom correspondence should be addressed. Phone: 785-532-6849. Fax: 785-532-6666. Email: kenjk@ksu.edu

Introduction

Since the discovery of water splitting by Fujishima and Honda in 1972, enormous efforts have been devoted to photocatalysis for its potential in the conversion of light energy into chemical energy.[1] Photocatalytic degradation and complete mineralization of toxic organic compounds in the presence of catalysts have also received much attention during the last two decades.[2,3,4] TiO_2 is considered the most promising photocatalyst for its high efficiency, chemical stability, nontoxicity, and relative cost. However, the use of TiO_2 is impaired by its wide band gap (3.2 eV), which requires ultraviolet (UV) irradiation for photocatalytic activation ($\lambda < 387$ nm).[5,6,7] UV light accounts for only about 5% of solar energy while 45% is visible light.[8] Thus, a lot of effort has been directed towards the development of visible light active photocatalysts.

One approach to synthesize visible light active photocatalysts is the substitution of Ti by V, Cr, Mn, Ni, or Fe.[7,9,10] Recently, it has been shown that the optical response of TiO_2 could also be shifted from UV to visible light by doping with nonmetal atoms, such as carbon, nitrogen, and sulfur.[5,11,12,13,14] In addition, the synergetic effect of transition metal and nonmetal atoms on the photocatalytic activity of TiO_2 was studied. Zhao et al.[15] found that the incorporation of boron atoms into TiO_2 extended the spectral response to the visible region and the loaded Ni_2O_3 species facilitated the charge separation. Wei et al.[16] reported that lanthanum and nitrogen led to significant enhancement in degrading of methyl orange under visible light ($350 < \lambda < 450$ nm); the substitution of N for O was responsible for the band gap narrowing of TiO_2 and La^{3+} doping prevented the aggregation of powder in the process of preparation. Sakatani et al.[17] found the doped La^{3+} hindered the crystal growth and the number of the paramagnetic N species seemed to control the absorption of visible light and the activity under visible light irradiation.

In this research, carbon and vanadium co-doped TiO_2 photocatalysts are prepared by a sol-gel process. The photocatalytic performance of the doped catalysts is investigated and demonstrated.

Experimental

Sample preparation

3.8 g thiourea (99.0-100.5%, Aldrich) and 3.0 g urea (99% Aldrich) are dissolved in 75 ml deionized water. While under vigorous stirring in an ice bath, the desired amount of vanadyl acetylacetonate is added to the solution. Then 0.0375 mole titanium isopropoxide (97%, Aldrich) is added drop-wise. The mixture is stirred for 24 h and aged for 48 h. The water is removed by drying in air at 80 °C. The dried powder is crushed and calcined at 500 °C for 2 h in air. Pure TiO_2 catalyst is obtained using a similar method without urea and thiourea.

Characterization

X-Ray powder diffraction (XRD) patterns are obtained with a Bruker D8 diffractometer, using Cu Kα radiation (1.5406Å) at 40 kV and 40 mA. X-ray photoelectron spectroscopy (XPS) data are recorded using a Perkin-Elmer PHI 5400 electron spectrometer. The spectrometer utilizes acrochromatic Al $K\alpha$ radiation (1486.6 eV). Diffuse reflectance spectra (DRS) are recorded with a Cary 500 Scan UV-vis NIR spectrophotometer with an integrating sphere attachment for their diffuse reflectance in the range of 200-800 nm. N_2

adsorption-desorption isotherms are obtained at -196°C on a NOVA 1000 series instrument (Quantachrome, USA). The specific surface areas are calculated according to the multi-point Brunauer-Emmet-Teller (BET) method.

Degradation of acetaldehyde

Acetaldehyde is chosen as probe molecule for testing the activity of the catalysts. Experiments are performed in a cylindrical glass reactor (305 ml), which has a quartz window. The light source is a 1000 W high pressure Hg arc lamp (Oriel Corporation). Filters are used to eliminate UV radiation during visible light experiments (>420 nm). 100 mg catalyst and 100 μl of liquid acetaldehyde are introduced into the reactor. The gaseous mixture of acetaldehyde and air inside the reactor is magnetically stirred constantly. Prior to analysis, the catalyst and acetaldehyde are kept in the reactor for about 30 min to obtain desorption-adsorption equilibrium. The reactor is cooled by water circulation and experiments are performed at 25 °C. Gaseous samples are periodically extracted from the reactor and analyzed by GCMS (QP5000 from Shimadzu).

Results and discussion

The evolution of CO_2 during acetaldehyde degradation is used to estimate the activity of the catalysts. Figure 1 shows the CO_2 evolution on different catalysts. It can be seen that the photo-catalytic activity of co-doped TiO_2 is greatly enhanced. Most important, the co-doped TiO_2 has high activity in the dark: the higher values at zero time are due to reaction in the dark which occurs during the 30 min period prior to sampling. In order to enable quantitative comparison, the initial rates for the initial 40 min are calculated to represent the activities. These values are shown in Table 1. In the dark, P25 TiO_2 shows no activity. However, the co-doped TiO_2 catalysts are highly active for the degradation of acetaldehyde, especially when the vanadium loading is 2%. Under visible light irradiation, co-doped TiO_2 for a 2.0% loading has 6.6 times higher activity than P25 TiO_2. Moreover, for co-doped TiO_2 with 2% vanadium content, the activity is comparable in the dark and under visible irradiation. The different activities among these catalysts likely result from different physical properties and reaction mechanisms, as discussed later.

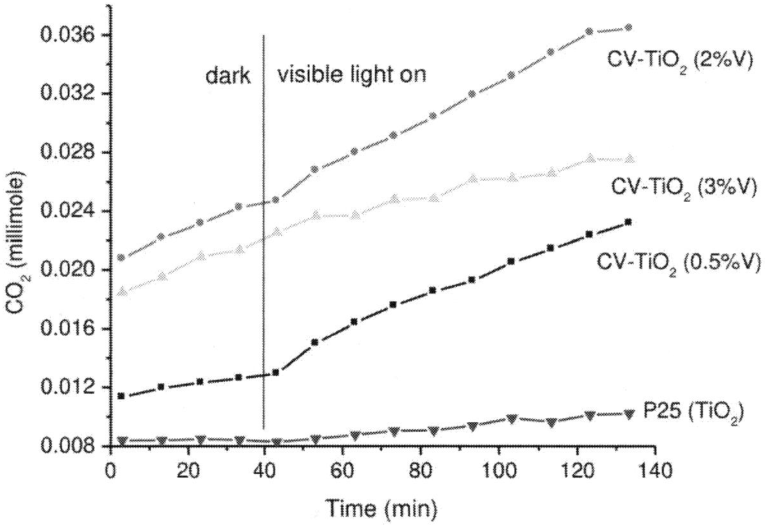

Figure 1 CO_2 evolution during CH_3CHO degradation on P25 TiO_2 and doped TiO_2

Table 1 Kinetic data and physical properties

Sample	Rate in the dark (mmol/min)	Rate under visible light (mmol/min)	Surface area (m^2/g)	Particle size (nm)
P25 TiO_2	0	2.13×10^{-5}	57	-
C, V-TiO_2 (0.5% V)	3.81×10^{-5}	1.38×10^{-4}	96	8.9
C, V-TiO_2 (2.0% V)	1.26×10^{-4}	1.37×10^{-4}	94	9.2
C, V-TiO_2 (3.0% V)	6.9×10^{-5}	5.57×10^{-4}	59	10.3

Figure 2 shows the XRD patterns of catalysts. The peak at $2\theta = 25.3°$ is representative of (101) anatase phase reflections. No other crystal phase (rutile or brookite) is detected. Compared with pure TiO_2, the peak position of the (101) plane of doped TiO_2 shifts slightly and the peak is broader, which suggests that the crystal lattice is distorted. The structures are distorted further with higher content of vanadium, indicating vanadium (V^{4+}) is incorporated into the lattice and substitutes for Ti^{4+} as suggested by the XPS spectra (see later discussion). From the full width at half maximum (FWHM) of the diffraction pattern, the particle sizes are calculated using Scherrer's equation: pure TiO_2 has the particle size of 14 nm and the values for doped TiO_2 are about 9 nm (shown in Table 1). No diffraction lines due to TiC, TiN, and TiS are found. Furthermore, no separate VO_x phases are observed possibly due to the high dispersion or the very small crystallite size of vanadia species that may be present.

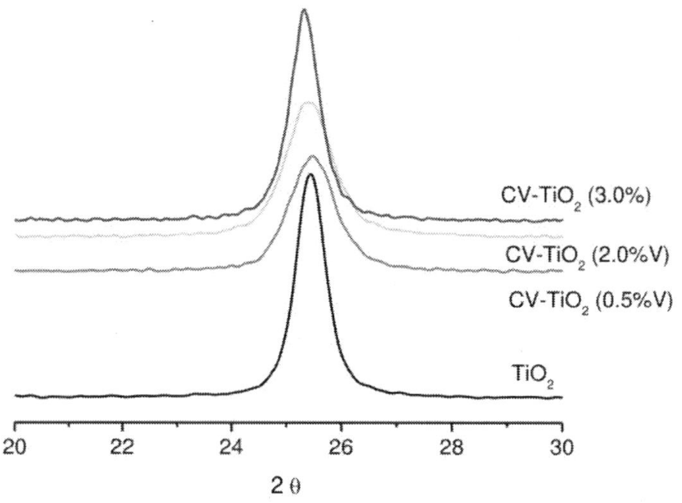

Figure 2 X-Ray diffraction patterns

XPS was used to investigate the chemical states of dopants incorporated into TiO_2. The results are shown in Figure 3. The C 1s XPS spectrum shows one peak at 284.8 eV and a shoulder around 288.6 eV. The feature at 284.8 eV is assigned to elemental carbon and the shoulder around 288.6 eV suggests the existence of carbonate species. After sputtering, the shoulder around 288.6 eV disappears and the peak at 284.8 eV still exists. These results strongly imply that carbonate species only exist on the surface and elemental carbon deposits not only on the surface but also in the bulk of the TiO_2 powder. A peak around 281 eV, which was reported to result from the Ti-C bond,[18] is not observed for our sample. It is obvious that the O 1s satellite interferes with the V 2p3/2 XPS peak. The magnitude of the V 2p3/2 peak is small. The shoulder at binding energy of 516.9 eV suggests V^{5+} species, while the shoulder at 516.3 eV is assigned to V^{4+}. These values are consistent with binding energy of V 2p3/2 measured for V_2O_5/TiO_2 catalyst.[19,20] We know that vanadium ions are incorporated into the crystal lattice of TiO_2 and no vanadium-titanium phase is present from the previous XRD discussion, so V^{4+} ions substitute for Ti^{4+} ions and form Ti-O-V bond. Obviously, some V^{4+} ions are oxidized into V^{5+} in the preparation process since vanadium only exists as V^{4+} in the precursor; the formation of V^{5+} possibly occurs during the annealing. The both disappearance of S 2p and N 1s peaks in XPS spectra (not shown here) indicated that sulfur and nitrogen species only existed on the surface.

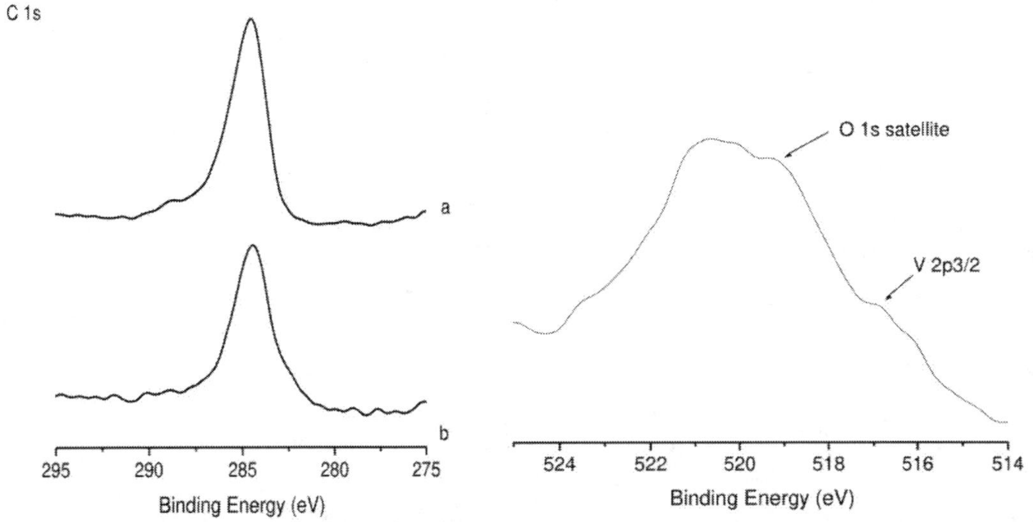

Figure 3 XPS spectra of C 1s (a: before sputtering; b: after sputtering) and V 2p3/2

Figure 4 shows the UV-Vis absorption spectra. As expected, commercial Degussa P-25 TiO_2 has absorption only in the UV region. However, the absorption curve of the co-doped TiO_2 extends to 800 nm. It was reported that V^{5+} shows an absorption at lower than 570 nm and that V^{4+} has an absorption band centered at 770 nm.[21] Therefore, the spectra suggest the co-existence of V^{5+} and V^{4+}, consistent with XPS spectra. The extended visible absorption obviously suggests modified electronic properties by incorporation of vanadium, forming impurity energy levels between the valence band and conduction band of TiO_2.[7]

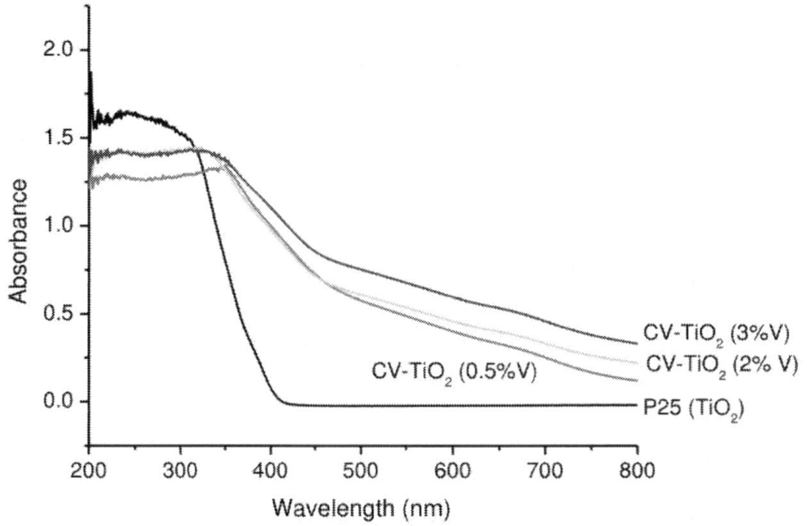

Figure 4 Diffuse reflectance spectra of P25 TiO_2 andC, V-doped TiO_2

Mechanistic studies

Figure 1 illustrates that P25 TiO_2 is inactive in the dark for the degradation of acetaldehyde. After the modification, catalysts show high activity in the dark. Based on the characterization results, one explanation could be considered for the high activity in the dark: it is possible that incorporated V^{4+} in the sublayers of TiO_2 is oxidized to V^{5+} during preparation, such as water-removal or annealing. If pentavalent V^{5+} ions are present at tetravalent sites, the substitution could induce a free electron.[22] V^{5+} ion only shares four valence electrons with four O^{2-} neighbor ions. The fifth electron cannot be shared and is delocalized around the V^{5+} positive center. Only a small amount of thermal energy is needed to delocalize this electron and promote it into the conduction band.[22] Therefore, the promoted electron can be transferred to surface adsorbed oxygen molecule to generate the superoxide radical anion $O_2^{\bullet -}$, which was known to degrade organic compounds.[14,23] When the loading of vanadium increases from 0.5 to 2.0%, more V^{4+} ions in the sublayer of TiO_2

are possibly oxidized into V^{5+}, enhancing the activity. However, further increased the vanadium content could deteriorate the dispersion and lead to the aggregation of vanadium, proven by the smaller surface area. That is why the activity decreases in Figure 1 when the vanadium loading increases from 2.0 to 3.0%.

Under visible light irradiation, two major reaction pathways could be considered: (1) the elemental carbon acts as a photosensitiser. In the presence of visible light, the excited photosensitiser injects an electron into the conduction band of titanium dioxide. Subsequently, the electron is transferred to oxygen adsorbed on the TiO_2 surface producing $O_2^{\bullet-}$.[14] After a series of reactions, acetaldehyde molecules are finally degraded into CO_2 and H_2O. (2) The incorporation of vanadium into the crystal lattice of TiO_2 modifies the electronic properties of TiO_2. Because the oxidation V^{4+} to V^{5+} occurs 2.1 eV below the TiO_2 conduction band, a 3d electron from a V^{4+} center is easily photoexcited into the TiO_2 conduction band under visible light irradiation, generating $O_2^{\bullet-}$ similarly as mentioned in parthway 1.

Conclusions

C-, V-doped TiO_2 catalysts are successfully synthesized by a sol-gel process. The doped elemental carbon acts as photosensitizer and increases the surface area, resulting in the high dispersion of vanadium for the co-doped TiO_2 catalysts which have high activity both in the dark and under visible light irradiation. For the first time, we report here that the activities are comparable in the dark and under visible light irradiation for the co-doped TiO_2, especially when the vanadium content is 2.0%.

Acknowledgment

The support of the Army Research Office and Kansas State University Targeted Excellence program is acknowledged with gratitude.

References

[1] M. A. Fox, M. T. Dulay, Chem. Rev. 93 (1993) 341.

[2] A. L. Linsebigler, G. Lu, J. T. Yates Jr, Chem. Rev. 95 (1995) 735.

[3] J. H. Carey, J. Lawrence, H. M. Tosine, Bulletin of Environmental Contamination and Toxicology 16 (1976) 697.

[4] J. C. Yu, J. Lin, R. W. M. Kwok, Journal of Photochemistry and Photobiology A: Chemistry 111 (1997) 199.

[5] C. Burda, Y. Lou, X. Chen, A. C. S. Samia, J. Stout, J. Gole, Nano Letters 3 (2003) 1049.

[6] X. Li, F. Li, C. Yang, W. Ge, Journal of Photochemistry and Photobiology A: Chemistry 141 (2001) 209.

[7] M. Anpo, S. Dohshi, M. Kitano, Y. Hu, M. Takeuchi, M. Matsuoke, Annu. Rev. Mater. Res. 35 (2005) 1.

[8] S. Yin, H. Yamaki, M. Komatsu, Q. Zhang, J. Wang, Q. Tang, F. Saito, T. Sato, J. Mater. Chem. 13 (2003) 2996.

[9] S. Klosek, D. Raftery, J. Phys. Chem. B 105 (2001) 2815.

[10] J. C.S. Wu, C. H. Chen, Journal of Photochemistry and Photobiology A: Chemistry 163 (2004) 509.

[11] S. Joung, T. Amemiya, M. Murabayashi, K. Itoh, Chem. Eur. J. 12 (2006) 5526.

[12] T. Umebayashi, T. Yamake, H. Itoh, K. Asai, Applied Physics Letters 81 (2002) 454.

[13] T. Ohno, T. Mitsui, M. Matsumura, Chemistry letters 32 (2003) 364.

[14] C. Lettmann, K. Hildenbrand, H. Kisch, W. Macyk, W. Maier, Applied Catalysis B: Environmental 32 (2001) 215.

[15] W. Zhao, W. Ma, C. Chen, J. Zhao, Z. Shuai, J. Am. Chem. Soc. 126 (2004) 4782.

[16] H. Wei, Y. Wu, N. Lun, F. Zhao, Journal of Materials Science 39 (2004) 1305.

[17] Y. Sakatani, J. Nunoshige, H. Ando, K. Okusako, H. Koike, T. Takata, J. N. Kondo, M. Hara, K. Domen, Chemistry Letters 32 (2003) 1156.

[18] H. Irie, Y. Watanabe, K. Hashimoto, Chemistry Letters 32 (2003) 772.

[19] M. Kobayashi, R. Kuma, S. Masaki, N. Sugishima, Applied Catalysis B: Environmental 60 (2005) 173.

[20] S. L. T. Andersson, Catalysis Letters 7 (1990) 351.

[21] H. Ozaki, S. Iwamoto, M. Inoue, Catalysis Letters 113 (2007) 95.

[22] J. M. Herrmann, J. Disdier, G. Deo, I. E. Wachs, J. Chem. Soc. Faraday Trans. 93 (1997) 1655.

[23] M. M. Mohamed, M. M. Al-Esaimi, Journal of molecular Catalysis A: Chemical 255 (2006) 53.

475b Application of Photocatalytic Oxidation Systems for Indoor Air Cleaning

Lynette de Ausen Vera[1], Xiangxin Yang[1], Larry E. Erickson[1], Ronaldo G. Maghirang[2], and Kenneth J. Klabunde[3]. (1) Chemical Engineering, Kansas State University, Manhattan, KS 66506, (2) Biological and Agricultural Engineering, Kansas State University, Manhattan, KS 66506, (3) Chemistry, Kansas State University, Manhattan, KS 66506

Photocatalytic oxidation (PCO) is a promising and emerging technique in controlling indoor air contaminants, including volatile organic compounds (VOCs). Numerous studies have been conducted to improve the effectiveness and performance of this technology. These include the development of new PCO configurations, energy-efficient catalysts (i.e., active under visible-light and dark conditions) and other parameters that control the process. However, most of the studies have been performed under laboratory conditions. Only limited research has been conducted under realistic indoor environmental conditions. This study investigated the possible application of novel PCO systems in cleaning indoor air. Mathematical modeling and simulation techniques were utilized to assess the potential use of some of these promising systems in the degradation of VOCs that can usually be found in offices, residential buildings, and other facilities.

Key words: catalyst, indoor air, mathematical model, photocatalytic oxidation, volatile organic compounds.

475c CO2 Reduction by Nanoscale Galvanic Couples

Kanchan Mondal, Department of Mechanical Engineering and Energy Processes, Southern Illinois University, Carbondale, IL 62901

Methanol, lower hydrocarbons, CO, and HCOOH formation have been by the reduction of CO2 with H2. Un-catalyzed electro-reduction requires a significant overvoltage. An alternate route has been conceptualized. Production of hydrogen from water can be achieved by chemical oxidation of a metal in water followed by oxidation of the hydronium ion by the released electrons to form hydrogen radical. It is hypothesized that circumventing the problem of the formation of hydrogen molecule is essential while having the hydrogen radical to react with CO2 forming methanol, etc. This may be possible by a bimetallic catalyst acting as a galvanic couple, wherein one metal acts as the electron donor for the production of hydrogen radical while the other acts as a catalyst for the reduction of CO2. This may enhance the reaction rates. The current concept circumvents the need for an external electric field unlike the electrochemical process. Utilization of carbon dioxide as a raw material, not only as carbon sources but also as useful chemical feedstock is a more holistic approach to reduce air pollution by flue gases. Several methods have been proposed to maintain the concentration of atmospheric CO2 or to reduce and recycle it. One of these is the chemical fixations using catalysts. The conversion of CO2 to methanol by hydrogenation over copper containing catalysts is one of the most widely studied subjects in this area. The limitation in this reaction is the low conversion of CO2. Methanol is currently produced from CO-rich synthesis gas (CO/CO2/H2) using a catalyst at 50¨C100 bar and 473¨C523 K. CO2 has been identified as a carbon source for methanol synthesis based on the fact that kinetic experiments, using isotope-labeled carbon oxides and spectroscopic methods, have demonstrated that methanol is produced from hydrogenation of CO2, whereby CO merely provides a source of CO2 and acts as a reducing agent by scavenging surface oxygen. Nevertheless, the catalyst that is active for CO-rich feedstock is not so active for CO2-rich sources. There have been many claims that the catalyst performance can be improved. The majority of the modified catalysts still contain Cu and Zn as the main components, together with different modifiers such as chromium, zirconium, vanadium, cerium, titanium, and gallium. The use of these catalysts leads to a more reduced state for the catalyst surface, a situation that may further facilitate surface hydrogenation reactions. Methanol synthesis from a CO2/H2 feed has already been achieved by either physically mixing a CuO-ZnO catalyst with a Pd catalyst or by using Pd-impregnated CuO-ZnO-Al2O3 catalysts instead of physical mixtures. In spite of the advances in direct hydrogenation technologies, the commercial implementation of these technologies is greatly hindered due to the cost prohibitive state of art of hydrogen production. In plants generating hydrogen as a fuel source, the utilization of this hydrogen in reducing CO2 is counter productive. Thus alternate methods need to be investigated that can convert CO2 to useful chemicals without the use of molecular hydrogen as the reducing agent. Two such processes currently under investigation are the electrochemical route of reducing CO2 at the cathode surface and the photo catalytic route. Both these routes utilize water as the source of hydrogen (that is generated in situ). Another method that utilizes water as the source of hydrogen to reduce CO2 is the use of zero valent iron. However, it was reported from the studies with zero valent iron that methane was the primary product. Direct methanol production from CO2 may be commercially feasible with an inexpensive source of H2. As mentioned earlier, use of hydrogen for this reduction chemistry is neither economically attractive nor is it the preferred route when the primary product is hydrogen. Hydrogen may be replaced by electrons and protons, available, for example, in electrochemical reduction in aqueous media. One-electron reduction of carbon dioxide to the radical anion CO, presents thermodynamic and kinetic barriers. In aqueous solutions the reduction potential is -1.9 V vs. SHE. The barrier to outer-sphere electron transfer for the couple is large because of the very different geometries of the linear, neutral carbon dioxide and the bent radical anion. Thus, direct (un-catalyzed) electro-reduction requires a significant overvoltage. This thermodynamic barrier can, however, be reduced by protonating the reduction product. [CO2 + 6H+ + 6e = CH3OH + H2O (-0.38 V)] An alternate route has been conceptualized. The production of hydrogen from water is achieved by chemical oxidation of a suitable metal in the presence of water followed by oxidation of the hydronium ion by the released electrons to form hydrogen radical. In the presence of a suitable catalyst, it may be very efficient in reducing carbon dioxide to methanol. If such a catalyst is not present in the vicinity of its production, it may combine with another hydrogen radical to form hydrogen molecule. To circumvent the problem of the formation of hydrogen molecule (which is less reactive than the radical), it is hypothesized that a bimetallic catalyst, wherein one metal acts as the electron donor for the production of hydrogen radical while the other acts as a catalyst for the reduction of CO2, may be utilized. If the bimetallic catalysts acted as galvanic couples such that the hydrogen radical was being produced on the catalysts while the other metal participated in maintaining the oxidation state of the catalyst as zero, then the reaction rates are further enhanced. The process is based on the ability of certain transition metals (M1) to oxidize in the presence of water giving of electrons that in turn converts the hydronium ion into hydrogen. This hydrogen is then adsorbed locally onto another metal (M2) of known catalytic activity towards CO2 hydrogenation. This process would follow the same route as that of an electrochemical route for CO2 hydrogenation with the added advantage of circumventing the need for an external electric field. Table 1 shows the mechanism for CO2 conversion to methanol. In this scheme, it is possible to have competing reactions that produce intermediates such as formats and bicarbonates as well as hydrocarbons as demonstrated by researchers working with a single metal catalyst. However, these intermediates would also be reduced by the highly active hydrogen radical/ metal hydride formed in situ and it is expected that hydrocarbon formation may be reduced/eliminated with proper design of the catalyst and the process. It must be noted here that the catalyst M2 may also be oxidize in a similar fashion as M1 depending on the choice of the catalyst. In this case the hydrogen radical would be formed on the M2 part of the bimetallic catalyst. If M1 is chosen such that it can form a galvanic couple with M2 , the reaction rates would further be enhanced. The selectivity of different galvanic products towards various end products will be presented. The presentation will also illustrate the mechanism of methanol production from carbon dioxide using galvanic couples. The thermodynamic approach adopted to identify suitable electron donors and catalysts will also be presented and the galvanic couples will be identified. The experimental procedure to synthesize the nanoscale galvanic couples will be shown along with the characterization data of the produced nanosized catalysts. We have found that the use of such bimetallic galvanic couples enable the reactions to process at a much faster rate than the corresponding electrochemical routes. In addition, synergistic activity was observed when the catalyst form a galvanic couple as opposed to the performance of the catalysts individually. Depending on the galvanic couple used, the end product may be methane or FT hydrocarbons or methanol. Results pertaining to the performance of the bimetallic catalysts in terms of their reactivity towards CO2 hydrogenation and their selectivity for methanol synthesis will be presented along with the future experimental plan to understand the mechanism of the reactions and activity of the catalysts. Both aqueous media and gas phase reactions are being considered. In the presented results, the effects of process parameters such as temperature, pressure, gas flow rates, pH (in aqueous media) will be shown. In addition, the effect of impurities present in the CO2 stream will also be shown. Table 1 Mechanism of CO2 Conversion to Methanol over Bimetallic Catalysts

475d Regeneration and Framework Stability of Heavy Metal Exchanged Nanoporous Zincosilicates

Tyler J. Selbe, Kansas State University, 1005 Durland Hall, Manhattan, KS 66506 and Jennifer L. Anthony, Chemical Engineering, Kansas State University, 1005 Durland Hall, Manhattan, KS 66506.

This presentation will report our findings of the framework stability and the regeneration properties of nano-structured porous zincosilicates that have previously captured mercury and lead from aqueous systems. Zincosilicates have similar framework structures to aluminosilicate zeolites, however they typically have a higher ion density due to the need for two counter cations per zinc in the framework as opposed to the one-to-one ratio of counter cation per aluminum of traditional zeolites. Molecular sieves containing zinc in the framework also have the ability to form three-membered rings (3MR) which lead to large void spaces in their structure. Interestingly, microporous zincosilicates do not have "cages" in their structure, which leads to all the cations being accessible. These features lead to high capture capacities especially for divalent heavy metals ions such as lead and mercury. Our results have shown that zincosilicates capture high capacities of heavy metals very rapidly indicating a highly favorable thermodynamic process. We will report the structural changes after ion exchange, quantify the reversibility of the ion exchange, and the subsequent structure after the heavy metal reversal process in aqueous systems for a series of zincosilicates, such as VPI-7, VPI-9, VPI-10, and CIT-6.

Synthesis and Characterization of Visible-Light-Active TiO_2-Based Photocatalysts

Xiangxin Yang[1], Chundi Cao[1], Keith Hohn[1], Larry Erickson[1], Ronaldo Maghirang[2], Kenneth Klabunde[3]*
[1]Department of Chemical Engineering, Kansas State University, Manhattan, Kansas 66506
[2]Department of Biological & Agricultural Engineering, Kansas State University, Manhattan, Kansas 66506
[3]Department of Chemistry, Kansas State University, Manhattan, Kansas 66506

Abstract

TiO_2 has gained great popularity for environmental treatment and purification purposes; however, it shows poor absorption of visible light and requires ultraviolet (UV) light for activation. Here we have successfully synthesized a visible-light-active carbon and nitrogen co-doped TiO_2 catalyst. The catalyst was characterized by Raman spectroscopy, X-ray diffraction (XRD), X-ray photoelectron spectroscopy (XPS), diffuse reflectance spectroscopy (DRS). Results showed that the optical response of modified TiO_2 was extended to the visible light region. Nitrogen substituted for some of the lattice oxygen atoms and most of the carbon was in the form of elemental carbon. The photocatalytic tests indicated that the modified catalyst demonstrated higher visible-light activity for the degradation of methylene blue (MB) than TiO_2.

Keywords: Visible-light-active; Nitrogen; Carbon; TiO_2; Photocatalyst;

* To whom correspondence should be addressed. Phone: 785-532-6849. Fax: 785-532-6666. Email: kenjk@ksu.edu

Introduction

Titanium dioxide (TiO_2) has been widely used as photocatalysts for solar energy conversion and environmental application because of good photoactivity, high chemical stability, low cost, and non-toxicity. However, the application of pure TiO_2 was limited because poor absorption of visible light. Therefore, efforts have been directed toward expanding the optical response of TiO_2-based photocatalysts from UV to the visible light region. One method is introduction of allowed energy states in the band gap of TiO_2 by doping of transition metal ions such V, W, or Fe.[1,2,3,4] However, these doped materials are thermally unstable and the introduction of these ions can also increase the carrier recombination rate, resulting in low photocatalytic efficiency.[5,6] Another approach is to modify TiO_2 with nonmetal atoms such as S, C, or N. It was assumed that these nonmetal atoms either substituted oxygen atoms or titanium in the lattice of TiO_2 or formed impurity level between the conduction and valence band of TiO_2, thus increased the visible light sensitivity.[7,8,9,10,11] In this paper, we synthesized carbon and nitrogen co-doped TiO_2 photocatalysts by the sol-gel method and their photocatalytic activities will be tested by using methylene blue (MB) as a probe molecule, which has been widely used in textile industry and as a standard compound in a test of photocatalysts.

Experimental

Preparation 3.0 g urea (99.0-100.5%, Aldrich) was dissolved in 30 ml deionized water. While under vigorous stirring in water-ice bath, 8 ml titanium isopropoxide (97%, Aldrich) was added drop-wise. The mixture was stirred for 12 h and aged for 24 h. Water removal was accomplished by drying in air at 80 °C. The as-obtained samples was calcined under NH_3 flow at 500 °C for 2 h then treated in air at 200 °C for 1 h, denoted as CN-TiO_2. For comparison purposes, pure TiO_2 was also prepared using the similar procedure without urea and the calcination was performed in air.

Characterization Raman spectrum is recorded using a Nicolet Nexus 670 Fourier transform IR (FT-IR) with a Raman module. X-Ray powder diffraction (XRD) patterns are obtained with a Bruker D8 diffractometer, using Cu Kα radiation (1.5406Å) at 40 kV and 40 mA. X-ray photoelectron spectroscopy (XPS) data are recorded using a Perkin-Elmer PHI 5400 electron spectrometer. Binding energies for the samples are normalized with respect to the position of the C 1*s* peak resulting from adsorbed hydrocarbon fragments. Diffuse reflectance spectra (DRS) are recorded with a Cary 500 Scan UV-vis NIR spectrophotometer with an integrating sphere attachment for their diffuse reflectance in the range of 200-800 nm.

Results and discussion

Figure 1 showed Raman spectrum of CN-TiO_2. Four peaks at 148, 399, 519, and 638 cm^{-1} were observed. It was reported that six modes A_{1g} (519 cm^{-1}), B_{1g} (399 and 519 cm^{-1}), and E_g (144, 197, and 639 cm^{-1}) were Raman-active of anatase phase. Four modes A_{1g} (612 cm^{-1}), B_{1g} (143cm^{-1}), B_{2g} (826 cm^{-1}), and E_g (447cm^{-1}) were Raman-active of rutile.[12,13,14] Therefore, anatase was the predominant phase structure. An additional peak appeared at 320 cm^{-1} in the spectrum and this feature was contributed to the first-order scattering of a non-stoichiometric titanium nitride.[15] In order to better understand the change in the Raman spectra, the spectra at the range of 300-700 cm^{-1} were fitted. The fitted results were shown in the inset of Figure 1. In the fitting curves, two peaks at 460 and 570 cm^{-1} were observed. The peak at 460 cm^{-1} was related to second-order acoustic mode of titanium nitride and peak at 570 cm^{-1} was related to transverse optical mode.[16] Consequently, nitrogen was doped and replaced oxygen atoms in the titanium oxide crystal lattice.

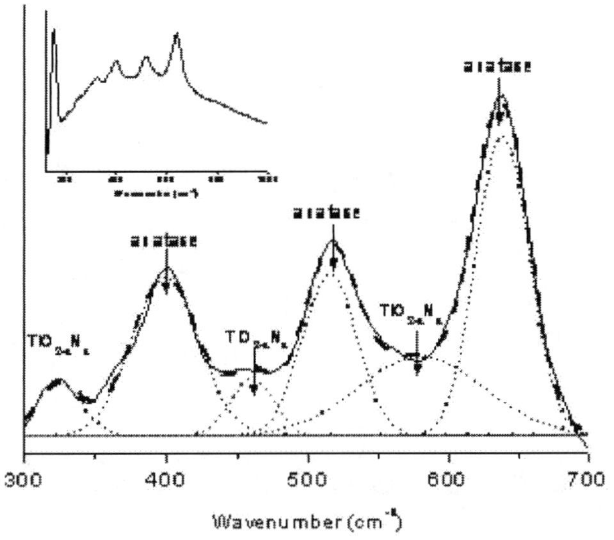

Figure 1 Raman spectra and fitting curves for CN-TiO_2

Figure 2 showed the XPS spectrum for CN-TiO_2. C 1s spectrum showed one strong peak at 284.6 eV and a weak shoulder around 288.5 eV. So most of the carbon incorporated in the TiO_2 matrix existed as elemental carbon and a small amount carbonate species existed on the surface. No peak at 281 eV resulting from Ti-C bond was observed, suggesting carbon did not substitute oxygen atom in the lattice of TiO_2. The N 1s spectra showed one peak at 396.2 eV and another at 399.5 eV. We attributed the peak at 396 eV to the presence of Ti-N bond resulting from substitution for oxygen sites by nitrogen atoms in the TiO_2 lattice, which was consistent with literatures.[17,18,19] The results also coincided with Ramam data in Figure 1. The feature at 399.5 eV was assigned to the adsorbed NH_3 based on our preparation method.

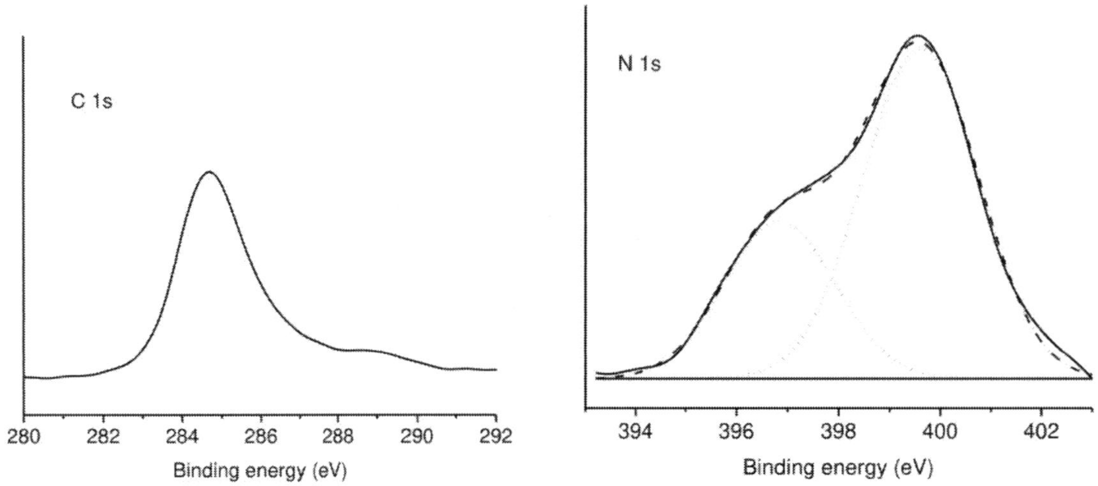

Figure 2 XPS spectra for C 1s, N 1s of of CN-TiO_2

Figure 3 showed the XRD patterns of CN-TiO_2 and TiO_2. Both samples only showed anatase phase ($2\theta = 25.3°$, representative of (101) anatase phase reflections). No other crystal phase (rutile or brookite) was detected. Compared with pure TiO_2, the peak position shifted toward higher 2θ value, which suggests that the crystal lattice of TiO_2 was distorted by the carbon or nitrogen dopants. From the full width at half maximum (FWHM) of the diffraction pattern, the particle sizes were calculated using Scherrer's equation: pure TiO_2 had the particle size of 14 nm and the values for doped TiO_2 are about 9 nm.

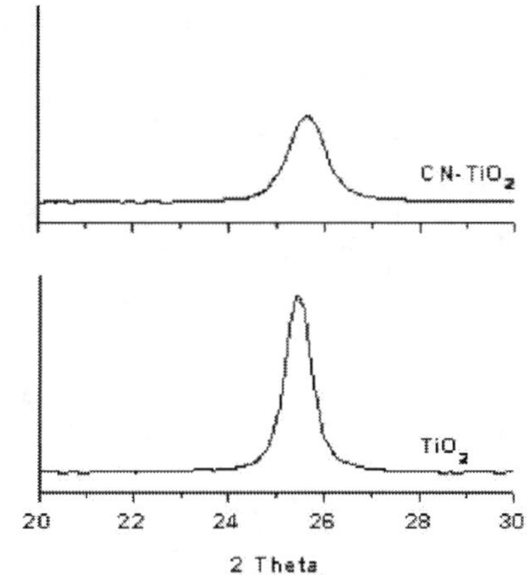

Figure 3 XRD pattern of $CN\text{-}TiO_2$ and TiO_2

Figure 4 showed the diffuse reflectance spectra of $CN\text{-}TiO_2$ and TiO_2. The band gap energies were calculated according to the following equation, $E_g = hc/\lambda$, where E_g is the band gap energy (eV), h is the Planck's constant (4.135667×10^{-15} eV·s), c is the light velocity (3×10^8 m/s), and λ is the wavelength (nm) of absorption onset. Therefore, the band gap energies are 3.05 and 2.12 eV for TiO_2 and $CN\text{-}TiO_2$, respectively. As expected, the optical response of $CN\text{-}TiO_2$ was extended into visible light because of narrower band gap, resulting from doped nitrogen and carbon.

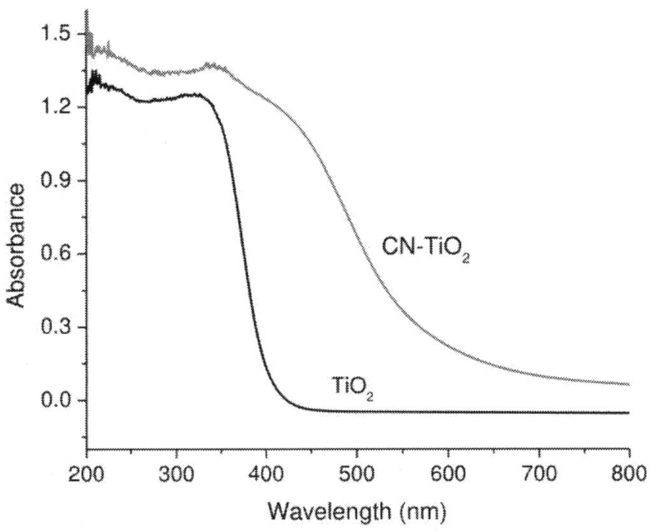

Figure 4 Diffuse reflectance spectra of TiO_2 and CN-TiO_2

Photocatalytic degradation of MB 30 mg of catalyst was suspended in 100 ml aqueous methylene blue dye solution (10 mg/l), contained in a water-jacketed reactor and its temperature was kept at 25 °C. The suspension was kept in the dark for 60 min to establish the adsorption-desorption equilibrium. Then the suspension was irradiation under visible light (>420 nm). Samples were withdrawn periodically and centrifuged and analyzed for the degradation of MB using a UV-vis Cary 500 Spectrophotometer. MB has the maximum absorbance at 664 nm and that was taken as a way for following up the MB degradation. Figure 5 depicts the results of degradation of MB in the presence of CN-TiO_2 and TiO_2 under visible light irradiation. It was clear that CN-TiO_2 showed much higher activity in the presence of visible light than pure TiO_2, which exhibited little activity as expected. If we assumed that the degradation of MB was the first order reaction ($dC/dt = -K \bullet C$), then the rate constant K for CN-TiO_2 was about 17.5 times higher than that for TiO_2. So doping C and N was an effective way to improve the visible-light-induced activity of TiO_2 for the degradation of MB.

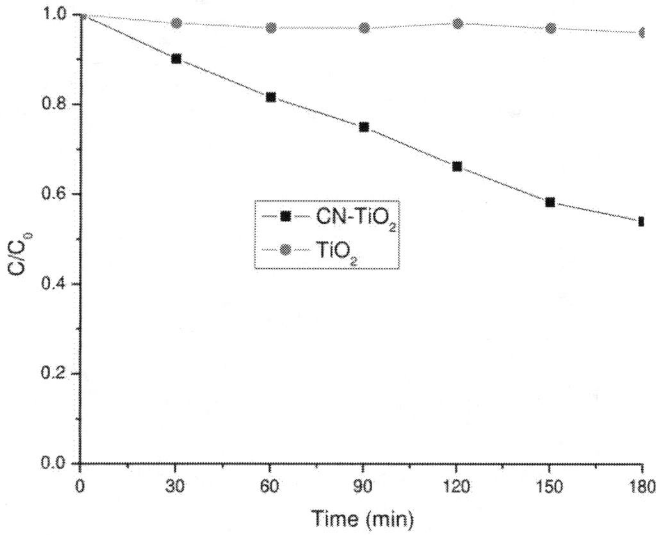

Figure 5 Degradation of MB on CN-TiO_2 and TiO_2 under visible light irradiation

Conclusions

Carbon and nitrogen co-doped TiO_2 catalysts were successfully synthesized by a facile method. Characterization results suggested that nitrogen substituted some of the oxygen atoms in the lattice of TiO_2. Carbon species existed in both elemental and carbonate forms. The co-doped TiO_2 showed higher activity than pure TiO_2 for the degradation of MB under visible light irradiation. The high activity was related to the several beneficial effects associated with the introduction of carbon and nitrogen.

References

[1] S. Klosek, D. Raftery, J. Phys. Chem. B 105 (2001) 2815

[2] P. K. Surolia, R. J. Tayade, R. V. Jasra, Industry & Engineering Chemical Research 46 (2007) 6196

[3] M. M. Mohamed, M. M. Al-Esaimi, Journal of Molecular Catalysis A: Chemical 255 (2006) 53

[4] K. E. Karakitsou, X. E. Verykios, J. Phys. Chem. 97 (1993) 1184

[5] S. Yang, L. Gao, J. Am. Ceram. Soc. 87 (2004) 1803

[6] Y. Cong, J. Zhang, F. Chen, M. Anpo, J. Phys. Chem. C 111 (2007) 6976

[7] T. Ohno, T. mitsui, M. Matsumura, Chemistry Letters 32 (2003) 364

[8] T. Ohno, T. Tsubota, K. Nishijima, Z. Miyamoto, Chemistry Letters 33 (2004) 750

[9] X. Yang, C. Cao, K. Hohn, L. Erickson, R. Maghirang, D. Hamal, K. Klabunde, Journal of Catalysis 252 (2007) 296

[10] R. Nakamura, T. Tanaka, Y. Nakato, J. Phys. B 108 (2004) 10617

[11] R. Asahi, T. Morikawa, T. Ohwaki, K. Aoki, Y. Taga, Science 293 (2001) 269

[12] H. Berger, H. Tang, F. Kevy, J. Cryst. Growth 130 (1993) 108

[13] S. P. S. Porto, P. A. Fbeury, T. C. Damen, Phys. Rev. 154 (1967) 522

[14] S. C. Pillai, P. Periyat, R. George, D. E. McCormack, M. K. Seery, H. Hayden, J. Colreavy, D. Corr, S. J. Hinder, J. Phys. Chem. C 111 (2007) 1605

[15] E. Gyorgy, A. P. D. PinoP. Serra, J. L. Morenza, Applied Surface Science 186 (2002) 130

[16] Y. Cheng, B. K. Tay, S. P. Lau, H. Kupfer, F. Richter, Journal of Applied Physics 92 (2002) 1845

[17] Q. Li, R. Xie, J. Shang, J. Am. Ceram. Soc. 90 (2007) 1045

[18] O. Diwald, T. L. Thompson, T. Zubkov, E. G. Goralski, S. D. Walck, J. T. Yates, J. Phys. Chem. B 108 (2004) 6004

[19] Y. Xie, Q. Zhao, X. Zhao, Y. Li, Catalysis Letters 118 (2007) 231

475f Particulate Methane Monooxygenase-Catalyzed Sustainable Methanol Production from Atmospheric Methane

Katherine K. Bearden and Daniela S. Mainardi. Louisiana Tech University, 911 Hergot Ave., Ruston, LA 71272

The ability of Methanotrophic bacteria to produce methanol from atmospheric methane by Methane Monooxygenase (MMO) enzyme, and further oxidize methanol to formaldehyde by Methanol Dehydrogenase (MDH) enzyme is an exciting phenomenon, since it implies that the organism can change product and substrate reversibly depending on the prevailing conditions. Understanding how these two enzymes, that are present in Methanotrophic bacteria, work in nature will provide clues to designing efficient synthetic catalysts (through biomimetics), which can mitigate the harmful effects of methane in the atmosphere, leading to a reduction in global warming.

Particulate Methane Monooxygenase(pMMO), a MMO form that is expressed when the growth medium contains copper at sufficient levels (>5μM) under high copper/biomass ratios, is the predominant methane oxidation catalyst in nature.[1] Lieberman et al.[2] suggested that the active site of pMMO is constructed partially (with 50% of the copper) of a multi-copper complex, although the nuclearity of the Cu cluster in pMMO is still unknown. Hence, clues on how pMMO activates the inert methane C-H bond are of fundamental chemical interest, and could lead to the development of new synthetic catalysts that could impact the use of methane (natural gas) as an alternative energy source.

This work focuses on fundamental research on the kinetics of an important catalyzed chemical reaction that relates to environmental biocatalysis, and involves atmospheric methane consumption (oxidation) for the production of fuel (methanol). Molecular Mechanics and Molecular Dynamics simulations are conducted in order to investigate the processes involved in methanol production by pMMO, looking at its interaction with MDH. In this work, models of MDH and pMMO are selected and geometry minimized first. Then, their interaction is investigated and tested upon the addition of oxidized forms of methane and methanol representing possible reaction steps in those oxidation mechanisms. The results from the simulations will provide clues on the location of the pMMO active site and its ability to break the methane C-H bond.

[1] S.K. Dubey, A.S.K. Sinha, and J.S. Singh, Differential inhibition of CH4 oxidation in bare, bulk and rhizosphere soils of dryland rice field by nitrogen fertilizers, Basic and Applied Ecology, 3. 2002. p. 347-355.

[2] R.L. Lieberman and A.C. Rosenzweig, Crystal structure of a membrane-bound metalloenzyme that catalyses the biological oxidation of methane, Nature, 434. 2005. p. 177-182.

551a Accelerated Demineralization of Reverse Osmosis Concentrate of High Gypsum Scaling Propensity

Anditya Rahardianto, Brian C. McCool, and Yoram Cohen. Chemical & Biomolecular Engineering, University of California, Los Angeles, 5531 Boelter Hall, 420 Westwood Plaza, Los Angeles, CA 90095-1592

Desalination at many inland locations has increasingly become a necessity for sustaining the quality of current traditional water resources, as well as for exploiting underutilized non-traditional brackish water. The technical and economical feasibility of RO desalting require operation at high product water recovery (>75-90%) to minimize the volume of generated concentrate (i.e., brine) and thus to improve the options for concentrate management. At high water recovery, however, dissolved mineral salts (e.g., $CaSO_4$, $BaSO_4$, $CaCO_3$) may become concentrated beyond their solubility limits and may crystallize in the bulk and onto surfaces (e.g., reverse osmosis membrane). Mineral crystallization may severely limit the attainable water recovery of various desalination processes, including reverse osmosis (RO).

In the present study, a novel demineralization process for desalination concentrate is proposed to enable water recovery enhancement. Chemical demineralization of RO concentrate by precipitation softening (i. e., calcium carbonate precipitation), followed by secondary RO desalting of the demineralized concentrate, has been previously demonstrated to enhance water recovery in Colorado River Water desalting up to 95% [1]. However, application of this desalination approach for brackish water of high gypsum scaling propensity, particularly those with low carbonate content (e.g., Agricultural Drainage water in California's San Joaquin Valley), would require uneconomical chemical usage. For this case, an alternative and more promising demineralization process is proposed, involving accelerated precipitation of dissolved mineral salts, which are kept in meta-stable supersaturation due to threshold inhibition effects of polymeric additives (i.e., antiscalants). The present demineralization process relies on sustainable recycle of crystal seeds and an economical antiscalant deactivation approach with minimal impact on subsequent membrane operations (e.g., does not contribute to fouling). In order to address these critical issues, precipitation experiments, using synthetic RO concentrate solution that contained a generic antiscaling agent, were conducted in a batch reactor. In addition to online monitoring of calcium activity and pH, the evolution of crystal size distribution was measured by an electrical zone sensing method. Experimental results revealed that significant "deactivation" of antiscalants (up to 86% was observed) can be achieved by the present approach. Furthermore, sustainable recycle of crystal seeds was enabled due to antiscalant deactivation. The relationships between antiscalants deactivation, seed crystals concentration, and demineralization kinetics were quantified experimentally. A mathematical kinetic model linking important process variables was proposed to provide a framework for optimization of the process. The results from this study indicate that demineralization of high sulfate RO concentrate by accelerated precipitation is technically feasible and is a promising option for enhancing product water recovery.

[1] C.J. Gabelich et al, High-recovery reverse osmosis desalination using intermediate chemical demineralization, J. Membr. Sci. 301 (2007) 131

Web Page: www.watercenter.ucla.edu

DEVELOPMENT AND OPTIMIZATION OF A HIGHLY EFFECTIVE AND LOW ENERGY INTENSIVE ELECTRO-DISINFECTION SYSTEM FOR BALLAST WATER TREATMENT

*K.G. Nadeeshani Nanayakkara; Yu-Ming Zheng; J. Paul Chen**
National University of Singapore, Singapore
**Corresponding author, email: esecjp@nus.edu.sg*

Abstract

Biological invasions through ships' ballast water create threats to the environment and to the human health and thus, need to treat prior to the discharge. In this study, an innovative electro-disinfection reactor system is developed and optimized for disinfection of *E. coli* in ballast water to achieve the International Maritime Organization's (IMO) guideline, 250 CFU/100ml. Optimized hydraulic retention time (HRT) and current to achieve the guideline are 30 s and 0.3 A respectively. The energy consumption is 0.004 kWh/m^3. Electrochemically produced chlorine is found to be the major disinfectant agent in the system. Electro-disinfection is effective and low energy intensive, thus possesses the ability to improve and use as a technology for ballast water treatment.

Keywords: Ballast water treatment, Chlorine, E. coli, Electro-disinfection, optimization.

1. Introduction

Ballast water is an essential component for the stability of the vessels during the voyage. Ballast water is taken in from the source port when the ship does not carry cargo and subsequently deballast at the destination port when it taken the cargo in. A wide variety of marine organisms can transport from one place to another through ship's ballast water and create environmental, economical and health crisis around the world. The only ballast water management technology practicing so far is the ballast water exchange at mid ocean. This process can either be sequential or flow through dilution (Endresen et al., 2004). Ballast water exchange is not efficient enough in removing the living organisms and not safe enough to the ship and to the crew (http://web.mit.edu/seagrant). Therefore, the need of a safe, efficient and cost effective alternative treatment technology is increasing. Heating technology using the waste heat from engine (Rigby et al., 1999; Thornton et al., 2004), Ultraviolet irradiation combined with pretreatment (Nilsen et al., 2001), ozonation (Herwig et al., 2006) and chemical biocides (Gregg et al., 2007) are some other technologies which have been researched for ballast water treatment. The abovementioned technologies have limitations such as reactivation of organisms, high cost, chemical storage, toxicity and low efficiency in disinfection.

Compared to the above disinfection technologies, electro-disinfection possesses the advantages such as no storage of chemicals and low foot print. In this technology, solution to be disinfected is passed through a reactor equipped with electrodes while applying direct or alternative current to the electrodes. Electro-disinfection has been found to be effective in killing wide variety of organisms in wastewater (Stoner et al., 1982; Patermarakis et al., 1990; Sarkka et al., 2008). Inactivation of living organisms can be due to chlorination (Stoner et al.,

1982), lethal oxidants (Patermarakis et al., 1990) or electric shock (Matsunaga et al., 2000). When the wastewater is mixed with seawater, the salinity increases and the electro-disinfection technology becomes more reliable with lower energy consumption (Li et al., 2002). Therefore, it is speculated that the electro-disinfection can be effectively incorporated to the purpose of ballast water treatment. Purpose of this study is to develop an electro-disinfection reactor which can fulfill the IMO guideline D-2 for living organisms in discharging ballast water. In the present discussion, bench scale electro-disinfection reactor has developed and optimized to achieve the IMO guideline for *E. coli* in discharging ballast water. Besides, the major disinfectant agent has investigated.

2. Experimental

Seawater was collected and artificially contaminated with *E. coli* (ATCC 10798) such a way that the initial *E. coli* concentration of the seawater solution is around 10^6 CFU/100 ml. The resulted seawater solution was used as the simulated ballast water in the experiments. The electrochemical disinfection reactor system mainly consisted of the designed reactor, a DC power supply (TDK-Lambda, Japan/Israel), a peristaltic pump (Cole Parmer Instrument Co., USA) and a container to fill seawater with an overhead stirrer. Volume of the disinfection reactor was 1.2 L and contained relatively inert metal electrodes. Experimental design for the optimization of designed reactor for disinfection of *E. coli* was done by statistical software, Minitab 15 (Minitab Inc., USA). Central composite design was selected, according to the number of factors to be handled. Range of HRT and current to be applied were selected as 15-45 s and 0.1-0.5 A respectively. The design resulted in 13 different combinations of applied current and HRT. Seawater solution and the experimental setup which are explained above were used in experiments. For each experiment, two samples were collected after the system became stable for chlorine analysis and detection of *E. coli*.

Na_3PO_4 was selected as the chloride free electrolyte. Concentration of the salt was adjusted such a way that the final conductivity of the Na_3PO_4 electrolyte was similar to the conductivity of seawater. Solution pH was adjusted using H_2SO_4 to reach the pH of seawater. This electrolyte was contaminated with cultured *E. coli* and the concentration of *E. coli* was set to 10^6 CFU/100ml. The experimental setup explained above was used for the experiments. Optimized HRT was used and the investigation was carried out at different energy inputs by varying the applied current. The experiments were repeated under the similar conditions using seawater as the electrolyte.

3. Results and discussion

3.1. System Optimization

Optimization was done based on the final *E. coli* concentration. The IMO guideline for *E. coli* in discharging ballast water, 250 CFU/100 ml, was chosen as the target to be achieved. Optimized HRT and current were found to be 30 s and 0.3 A respectively. To achieve the above results, energy consumption at the electro-disinfector was as low as 0.004 kWh/m^3 (see Fig. 1 (a) & (b)). Further, it was found that the developed electro-disinfection system is capable of achieving 4 log inactivation of *E. coli* which in-turn fulfill the IMO guideline at total residual chlorine concentration of 1 mg/L.

(a)

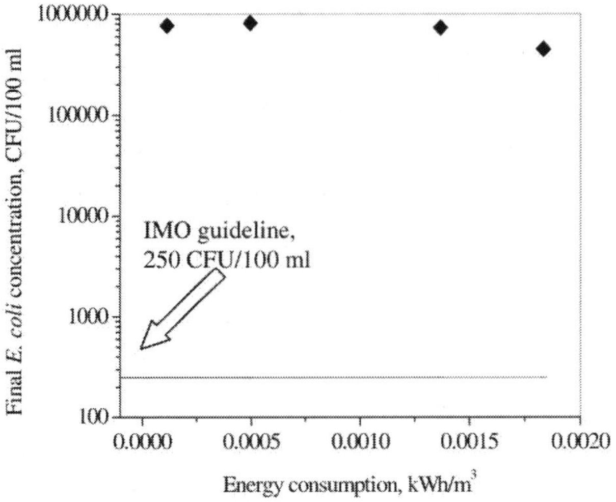

(b)

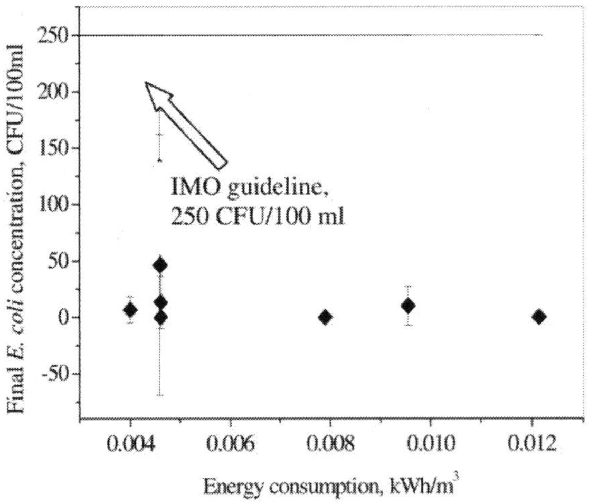

Fig. 1: Comparison of Final *E. coli* concentration with IMO guideline vs. energy consumption. (a) Energy consumptions < 0.004 kWh/m^3.(b). Energy consumption > 0.004 kWh/m^3.

3.2. Effect of chloride ion

Electro-disinfection process can produce several disinfectant agents such as chlorine, ozone and free radicals. Further, there is a possibility of disinfection due to the direct electric shock (Patermarakis et al., 1990; Matsunaga et al., 2000). In this study, the electrolyte is rich in chloride ion which can oxidize to chlorine at the anode. In order to investigate the effect of chlorine and the collective impact of other possible disinfectants, Na_3PO_4 was used as the electrolyte. It was found that the disinfection efficiency when use Na_3PO_4 as the electrolyte is far behind that of the seawater electrolyte. For example, at an energy consumption of 0.004 kWh/m^3, (the value of interest in the present system) disinfection efficiency of the system with Na_3PO_4 was only 22.9% while for the similar energy consumption, disinfection efficiency of the system with seawater was 99.99%. Therefore, it is anticipated that the major disinfection agent of the present system is chlorine, while the contribution of the other mechanisms is about 20%.

4. Conclusions

Electro-disinfection reactor was designed and optimized for disinfection of *E. coli* in seawater and the possibility of using the system in ballast water treatment was evaluated. Ability to meet the IMO regulation for *E. coli* in discharging ballast water was used as the major criterion of optimization. Optimized HRT and the current of the reactor were found to be 30 s and 0.3 A respectively. Specific energy consumption to achieve the optimized reactor performance was 0.004 kWh/m^3. The total residual chlorine concentration at the above conditions was 1 mg/L and the disinfection efficiency was 99.99%. Electrochemically produced chlorine was found to be the major disinfectant agent while the other disinfection mechanisms contributed to about 20% of the inactivation.

Acknowledgement

This research was co-funded by Maritime and Port Authority of Singapore and the National University of Singapore.

References

1. Endresen, Ø., Behrens, H.L., Brynestad, S., Andersen, A.B., Skjong, R. (2004), Challenges in global ballast water management, *Marine Pollution Bulletin,* 48, 615-623.
2. Gregg, M.D., Hallegraeff, G.M. (2007), Efficacy of three commercially available ballast water biocides against vegetation microalgae, dinoflagellate cysts and bacteria, *Harmful algae*, 6, 567-584.
3. Herwig, R.P., Cordel, J.R., Perrins, J.C., Dinnel, P.A., Gensemer, R.W., Stubblefield, W.A., Ruiz, G.M., Kopp, J.A., House, M.L., Cooper, W.J. (2006), Ozone treatment of ballast water on the oil tanker S/T Tonsina: chemistry, biology and toxicity, *Marine Ecology Progress Series*, 324, 37-55.
4. http://web.mit.edu/seagrant/, Marine bioinvasion fact sheet; ballast water treatment options, MIT Sea Grant.
5. Li, X.Y., Ding, F., Lo, P.S. Y., Sin, S.H.P. (2002), Electrochemical disinfection of saline wastewater effluent, Journal of environmental engineering, 128, 697-704.
6. Matsunaga, T., Okochi, M., Takahashi, M., Nakayama, T., Wake, H., Nakamura, N. (2000), TiN electrode reactor for disinfection of drinking water, Water research, 34, 3117-3122.
7. Nilsen, B., Nilsen, H., Mackey, T. (2001), The OptiMar ballast system, Proceedings of 1st International Ballast Water R&D symposium, IMO- London, 126-126.
8. Patermarakis, G., Fountoukidis, E. (1990), Disinfection of water by electrochemical treatment, Water research, 24, 1491-1496.
9. Rigby, G.R., Hallegraeff, G.M., Sutton, C. (1999), Novel ballast water heating technique offers cost effective treatment to reduce the risk of global transport of harmful marine organisms, Marine Ecology Progress Series, 191, 289-293.

10. Sarkka, H., Vepsalainen, M., Pulliainen, M., Sillanpaa, M. (2008), Electrochemical inactivation of paper mill bacteria with mixed metal oxide electrode, *Journal of hazardous materials*, 156, 208-213.
11. Stoner, G.E., Cahen, Jr., G.L., Sachyani, M., Gileadi, E. (1982), The mechanism of low frequency A.C. electrochemical disinfection, *Bioelectrochemistry and bioengineering*, 9, 229-243.
12. Thornton, G., Prentice, B. (2004), Biocidal heat treatment, *Ballast water treatment R&D directory*, 2nd Edition, Global Ballast Water Treatment Program, 17-17.

551c Treatment of Reverse Osmosis Concentrate to Improve Overall Recovery: The Effect of Prior Antiscalant Oxidation on Particle Characteristics and the Extent of Precipitation

Lauren F. Greenlee[1], Elise Barbot[2], Benny D. Freeman[1], Benoit Marrot[2], Philippe Moulin[2], and Desmond F. Lawler[3]. (1) Chemical Engineering, The University of Texas at Austin, 1 University Station C0400, Austin, TX 78712, (2) Laboratoire Mécanique, Modélisation et Procédés Propres, Aix en Provence, France, (3) Civil, Environmental and Architectural Engineering, The University of Texas at Austin, 1 University Station C1786, Austin, TX 78712

In many locations, fresh water resources are insufficient for local needs, and alternative sources with lesser water quality are being considered as drinking water supplies. In particular, the United States has many inland regions with untapped brackish water (500–10,000 mg/L total dissolved solids (TDS)) resources. Interest has increased for using these brackish water resources to produce drinking water. The two primary process choices for water desalination are thermal distillation or membrane filtration. Reverse osmosis (RO) desalination has become the primary choice for brackish water desalination due to lower energy requirements and a smaller footprint. However, from brackish water, the product recovery (volume of product water per volume of feed water) range is only 75 – 90%; *i.e.*, at least 10% of the feed water becomes the RO waste stream, or concentrate. The costs and technical feasibility of concentrate disposal are the key limitations to widespread application of inland RO. This research focuses on the development of a novel process to reduce the volume of brackish water RO concentrate.

Several critical differences between brackish and sea waters are the relative metal ion concentrations and the typical RO recovery ranges. While a typical brackish water (1,500 mg/L TDS) has a chloride ion concentration that is 30 times smaller than sea water and a sodium ion concentration that is 90 times smaller, the calcium concentration is only 2 – 5 times smaller and the carbonate concentration is equal to or larger than that of sea water. In addition, brackish water RO recovery is higher than sea water RO recovery (40-60%). These factors cause the recovery in brackish water RO systems to be limited by salt precipitation.

Chemicals called antiscalants are often used to complex with problematic salts ($CaCO_3$, $CaSO_4$, $BaSO_4$, $SrSO_4$, silica), delaying precipitation. However, salt concentration increases with recovery, and antiscalants work successfully within a limited concentration range. Therefore, eventually precipitation control is overcome. To increase system recovery and decrease the concentrate volume, a new approach is required.

Previous research using precipitation and separation to treat concentrate has shown that significant increases in total system recovery are possible [1]. However, the presence and influence of antiscalants and natural organic matter (NOM) during RO concentrate treatment have not been investigated.

This paper presents the development of a novel three-stage process to treat the concentrate from a brackish water RO system. The process achieves problematic salt removal through (I) antiscalant deactivation, (II) precipitation, and (III) solid/liquid separation. Antiscalant deactivation is performed using ozone (O_3) and hydrogen peroxide (H_2O_2). pH elevation is used to precipitate salts, and solid/liquid separation is achieved through sedimentation and filtration. While technologies for solid/liquid separation are well-established, the combination of antiscalant oxidation and precipitation represents a new system; research on antiscalant oxidation has been limited [2], and the effect of ozonation on precipitation has not been investigated.

Specifically, this study focused on changes to the precipitate particles and the extent of precipitation caused by prior antiscalant oxidation. Experiments were performed on a series of increasingly complex waters. First a simplified concentrate (~ 8,000 mg/L TDS), containing only sodium chloride (NaCl), sodium bicarbonate ($NaHCO_3$), and calcium chloride ($CaCl_2*2H_2O$) was used. Subsequently, magnesium and sulfate were added, and finally, real water samples were tested. Four different antiscalants, including several phosphonates and one acrylic polymer blend were used during experiments.

Several oxidation parameters, such as ozonation time and antiscalant concentration, were varied. Ozonation times of 1, 10, and 30 minutes were tested. For the first antiscalant tested (amino tri (methylene phosphonic acid), or AMPA) and the simplified concentrate, results show increased calcium precipitation for all ozonation times and all antiscalant concentrations (4 – 85 mg/L). The simplified concentrate with 85 mg/L AMPA, treated with only 1 minute of ozonation and subsequent precipitation at pH 10.5 for 1 hour, had a final dissolved calcium concentration of 5.2 mg/L Ca^{2+}. In comparison, the same precipitated solution with no antiscalant resulted in a final dissolved calcium concentration of 3.7 mg/L, while the same precipitated solution with 85 mg/L AMPA and no prior ozonation resulted in a final dissolved calcium concentration of 77 mg/L.

A laser granulometer Mastersizer S (Malvern Instruments) and a laser particle counter (Met One) were used to evaluate the effect of the ozonation step on precipitate particle size and number. Previous results have shown that certain antiscalants, such as AMPA (40 - 85 mg/L), can change the particle size range and the modality of the size distribution for calcium carbonate ($CaCO_3$) precipitation. A solution of precipitated simplified concentrate with 40 – 85 mg/L AMPA is bimodal and has particle size ranges of 0.2 – 2 microns and 2 - 50 microns, whereas the same solution without antiscalant has a single particle size range between 10 and 100 microns. Results from combined ozonation-precipitation experiments show that ozonation causes the particle size distribution to shift towards that of a solution containing no antiscalant. Light microscope photos (25x) are consistent with the granulometer results and show particles that resemble a precipitated solution with no antiscalant present.

The fouling potential of precipitated solutions was evaluated through dead end filtration experiments (MWCO = 0.1 micrometers, Millipore nitrocellulose membranes). For antiscalant AMPA (85 mg/L), ozonation times of 1 and 10 minutes increase the permeate flux. An ozonation time of 30 minutes causes a higher initial permeate flux but the flux decline is greater overall and results in a lower final permeate flux. These differences in flux decline indicate different membrane fouling mechanisms. Moreover, this ozonation duration is not in agreement with an industrial development.

Future work includes manipulation of flux data to determine the type of membrane fouling occurring and continuing experiments with more complex waters.

[1] Rahardianto, A.; Gao, J.; Gabelich, C.J.; Williams, M.D.; Cohen, Y., High recovery membrane desalting of low-salinity brackish water: Integration of accelerated precipitation softening with membrane RO. *Journal of Membrane Science* **2007,** 289, 123-137.

[2] Yang, Q.; Ma, Z.; Hasson, D.; Semiat, R., Destruction of Anti-Scalants in RO Concentrates by Electrochemical Oxidation. *Journal of Chemical Industry and Engineering (China)* **2004,** 55(2), 339-340.

A Hybrid Membrane-Biofilm Process for Concurrent Nitrification and Denitrification of Wastewater

Leon S. Downing and Robert Nerenberg
University of Notre Dame, Notre Dame, IN

Introduction

Total nitrogen (TN) removal from wastewater is an increasingly important treatment objective, yet many wastewater treatment facilities were designed with low solids retention times, and cannot achieve biological nitrification, i.e., conversion of ammonium to nitrate. For nitrifying plants, upgrading to denitrification, i.e., reduction of nitrate to nitrogen gas, often requires the use of exogenous electron donors, increasing operating costs.

We proposed a new process, the hybrid (attached and suspended-growth) membrane biofilm process (HMBP), for TN from wastewater. The HMBP (Figure 1) incorporates air-supplied hollow-fiber membranes into an activated sludge tank, achieving nitrification with a membrane-aerated biofilm. By maintaining completely-mixed conditions, the BOD concentration is kept low, thus denitrifying heterotrophs are mainly in the suspended phase (Downing and Nerenberg 2007). This allows for TN removal with a single tank, short bulk-liquid solids retention times (bSRTs), and minimal or no exogenous electron donor addition. Also, by utilizing passive aeration via the membranes, rather than bubbled aeration, and maximizing BOD use for denitrification, up to 70 percent of a plant's energy needs for aeration can be saved (Semmens 2005).

The HMBP combines the benefits of the integrated fixed-film activated sludge (IFAS) process, a hybrid process where nitrifiers are retained on suspended or fixed carriers, with membrane aerated biofilm reactors (MABRs), a fixed-film process that can concurrently nitrify and denitrify, but performs poorly when influent BOD is present.

A potential added benefit of membrane aeration is shortcut TN removal, which has been suggested for MABR systems (Hibiya, Terada et al. 2003; Terada, Hibiya et al. 2003; Terada, Yamamoto et al. 2006). Shortcut TN removal is the reduction of nitrite rather than nitrate during denitrification, which results in a reduced carbon source requirement of 40% and oxygen requirements by 25% (van Loosdrecht and Salem 2006). Nitrite accumulation is based on selection of ammonia oxidizing bacteria (AOB) over nitrite oxidizing bacteria (NOB), thus limiting the amount of nitrite oxidized to nitrate. Conditions that favor AOB over NOB include high temperature (Hellinga, Schellen et al. 1998), high free ammonia (FA) (Anthonisen, Loehr et al. 1976), high pH (Villaverde, Garcia-Encina et al. 1997), and low dissolved oxygen (DO) (Munich, Lant et al. 1996). Inhibition of NOB via low DO may occur when membrane-aerated biofilms operate in an anoxic environment, where the outer biofilm regions have low or zero oxygen concentrations. AOB can outcompete NOB at lower oxygen concentrations, as they have a lower K value for oxygen (Schramm, DeBeer et al. 1998). In a nitrifying MABR, microsensor studies indicated that nitrite was the main product of nitrification (Schramm, De Beer et al. 2000). MABRs achieving TN removal were shown to have low levels of nitrate in the biofilm (Hibiya, Terada et al. 2003), and an oxygen mass balance indicated the majority of oxygen transferred to a MABR was used for ammonia oxidation, with little oxygen remaining for nitrite oxidation (Terada, Hibiya et al. 2003).

We examined performance of the HMBP at several nitrogen and organic loading rates. Microsensor and FISH measurements were used to provide new insights into the structure and function of the HMBP biofilm, including evidence of short-cut denitrification (Downing and Nerenberg 2008).

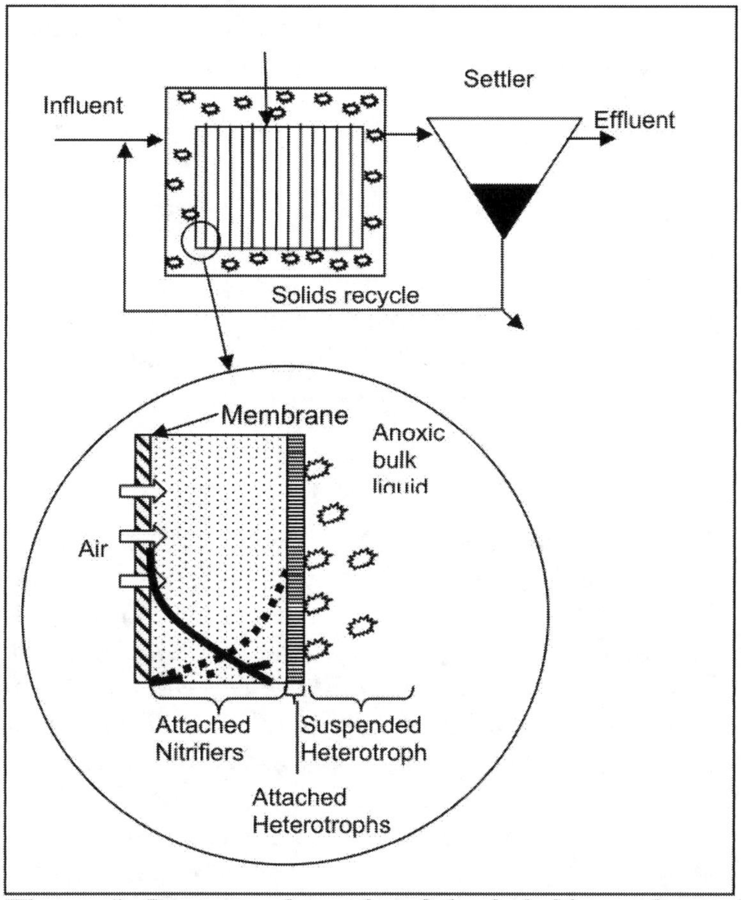

Figure 1. Process schematic of the hybrid membrane biofilm process (HMBP). Gradients of oxygen (—), ammonia (▪▪▪), and BOD (▪—▪) are shown through the biofilm.

Methods

Bench-scale reactor

The bench-scale HMBP reactor was designed following Downing and Nerenberg (2007), and was intended to simulate an activated sludge system. The reactor influent flow rate was 7.9 or 15.7 mL/min, and the sludge recycle ratio was 1.0. The activated sludge tank volume was 6 liters, resulting in a hydraulic retention time (HRT) of 10 or 20 hours. A bulk liquid sludge retention time (bSRT) of 2.5 days was maintained. The settler volume was 5 liters. The aeration tank included an axial mixer with a 3.5-inch blade operated at 1000 rpm. An open rectangular chamber contained a bank of hollow fiber membranes, referred to as the membrane bank. The membranes were a composite, with microporous polyethylene encasing a dense, polyurethane core (HFM200TL, Mitsubishi Rayon, Japan). The membrane outside diameter was 280 μm, and the total surface area of fiber was 1,900 cm^2 or 1,400 cm^2, depending on the loading rate being tested. The membranes were pressurized with compressed air at 35 kPa. Measurement of suspended solids in the influent, reactor tank, effluent, and settled sludge determined the volume of wasted sludge necessary to maintain the 2.5 day sbSRT.

A synthetic wastewater was prepared from distilled water amended with 1.386 g Na_2HPO_4, 0.849 g KH_2PO_4, 0.05 g $MgSO_4 \cdot 7H_2O$ per liter, as well as a trace mineral solution (Nerenberg, Rittmann et al. 2002). $(NH_4)_2SO_4$ was added to achieve either 20 or 12 mg/L NH_3-N. Various BOD concentrations were achieved via magnesium acetate addition. BOD was calculated from the acetate by multiplying the acetate concentration by 1.08 mg BOD/mg acetate.

Analytical methods

For the bench-scale studies, nitrate, nitrite, and acetate were analyzed by ion chromatography (IC2500 with AS11/AG11 column, Dionex Corp, Sunnyvale CA.). The eluant was sodium hydroxide. NH_3-N was measured using a colorimetric method (High Range, Hach, Loveland, CO). pH was monitored using a glass electrode pH meter. Bulk liquid dissolved oxygen (DO) concentration was measured using a DO probe (YSI Model No. 55/25 FT). Suspended solids were measured according to *Standard Methods in Water and Wastewater*(Rand, Greenberg et al. 1978).

Microsensors

Microsensor measurements were completed at the bench-scale. Oxygen profiles through the biofilm were measured with Clark type oxygen microelectrodes (ox10, Unisense A/S, Aarhus, Denmark). Ammonium, nitrite, and nitrate profiles were measured using LIX microsensors. These microsensors were constructed as described elsewhere (De Beer and van den Heuvel 1988; De Beer and Sweerts 1989; De Beer, Schramm et al. 1997). Measurements were taken at intervals of 20 μm through the biofilm when influent concentrations of ammonium were 20 mgN/L and BOD were 150 mg/L. A micromanipulator was used to position the microsensors through the biofilm (Unisense A/S, Aarhus, Denmark). A total of three profiles were taken at three different locations for each sensor, resulting in nine profiles each for oxygen, ammonium, nitrite, and nitrate. Net fluxes into, or out of, the biofilm for ammonium, nitrite, and nitrate were calculated as described elsewhere (Schramm, De Beer et al. 1999).

FISH

The ecology of the biofilm was analyzed using FISH. Oligonucleotide probes used were: Nso190, which targets the majority of AOB (Mobarry, Wagner et al. 1996); NIT3, which targets *Nitrobacter* spp. (Wagner, Rath et al. 1996); NSR1156, which targets *Nitrospira* spp. (Schramm, DeBeer et al. 1998); and EUB338, which targets general bacteria (Amann, Krumholz et al. 1990). Membranes with attached biofilm were removed from the reactor, fixed in 4% paraformaldehyde for 60 minutes at 4°C, embedded in OCT compound, and sectioned using a cryostat, as described elsewhere (Schramm, De Beer et al. 2000). Hybridizations were performed as previously described (Manz, Amann et al. 1992).

Results

Bench-scale

A major advantage of the HMBP, as compared to previous membrane aeration systems, is the ability to treat increased BOD loadings without losing nitrification efficiency. As shown in Table 1, the HMBP has a similar nitrification rate as MABR systems. However, at high BOD loading rates, the HMBP maintains the same nitrification rate, whereas MABR systems have been shown to have a decreasing nitrification rate at increasing BOD loadings (Walter, Haase et al. 2005).

Table 1. Nitrification rates in the bench-scale HMBP and previous MABR systems

Nitrogen Loading gN m^{-2} day^{-1}	Organic Loading gBOD m^{-2} day^{-1}	Nitrification Rate g m^{-2} day^{-1}	Denitrification %	Reference
2	8	0.4	>90	Timberlake et al., 1988
3	13	2.2	39	Yamagiwa et al., 1994
1.9	8.3	1.71	>90	Semmens et al, 2003
2	7	0.77	>95	Hibiya et al, 2003
1.65	3.3 6.6 9.9	1.1 1.1 1.1	24% 29% 67%	This study
2.24	4.5 9.0 12.4 16.8	1.1 1.1 1.1 1.1	10% 40% 70% 80%	This study
1.46	16.8	1.0	99%	This study

The bulk BOD concentration was shown to have a major impact on nitrification in the HMBP (Figure 2). This may be explained by greater BOD penetration into the biofilm, allowing heterotrophic bacteria to compete for oxygen from the membranes. Although the BOD spikes were below 2.5 mg/L, the BOD source was acetate, which is readily degradable and consequently can stimulate growth at very low concentrations. A typical municipal wastewater contains more complex BOD, and significant effects on nitrification would not be expected at these low concentrations.

The mechanisms for nitrification and denitrification in the HMBP were also explored at the bench scale using microsensors. Shortcut nitrification, where nitrite is the main product of nitrification rather than nitrate, has been suggested in previous MABR studies (Satoh, Ono et al. 2004; Terada, Yamamoto et al. 2006). The use of microsensors during these experiments quantified the amount of nitrite and nitrate produced by the biofilm and consumed by the suspended heterotrophs. Results of microsensor profiles are summarized in Figure 3.

Under the three loading conditions in which microsensor measurements were taken, nitrite was the dominant form of oxidized nitrogen produced by the biofilm, confirming shortcut nitrogen removal. Effluent BOD was below 0.2 mg/L for the first two loading conditions, but spiked to approximately 2.5 mg/L for the final loading condition. This spike in effluent BOD resulted in decreased nitrification rates due to competition of AOB and heterotrophs in the biofilm.

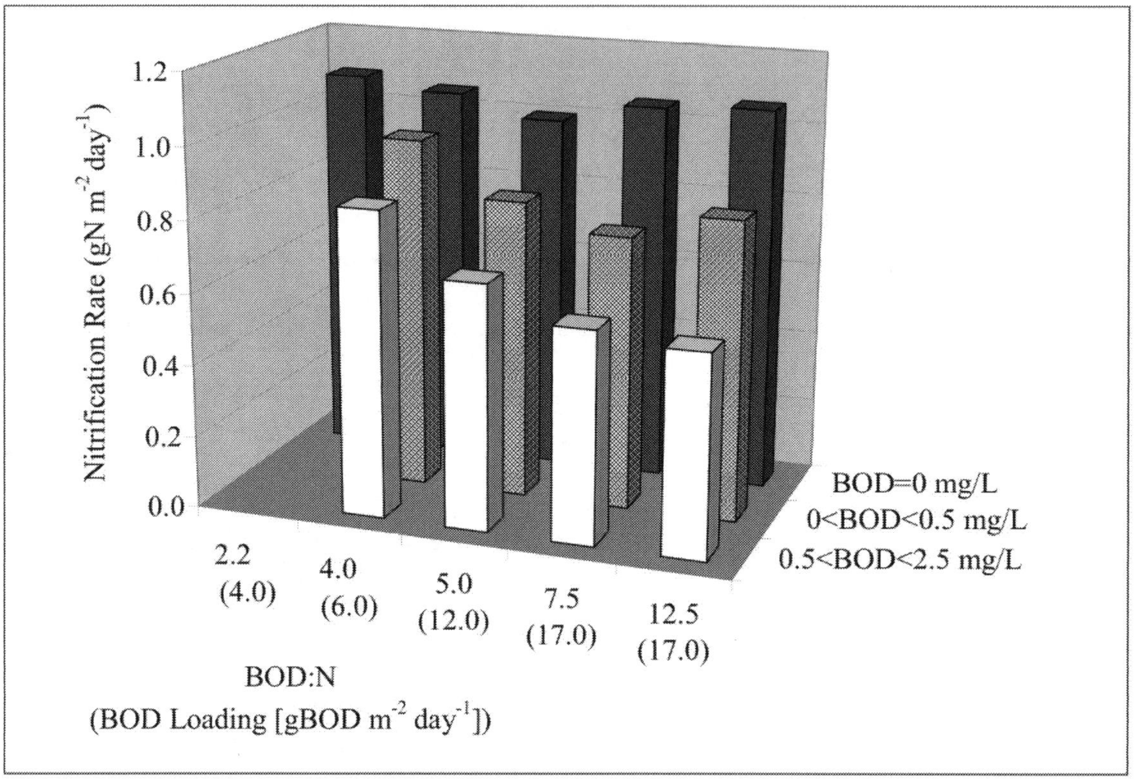

Figure 2. Impact of effluent BOD concentration of average nitrification rates at varying influent BOD:N ratios. BOD concentrations indicated on the right axis are for the bulk-liquid

All denitrification occurred in the bulk liquid under BOD to nitrogen (BOD:N) loading ratios of 5 and 7.5, where 70% and 80% denitrification was achieved, respectively. Under a BOD:N loading ratio of 12.5, full denitrification was consistently achieved. Under this condition, about 25% of the denitrification occurred in the biofilm, with 75% occurring in the bulk liquid. The increased biofilm denitrification is due to the increased bulk liquid BOD concentration, which allows effective diffusion of BOD into the biofilm.

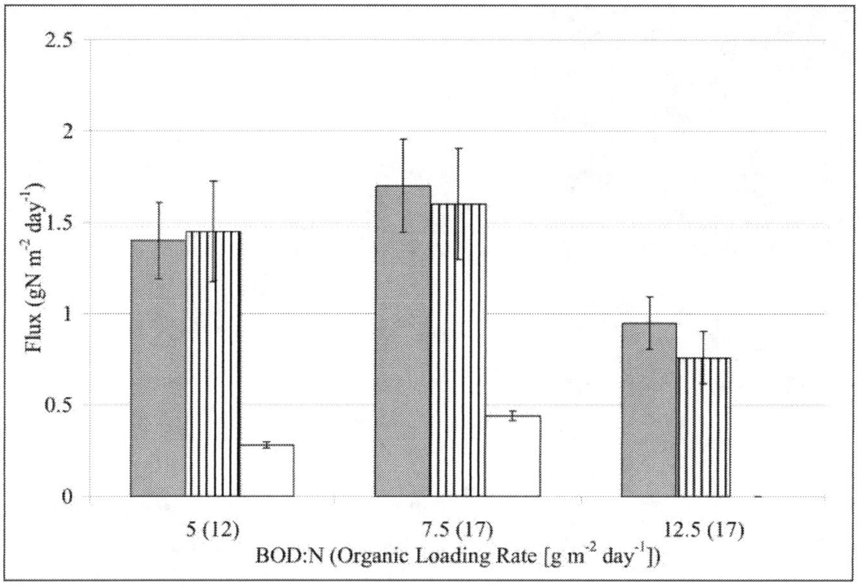

Figure 3. Rates of nitrification (solid), nitrite production (vertical lines), and nitrate production (white) by the biofilm in the bench-scale HMBP at varying loading conditions, as calculated from biofilm microgradients.

FISH results from the bench-scale study indicated a uniquely stratified biofilm. With 150 mg/L BOD and 20 mg/L NH_4^+-N in the influent, AOB were present throughout the biofilm, whereas two distinct layers of NOB and heterotrophs were present (Figure 4). Starting from the membrane biofilm interface, a mixture of AOB and NOB were observed, consistent with Schramm et al. (2000). At a distance of approximately 50 μm from the membrane biofilm interface, NOB populations decreased, and only AOB were present. A section of biofilm at 50 μm to 75 μm from the membrane contained exclusively AOB. The outer edge of the biofilm was composed of putative heterotrophic bacteria. Ecology studies of MABR system revealed some denitrifying heterotrophs throughout the biofilm thickness (Cole, Semmens et al. 2004). The existence of a strictly nitrifying layer in the HMBP results in increased nitrification rates by limiting heterotrophic competition in the aerobic portions of the biofilm.

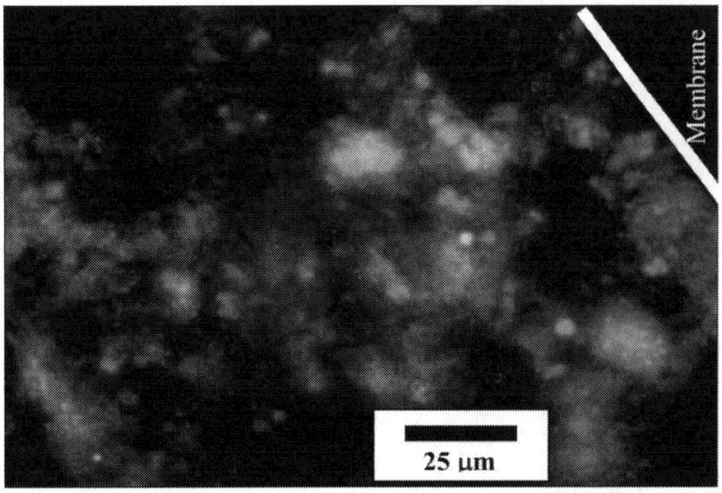

Figure 4. FISH image with an ammonium loading of 2 gN m^{-2} day^{-1} (20 mgN/L) and an organic loading of 17 gBOD m^{-2} day^{-1} (150 mg/L). Biofilm thickness was approximately 120 µm. AOB are shown in red/orange, while other bacteria are shown in green. Green cells near the membrane surface were proven to be NOB in separate hybridizations, while green cells near the biofilm edge are putative heterotrophs.

FISH results also indicated an increasing heterotrophic presence in the biofilm under the last loading condition (BOD:N of 12.5 in the influent). As previously stated, increased effluent BOD occurred on a more regular basis at this loading rate. The increased effluent BOD allowed BOD to diffuse into the biofilm. This increased BOD concentration in the biofilm results in the increased heterotrophic presence.

Conclusions

The HMBP achieves sustained nitrification rates despite increasing BOD loadings. The results suggest that nitrite is the main product of nitrification, and is preferentially reduced by heterotrophic bacteria. Given the HMBP's ability to achieve short-cut denitrification with short bSRTs, to aerate passively, and to maximize use of endogenous donors for denitrification, it has great potential to contribute to sustainable wastewater treatment. It also may be ideal to retrofit into existing plants that were designed with short SRTs and are unable to achieve nitrogen removal.

References

Amann, R., L. Krumholz, et al. (1990). "Fluorescent-oligonucleotide probing of whole cells for determinative, phylogenetic, and environmental studies in microbiology." Journal of Bacteriology **172**: 762-770.

Anthonisen, A. C., R. C. Loehr, et al. (1976). "Inhibition of nitrification by ammonia and nitrous acid." Journal of Water Pollution Control Federation **48**(5): 835-852.

Cole, A. C., M. J. Semmens, et al. (2004). "Stratification of activity and bacterial community structure in biofilms grown on membranes transferring oxygen." Applied and Environmental Microbiology **70**(4): 1982-1989.

De Beer, D., A. Schramm, et al. (1997). "A Nitrite Microsensor for Profiling Environmental Biofilms." Applied and environmental microbiology **63**(3): 973-977.

De Beer, D. and J.-P. R. A. Sweerts (1989). "Measurement of nitrate gradients with an ion-selective microelectrode." Analytica Chimica Acta **219**: 351-356.

De Beer, D. and J. C. van den Heuvel (1988). "Response of ammonium-selective microelectrodes based on the neutral carrier nonactin." Talanta **35**(9): 728-730.

Downing, L. and R. Nerenberg (2007). "Microbial Ecology and Performance of a Hybrid Membrane Biofilm Process for Concurrent Nitrification and Denitrification." Water science and technology **55**(8-9): 355-362.

Downing, L. and R. Nerenberg (2008). "Total Nitrogen Removal in a Hybrid, Membrane-Aerated Activated Sludge Process. ." Water Research **42**(14): 3697-3708.

Hellinga, C., A. Schellen, et al. (1998). "The SHARON process: An innovative method for nitrogen removal from ammonium-rich wastewater." Water Science and Technology **37**(9): 135-142.

Hibiya, K., A. Terada, et al. (2003). "Simultaneous nitrification and denitrification by controlling vertical and horizontal microenvironment in a membrane-aerated biofilm reactor." Journal of Biotechnology **100**(1): 23-32.

Manz, W., R. Amann, et al. (1992). "Phylogenetic oligodeoxynucleotide probes for the major subclasses of proteobacteria: problems and solutions." Systems for Applied Microbiology **15**: 593-600.

Mobarry, B. K., M. Wagner, et al. (1996). "Phylogenetic Probes for Analyzing Abundance and Spatial Organization of Nitrifying Bacteria." Applied and Environmental Microbiology **62**(6): 2156-2162.

Munich, E. V., P. Lant, et al. (1996). "Simultaneous nitrification and denitrification in bench-scale sequencing batch reactors." Water Research **30**(2): 277-284.

Nerenberg, R., B. Rittmann, et al. (2002). "Perchlorate reduction in a hydrogen-based membrane-biofilm reactor." Journal AWWA **95**(2): 10.

Rand, M., A. E. Greenberg, et al. (1978). Standard methods for the examination of water and wastewater. New York, American Public Health Association.

Satoh, H., H. Ono, et al. (2004). "Macroscale and microscale analyses of nitrification and denitrification in biofilms attached on membrane aerated biofilm reactors." Water Research **38**(6): 1633-1641.

Schramm, A., D. De Beer, et al. (2000). "Microenvironments and distribution of nitrifying bacteria in a membrane-bound biofilm." Environmental microbiology **2**(6): 680-686.

Schramm, A., D. De Beer, et al. (1999). "Microscale distribution of populations and activities of Nitrosospira and Nitrospira spp. along a macroscale gradient in a nitrifying bioreactor: quantification by in situ hybridization and the use of microsensors." Applied and environmental microbiology **65**(8): 3690-3696.

Schramm, A., D. DeBeer, et al. (1998). "Identification and Activities In Situ of Nitrosospira and Nitrospira spp. as Dominant Populations in a Nitrifying Fluidized Bed Reactor." Applied and environmental microbiology **64**(9): 3480-3485.

Semmens, M. I. (2005). Membrane Technology: Pilot Studies of Membrane Aerated Bioreactors, WERF.

Terada, A., K. Hibiya, et al. (2003). "Nitrogen removal characteristics and biofilm analysis of a membrane-aerated biofilm reactor applicable to high-strength nitrogenous wastewater treatment." Journal of Bioscience and Bioengineering **95**(2): 170-178.

Terada, A., T. Yamamoto, et al. (2006). "Feasibility of a membrane-aerated biofilm reactor to achieve controllable nitrification." Biochemical engineering journal **28**(2): 123-130.

van Loosdrecht, M. C. M. and M. Salem (2006). "Biological treatment of sludge digestor liquids." Water Science and Technology **53**(12): 11-20.

Villaverde, S., P. A. Garcia-Encina, et al. (1997). "Influence of pH over nitrifying biofilm activity in submerged biofilters." Water Research **31**(5): 1180-1186.

Wagner, M., G. Rath, et al. (1996). "In situ analysis of nitrifying bacteria in sewage treatment plants." Water Science and Technology **34**: 237-244.

Walter, B., C. Haase, et al. (2005). "Combined nitrification/denitrification in a membrane reactor." Water Research **39**(13): 2781-2788.

551e Removal of Arsenate Ions from Water Using Micellar Enhanced Ultrafiltration

Miklós Szidarovszky, Kyle Heckel, A. Eduardo Sáez, and Wendell Ela. Department of Chemical and Environmental Engineering, University of Arizona, 1133 E. James E. Rogers Way, Tucson, AZ 85721-0011

Arsenic is a naturally occurring element that can cause vascular disease, non-malignant skin alterations, and internal malignancies cancers and is thus considered to be a serious hazard to human health. It is commonly found in mine tailings, industrial wastewater, and even natural river systems. Due to its toxicity, the EPA reduced the maximum contaminant limit for arsenic to 10 ìg/L in 2006. Current techniques for the removal of arsenic from drinking water and high concentration brines include ion exchange, adsorption and reverse osmosis. Previous works have shown that negatively charged ultrafiltration membranes, such as regenerated cellulose, can separate arsenic oxyanions from water as long as no other ions are present, and that the addition of cationic surfactants improves separation. This is accomplished by arsenic oxyanions binding to the micelles, which can then be retained using relatively coarse ultrafiltration membranes. The main benefit of this method is that ultrafiltration can be operated at relatively low pressures and that the surfactant in the permeate can potentially be recycled by processes such as foam fractionation. The micelle-driven separation process is known as micellar-enhanced ultrafiltration.

This research focuses on the quantification of ionic strength effects on micellar ultrafiltration of arsenate ions, the optimization of the membrane pore size, and the quantification of the concentration polarization effect apparent at the membrane surface. The surfactant used was cetyl pyridinium chloride, and the membranes used were regenerated cellulose membranes with pore sizes ranging from 1 kDa to 100 kDa. Results show that the addition of surfactant below the critical micelle concentration (CMC) lowers separation by the membrane as the surfactant monomers screens the surface charges of the membrane, making the pores more permeable to arsenate ions. The addition of surfactant above the CMC has to overcome this negative effect before actual separation enhancement can be observed. The addition of surfactant does not improve separation if the membrane pore size is larger than the surfactant micelle. The average molecular weight of a CPC micelle is 32 kDa. Results show that CPC concentrations that are higher than the CMC lead to arsenic rejections of 85% and higher for membranes with pore sizes less than or equal to 30 kDa. The number of arsenate ions bound to each micelle varies between 28 and 0.5 as the concentration of surfactant is varied in the range from 1.0 mM to 15 mM. Due to pumping costs and pressure requirements, the ideal membrane size would be somewhere between 10 kDa and 30 kDa. Concentration polarization at the membrane surface was found to increase arsenic rejection due to a presieving effect; however, the surfactant rejection decreased due to a build up of surfactant micelles and monomers at the surface, creating a sizeable surfactant concentration gradient across the membrane. Increase in the ionic strength of the solutions by means of additional ionic species led to a decrease in arsenic rejection due to ion competition and charge screening, but also to a decrease in surfactant rejection due to the lowering of the surfactant CMC. The results support the use of micellar-enhanced ultrafiltration for the removal of arsenate ions within a specific range of ionic strengths.

551f Crossflow Nanofiltration (NF) of Silica Nano-Particles and Natural Organic Matter (NOM): Fouling, Transport and Interaction Effects

James (Chip) Kilduff, Civil and Environmental Engineering, Rensselaer Polytechnic Institute, 110 8th Street, Troy, NY 12180 and Yanxiao Yuan, Cee, Rensselaer Polytechnic Institute, 110 8th Street, Troy, NY 12180.

Applications of nanofiltration membrane processes to drinking water treatment are increasing. A typical objective is to control natural organic matter (NOM) to reduce disinfection by-product formation. Major considerations include NOM rejection and flux decline during filtration. The objective of this study was to investigate the influence of interactions between natural nanoparticles (silica colloids) and NOM on membrane fouling, particle deposition, and solute (NOM and salt) selectivity. A reduction in NOM selectivity by polyamide NF membrane was observed when the concentration of colloids was high and/or when a thick particle cake formed on the membrane. Moreover, the presence of NOM in the cake layer matrix substantially attenuated the enhancement of salt concentration polarization and increased its retention. The transport processes of both NOM and salt were successfully simulated using a multi-layer hydrodynamic solute transport model. Properties of the first deposited colloidal cake layer were critical in determining permeate quality.

560a Overview of Nanotechnology Activities at EPA Under the Toxic Substances Control Act (TSCA)

Jim Alwood, Chemical Control Division, U.S. EPA (OPPT), Washington DC, DC 20460

EPA will give an overview of nanotechnology activities at EPA under the Toxic Substances Control Act (TSCA). This will include components of the review process for Premanufacture Notices of new chemical substances, the goals and current outcomes for the Nanoscale Materials Stewardship Program, other potential regulatory activities under TSCA, and related EPA, federal, and international activities regarding nanotechnology.

Nanotechnology: Chemical and Toxicological Risk Assessment Issues with Antimicrobials

By

A. Najm Shamim, Jenny Tao and Jonathan Chen

US EPA, Antimicrobials Division, Potomac Yard, Arlington, VA, USA

Shamim.Najm@epa.gov, Tao.Jenny@epa.gov, and Chen.Jonathan@epa.gov

EXTENDED ABSTRACT

The United States Environmental Protection Agency (U.S. EPA) Office of Pesticide Programs (OPP) is a regulatory office that is responsible for pesticides registration for distribution and sale in the U.S. OPP is mandated by the Federal Insecticide, Fungicide, and Rodenticide Act (FIFRA) to conduct hazard and exposure risk assessments to ensure that agricultural and antimicrobials pesticides and biopesticides, when become commercially available, do not pose unreasonable adverse effects to humans or the environment (air, soil, water, terrestrial and aquatic organisms). OPP reviews enormous amount of scientific and technical data submitted by the registrants as well as has the burden of making regulatory decisions on these pesticides. The new nanoscience and nanotechnology, which may be used in pesticides, present an enormous challenge to the Agency in both domains.

The antimicrobial pesticides are handled in the Antimicrobial Division (AD) in OPP. Uses, application rates, and application scenarios of antimicrobial pesticides are relatively less than those of agricultural pesticides. However, antimicrobials are more used at homes, institutions, hospitals, and other residential-related fields, such as, construction materials, playground and recreational equipments, swimming pool and spa, etc. There are 12 categories of antimicrobial pesticides based on their use patterns. Of these use patterns, eight are categorized as indoor/residential. Some outdoor antimicrobial uses include heavy duty wood preservatives and antifoulants. Each use pattern triggers a set of data requirements.

Manufactured nanomaterials (MN) can be broadly divided into two major types: (1) carbon nanotubes, fullerenes, and functionalized carbon nanotubes or functionalized fullerenes; (2) manufactured nano-metals and nano-metal oxides. Both types can have different physical shapes like cylindrical or spherical etc. From these two types of MNs, a series of nanoproducts can be made. Mike Roco[1] believes that five generations of nanoproducts can be devised from these MNs. These are:

1) Passive nanostructures which include: polymers, ceramics, nanostructures of metals, nanoparticles, coatings, and possible applications are in the food industry, consumer products, cosmetics and pharmaceuticals.

2) Active nanostructures which include amplifiers, targeted drugs, actuators, adaptive structures and 3D transistors and possible applications are in the fields of

nanobiotechnology, neuroelectronics interfaces, hybrid nanomanufacturing (nanocomposites).

3) Systems of nanosystems which include guided assembling, 3D networking robotics, and possible applications are nanorobotics, regenerative medicine, brain-machine interface, and engineered agriculture products.

4) Molecular nanosystems which include molecular devises 'by design', atomic design, emerging functions, and possible applications are neuromorphic engineering, complex systems, human-machine interface

5) Converging Technology which include nano-bio-info from nanoscale, cognitive technologies, large complex systems from nanoscale and applications include hybrid nano-bio-info-medical-cognitive systems. At the present time, research and applications are at the third generation stage.

OPP mandates, through FIFRA, to conduct risk assessment for pesticides. The following data are required and acquired for a) physical/chemical characterization of a pesticide; b) data to show that the intended use of pesticide is efficacious; c) ecological effects, fate and transport of a pesticide in air, water, soils, terrestrial and aquatic organisms, bioaccumulation; d) human exposure risks for workers and post-applicators; e) mammalian toxicity.

OPP requires the submission of physical/chemistry characteristics for all pesticides which are submitted for registration. These are presented Table 1:

Table 1. OPPTS Physical/Chemical Properties Series 830 Test Guidelines

Series#	Title	TGAI[1]	MUP[2]	EUP[3]
830.1000	**Background for prod properties test guidelines**	**X**	**X**	**X**
830.1550	**Prod. identity and composition**	**X**	**X**	**X**
830.1600	**Description of materials used to produce the products**	**X**	**X**	**X**
830.1620	**Description of production process**	**X**		
830. 1650	**Description of formulation process**		**X**	**X**
830.1670	**Discussion of formation of impurities**	**X**		
830.1700	**Preliminary analysis**	**X**		
830.1750	**Certified limits**	**X**	**X**	**X**
830.1800	**Enforcement analytical method**	*	*	*
830.1900	**Submittal of samples**	*	*	*
830.6302	**Color**	**X**	**X**	#
830.6303	**Physical state**	**X**	**X**	**X**
830.6304	**Odor**	**X**	**X**	#
830.6313	**Stability to sunlight at normal and elevated temp.**	**X**		
830.6314	**Oxid/reduc chemical incompatibility**	**X**	**X**	**X**
830.6315	**Flammability**		**X**	**X**
830.6316	**Explodability**		**X**	**X**
830.6317	**Storage stability**		**X**	#
830.6319	**Miscibility**		**X**	**X**
830.6320	**Corrosion characteristics**		**X**	**X**
830.6321	**Dielectric breakdown voltage**		**X**	**X**

Series#	Title	TGAI[1]	MUP[2]	EUP[3]
830.7000	pH of aqueous solutions and suspensions	X	X	X
830.7050	UV/visible absorption	X		
830.7100	Viscosity		X	X
830.7200	Melting point/melting range	X		
830.7220	Boiling point/boiling range	X		
830.7300	Density/relative density/bulk density	X	X	X
830.7370	Dissociation constant in water	X		
830.7520	Particle size, fiber length, and diameter distribution	X		
830.7550	Partition coefficient (n-octanol/water), shake flask method	X		
830.7560	Partition coefficient	X		
830.7570	Partition coefficient (n-octanol/water) estimation by liquid chromatography	X		
830.7840	Water solubility: column elution method; shake flask method	X		
830.7860	Water solubility, generator column method	X		
830.7950	Vapor pressure	X		

[1] Technical grade cctive ingredient

[2] Manufacture use product

[3] End use product

Open literature search on nanotechnology and nanoproducts indicates that the determination of the physical/chemical characteristics of nanomaterials is beset with difficulties and uncertainties. This means that the uncertainties will make the data, results and risk assessments less reliable for nanomaterials.

Some of the problems, difficulties and uncertainties discussed here by no means represent a complete and comprehensive list:

1. Metal M or metal oxide MO as a MN is brought for registration. Should OPP consider it a NEW chemical or an old chemical but new use? The second approach will require very little new data. It is becoming quite clear that nanomaterials have some unique and new characteristics not possessed by same material if it is the 'normal' or 'conventional' state of existence. Some of the new and novel characteristics are: surface area, surface activity, size, and size distribution, shape, morphology, enhanced catalytic activity. Therefore treating a nanomaterial formed from a conventional substance as a new chemical appears a better approach.
2. a) Product identification and composition is of critical importance. Evidence abounds that most of the MNs suffer from the presence of impurities, particularly the presence of metals as impurity. How can a nano Technical Grade Active Ingredient (TGAI) be defined? What percent (%) or number of ppm level of impurity will the Agency regard as acceptable of unacceptable? b) Which analytical techniques would be acceptable for the Agency to be standard enforcement method for the full characterization, identification of nanoproducts? SEM, TEM, AFT is expensive techniques. Moreover, a single technique will not provide a complete characterization of a nanoproduct. It should be pointed out that recent research shows that some traditional

techniques can yield good and reliable data in identification and characterization of nanoproduct, as shown by the recent work by Xu,[2] Guar-Tzo Wei[3], C. Degueldre [4, 5], and Jorg Bettmer[6]

3. Most of the nanomaterials are insoluble in water. One simple technique used to identify chemicals is the uv-vis spectra. However, because of the solubility issue, should the Agency abandon asking for this data?
4. Similarly using the existing guideline methods as shown in Table 1, some of the characteristics like boiling point, partition coefficients (Kow), vapor pressure, and density can not be determined. Should Agency not require these characteristics? OR should the Agency develop new methods and employ new guidelines for the determination of these and other characteristics?
5. In addition to the date requirements listed in Table 1, should the Agency require new data, and therefore develop new guidelines due to the nanomaterials having new and unique characteristics? Some of the new characteristics that will play pivotal role in the characterization as well as help in the risk assessments are: specific surface area of the MNs, determination of zeta potentials of the MNs, surface charge on MNs, agglomeration/aggregation conditions of the MNs, crystalline phase of metal/metal oxides MNs, grain size, hydrodynamic size, size, length, and shape of MNs etc.

Metal Risk Assessment and Uncertainties:

Scientific community as well as the Agency has recognized that the risk assessment methodology used for conventional pesticides (agricultural as well as antimicrobials) can not be used for the risk assessment for metals. For example, fate and transport studies based on the conventional test guidelines to determine hydrolysis, photolysis, degradation (metabolism) studies under aerobic and anaerobic conditions, bioavailability, bioaccumulation etc. can not be investigated for metals[7]. Some of the reasons are: metals are naturally occurring substances, and are found in nature as mixtures in the environmental media like soils, water, air, plants, humans, animals and microorganisms. Metals do not degrade, biodegrade in the traditional sense, but undergo speciation (exist as different forms of chemical moieties at various PHS, redox conditions, in water, soils sediments, plants, humans and animals. In various organisms all important processes like absorption, distribution, transformation, and excretion of metals depend on the metal, organ's ability to absorb, transform, and excrete or to store the metal. This difficulty gets compounded with metals that are MNs[8]. This gives rise to the following uncertainties and concerns:

- Silver meets the PBT criteria in environmental media like water and soils. The exact hazard associated with nano-silver is not well-defined and hence precise hazard and exposure risk assessments can not be done for nano-silver.
- Despite the claim of nano-silver uses in numerous consumer products (and actual antimicrobial uses have not been figured in), the mass loadings into the environmental media is not known. This is a major barrier to conducting risk assessments. No adverse effects of nano-silver in the environmental media have been substantiated yet.

- Toxicity of silver to bacteria is well established. And some reports indicate that toxicity to bacteria from nano-silver is enhanced. But dose-responses in various environmental media have not been systematically investigated.
- In its bioavailable form, silver is known to be highly toxic to aquatic organisms. A similar database for nanosilver is not yet established.
- Normal silver is generally not a systemic toxic substance to humans. However, nanosilver has shown increased toxicity. Full blown research must be conducted to establish the possibility if nanosilver is a systemic toxic substance.

These issues are for one metal-silver in this case. The trend in the nanoscience and nanotechnology are clear for quite sometime, that the manufacture and applications of new products will include metals and metal oxides. As the recent document has established that a new framework is needed to conduct the risk assessments for metals, perhaps it is true that a brand new framework will be needed to conduct the risk assessment on nano-metals and nano-metal oxides.

Because of their nanometer-range size and high surface area and/or reactivity, nanomaterials possess unique physiochemical properties, and in turn, the toxicity of nanomaterials might be different from conventional materials. Some of the critical issues that EPA is struggling with and may continue to do so for the foreseeable future are:

How useful are the data generated on conventional size pesticides? Is data bridging between conventional size chemical and naomaterial possible?

Would current data requirement be adequate to characterize the hazards and to assess the risks related to the exposure of a nanopesticide?

Would the Agency need nanopesticide-specific data?

Would the current test guidelines be suitable and adequate for testing nanopesticide or existing guidelines be modified? Or would new test guideline(s) be needed?

For toxicity risk assessment would traditional animal testing model be appropriate for testing nanopesticide? What kind of dosimetrics would be appropriate for testing nanomaterial, surface area/activity-based and/or particle size/number-based in addition to mass-based?

Could the results derived from "non-traditional" *in vivo* method (e.g., intratracheal) be used in hazard characterization and risk assessment?

Would the data generated on 10 nm nanomaterials be the same as those of 30 nm ones?

How nanomaterials enter, travel through, and deposit in the body? - PK/PD of nanomaterials in animals

How would aggregation/agglomeration affect the toxicity of nanomaterials in organisms and environmental media?

The current data requirement for health effects (OPPTS mammalian toxicity series 870 test guidelines) for antimicrobial agents is a "tiered" approach. The Agency requires the Tier I data for all antimicrobial agents unless a waiver is granted based on its use pattern (e.g., in so-called "closed system" use that poses a minimum or negligible concern of exposure) or the physical/chemical properties (e.g., a corrosive agent). The

Tier I data requirement is listed below and all studies are required to test on the technical grade of active ingredient (TGAI). The product-specific acute 6 package may be used for the acute toxicity testing requirement.

Tier I data requirement:

Acute Toxicity Testing:
870.1100: Acute oral (one species) (TGAI/MP and EP)
870.1200: Acute dermal toxicity (one species) (TGAI/MP and EP)
870.1300: Acute inhalation toxicity (one species) (TGAI /MP and EP)
870.2400: Primary eye irritation (one species) (TGAI/MP and EP)
870.2500: Primary dermal irritation (TGAI/MP and EP)
870.2600: Dermal sensitization (TGAI/MP and EP)

Subchronic Toxicity Testing:
870.3100: 90-Day oral – rodent (TGAI)

Developmental Toxicity and Reproduction:
870.3700: Prenatal developmental toxicity - rodent (TGAI)

Mutagenicity:
Subdivision F, App. 9, 870.5100: Bacterial reverse mutation assay (TGAI)
Subdivision F, App. 9, 870.5300: *In vitro* mammalian gene mutation (TGAI)
Subdivision F, App. 9 870.5380, 870.5385/870.5395: *In vivo* cytogenetics (mutagenicity) (TGAI)

Depending upon the use pattern (e.g., food/feed-contact), the exposure concerns (e.g., long-term/high human exposure), or chemical-specific toxicity (e.g., a nuerotoxicity- or immunotoxicity-causing agent), additional testing including route-specific (dermal and/or inhalation), chronic toxicity and/or carcinogenicity studies, and other specific testings may also be required.

The conventional toxicity testing is to determine if there are any health effects in humans and/or animals exposed to a substance via specific exposure route. The health effects and exposure route are defined by the physiochemical properties of the substance as well as the likelihood of human exposure. The commonly used duration of testing includes acute, subchronic, and chronic. The dosage used in conventional toxicity testing is mass-based. Although toxicological endpoints vary among the studies, the lethal concentration level of 50% testing animal die after one exposure in an acute test (LD50/LC50), the maximum tolerated dose (MTD), and the none-observed-adverse-effect level (NOAEL) are defined endpoints for an acceptable level of human exposure. In addition, the histopathological changes of tissues are normally examined by light microscopy techniques.

Nanomaterials characterized by its well-known unique nano size and large surface area as well as surface reactivity would likely manifest a unique toxicity both in humans and environment. The mass concentration of doses used in the conventional toxicity testing would most likely result in non-representative exposure of nanomaterial particles. In turn, the toxic effects observed in the study are unlikely to be the potential toxic effects of the tested nanomaterial; but rather, the effect of massive blockage, e.g., preventing the ingestion of nutrients due to the amount of foreign material in the stomach in an oral study, or clogging the airways in an inhalation study. In addition, the light microscopic examination used in the conventional testing will not be able to detect any nanomaterials presented in the organs/tissues[8].

The short-term duration, such as those defined in acute toxicity testing, is unlikely to occur in human exposure to small amount of nanomaterials in reality unless a massive release happens as a result of accident. Therefore, subchronic and/or chronic toxicity testing would be more appropriate to determine the toxic effects of nanomaterials.

The OECD Working Party on Manufactured Nanomaterials report (ENV/CHEM/NANO 2008/7)[9] indicates that the current OECD testing guidelines are generally appropriate for investigating the human health effects of nanomaterials. However, it is recommended that certain important considerations need to be borne in mind, particularly those related to the physiochemical characteristics, including such characteristics in the actual dosing solution. Considerations should be given as well to what the most appropriate dose metrics would be to the tested nanomatereial. If not known, a serial of measurements including mass, particle number and surface area need to be considered. Modification to the testing guideline may also be necessary, particularly via inhalation route, which is considered to be the primary exposure route for nanomaterials. Detailed histopathological examination of the entire respiratory tract is needed.

There are many *in vitro* and "unconventional" *in vivo* (e.g., intratracheal) studies on nanomaterials. It is important to establish a scientifically sounding approach, based on current knowledge and practical solutions, when using these test methods. Nanomaterials have a tendency of translocating to whatever the exposure conditons are, including inside the organism after administration and will likely to show systemic effects. Therefore, it is almost impossible to determine such property of the nanomaterials and their systemic health effects by using single cell *in vitro* testing environment, until more valid screening tests or the relationship between the physiochemical properties of nanomaterials and their toxic effects are established.

The effects of chemical on human health and environment are dependent upon physiochemical and toxicological properties of the substance. To understand how the substance behaves in humans and animals, including persistence, bioavailability, internal distribution and bioaccumulation, are the keys to accurately assess and characterize the toxicological properties of the substance. No exceptions for nanomaterials. The toxicokinetics studies, which are defined to study the absorption, distribution and elimination of nanomaterials in humans and animals, are fundamentally important in assessing their potential health effects. While it is technically challenging, studies tracking the distribution of nanomaterials *in vivo* at realistic exposure scenarios (dose level and exposure route) seems necessary.

For what have been said, there are knowledge gaps on whether or not or to what extent that extrapolation would be possible from the toxicology of non-nanomaterials and other physical forms (e.g., fiber of the same substance) to the toxicology of nanomaterials and between nanomaterials of different size, range, and shape.

Given the complexity and variety of antimicrobial pesticides (twelve use pattern categories, different use sites and dual jurisdiction with FDA), the data requirement and policy on regulating pesticides containing nanomaterials are a "case-by-case" approach, although there are no registrations of pesticides containing nanomaterials at present time. From the toxicity perspective, the uncertainties of interpretations between *in vitro* and *in vivo* mechanistic studies as well as the limitations of the existing regulatory data requirements are presented.

Research and applications in the field of nanotechnology is fast-paced. Virtually all disciplines of science are or will be impacted by nanotechnology. It has been pointed out[10] recently that from the science and regulatory perspectives, short-term and long policies are needed to have strong grip on the development of the new field.

Some of these short-term policies are:

- Environmental health and safety (EHS) related research funds should be increased, and a peer-reviewed EHS research plan should be instituted.
- A stand-alone Nanotechnology Effects Institute should be established. This Institute should handle all hazards, exposure and safety related issues.
- A similar stand-alone interagency nanotechnology Regulatory Group should form.
- Each Regulatory agency should have its own unique nanotechnology plan

Some of the long-term policies include:

- Existing regulatory laws such as Toxic Substance Control Act (TSCA), Federal Food, Drug, and Cosmetic Act (FFDCA), and Federal Insecticide, Fungicide and Rodenticide Act (FIFRA), should be amended to incorporate nanotechnology related guidelines and make these laws more stringent.
- Introduce new laws concerning the data requirements for nanoproducts
- Enact new regulatory law specific to nanoproducts

REFERENCES

1. Mike Roco, June 2008, A talk presented at the: International Food Policy Research Institute , Washington, DC
2. P.D. Nallathamby, Kerry J. Lee, and Xiao-Hong Nancy Xu, 2008, ACS NANO, Volume 2(No.7), pp 1371-1380
3 Guor-Tzo Wei, and Fu-Ken Liu, 1999, Journal of Chromatography A, 836, pp 253-260
4 C. Degueldre et.al, 2004, Analytica Chimica acta, 519, pp137-142
5 C. Degueldre et al., 2006, Analytica Chimica Acta, 555, pp 263-268
6 Andreas Helfrich, Wolfram Bruchert, and Jorg Bettmer, 2006, Journal of Analytical Atomic Spectrometry, Vol.21, pp 431-434
7 Risk Assessment Framework for Metals, 2007/ US EPA, Office of Research and Development.
8 Rocks. S, Pollard, S, and Dorey, R., et al. 2008. Comparison of Risk Assessment Approaches for Manufactured Nanomaterials. Final report, Defra Project CB403, Cranfield University. May 30, 2008.
9 Working Party on Manufactured Nanomaterials. 2008. Progress Report for Project 4 "Manufactured Nanomaterials and Test Guidelines". ENV/CHEM/NANO (2008)7. Organization for Economic Co-operation and Development (OECD). May 30, 2008.
10 Davies, J. C. 2008. Woodrow Wilson International Center for Scholars, Project on Emerging Nanotechnologies (PEN) # 13.

560c Human Health Effects of Nanomaterials

David A. Jones and **Megan J. Winter**. Engineering and Production, NanoScale Corporation, 1310 Research Park Dr., Manhattan, KS 66502

NanoScale Corporation in Manhattan, Kansas manufactures high surface area specialty metal oxide nanoparticles. A proprietary blend of these powders called FAST-ACT has been developed to be highly reactive against chemical warfare agents and toxic industrial chemicals. This product was developed for use by the military against chemical warfare threats and is currently being used by first responders to neutralize harmful vapor and liquid spills. An intensive study is in progress in conjunction with the Kansas State University School of Veterinary Medicine Department of Diagnostic Medicine/ Pathobiology to determine the dermal penetration ability and effect on human epithelial cells of FAST-ACT and its components.

The first area of focus for the project is to determine the ability of the nanocrystalline materials to pass through human skin. Quantitative testing is being completed to determine the penetration of FAST-ACT, a FAST-ACT/water mixture and a FAST-ACT/water/surfactant mixture through human skin. The FAST-ACT/water mixture and the FAST-ACT/water/surfactant mixture are being tested to simulate the application of water and FAST-ACT to the skin or respectively water, soap and FAST-ACT. Additionally, the affinity of the lipids in skin for the nanocrystalline material is being tested by an octanol/water partitioning studies.

The second area of focus of the project is the effect of inhaled FAST-ACT on the human lung. The solubility of FAST-ACT in various body fluid simulants is being tested. Additionally, the effect of FAST-ACT as well as FAST-ACT with adsorbed chemical warfare agent reaction products on lung epithelial cells will be determined.

The project will be completed at the end of July, 2008 and at that time it is expected that the dermal penetration ability of FAST-ACT and the effect of FAST-ACT on lung epithelial cells will be determined.

560d Classification of Engineered Nanoparticles According to Their Long Time Risk Potential for Human Health and Environment

Wendelin J. Stark and Ludwig K. Limbach. Chemistry and Applied Biosciences, ETH Zurich, Wolfgang-Pauli-Str. 10, ETH Hönggerberg, Zurich, 8093, Switzerland

Time- and cost-effective research instruments are currently required as tools for a sustainable and safety oriented product development both in academia and industry. The present contribution investigates a very rapid method to assess critical materials in an early stage of development or product design. The use of physico-chemical properties as primary decision parameters allow a first step of material risk evaluation at the design stage using predominantly accessible data. The method has been validated on the basis of the currently available toxicity data and further supported by in vitro experiments.

We propose a mapping of nanomaterials along their persistence, catalytic and chemical activity within biological systems and matrix related effects. This contribution will first demonstrate a justification for the chosen risk evaluation parameters. In a second step, existing materials with known risk profile are used to validate the approach. The last part will guide the audience through the evaluation of currently used industrial materials and novel products based on nanoparticles or carbon nanotubes.

The Swiss Departments of Public Health (BAG) and environment (BAFU) are currently implementing the here presented proposal as part of chemical and materials legislation. It is expected that significant parts of the Swiss proposal will be transferred to other European Countries.

References:

[1] C. Meili, M. Widmer, F. Husmann, P. Gehr, F. Blank, M. Riediker, K. Schmid, L. Limbach, W. Stark, Synthetische Nanomaterialien: Risikobeurteilung und Risikomanagement. Grundlagenbericht zum Aktionsplan. Umwelt-Wissen Nr. 0721. Bundesamt für Umwelt und Bundesamt für Gesundheit, Bern. 284 S. (2007)

[2] L.K. Limbach, P. Wick, P. Manser, R. Grass, A. Bruinink, W.J. Stark, Exposure of Engineered Nanoparticles to Human Lung Epithelial Cells: Influence of Chemical Composition and Catalytic Activity on Oxidative Stress. Env. Sci. Technol, 41, 4084-9 (2007)

[3] T.J. Brunner, P. Wick, P. Manser, P. Spohn, R.N. Grass, L.K. Limbach, A. Bruinink, W.J. Stark, In Vitro Cytotoxicity of Oxide Nanoparticles: Comparison to Asbestos, Silica, and the Effect of Particle Solubility, Env. Sci. Technol., 40 (14), 4374-81 (2006).

[4] P. Wick, P. Manser, L.K. Limbach, U. Dettlaff-Weglikowska, F. Krumeich, S. Roth, W.J. Stark, A. Bruinink, The degree and kind of agglomeration affect carbon nanotube cytotoxicity, Toxicol. Lett., 168 (2), 121-31 (2007).

560e Engineered Nanomaterial Interactions with Suspended Lipid Bilayers

Steven A. Klein, Mechanical Engineering, Arizona State University, 1711 S Rural Rd ECG 346, Tempe, AZ 85287-6106, Trevor J. Thornton, Electrical Engineering, Arizona State University, 1711 S Rural Rd., Tempe, AZ 85287-6206, and Jonathan D. Posner, Mechanical Engineering, Chemical Engineering, Arizona State University, 1711 S Rural Rd ECG 346, Tempe, AZ 85287-6106.

Over the past five years there has been a growing interest in the interactions between engineered nanomaterials and cellular constituents. Quantitative measures of nanomaterial bioavailability and toxicity must be made so that the impact of nanotechnology on human health and the environment can be addressed. Recent research focuses on either collection of empirical epidemiological data (e.g. uptake of nanoparticles (NP) by cells, toxicity to organisms such as rats or fish, mutation of SNP) or precise NP characterization (e.g. size, shape, degree of aggregation, charge, and surface chemistry). However, it is difficult to transition from these measurements to a rapid assessment of emerging nanomaterials toxicity or to understand the fundamental interactions of NM with biomolecules or biological interfaces (e.g. cellular uptake, infusion in tissue, etc.) In this talk, a novel nanoparticle platform is presented which can rapidly provide information about the interaction of nanomaterials at the lipid bilayer-fluid interface. Lipid bilayers are arguably the most important interface between life and its environment. The platform uses lipid bilayers suspended over apertures, low-noise patch clamp type amplifiers and high speed, spinning disk epi-fluorescence microscopy for real-time measurements of NP-bilayer interactions. These measurements can help determine the possible types of interactions taking place: unmediated transport of NP through lipid membranes, aggregation of NP in solution and on the lipid bilayer, as well as degradation of the membrane due to NP influence. This system operates over a wide range of conditions (NP composition, charge, shape and size) and physicochemical conditions (e.g. EDL thickness, aggregation, etc.). Here, results of CdSe quantum dot, polystyrene, and silver NP interactions with DOPC:DOPE suspended membranes will be presented. Nanoparticle characteristics (composition, charge, shape and size) are investigated to determine parameters that induce interactions and better understand underlying processes. Optical images and low noise current signatures are correlated to provide insight into the physical mechanisms of lipid bilayer-NP interaction.

Web Page: microfluidics.asu.edu

560f In Vitro Toxicity Testing of CdTe Nanoparticles In Three-Dimensional Cell Culture

Jungwoo Lee, Biomedical Engineering, University of Michigan, 2300 Hayward St., 3074 H. H. Dow Bldg., Ann Arbor, MI 48109-2136, G. Daniel Lilly, Chemical Engineering, University of Michigan, 2300 Hayward, HH Dow Building #3074, Ann Arbor, MI 49109, and Nicholas A. Kotov, Chemical Engineering Department, Materials Science Department, Biomedical Engineering Department, University of Michigan, 2300 Hayward, Ann Arbor, MI 48109.

Nanoparticles (NPs) have been actively investigated due to their great benefits in numerous applications; however, less effort has been committed to explore their toxic effects despite high hazardous potential. Until now all in vitro NPs toxicity testing have been performed on the 2D cell culture. However, it has become increasing apparent the discrepancy of toxicity testing results between 2D cell culture and animal models because simplified 2D cell culture is far too limited to recapitulate appropriate level of complexity in vivo tissues. To achieve more realistic toxic effects of NPs, we developed a 3D spheroid culture based NPs toxicity testing system which represents better physiology of natural tissues.

As an initial step, we prepared spheroid formation of human hepatocarcinoma cell (HepG2) utilizing hydrogel inverted colloidal crystal scaffolds. The extensive cell-cell interactions in spheroid culture promoted the recovery of 3D structure and partial function of the liver tissue. In addition, highly regulated and homogeneous spheroid culture allowed systematic and reproducible toxicology assays. Toxic effects of lcysteine-stabilized cadmium telluride (CdTe) nanoparticles were tested in multiple aspects including morphology, membrane integrity, metabolic activity and cell death mechanism. The spheroid culture showed significantly reduced CdTe NPs toxic effects compared to the conventional 2D culture.

Overall this standardized spheroid culture system extends current cellular level NPs cytotoxicity testing on 2D culture to the tissue level toxicity, which is closer to the real human body responses. It can significantly contribute to the realization of emerging opportunities of NPs by providing their toxic information accurately as well as easily.

619a Regulation and Societal Considerations of Nanotechnology in Commerce

Michael E. Heintz, Porter Wright Morris & Arthur LLP, 41 South High Street, Columbus, OH 43215

With an estimated 600 consumer products on the market containing nanomaterials, protection of human health and the environment is becoming a paramount consideration. Environmental, occupational health, and consumer protection laws are coming under close scrutiny to determine whether each are appropriate methods for regulating the widespread use of nanomaterials. In addition, support businesses, such as insurance companies, and consumer safety organizations are also releasing information concerning the impacts and considerations surrounding nanotech. This presentation will provide an overview of how the major regulatory schemes at all levels of government are addressing the unique and new concerns raised by nanotechnology in commerce. In addition, ethical and societal considerations concerning the use of nanotechnology in a time of limited information concerning the impacts of such materials, including reactions by insurance companies and public interest groups, will be discussed as well.

The rapidly advancing field of nanotechnology offers potential benefits to almost all industries and products. According to the Woodrow Wilson International Center for Scholars' Project on Emerging Nanotechnology's *Nanotechnology Consumer Product Inventory*, nanomaterials are currently being used in over 500 consumer products. However, the environmental and human health risks associated with these materials are largely unknown. Before the environmental and general population risk of these materials can be measured or monitored, methods will be needed to collect, separate, detect, identify, and quantify the occurrence of nanomaterials in the environment over time. The rate of product development has outpaced the rate of methods development, and, unfortunately, the methods needed to monitor the environmental occurrence of nanomaterials are very limited or nonexistent. Only limited research has been conducted, in part due to the immense challenges of sampling nanomaterials in the environment (e.g., distinguishing between naturally-occurring nanoparticles and engineered nanoparticles). While EPA's Office of Research and Development (ORD) may eventually sponsor research to develop remote, in situ, and continuous monitoring devices to detect engineered nanomaterials at very low concentrations, the purpose of this initial effort is to identify and assess currently-available sampling and analysis methods. This state-of-the-science review focuses on soil, sediment, and groundwater environmental sampling. A companion research effort is focusing on air sampling and analytical techniques.

Due to the limited data available and the rapid advancements that are being made, a traditional literature search (i.e., using an abstract database to search peer-reviewed journals) is not expected to effectively capture the cutting-edge information on environmental sampling and analysis methods for nanomaterials, particularly those under development and those that have not been fully peer-reviewed. Because this research is current and ongoing, much of the applicable information is found in gray literature (e.g., conference proceeding, communications with research scientists and other experts). Additionally, many of the techniques for environmental sampling and analysis of nanomaterials may need to be adapted from trace sampling methods and methods used in quality assurance or characterization of manufactured nanomaterials; therefore, peripheral literature has also been identified and reviewed.

The approach to this review included three separate strategies:

1) Collection of available published literature using Dialog®;

2) Review of information from targeted sources, such as nano-specific journals, conference proceedings, grants databases, and research databases; and

3) Contacts with industry and academic experts.

This review summarizes the key characteristics that must be considered when collecting and analyzing samples in various environmental media. Based on results of the literature review, and personal communication with researchers, typical analytical methods and techniques that are currently used for nanomaterials are identified and briefly discussed.

The review also identifies several sources that provided information on analytical techniques and equipment for nanomaterials. However, little information was obtained for sampling techniques that are specific for analysis of nanomaterials. Information obtained indicated that nanomaterial-specific sampling techniques have not yet been developed. Another area for which the search results provided little information is for differentiating anthropogenic (man-made) nanomaterials from natural nanomaterials. A number of potential sources that may provide additional information upon a more in-depth review were identified. As this review provides a current state-of-the-science review of active research topics at the time of writing, this review may require modifications as additional research is conducted.

619c Removal of Oxide Nanoparticles In a Model Waste Water Treatment Plant

Ludwig K. Limbach[1], Robert Bereiter[2], René Gälli[2], and Wendelin J. Stark[1]. (1) Chemistry and Applied Biosciences, ETH Zurich, Wolfgang-Pauli-Str. 10, ETH Hönggerberg, Zurich, 8093, Switzerland, (2) BMG Engineering, Schlieren / Zurich, 8952, Switzerland

The rapid growth of nanotechnology has resulted in various implementations of nanomaterials in advantageous products or as process enhancers in manufacturing. As a result of the increasing production it is important to obtain a better insight in possible risks related to these ultra small particles in the environment. Of particular interest is the clearing efficiency of nanomaterials within sewage plants that could avoid the propagation of nanoparticulate contamination into the aquatic environment.

However, little is known so far about the clearing efficiency of nanomaterials in sewage treatment. Reijnders (Reijnders, 2006) supposes that the standard waste water treatment seems to be poorly suited to the capture of nanomaterials, whereas others (Wiesner et al., 2006) did not support scenarios where containment of nanoparticles in current water treatment infrastructure may become problematic. Due to the present lack of experimental data, the disposal of nanomaterials is not regulated. This appears problematic as it is not clear what happens when inorganic nanomaterials are dispersed in the environment or brought into public sewage treatment plants.

Efficient removal is particularly important in view of increasing long term persistence and evidence for considerable ecotoxicity of specific nanoparticles. The present work investigates the use of a model waste water treatment plant for removal of oxide nanoparticles. (Limbach et al., 2008) While a majority of the nanoparticles could be captured through adhesion to clearing sludge, a significant fraction of the engineered nanoparticles escaped the waste water plant's clearing system and was found in the exit stream of the model plant. Our study demonstrates a significant influence of surface charge and the addition of dispersion stabilizing surfactants as routinely used in the preparation of nanoparticle derived products. A detailed investigation on the agglomeration of oxide nanoparticles in waste water streams revealed an astonishingly high stabilization of the particles against clearance (adsorption on the bacteria from the sludge) presumably through peptide adsorption (Figure 1).

Since cerium oxide is only one representative of the currently rapidly growing material class of nanoparticles, we have additionally compared stability measurements with titania, silica, iron oxide, zinc oxide and calcium phosphate nanoparticles to the here investigated cerium oxide.

This unexpected finding suggests a need to investigate nanoparticle clearance in more detail and demonstrates the complex interactions between dissolved species and the nanoparticles within the continuously changing environment of the clearing sludge

Figure 1. Scanning Transmission electrograph of ceria nanoparticles exposed to clearing sludge. In the Z-contrast images, the high density cerium oxide nanoparticles are bright.

Limbach, L.K., Bereiter,R., Müller, R., Krebs, R., Gälli, R., Stark, W.J., Removal of oxide nano-particles in a model waste-water treatment plant: Influence of agglomeration and surfactants on clearing efficiency (2008) Envrion. Sci. Technol, accepted.

Reijnders, L., Cleaner nanotechnology and hazard reduction of manufactured nanoparticles, (2006), J. Clean Prod. 14 (2), 124-133.

Wiesner, M. R., Lowry, G. V., Alvarez, P., Dionysiou, D., Biswas, P., Assessing the risks of manufactured nanomaterials (2006), Environ. Sci. Technol., 40 (14), 4336-4345.

Limbach, L.K, Li, Y., Grass, R.N., Brunner, T.J., Hintermann, M.A., Muller, M., Gunther, D., Stark, W. J, Oxide nanoparticle uptake in human lung fibroblasts: Effects of particle size, agglomeration, and diffusion at low concentrations, (2005) Env. Sci. Technol 39, 9370-9376.

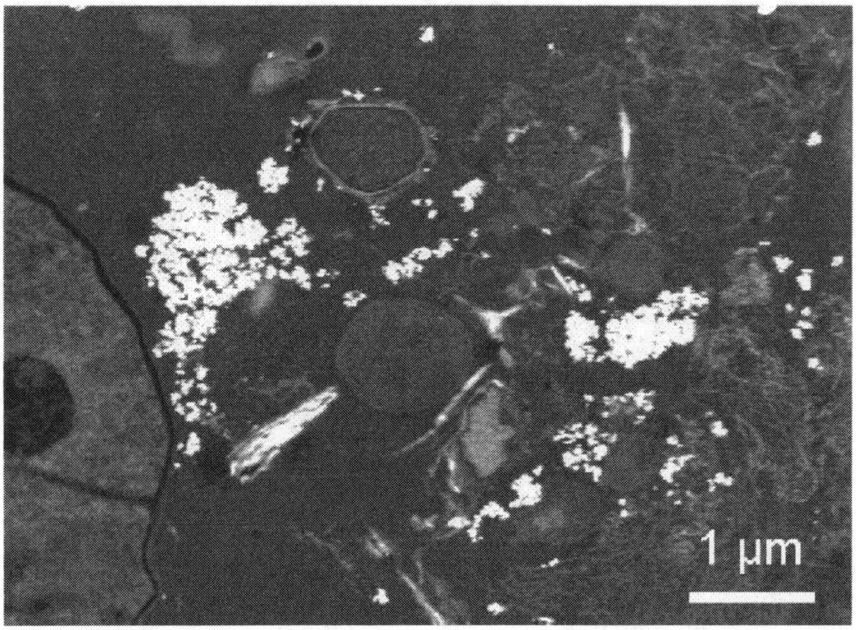

619d Aggregation of Fullerene (C_{60}) Nanoparticles in Monovalent and Divalent Electrolytes: Implications for Fate, Transport, and Bioavailability

Kai Loon Chen, Department of Geography and Environmental Engineering, Johns Hopkins University, Baltimore, MD 21218 and Menachem Elimelech, Department of Chemical Engineering, Yale University, Mason Lab, 313A, 9 Hillhouse Avenue, New Haven, CT 06520.

Buckminsterfullerene C_{60} has been attracting attention in recent years both for its potential as a nanomaterial as well as its toxicity towards certain microorganisms and human cells. In aquatic environments, fullerene C_{60} molecules self-assemble to form negatively charged nanoparticles. The objective of this study is to investigate the aggregation behavior of fullerene nanoparticles in the presence of monovalent (NaCl and KCl) and divalent ($CaCl_2$) electrolytes prevalent in natural aquatic systems. A comprehensive knowledge of the aggregation behavior of the nanoparticles will facilitate the prediction of their fate, transport, and bioavailability in natural aquatic systems. Furthermore, the aggregation state of these nanoparticles is expected to play a critical role in determining the degree of fullerene toxicity. In this study, the electrophoretic mobilities of fullerene nanoparticles synthesized through two methods are shown to become less negative when electrolyte concentrations are increased, indicative of charge screening effects. By conducting time-resolved dynamic light scattering measurements, the aggregation kinetics of both fullerene nanoparticles are found to increase with increasing salt concentrations. At high enough salt concentrations, the charges of the nanoparticles are completely screened and the nanoparticles undergo diffusion-limited aggregation. Even though the stabilities of both fullerene nanoparticles are very different, their Hamaker constants are experimentally derived and demonstrated to be similar, indicating their bulk material to be the same despite different synthesis methods. The influence of particle size and solution pH on the stability of fullerene nanoparticles will also be discussed.

624a Desalination Membranes Based on Directly Sulfonated Poly(arylene ether sulfone) Copolymers

Ho-Bum Park[1], Wei Xie[2], Joe Cook[2], James McGrath[3], and **Benny D. Freeman**[2]. (1) Chemical Engineering , Hanyang University, Seoul, 680-749, South Korea, (2) Chemical Engineering, The University of Texas at Austin, 10100 Burnet Rd., Bldg 133, Center for Energy and Environmental Resources, Austin, TX 78758, (3) Chemical Engineering, Virginia Polytechnic Institute and State University, Blacksburg, VA 24060

Reverse osmosis and nanofiltration membranes are susceptible to biofouling, in particular when treating municipal wastewater, and chlorine compounds are widely used to successfully control the bacteria that causing biofouling. However, chlorine-containing disinfectants chemically degrade current commercial RO and NF membranes, especially the polyamide (PA)-thin film composite membranes that dominate the current market. The low chlorine tolerance of conventional PA membranes leads to irreversible damage to membrane performance due to chlorine attack beginning at the amide nitrogen in PA membranes. This presentation presents the synthesis and characterization of a systematic series of highly chlorine-tolerant, sulfonated copolymer membranes, for use as desalination membranes, based on chemistry that is entirely different from the conventional post-polymerization sulfonation technique. Using direct copolymerization of sulfonated and other monomers, reproducible sulfonated copolymer membranes can be prepared as various polymer structures and compositions at different sulfonation levels. This synthesis method overcomes the problems of conventional sulfonation technology such as molecular weight reduction during sulfonation. This study will discuss the preparation and evaluation of several families of sulfonated copolymers such as random or segmented multiblock copolymers in terms of fundamental desalination characteristics (e.g., water permeability (or permeance), salt permeability, and salt rejection). In addition, thin-film composite membranes were prepared via brush coating of dilute sulfonated polymer solutions on commercial microporous polysulfone membrane. These sulfonated copolymers or their thin-film composite membranes exhibit high water permeability (or permeance) and good salt rejection. To delineate structure-property relations for these materials, solubility and diffusivity of water and salt (e.g., NaCl) were also evaluated for a series of sulfonated copolymers with various sulfonation levels in acid or salt forms. These intrinsic properties were compared with those of commonly used cellulose acetate and polyamide membranes in the literature. This fundamental and systematic study of structure-property relations regarding newly synthesized sulfonated copolymer membranes provides guidelines regarding material selection for new desalination membranes.

Energy Cost Optimization In Membrane Desalination and the Thermodynamic Restriction*[‡,†]

Aihua Zhu, Panagiotis D. Christofides and Yoram Cohen
Department of Chemical and Biomolecular Engineering
Water Technology Research Center
University of California, Los Angeles, CA 90095-1592, U.S.A.

September 21, 2008

Reverse osmosis (RO) membrane water desalination is now well established as a mature water desalination technology. However, there are intensive efforts to reduce the cost of RO water desalination in order to broaden the appeal and deployment of this technology. The water production cost in a typical RO desalination plant generally consists of the cost of energy consumption, equipment, membranes, labor and maintenance and financial charges. Energy consumption is a major portion of the total cost of water desalination[1-3] and can reach as high as about 58% of the total permeate production cost as shown in Fig. 1. The energy cost per volume of produced permeate (i.e., the Specific Energy Consumption or SEC) is significant in RO operation due to the high pressure requirement (up to about 1000 psi for seawater and in the range of 100-600 psi for brackish water desalting). Considerable effort has been devoted to find means for reducing the transmembrane pressure required for a given water permeate productivity level dating back to the initial days of RO development in the early 1960's.

*American Institute of Chemical Engineers (AIChE),
†Annual Meeting,
‡November 16-21, 2008,
§Loews Philadelphia Hotel,
¶Philadelphia Marriott Downtown,
‖Pennsylvania Convention Center,
**Philadelphia, Pennsylvania, USA,
††Paper 624b

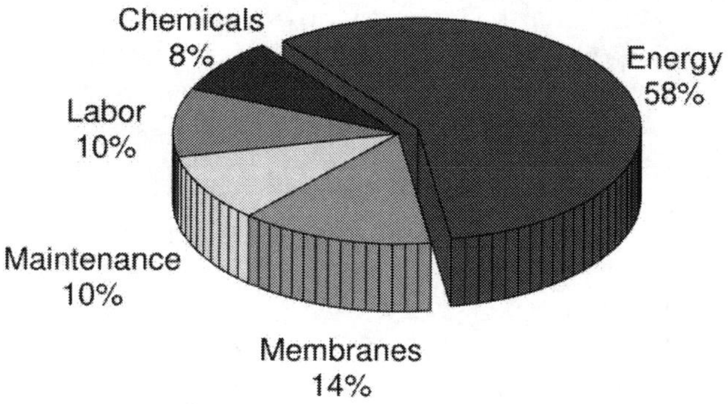

Figure 1: Annual operating cost distribution of a seawater reverse osmosis facility[4].

Early research in the 1960's[5–8] focused on unit cost optimization with respect to water recovery, energy recovery system efficiency, feed flow rate and the applied transmembrane pressure. Efforts to reduce the SEC also considered increasing the permeate flow rate, at a given applied pressure and feed flow rate, by either optimizing the membrane module with respect to its permeate flux[9–16] and/or by using more permeable membranes[17–20]. For example, studies have shown that specific permeate productivity of spiral wound RO and nanofiltration modules could be improved by optimizing module configuration (e.g., feed channel height, permeate channel height, and porosity)[13].

The introduction of highly permeable membranes in the mid 1990's with low salt passage[17] has generated considerable interest given their potential for reducing the pumping energy required to attain a given permeate[17–20]. Wilf[17] and later Spiegler[21] reported that operation close to the minimum level of applied pressure (i.e., pressure approaching the concentrate osmotic pressure plus frictional pressure losses), would result in the lowest energy cost. Clearly, in the absence of pressure drop in the membrane module, the minimum required applied pressure when a highly permeable membrane is used would be very close to the osmotic pressure of the RO concentrate that would be reached at the membrane outlet[17,22–24]. As illustrated in Fig. , in order to achieve a given water recovery and utilize the entire membrane area, there is a minimum pressure that must be applied and this pressure must be greater than the osmotic pressure of the concentrate exiting the process, but this applied pressure can approach the osmotic pressure of the brine stream when highly permeable membranes are used. It is noted, that the requirement of a minimum pressure, for the lowest energy cost, will apply even when one considers concentration polarization, albeit the required pressure will be based on the osmotic pressure at the membrane surface at module exit[22].

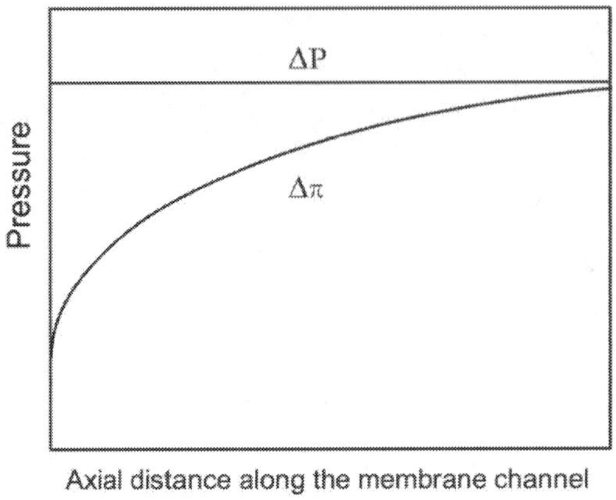

Figure 2: Schematic illustration of the thermodynamic restriction for cross-flow RO desalting[22].

In order to reduce energy consumption, energy recovery from the concentrate stream has been implemented using a variety of energy recovery devices (ERDs), in addition to optimization of the configurations of the RO membrane arrays. The effect of an energy recovery device (ERD) on the SEC was first studied in the early 1960's[6,7]. Avlonitis et al.[25] discussed four kinds of ERDs (i.e., Pelton wheel, Grundfos Pelton wheel, Turbo charger, Pressure exchanger) and reported that the pressure exchanger was the most efficient energy recovery device. More recently, Manth et al.[1] proposed an energy recovery approach, in which a booster pump is coupled with a Pelton turbine (instead of a single-component high-pressure feed pump), or is used as an interstage booster for dual-stage brine conversion systems.

Simplified process models to optimize the structure of RO membrane desalination plants have been proposed in the literature[26–33]. Early studies have shown that the "Christmas tree" configuration developed in the early 1970's was suitable for the early generation of RO spiral-wound membranes. However, with the emergence of higher permeability membranes, it is unclear if the above configuration of membrane modules is also optimal for ultra low pressure RO modules[26]. It has been argued that the SEC can be lowered by utilizing a large number of RO membrane units in parallel so as to keep the flow and operating pressure low[29]. It has also been claimed that the SEC decreases upon increasing the number of membrane elements in a vessel[3]. In the mid 1990's researchers have suggested that a single-stage RO process would be more energy efficient[34]. However, it has also been claimed that two-stage RO is more energy efficient than single-stage RO[29]. The above conflicting views suggest that there is a need to carefully compare the energy efficiency of RO desalination by appropriately comparing single and multiple-stage RO on the basis of appropriately normalized feed flow rate and SEC taking into consideration the feed osmotic pressure, membrane permeability and membrane area.

Optimization of RO water production cost with respect to capital cost has also been addressed in order to explore means of reducing the total specific cost of water production[29,34]. Such optimization studies have considered the costs associated with feed intake (primarily for seawater) and pretreatment, high pressure pumps, energy recovery system, and membrane replacement[34]. The problem of maximizing RO plant profit, considering energy cost, amortized membrane plant cost, cleaning and maintenance cost, and amortized cost of process pumps in the absence of energy recovery devices has also been addressed[29]. The majority of the existing studies have accepted the standard operating procedure whereby the applied pressures is set to be significantly higher than the minimum required pressure limit that would correspond to the lowest SEC. Moreover, a formal mathematical approach has not been presented to enable an unambiguous evaluation of the optimization of the RO water production cost with respect to the applied pressure, water recovery, pump efficiency, membrane cost and the use of energy devices.

It is important to recognize that previous studies that focused on optimization of the SEC have only evaluated the SEC dependence on water recovery at one or several normalized feed and permeate flow rates. Previous researchers have reported the minimum SEC for one or several flow rates or a range of product water recoveries[5–20]. However, the global minimum SEC has not been identified along with SEC optimization via a general theoretical framework. Motivated by the above considerations, the current study revisits the problem of RO energy cost optimization when highly permeable membranes are used, via a simple mathematical formalism, with respect to the applied pressure, water recovery, feed flow rate, and permeate flow rate and accounting explicitly for the limitation imposed by the minimum required applied pressure. Subsequently, the impact of using an energy recovery device, brine disposal cost, membrane hydraulic permeability and pressure drop within the membrane module are discussed for one-stage RO. Additionally, an analysis is presented of the energy efficiency of a two-stage RO relative to one-stage RO following the formalism proposed in the present study.

In previous work[35], we systematically studied the effect of the thermodynamic restriction (i.e., the fact that the applied pressure cannot be lower than the osmotic pressure of the exit brine stream plus pressure losses across the membrane module) on the optimization of the specific energy consumption of an RO process[35]. Specifically, we computed the optimum SEC, corresponding water recovery, and permeate flux for single-stage and two-stage RO membrane desalination systems. We also studied the effect of energy recovery device, membrane cost and brine disposal costs on SEC. The developed approach can also be utilized to evaluate the energy savings of a two-stage RO system over single-stage RO and the drawback of extra membrane area consumption of two-stage over single-stage. In the present work, we extend our previous results[35] to include the effect of membrane salt rejection on SEC and to study the energy consumption optimization of a two-pass membrane desalination process as shown in Fig. . The two-pass configuration is a relatively new configuration used in seawater desalination in which the permeate water from the first-pass goes through a second-pass. Previous work on energy consumption optimization of two-pass membrane desalination has addressed a number of issues; specifically, Noronha *et al.* first studied the specific energy consumption optimization of a two-pass (called "product-staging" in their work) membrane desalination process with recirculation pumps for each pass' retentate stream but without energy recovery devices[36]. Based on their study, they argued that the lower the water recovery in the first-pass, the lower the specific energy consumption of the two-pass system. Later on, Cardona *et al.* compared the energy consumption of a two-pass membrane desalination process (called "double-stage" in their work) without energy recovery devices to a single-stage RO process without an energy recovery device and reported that two-pass has a potential for energy savings on the order of 13-15% when the overall water recovery is less than 50% and the salt rejection is 98.3%[37]. Both papers did not address the effect of thermodynamic restriction on the computation of the optimal solution.

The wide application of low pressure membrane modules, owing to the development of high permeability RO membranes, has enabled the applied pressure in RO processes to approach the osmotic pressure limit. Therefore, it is now possible to optimize RO mem-

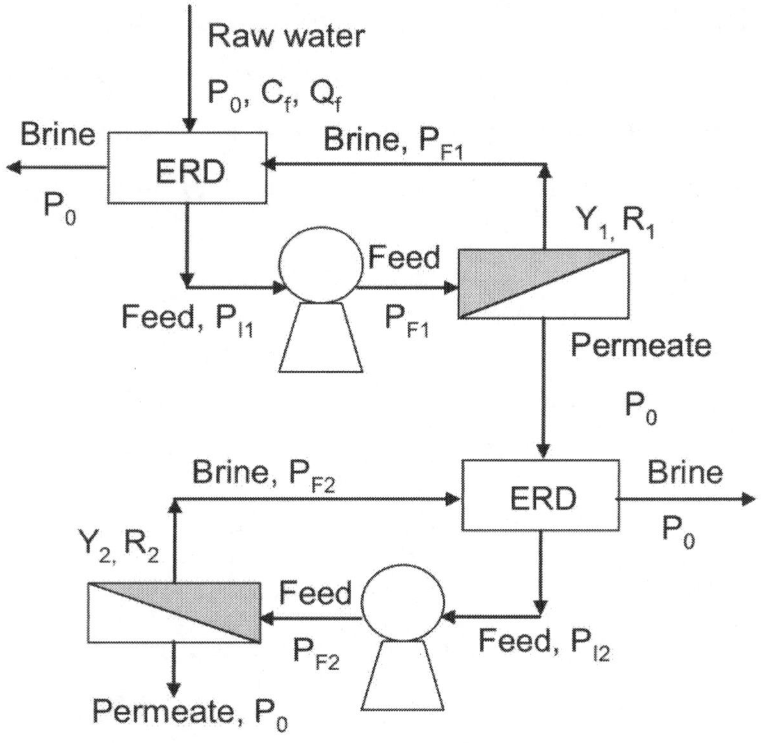

Figure 3: Schematic of a two-pass RO/NF process with ERDs.

brane processes with respect to product water recovery, with the goal of minimizing energy consumption, while considering constraints imposed by the thermodynamic cross-flow restriction and feed or permeate flow rate. In the present study, an approach to optimization of product water recovery in RO membrane desalination when highly permeable membranes are utilized was presented via a number of simple RO process models. The current results suggest that, it is indeed feasible to refine RO desalting so as to target the operation at the condition of minimum energy consumption, while considering the constraint imposed by the osmotic pressure as specified by the thermodynamic cross-flow restriction. The impact of energy recovery devices, membrane permeability, process configuration, brine management cost, pump efficiency, and frictional pressure drop can all be considered using the proposed approach as shown in a series of illustrations. Overall, as process costs above energy costs are added, the operational point for achieving minimum energy consumption shifts to higher recoveries. Although the newer generation of highly permeable RO membranes can allow high recovery operations, limitations due to mineral scaling and fouling impose additional constraints. The incorporation of these phenomena in an expanded optimization framework is the subject of ongoing research. To provide a clear picture to the problem of energy consumption optimization of two-pass reverse osmosis membrane desalination, the present work considers a systematic comparison of the SEC between single-pass and two-pass reverse osmosis systems and accounts for key practical issues like the effect of thermodynamic cross-flow restriction, energy recovery devices and concentration polarization.

References

[1] T. Manth, M. Gabor, and E. Oklejas. Minimizing RO energy consumption under variable conditions of operation. *Desalination*, 157(1-3):9–21, 2003.

[2] M. Busch and W.E. Mickols. Reducing energy consumption in seawater desalination. *Desalination*, 165:299–312, 2004.

[3] M. Wilf and C. Bartels. Optimization of seawater RO systems design. *Desalination*, 173(1):1–12, 2005.

[4] Srinivas (Vasu) Veerapaneni, Bruce Long, Scott Freeman, and Rick Bond. Reducing energy consumption for seawater desalination. *Journal AWWA*, 99:6:95, 2007.

[5] L. T. Fan, C. Y. Cheng, L. E. Erickson, and C. L. Hwang. The optimal design of desalination systems. *Desalination*, 3(2):225–236, 1967.

[6] L. T. Fan, C. Y. Cheng, L. Y. S. Ho, C. L. Hwang, and L. E. Erickson. Analysis and optimization of a reverse osmosis water purification system–part I. Process Analysis and Simulation. *Desalination*, 5(3):237–265, 1968.

[7] L. T. Fan, C. Y. Cheng, L. Y. S. Ho, C. L. Hwang, and L. E. Erickson. Analysis and optimization of a reverse osmosis water purification system–part II. Optimization. *Desalination*, 6(2):131–152, 1969.

[8] G. B. Hatfield and G. W. Graves. Optimization of a reverse osmosis system using nonlinear programming. *Desalination*, 7(2):147–177, 1970.

[9] R. Rautenbach and W. Dahm. Design and optimization of spiral-wound and hollow fiber RO-modules. *Desalination*, 65(1-3):259–275, 1987.

[10] S. Avlonitis, W. T. Hanbury, and M. B. Boudinar. Spiral wound modules performance. an analytical solution, part I. *Desalination*, 81(1-3):191–208, 1991.

[11] M. Ben Boudinar, W. T. Hanbury, and S. Avlonitis. Numerical simulation and optimisation of spiral-wound modules. *Desalination*, 86(3):273–290, 1992.

[12] S. Avlonitis, W. T. Hanbury, and M. B. Boudinar. Spiral wound modules performance an analytical solution: Part II. *Desalination*, 89(3):227–246, 1993.

[13] W. G. J. van der Meer and J. C. van Dijk. Theoretical optimization of spiral-wound and capillary nanofiltration modules. *Desalination*, 113(2-3):129–146, 1997.

[14] V. Geraldes, N. E. Pereira, and M. N. de Pinho. Simulation and optimization of medium-sized seawater reverse osmosis processes with spiral-wound modules. *Industrial & Engineering Chemistry Research*, 44(6):1897–1905, 2005.

[15] A. Villafafila and I. M. Mujtaba. Fresh water by reverse osmosis based desalination: simulation and optimisation. *Desalination*, 155(1):1–13, 2003.

[16] S.A. Avlonitis, M. Pappas, and K. Moutesidis. A unified model for the detailed investigation of membrane modules and RO plants performance. *Desalination*, 203(1-3):218–228, 2007.

[17] M. Wilf. Design consequences of recent improvements in membrane performance. *Desalination*, 113(2-3):157–163, 1997.

[18] J. E. Nemeth. Innovative system designs to optimize performance of ultra-low pressure reverse osmosis membranes. *Desalination*, 118(1-3):63–71, 1998.

[19] K. Fethi. Optimization of energy consumption in the 3300 m(3)/d RO kerkennah plant. *Desalination*, 157(1-3):145–149, 2003.

[20] J. Beca. Pharmaceutical discharge: Zero discharge for pharma plant. *Filtration & Separation*, 44(6):40–41, 2007.

[21] K. S. Spiegler and Y. M. El-Sayed. The energetics of desalination processes. *Desalination*, 134(1-3):109–128, 2001.

[22] L. Song, J. Y. Hu, S. L. Ong, W. J. Ng, M. Elimelech, and M. Wilf. Performance limitation of the full-scale reverse osmosis process. *Journal of Membrane Science*, 214(2):239–244, 2003.

[23] L. Song, J. Y. Hu, S. L. Ong, W. J. Ng, M. Elimelech, and M. Wilf. Emergence of thermodynamic restriction and its implications for full-scale reverse osmosis processes. *Desalination*, 155(3):213–228, 2003.

[24] L. Song and K. G. Tay. Performance prediction of a long crossflow reverse osmosis membrane channel. *Journal of Membrane Science*, 281(1-2):163–169, 2006.

[25] S.A. Avlonitis, K. Kouroumbas, and N. Vlachakis. Energy consumption and membrane replacement cost for seawater RO desalination plants. *Desalination*, 157(1-3):151–158, 2003.

[26] W. G. J. van der Meer, M. Riemersma, and J. C. van Dijk. Only two membrane modules per pressure vessel? hydraulic optimization of spiral-wound membrane filtration plants. *Desalination*, 119(1-3):57–64, 1998.

[27] W. G. J. van der Meer, C. W. A. Averink, and J. C. van Dijk. Mathematical model of nanofiltration systems. *Desalination*, 105(1-2):25–31, 1996.

[28] L. P. Wessels, W. G. J. Van der Meer, W. C. Van Paassen, and G. Vos. Innovative design of nano- and ultrafiltration plants. *Desalination*, 119(1-3):341–345, 1998.

[29] F. Maskan, D. E. Wiley, L. P. M. Johnston, and D. J. Clements. Optimal design of reverse osmosis module networks. *AIChE Journal*, 46(5):946–954, 2000.

[30] Y. Lu, Y. Hu, X. Zhang, L. Wu, and Q. Liu. Optimum design of reverse osmosis system under different feed concentration and product specification. *Journal of Membrane Science*, 287(2):219–229, 2007.

[31] N. G. Voros, Z. B. Maroulis, and D. Marinos-Kouris. Short-cut structural design of reverse osmosis desalination plants. *Journal of Membrane Science*, 127(1):47–68, 1997.

[32] N. Voros, Z. B. Maroulis, and D. Marinos-Kouris. Optimization of reverse osmosis networks for seawater desalination. *Computers & Chemical Engineering*, 20(Supplement 1):S345–S350, 1996.

[33] H. J. See, V. S. Vassiliadis, and D. I. Wilson. Optimisation of membrane regeneration scheduling in reverse osmosis networks for seawater desalination. *Desalination*, 125(1-3):37–54, 1999.

[34] A. Malek, M. N. A. Hawlader, and J. C. Ho. Design and economics of RO seawater desalination. *Desalination*, 105(3):245–261, 1996.

[35] A. Zhu, P. D. Christofides, and Y. Cohen. Effect of thermodynamic restriction on energy cost optimization of RO membrane water desalination. *Ind. Eng. Chem. Res.*, 2008.

[36] M. Noronha, V. Mavrov, and H. Chmiel. Simulation model for optimisation of two-stage membrane filtration plants–minimising the specific costs of power consumption. *Journal of Membrane Science*, 202(1-2):217–232, 2002.

[37] E. Cardona, A. Piacentino, and F. Marchese. Energy saving in two-stage reverse osmosis systems coupled with ultrafiltration processes. *Desalination*, 184(1-3):125–137, 2005.

UTILIZATION OF BLOCK COPOLYMERS AS ULTRAFILTRATION MEMBRANES.

Background. Block copolymers offer a potential route to superior ultrafiltration membranes. Such membranes are made of a polymer with two chemically distinct segments which self-assemble into a wide variety of structures[1]. Such nanometer-sized structures can include hexagonally packed cylinders of one component oriented perpendicularly to the surface of the membrane, if properly aligned[2]. When these cylinders are removed by selective etching, the resulting pores are closely packed and monodisperse.

Such membranes have shown promise for ultrafiltration[3,4]. These monodispersed pores should give high selectivity, and their close packing should allow high fluxes. The membranes' flat surface suggests that fouling will be reduced. However, even starting to evaluate these possible advantages means that the cylinders in the self-assembled structure must be carefully oriented perpendicular to the membrane surface. In this research, we avoid the formidable problem of alignment by using a bicontinuous network structure. Where a "doubly reactive" block copolymer polynorbornenylethyl styrene-polylactide is used as a structural template during the crosslinking of dicyclopentadiene[5]. After crosslinking and subsequent etching with dilute base, the dicyclopentadiene membrane which remains has a nanoporous structure. Such a membrane does have nearly monodisperse pores and so should give high selectivity. Additionally, the membrane's pores are closely packed and so should be capable of high flux.

This research explores how much of the promise of these membranes is real. We explore the details of the geometry by measuring the diffusion of a variety of gases across these membranes. We measure the degree of etching and wetting by determining the flow of water at various values of pH and ionic strength. We demonstrate membrane selectivity by studying transport of a series of polyethylene glycols as well as mixtures of dextrans. The research shows that the promise of these membranes is real, but it also identifies additional barriers which need to be overcome to realize practical value.

Results. The experiments in this work are based on membranes 100 microns thick. For gas diffusion experiments they are mounted in a diaphragm cell[6]. This cell consists of two compartments separated by a membrane. The top, "donating" compartment initially contains gases at high pressure which diffuses across the membrane into the bottom, "receiving" compartment. The change in pressure is recorded a function of time, the results for a variety of gases are shown in Figure 1. The slope of the curves is used to calculate the effective diffusion coefficient for each gas. Table I presents the selectivities for the gases relative to helium diffusion. Helium is chosen as the reference gas because it is not expected to interact with the solid membrane (Yang book). The experimental selectivities are within an average of 5 percent of those predicted demonstrating they have molecular weight dependence consistent with Knudsen diffusion, i.e. $\sqrt{1/\tilde{m}}$[7].

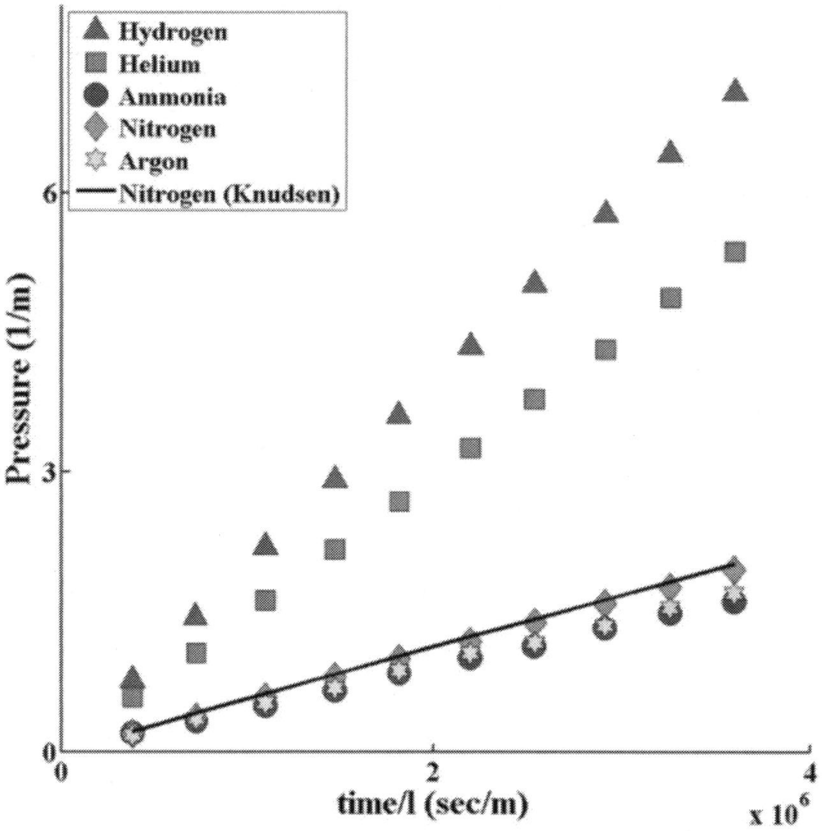

Figure 1: Gases Diffuse Across the Membrane by a Knudsen Mechanism. The slope of the experimental data is proportional to the gas permeability. Knudsen theory, shown for N_2 as a solid line, predicts a $\sqrt{\frac{1}{M}}$ dependence.

Table 1: Summary of Gas Diffusion Data.

Solute	D_{eff}	α_{exp}	α_{Kn}
H_2	0.0184	0.84	0.71
He	0.0154	1.00	1.00
NH_3	0.0043	3.58	2.06
N_2	0.0055	2.80	2.66
Ar	0.0047	3.27	3.15

Liquid flow measurements were made using an Amicon stirred cell set up. A pressure head was applied using compressed nitrogen and the mass of permeating solution recorded as a function of time. The slope of these plots were used to calculate the flux through the membrane at a given pressure drop. The flux, in m^3 per m^2 second, is plotted in Figure 2 *vs* the pressure drop, in kPa. A linear relation as expected from the Hagen-Poiseuille relationship is observed. The fluxes are small, around $1 \cdot 10^{-6}$ m^3/m^2 sec (4 gal/ft^2 day) at a pressure drop of 30 kPa. However, if the membrane thicknesses were decreased from 100 μm to 0.1 μm and the pressure drop was increased to 200 kPa, the flux would be $1.4 \cdot 10^{-2}$ m^3/m^2 sec (28,000 gal/ft^2 day). These experiments do not show these high fluxes, though they do suggest the potential of these membranes.

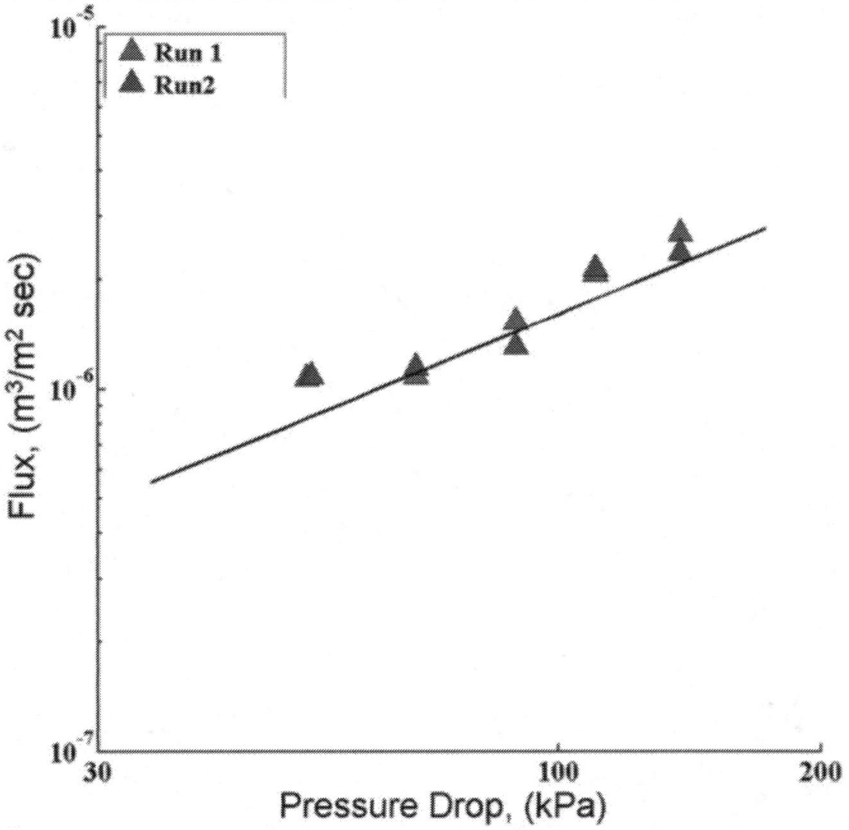

Figure 5: Water Flux is Proportional to Pressure Drop. Experimentally observed fluxes are compared to a theoretical prediction based on a pore diameter of 14.2 nm and a tortuosity of 1.56.

Hexagonally packed right cylinders would give a higher flux but they are difficult to align. Because the bicontinuous structure used here is isotropic, it requires no special alignment. The price paid is the lower flux and the less well-defined pore structure. These results of the gas diffusion and liquid flow show that the membranes have reproducible pores, which span the membrane thickness.

The ultrafiltration experiments begin to explore the practical value of these membranes. A series of experiments were made with a mixed feed of dextrans. These experiments, performed at GE Osmonics, imitate quality control studies carried out industrially[8]. In experiments like these, all solutes are fed simultaneously, and the flux of each out of the membrane is determined using SEC with refractive index. The ratio of the appropriate peak areas from the RI detector output is used to calculate the rejection curves shown in Figure 3.

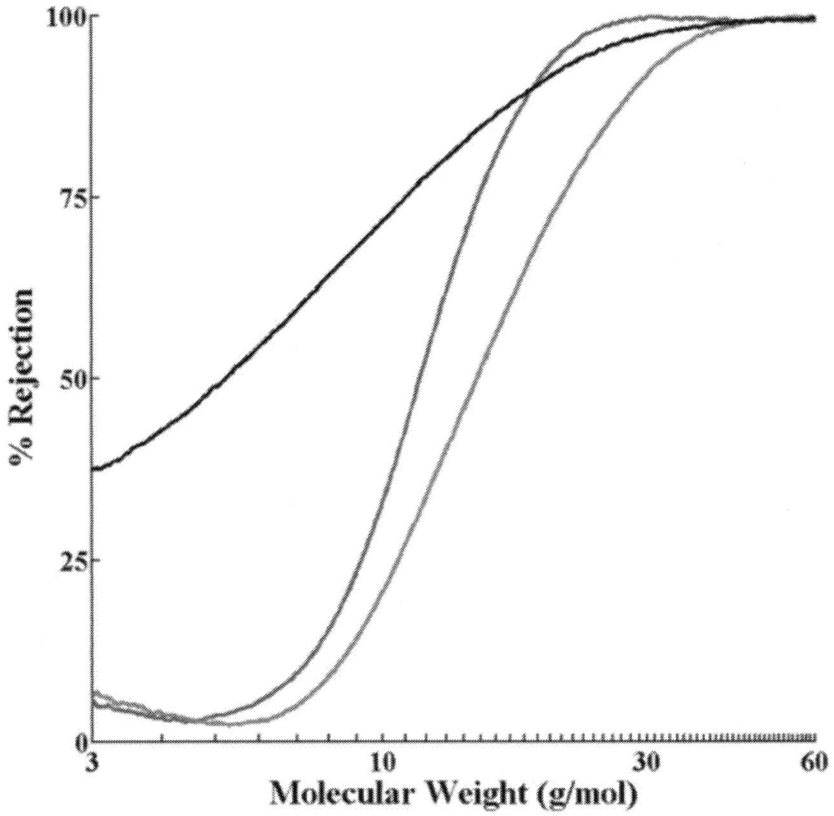

Figure 3. PDCPD Membranes Show a Sharper MWCO Than a Membrane Made by Phase Inversion. The green (32 kDa) and red (65 kDa) lines are rejection

curves for the membranes developed in this work. The black line is the rejection curve for a representative phase inversion membrane.

We would like these curves to be step functions with no rejection until a critical molecular weight and then complete rejection above this critical weight. However, this is inconsistent with the underlying theory of Brenner and Bungay[9], where the rejection is gentler because of altered free volume and drag of the solute within the pores. Even though step functions are not possible the results in Figure 3 are encouraging. The data for two membranes based on etched, self-assembled, block copolymer-based membranes show a sharper rejection than those of the commercial membrane. Moreover, the data for a membrane with a 32 kD chain reassuringly show a cut-off at lower molecular weight for a 65 kD chain. This suggests that block copolymer membrane properties can be further tuned to perform a specific desired separation. These membranes show promise as ultrafilters. The sharp rejection curves demonstrate the membranes are able to perform separations better than current technologies. The flux data suggest that if thin films can be made these membranes will be comparable to current membranes. The final area that must be explored is the tendency of theses membranes to foul during operation. These studies have begun and seem promising.

Reference.

(1) F. Bates and G. Fredrickson, Block copolymer thermodynamics: theory and experiment, *Ann. Rev. Phys. Chem.* **41** (1990), pp. 525–557.
(2) A.S. Zalusky, R. Alayo-Valles, J.H. Wolf and M.A. Hillmyer, Ordered nanoporous polymers from polystyrene–polylactide block copolymers, *J. Am. Chem. Soc.* **124** (2002), pp. 12761–12767.
(3) Phillip, W.A.; Rzayev, J.; Hillmyer, M. A.; Cussler, E. L., Gas and water liquid transport through nanoporous block copolymer membranes, *J. Membrane Sci.* **2006**, *286*, 144–152.
(3) S.Y. Yang, I. Ryu, H.Y. Kim, J.K. Kim, S.K. Jang and T.P. Russell, Nanoporous membranes with ultrahigh selectivity and flux for the filtration of viruses, *Adv. Mater.* **18** (2006), pp. 709–712.

(4) Chen, L.; Phillip, W. A.; Cussler, E. L.; Hillmyer, M. A., Robust Nanoporous Membranes Templated by a Doubly Reactive Block Copolymer, *J. Am. Chem. Soc.* **2007**, *129*, 13786–13787.
(5) C.F. Yang and E.L. Cussler, Oxygen barriers that use free radical chemistry, *AIChE J.* **47** (2001), pp. 2725–2732.
(6) E.L. Cussler, Diffusion (2nd ed.), Cambridge University Press, Cambridge (2002).
(7) ASTM Standard E 1343 – 97, Standard Test Method for Molecular Weight Cutoff Evaluation of Flat Sheet Ultrafiltration Membranes.
(8) Bungay, P. M., and H. Brenner, "The Motion of a Closely Fitting Sphere in a Fluid-Filled Tube," *Int.* J. *Multiph. Flow*, **25** (1973).

624d Kinetics of Mineral Precipitation and Organics Removal In Brackish Water

Brian C. McCool, Anditya Rahardianto, and Yoram Cohen. Chemical & Biomolecular Engineering, University of California, Los Angeles, 5531 Boelter Hall, 420 Westwood Plaza, Los Angeles, CA 90095-1592

Natural waters often contain natural organic matter (NOM). In many water treatment processes synthetic organic polymer additives are often used to enhance system performance (e.g. coagulants, flocculants, and dispersants). However, it is often desirable to reduce these organic concentrations prior to use. Recent studies of high recovery brackish water desalination suggested that chemical demineralization (an inter-stage process integrated with RO) has the added benefit of reducing NOM concentrations by up to 30%. The removal of organics from water is believed to occur via adsorption during precipitation of mineral salts during enhanced softening via high alkaline dosing. Mineral salt precipitation has been shown to be accompanied by the removal of organics primarily via adsorption. The interactions between organics and crystal surfaces during precipitation processes control the organic removal efficiency and these interactions merit a fundamental study of the removal process and its implication for attaining high product water recovery in RO desalting of brackish water of high scaling propensity.

In order to better understand the phenomenon of organic removal during mineral salt precipitation of RO concentrate, the removal of organics during mineral salt precipitation was studied in a model brackish water system. Laboratory precipitation studies were carried out in batch reactors with on-line monitoring of pH and mineral salt ion concentrations via ion-selective electrodes. Grab samples were taken periodically for DOC analysis, crystal size distribution measurements and imaging of mineral crystals. The results of the study served as a basis for assessing the integration of mineral precipitation with RO desalting so as to attain high recovery and thereby minimize the volume of generated concentrate.

The present results demonstrate that the concentration of organics can be significantly reduced by seeded mineral crystallization. The interaction between the organic and mineral salt precipitation was quantified via the effective rate constants for mineral salt crystallization, while varying the organic concentrations. The results indicate that organic removal during mineral salt precipitation is affected by the evolution of mineral salt crystal size distribution. In addition, the removal of mineral salt precursors was found to be sufficiently high to allow operation of secondary RO desalting while eliminating the problem of membrane scaling.

Title

Passive Vacuum Solar Flash Desalination

Authors

1. Mohammad Abutayeh, Department of Chemical Engineering, University of South Florida, Tampa, FL 33620, USA

2. Yogi Goswami, Clean Energy Research Center, University of South Florida, Tampa, FL 33620, USA

Correspondence

Correspondence concerning this article should be addressed to Mohammad Abutayeh at abutayeh@mail.usf.edu

Citation

M. Abutayeh and Y. Goswami, *Passive Vacuum Solar Flash Desalination*, AIChE National Meeting, Philadelphia, PA, November 2008.

Abstract

A model depicting a sustainable desalination process has been developed. The simulated process consists of pumping seawater through a solar heater before flashing it under vacuum in an elevated chamber. The vacuum enhances evaporation and is maintained by the balance between the hydrostatic pressure inside the elevated flash chamber and the outdoor atmospheric pressure. The developed model employs theoretical thermodynamic relations to describe the process setting it apart from previous empirical correlations.

Area

Environmental Division (09)

Session

Water Desalination and Purification (09016)

Keywords

Desalination, Flash Desalination, Solar Desalination

Introduction

The fresh water demand is persistently increasing as populations around the world keep growing and as existing fresh water reserves keep declining due to consumption and pollution.

Marine waters represent an infinite water source since almost 98% of global water is present in oceans; therefore, desalination is the logical approach to meet the rising fresh water demand.

The energy demand is continually increasing due to the relentless global industrialization. Oil and gas remain the principal source of energy for most of the world; however, their reserves are dwindling, their production is peaking, and their consumption is harming the environment. Serious economic and social disruptions are unfolding over those finite energy resources; hence, employing renewable energy resources will help avoid catastrophic conflicts, continue modern lifestyles, and preserve an increasingly warming and polluting environment. Solar radiation is a very appealing source of energy because it is available at no cost; furthermore, exploiting it has no notable adverse effect on the environment. Plenty of research and development have been undertaken to better utilize this free form of energy to develop more efficient sustainable processes such as water desalination and power generation [1].

The objective of this study is to simulate a solar flash desalination process under a hydrostatically sustained vacuum and analyze its controlling variables. Good quality simulation can be achieved with a rigorous model built to depict the proposed unit by employing the fundamental physical and thermodynamic relationships to describe the process. The model will be complimented by reliable empirical correlations to estimate the physical properties of the involved species and the operational parameters of the proposed system needed to perform accurate mass and energy balances. This approach is scientifically more valid than exclusive empirical methods and will increase the applicability of the developed model.

Background

Low-grade solar heat desalination under passive vacuum was devised and simulated theoretically and experimentally by Goswami and Kharabsheh [2]. A simplified process flow diagram of their desalination process is shown in Figure 1.

Atmospheric pressure moves cool seawater from a ground level tank into an elevated vacuumed evaporator through an injection pipe where water starts to evaporate due to solar or other low grade heat supplied to the chamber. The concentrated brine is then withdrawn through a withdrawal pipe annulus to the injection pipe to recover heat while the vapor moves towards a condenser due to a vapor pressure gradient through a finned pipe. Vapor then condenses by losing its latent heat of condensation to the ambient and flows down to a ground level tank due to gravity. The vacuum is naturally maintained by the hydrostatic balance between all of the connected vessels and the atmospheric pressure.

The experimental results of Goswami and Kharabsheh matched well with the predictions of their developed model [3]; however, the model was largely empirical thus limiting its applicability. In addition, the size of the evaporator of the desalination unit had to be very large to produce a practical output.

The above desalination system was later modified by Goswami and Maroo [4] to overcome the big size and the large level fluctuations of the evaporator. A simplified process flow diagram of the modified desalination process is shown in Figure 2.

Seawater is pumped through a condenser to preheat it before it enters a solar heater and then flashes into a vacuumed evaporator through an expansion orifice producing water vapor and concentrated brine. The flashed vapor then moves towards the condenser due to a vapor pressure gradient and condenses by losing its latent heat of condensation to the entering seawater in the condenser. The condensate and the concentrated brine flow down to ground level tanks by gravity while the vacuum is naturally sustained by the hydrostatic balance between all of the connected vessels and the atmospheric pressure.

Single and double stage configurations were simulated; however, the model was largely empirical which limited its applicability. In particular, no fundamental thermodynamic relations were used to characterize the flashing process which is the most fundamental operation of the entire desalination system.

Schematics

Reducing the pressure above the surface of liquids enhances their evaporation. This phenomenon can be integrated into a continuous desalination process by flashing seawater in a vacuumed chamber to produce water vapor that is condensed producing fresh water. Gravity can be used to balance the hydrostatic pressure inside the elevated flash chamber with the outdoor atmospheric pressure to maintain that vacuum, while low grade solar radiation can be used to heat seawater before flashing.

The proposed passive vacuum solar flash desalination system consists of a saline water tank, a concentrated brine tank, and a fresh water tank placed on ground level plus an evaporator and a condenser located at least 10 meters above ground, as shown in Figure 3. The ground level tanks are open to the atmosphere, while the evaporator-condenser assembly, or flash chamber, is insulated and sealed to retain both heat and vacuum.

The process begins by pumping the flash chamber with seawater while its top valves are open and its bottom ones are closed until it is completely full with water and free of air. The valve positions of the flash chamber are then switched around to let the water drop under gravity creating the desired vacuum without a vacuum pump.

In a continuous operation, cool saline water is pumped through a condenser to preheat it before it enters a solar heater and then flashes into a vacuumed evaporator through an expansion orifice or a pressure-reducing valve producing water vapor and concentrated brine. The water vapor then moves towards the condenser due to a vapor pressure gradient and condenses by losing its latent heat of condensation to the entering saline water. The condensate and the concentrated brine flow down to a fresh water and a brine water tanks respectively due to gravity through discharge pipes. Each of the fresh water and the brine water tanks have a discharge pipe located a few centimeters above the level of the inlet water pipes keeping their levels constant to hydrostatically maintain the vacuum in the flash chamber as well as retrieving the fresh water product and rejecting the concentrating brine.

Multi-stage solar flash desalination can be implemented by flashing saline water in sequentially lower pressure flash chambers, as shown in Figure 4. Applying the multi-stage strategy to solar flash desalination will result in more evaporation and better recovery of the heat of condensation resulting in more fresh water output.

Model

The following model is built to depict the continuous desalination process outlined above in Figure 3 to analyze the controlling variables of the fresh water output of the proposed desalination unit. The model will assume a semi steady state operation accounting for the build up of non-condensable gases in the flash chamber.

Salt balances around the transitional process equipment are given by

$$\phi_S \cdot M_S = \phi_P \cdot M_P \tag{1}$$
$$\phi_P \cdot M_P = \phi_X \cdot M_X \tag{2}$$
$$\phi_X \cdot M_X = \phi_H \cdot M_H \tag{3}$$
$$\phi_H \cdot M_H = \phi_W \cdot M_W \tag{4}$$

Overall energy balances around the transitional process equipment are given by

$$Q_P - W_P + E_P^{in} - E_P^{out} = 0 \tag{5}$$
$$Q_C - W_C + E_C^{in} - E_C^{out} = E_C^{a} \tag{6}$$
$$Q_H - W_H + E_H^{in} - E_H^{out} = 0 \tag{7}$$
$$Q_E - W_E + E_E^{in} - E_E^{out} = 0 \tag{8}$$

Energy inputs to the transitional process equipment are given by

$$E_P^{in} = M_S \cdot H_S \tag{9}$$
$$E_C^{in} = M_P \cdot H_P + M_E \cdot \left(H_E + H_E^L\right) \tag{10}$$
$$E_H^{in} = M_X \cdot H_X \tag{11}$$
$$E_E^{in} = M_H \cdot H_H \tag{12}$$

Energy outputs from the transitional process equipment are given by

$$E_P^{out} = M_P \cdot H_P \tag{13}$$
$$E_C^{out} = M_X \cdot H_X + M_C \cdot H_C \tag{14}$$
$$E_H^{out} = M_H \cdot H_H \tag{15}$$
$$E_E^{out} = M_E \cdot \left(H_E + H_E^L\right) + M_W \cdot H_W \tag{16}$$

Energy accumulation in the condenser is given by

$$E_C^{a} = M_C^{a} \cdot H_C^{a} \tag{17}$$

Adding heat to seawater upon flashing is accomplished by a solar heater. The present model assumes the heater to be a single-glazed flat-plate solar collector directly heating seawater flowing through its absorbing tubes. Solar insolation is geographically referenced and continually varying due to dynamic solar angles. In addition, solar insolation incident on the collector varies with plate geometry, sky clearness, ground reflectivity, and other factors. Average values for a generic single-glazed flat-plate solar collector will be used to simplify the comparison between the different simulation scenarios. The solar insolation area needed to meet the required heating load can be found using the Hottel-Whillier-Bliss equation [5]:

$$A_{SC} = \frac{Q_H}{F_{SC} \cdot \left[\tau_{SC} \cdot \alpha_{SC} \cdot I_{SC} - U_{SC} \cdot (T_X - T_S)\right]} \tag{18}$$

Seawater is a solution of many salts and contains a small amount of dissolved gases. To simplify calculations, seawater salt will be treated as one substance with Nitrogen, Oxygen, Argon, and Carbon Dioxide making up the dissolved gases.

The average molecular weight of seawater salt can be estimated by considering its major components [6] as follows

$$\frac{1}{MW_{Salt}} = \frac{\omega_{Cl}}{MW_{Cl}} + \frac{\omega_{Na}}{MW_{Na}} + \frac{\omega_{SO_4}}{MW_{SO_4}} + \frac{\omega_{Mg}}{MW_{Mg}} + \frac{\omega_{Ca}}{MW_{Ca}} + \frac{\omega_K}{MW_K} + \frac{\omega_{HCO_3}}{MW_{HCO_3}} + \frac{\omega_{Br}}{MW_{Br}} + \frac{\omega_{BO_3}}{MW_{BO_3}} + \frac{\omega_{Sr}}{MW_{Sr}} + \frac{\omega_F}{MW_F} \tag{19}$$

The average molecular weight of streams around the flash chamber can be estimated by considering their major components as follows

$$MW_H = z_{N_2} \cdot MW_{N_2} + z_{O_2} \cdot MW_{O_2} + z_{Ar} \cdot MW_{Ar} + z_{CO_2} \cdot MW_{CO_2} + z_{Salt} \cdot MW_{Salt} + z_{H_2O} \cdot MW_{H_2O} \tag{20}$$

$$MW_E = y_{N_2} \cdot MW_{N_2} + y_{O_2} \cdot MW_{O_2} + y_{Ar} \cdot MW_{Ar} + y_{CO_2} \cdot MW_{CO_2} + y_{H_2O} \cdot MW_{H_2O} \tag{21}$$

$$MW_W = x_{N_2} \cdot MW_{N_2} + x_{O_2} \cdot MW_{O_2} + x_{Ar} \cdot MW_{Ar} + x_{CO_2} \cdot MW_{CO_2} + x_{Salt} \cdot MW_{Salt} + x_{H_2O} \cdot MW_{H_2O} \tag{22}$$

$$MW_C = v_{N_2} \cdot MW_{N_2} + v_{O_2} \cdot MW_{O_2} + v_{Ar} \cdot MW_{Ar} + v_{CO_2} \cdot MW_{CO_2} + v_{H_2O} \cdot MW_{H_2O} \tag{23}$$

The gas and salt content of process streams is assumed constant prior to flashing

$$MW_S = MW_P = MW_X = MW_H \tag{24}$$

$$\phi_S = \phi_P = \phi_X = \phi_H \tag{25}$$

The distribution of non-condensable gases between the flashed vapor, the concentrated brine, and the condensed water in the flash chamber can be estimated by assuming equilibrium between the three phases. Salt is considered non-volatile and therefore is not present in the flashed vapor or condensed water streams. Henry's constants for the non-condensable gases and the saturated pressure of water are needed to describe this assumed equilibrium.

Henry's constants for the non-condensable gases are given by [7]

$$HC_i = HC_i^o \cdot exp\left[- HF_i \cdot \left(\frac{1}{T + 273.15} - \frac{1}{298.15} \right) \right] \tag{26}$$

The saturated pressure of water is given by [8]

$$P_{H_2O}^{sat} = exp\left[PA - \left(\frac{PB}{T + PC}\right)\right] \tag{27}$$

The vapor liquid equilibrium distribution coefficient of species i is defied as $K_i = y_i / x_i = P_i^{sat} / P$, and it is widely used to determine the distribution of chemicals between phases in equilibrium [9]. Vapor liquid equilibrium distribution coefficients are also known as partition coefficients in literature or simply as K-values. The K-value of salt is zero due to its non-volatility and the K-values of the non-condensable gases can be approximated as follows

$$K_i = \frac{y_i}{x_i} = \frac{\frac{P_i}{P_V}}{x_i} = \frac{\frac{HC_i \cdot x_i}{P_V}}{x_i} = \frac{HC_i}{P_V} \tag{28}$$

The K-value of water is given by

$$K_{H_2O} = \frac{P_{H_2O}^{sat}}{P_V} \tag{29}$$

The vapor liquid equilibrium distribution coefficients were obtained using SUPERTRAPP™, a computer code distributed by National Institute of Standards and Technology that calculates the thermodynamic properties of mixtures based on the Peng-Robinson equation of state. It was utilized to perform isobaric phase equilibria flash calculations at various temperatures to produce vapor liquid equilibrium distribution coefficient data. Least squares regression was then used to fit the produced data to the above equilibrium equations by manipulating the values of $HC°_i$, HF_i, PA, PB, and PC. SUPERTRAPP™ simulations are fresh water based and no salts are included in its flash calculations. To adjust the calculations for saline water, the K-values are multiplied by a correction factor which can be defined as α_i = *solubility in fresh water / solubility in seawater* for solutes and α_{H2O} = *seawater saturated pressure / fresh water saturated pressure* for water. The correction factor is a single constant obtained by averaging literature solubility data given over the operating temperature range to simplify calculations [10]. Table 1 summarizes the generated equilibrium parameters used in this model.

Two sets of K-values will be used in this model. One set is evaluated for the evaporator at a specified temperature and will be used in flash calculations. The other set is designated for the condenser and is used to solve for the temperature of the condensed water needed to establish equilibrium with the vapor. Flash calculations can now be carried out using the Rachford-Rice equation [11].

The molar composition of the stream entering the flash chamber is needed to proceed with the flash calculations and can be calculated from the average composition of seawater reported in mass basis [6] as follows

$$z_i = \frac{\frac{\phi_i}{MW_i}}{\sum_{All} \frac{\phi}{MW}} \tag{30}$$

The molar composition of the concentrated brine is given by

$$x_i = \frac{z_i \cdot N_H}{N_W + N_E \cdot \alpha_i \cdot K_i^E} \tag{31}$$

Similarly, the molar composition of the flashed vapor is given by

$$y_i = x_i \cdot \alpha_i \cdot K_i^E \tag{32}$$

The molar composition of the condensed water in the condenser can be determined by

$$v_i = \frac{y_i}{K_i^C} \tag{33}$$

The operating pressure inside the flash chamber has to be between the dew point and the bubble point to carry out a successful flash separation. The dew point and the bubble point pressures are estimated by

$$\frac{P_{BP}}{P_V} = \sum z_i \cdot \alpha_i \cdot K_i^E \tag{34}$$

$$\frac{P_V}{P_{DP}} = \sum \frac{z_i}{\alpha_i \cdot K_i^E} \tag{35}$$

The low vacuum pressure marginalizes the effect of molecular size and the fairly high flash temperature weakens the relative importance of intermolecular attractions. As a consequence, the ideal gas law becomes a very suitable equation of state to express the rising vacuum pressure inside the flash chamber as follows

$$P_V = \frac{P_V^i \cdot V_V^i + R \cdot (T_E + 273.15) \cdot \int (N_C^a) d\theta}{V_V} \tag{36}$$

The initial vacuum pressure will be assumed equal to the vapor pressure of water at the flash temperature and will be evaluated using the above given saturated pressure relation. The initial vacuum volume can be determined from the geometry of the system and the initial tank levels. The hydrostatic balance relations between the pressure inside the flash chamber and the outdoor atmospheric pressure determine the dynamic tank levels allowing for the calculation of the dynamic vacuum volume. The hydrostatic balance relations are written using Bernoulli's fluid equation [12] and correlations for stream pressure drop [8] are included as well. The integral part can be numerically evaluated since time is an input variable.

Summation of fractions are included to complete the flash calculations

$$\sum \phi_i = \sum z_i = \sum x_i = \sum y_i = \sum v_i = 1 \tag{37}$$

The vapor pressure of seawater is 1.84% lower than that of pure water at the same temperature due to the non-volatile salts and therefore the boiling point of seawater is slightly higher than that of fresh water. This phenomenon is known as the boiling point elevation or the vapor pressure depression. Boiling point elevation is a function of salinity and does not depend on the properties of the solute or the solvent [13].

The temperature of seawater drops upon entering the flash chamber to attain equilibrium; however, equilibrium is not always fully achieved. This phenomenon is known as the non equilibrium allowance and it depends on several factors such as flash temperature, flow rates, concentrated brine depth, and chamber geometry.

Correlations for boiling point elevation [14] and non equilibrium allowance [15] are incorporated into the model as follows

$$T_W = T_E + BPE + NEA \tag{38}$$

$$BPE = 8.77 \cdot \phi_H \cdot \left(95.76 \cdot 10^{-2} + 81.89 \cdot 10^{-4} \cdot T_E + 16.47 \cdot 10^{-6} \cdot T_E^2\right) - 0.05 \tag{39}$$

$$NEA = \frac{33 \cdot (T_H - T_E)^{0.55}}{T_E - BPE} \tag{40}$$

A semi steady state operation is assumed to account for the accumulation of non condensable gases in the flash chamber. This is accomplished by including the accumulation rate in the total mole balance around the flash chamber

$$N_H = N_E + N_W \tag{41}$$

$$N_E = N_C + N_C^a \tag{42}$$

$$M_E = M_C + M_C^a \tag{43}$$

Assuming total condensation of the flashed water vapor

$$v_{H_2O} \cdot N_C = y_{H_2O} \cdot N_E \tag{44}$$

The average molecular weights of process streams are used to relate mass and molar flow rates that were interchangeably used throughout this model and can be estimated by considering their major components. Mass and molar flow rates are related by the following relations

$$M_S = N_S \cdot MW_S \tag{45}$$

$$M_P = N_P \cdot MW_P \tag{46}$$

$$M_X = N_X \cdot MW_X \tag{47}$$

$$M_H = N_H \cdot MW_H \tag{48}$$
$$M_E = N_E \cdot MW_E \tag{49}$$
$$M_W = N_W \cdot MW_W \tag{50}$$
$$M_C = N_C \cdot MW_C \tag{51}$$
$$M_C^a = N_C^a \cdot MW_C^a \tag{52}$$

Correlations for enthalpy [16], density [17], and viscosity [18], are included in the model to estimate stream properties as functions of temperature and salinity.

The model is now complete and a corresponding computer code will be used to carry out different simulations of the process using several inputs of the controlling variables.

Results

A TK Solver™ computer code was developed to simultaneously solve the above equations in conjunction with the above mentioned correlations to estimate the physical properties of the involved species and other operational parameters of the proposed system. Solution algorithm is iterative with one minute long iterations. The following inputs were supplied to the code:

F_{SC} = 0.82
I_{SC} = 600 W/m^2
M_S = 500 g/min
T_E = 50, 60, 70, 80°C (alternating input)
T_S = 25 °C
U_{SC} = 0.92 W/m^2-°C
α_{SC} = 0.92
Θ = 0 – 180 min (array input)
τ_{SC} = 0.90
Φ_S = 0.035

In addition, conservative estimates of geometrical specifications such as pipe length, pipe diameter, tank length, tank diameter, equipment elevations, and initial tank levels were supplied to the code. The molecular weight of the involved species and the initial composition of seawater were also supplied to the code. The feed pump was assumed to be 1 HP and 70% efficient. Initial guesses for T_P, T_X, T_C, M_W, N_W, and NEA were also supplied. General solution algorithm is outlined in Figure 5.

Discussion

Operating the unit under elevated flash temperatures will increase the energy of all the streams around the flash chamber according to the energy balance above. The higher energy demand is supplied by the solar heater; therefore, increasing the flash temperature will increase the heater load, as shown in Figure 6, and will increase the required solar collection area, as shown in Figure 7. In addition, the heater load is gradually decreasing with time due to a declining energy demand for flashing caused by the slowly eroding vacuum; therefore, the heater load and the required solar collection area slightly decrease with time at a given constant flash temperature, as shown in Figure 6 and Figure 7 respectively.

System pressure is directly proportional to temperature as declared by any equation of state; accordingly, increasing the flash temperature will also increase the vacuum pressure as shown in Figure 8. In addition, flash chamber vacuum is continually eroding due to the build up of non-condensable gases. Continuous flash desalination systems include procedures to halt the operation, vent out the non-condensable gases, and resume under the regenerated vacuum.

Operating the unit under elevated flash temperatures will increase overall vaporization of seawater in compliance with thermodynamic phase equilibria. The increased vaporization gained from higher flash temperatures leads to more fresh water production rates as shown in Figure 9. In addition, fresh water production rates are progressively declining due to the reduced flash vaporization attributed to the eroding vacuum. The cumulative effect of flash temperature on the total fresh water output, $M_C \times \theta$, is shown in Figure 10.

Conclusion

The proposed passive vacuum solar flash desalination process was simulated with a comprehensive computer model utilizing the fundamental physical and thermodynamic relationships to depict the process and several empirical correlations to estimate the physical properties of the involved species and the operational parameters of the proposed system.

The proposed desalination system is very energy efficient because it does not require high flash temperatures since it operates under vacuum; moreover, renewable solar radiation is used in the heating module. In addition, the vacuum is naturally created by the hydrostatic forces without the need for vacuum pumps making the unit even more energy efficient.

Running the system under higher flash temperatures will increase the heater load demanding more solar collection area; however, it will increase the fresh water output. Higher flash temperatures also entail faster vacuum erosion and shorter run time since the vacuum has to be recreated more often. Experimental replication of the proposed system is currently underway to validate model assumptions and predictions.

Notation

A	Area (m^2)
BPE	Boiling point elevation (°C)
E	Energy flow (J/min)
F	Solar collector heat removal factor
H	Enthalpy (J/g)
HC	Henry's Constant (bar)
HF	Henry's Coefficient (°C)
I	Incident insolation on solar collector (W/m^2)
K	Vapor-Liquid equilibrium distribution coefficient
M	Mass flow rate (g/min)
MW	Molecular weight (g/mol)
N	Mole flow rate (mol/min)

NEA Non equilibrium allowance (°C)
P Pressure (bar)
PA Vapor Pressure Coefficient
PB Vapor Pressure Coefficient (°C)
PC Vapor Pressure Coefficient (°C)
Q Heat input (J/min)
R Universal gas constant (bar-cm^3/mol-°C)
T Temperature (°C)
U Solar collector heat loss conductance (W/m^2-°C)
v Mole fraction in condensed water
V Volume (cm^3)
W Work output (J/min)
x Mole fraction in concentrated brine
y Mole fraction in flashed vapor
z Mole fraction before flash
α K-value correction factor / solar collector absorptance
θ Run time (min)
τ Solar collector transmittance
φ Stream mass fraction
ω Salt mass fraction

Subscripts

BP Bubble point
C Condenser
DP Dew point
E Evaporator
H Heater
P Pump
S Seawater
Salt Seawater salt
SC Solar Collector
V Vacuum
W Brine
X Preheat

Superscripts

° Reference State
a Accumulated
C Condenser
E Evaporator
i Initial state
in Entering
L Latent
out Existing
sat Saturation

Literature Cited

1. Culp A. *Principles of Energy Conversion*. New York, NY: McGraw Hill; 1991.

2. Al-Kharabsheh S, Goswami DY. Analysis of an Innovative Water Desalination System Using Low-Grade Solar Heat. *Desalination*. 2003; 156(1-3): 323-332.

3. Al-Kharabsheh S. *Theoretical and Experimental Analysis of Water Desalination System Using Low Grade Solar Heat*. PhD Dissertation. Gainesville, FL: University of Florida; 2003.

4. Maroo SC. *Theoretical Analysis of Solar Driven Flash Desalination System Based on Passive Vacuum Generation*. MS Thesis. Gainesville, FL: University of Florida; 2006.

5. Goswami DY, Kreith F, Kreider JF. *Principles of Solar Engineering*. Second Edition. Philadelphia, PA: Taylor & Francis; 2000.

6. Turekian KK. *Oceans*. Englewood Cliffs, NJ: Prentice Hall; 1968.

7. Sander R. *Compilation of Henry's Law Constants for Inorganic and Organic Species of Potential Importance in Environmental Chemistry*. www.henrys-law.org; 1999.

8. Geankoplis CJ. *Transport Processes and Separation Process Principles*. Englewood Cliffs, NJ: Prentice Hall; 2003.

9. Perry RH, Green D. *Perry's Chemical Engineers' Handbook*. New York, NY: McGraw Hill; 1984.

10. Thibodeaux LJ. *Environmental Chemodynamics*. New York, NY: John Wiley & Sons; 1996.

11. Rachford HH, Rice JD. Procedure for Use of Electronic Digital Computers in Calculation Flash Vaporization Hydrocarbon Equilibrium. *Petroleum Technology*. 1952; 4(1): 19 & 4(2): 3.

12. Granet I. *Fluid Mechanics*. Englewood Cliffs, NJ: Prentice Hall; 1996.

13. Bemporad GA. Basic Hydrodynamic Aspects of a Solar Energy Based Desalination Process. *Solar Energy*. 1995; 54(2): 125-134.

14. El-Nashar AM, Qamhiyeh, AA. Simulation of the Steady-State Operation of a Multi-Effect Stack Seawater Distillation Plant. *Desalination*. 1995; 101(3): 231-243.

15. Miyatake O, Murakami K, Kawata Y, Fujii T. Fundamental Experiments with Flash Evaporation. *Heat Transfer - Japanese Research*. 1973; 2(4): 89-100.

16. Connors DN. On the Enthalpy of Seawater. *Limnology and Oceanography*. 1970; 15(4): 587-594.

17. Millero FJ, Poisson A. International One-Atmosphere Equation of State of Seawater. *Deep Sea Research*. 1981; 28(6): 625-629.

18. Sündermann J. *Landolt-Börnstein: Numerical Data and Functional Relationships in Science and Technology*. Group V: Geophysics and Space Research. Third Volume: Oceanography. Sub-Volume: A. Berlin, Germany: Springer; 1986.

19. Chopey NP. *Handbook of Chemical Engineering Calculations*. New York, NY: McGraw Hill; 1994.

Table Legends

1. Equilibrium parameters

	HC°	*HF*	*PA*	*PB*	*PC*	*α*
N_2	8067573	-3546				1.21
O_2	358815	-2209				1.22
Ar	384073	-2308				1.23
CO_2	10915	-445				1.17
H_2O			12.76	4391	245	0.98

Figure Legends

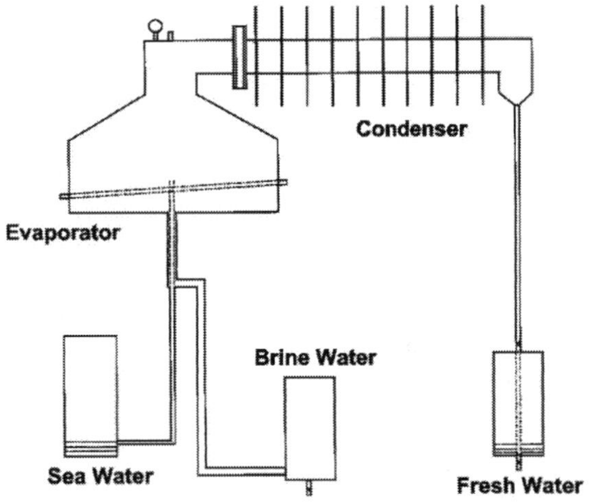

1. Passive vacuum desalination system

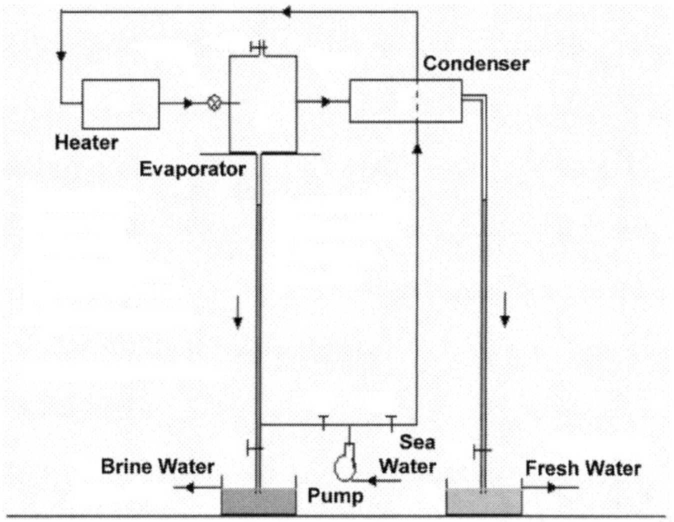

2. Passive vacuum flash desalination system

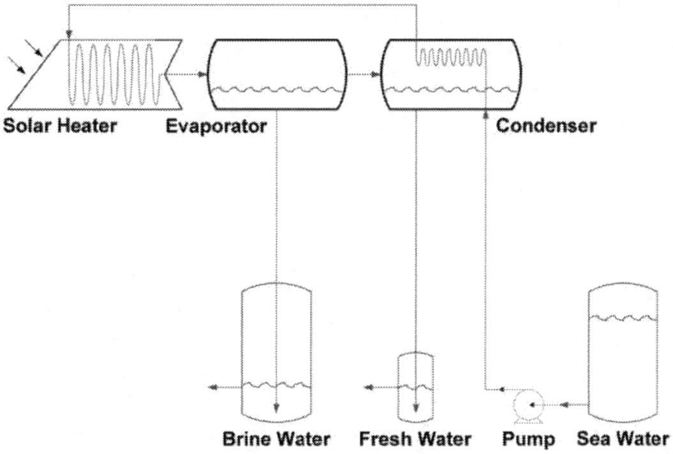

3. Proposed desalination system

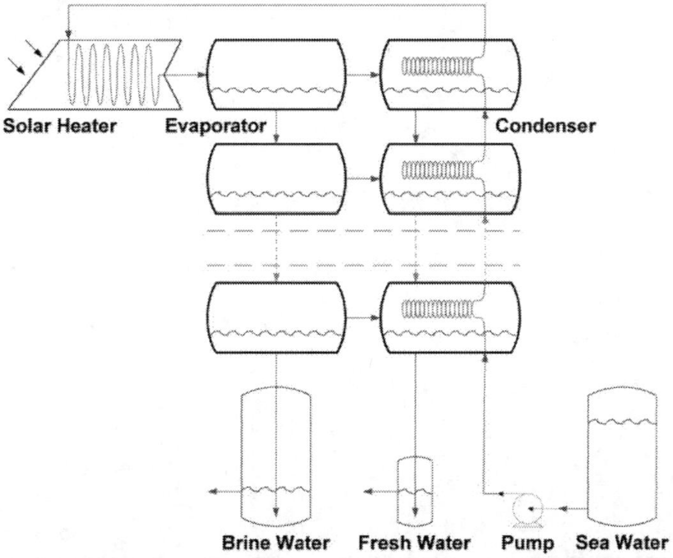

4. Multi-stage configuration of the proposed system

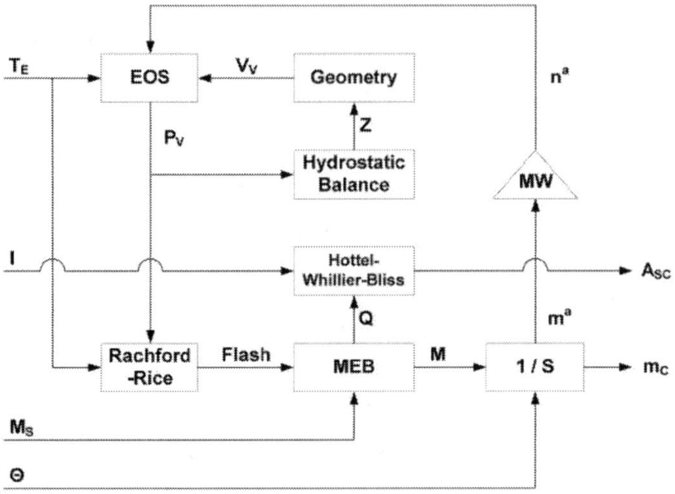

5. Solution Algorithm

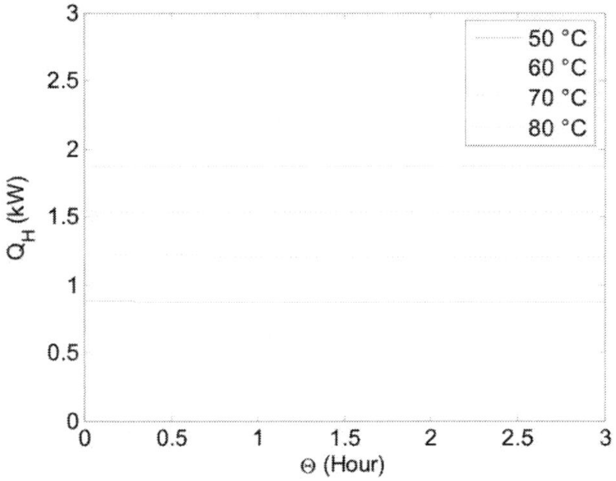

6. Heater load at several flash temperatures

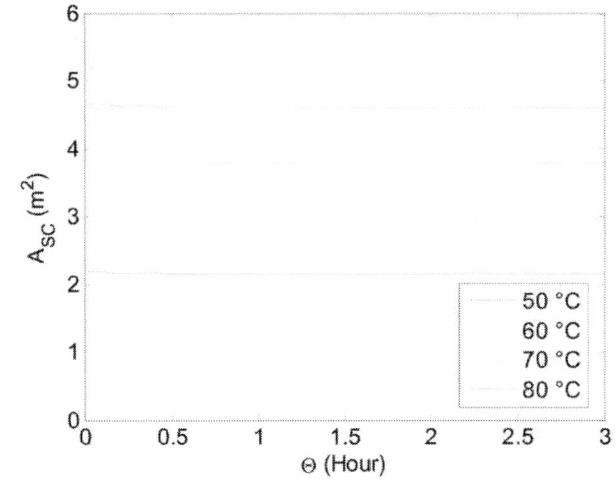

7. Solar collection area at several flash temperatures

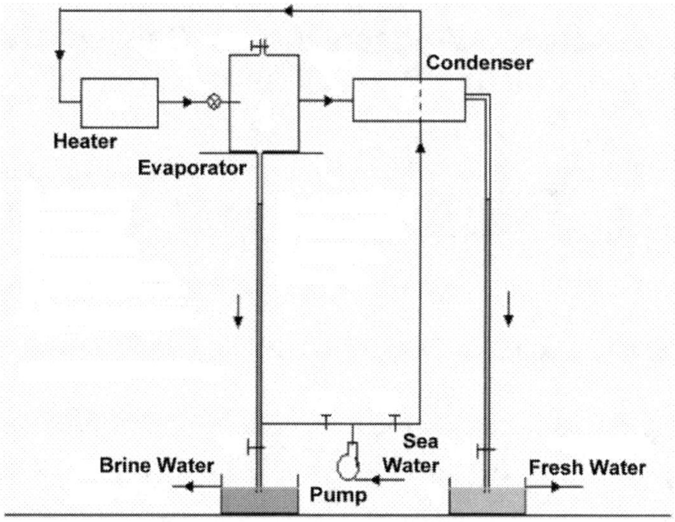

2. Passive vacuum flash desalination system

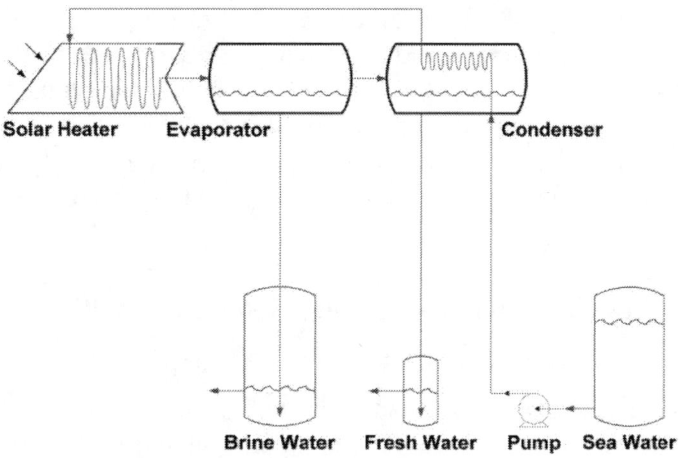

3. Proposed desalination system

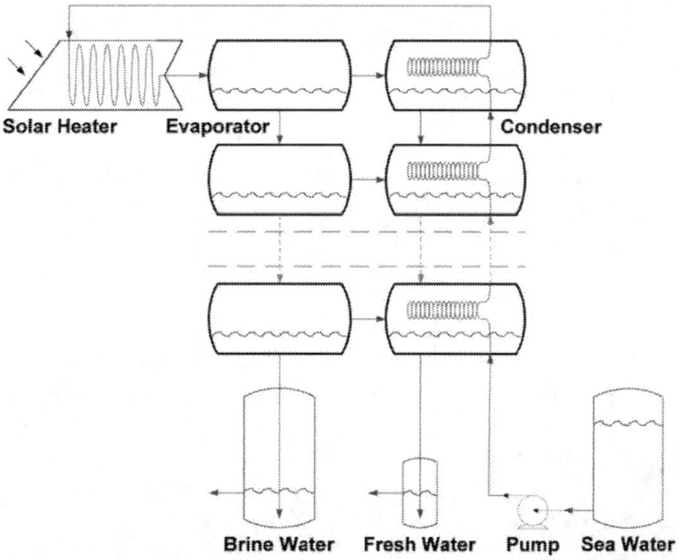

4. Multi-stage configuration of the proposed system

624f The Effect of the Presence of Antiscalants on Calcium Carbonate Precipitation: Evaluation of Calcium Precipitated, Particle Characteristics, and Fouling Potential

Lauren F. Greenlee[1], Elise Barbot[2], Benny D. Freeman[1], Desmond F. Lawler[3], Benoit Marrot[2], and Philippe Moulin[2]. (1) Chemical Engineering, The University of Texas at Austin, 1 University Station C0400, Austin, TX 78712, (2) Laboratoire Mécanique, Modélisation et Procédés Propres, Aix en Provence, France, (3) Civil, Environmental and Architectural Engineering, The University of Texas at Austin, 1 University Station C1786, Austin, TX 78712

Reverse osmosis (RO) membrane desalination has been increasingly used to produce drinking water from salt water. Most of the United States has relied on fresh water resources to produce drinking water, but even water-rich regions, such as states in the northeast, are experiencing a strain on traditional resources. Furthermore, many regions lack sufficient fresh water resources and are turning to alternate resources, such as seawater or brackish water, to sustain water needs. In particular, a growing number of inland communities have both insufficient fresh water and unused brackish water (500 – 10,000 mg/L total dissolved solids) resources. A key financial and technical limitation to inland RO desalination is disposal of the waste stream (concentrate); typically, 10 – 25% of the influent volume becomes the RO concentrate stream. This waste volume is large when compared to the waste volume produced by traditional fresh water treatment (less than 1%). To improve the feasibility of RO desalination, RO system recovery (volume of product water per volume of feed water) must be increased to decrease the concentrate volume.

Brackish water RO treatment recovery is limited by sparingly soluble salt ($CaCO_3$, $CaSO_4$, $BaSO_4$, $SrSO_4$, silica) precipitation. Specifically, calcium carbonate ($CaCO_3$) is known to be a key, omnipresent precipitate. Salt precipitation can be limited or prevented through a combination of chemical addition, pH control, and decreased recovery. Chemicals called antiscalants are dosed prior to membrane treatment and delay precipitation through association with and modification of the salt crystals in solution. Antiscalants are typically synthetic organic phosphonates, acrylic polymers, or polymer blends. Precipitation prevention through antiscalants can be achieved within a limited range of specific salt concentrations. As recovery is increased, antiscalant control is overcome and precipitation occurs. An alternate approach is thus required for further recovery augmentation.

This research focuses on the development of a novel process to treat brackish water RO concentrate and increase overall recovery. This research focuses on a novel process and includes three stages: antiscalant degradation, salt precipitation, and solid/liquid separation.

This study focused on the second stage, salt precipitation, using model synthetic concentrates and several antiscalant types to evaluate the precipitation of calcium carbonate. The model concentrate is based on data from a brackish water source in Arizona, assuming an 80% recovery and 100% ion retention. First, the precipitation of calcium carbonate as a function of pH was determined for solutions with a phosphonate antiscalant and with no antiscalant. For the model concentrate used (Saturation Index, SI, for $CaCO_3$ = 2.3 at pH 8), samples without antiscalant resulted in 87% calcium precipitation, while, as expected, samples with antiscalant showed 0% calcium precipitation. This result corresponds to the recommended SI limit for successful precipitation prevention by an antiscalant (2.5 for $CaCO_3$) [1]. As pH increased, samples with antiscalant showed increased calcium precipitation (92 – 93% at pH 10.5). However, the same sample without antiscalant always resulted in a higher calcium precipitation (99.7% at pH 10.5). Subsequent experiments with varied water composition showed a decrease in calcium precipitation with the addition of magnesium and sulfate.

The effect of antiscalants on $CaCO_3$ precipitation was also evaluated through particle size and particle number distributions, using a laser granulometer Mastersizer S (Malvern Instruments) and a laser particle counter (Met One). Several antiscalants, including organic phosphonates and one acrylic polymer blend, were tested. Most samples showed particle size distributions similar to samples without antiscalant. However, two antiscalants (amino tri(methylene phosphonic acid), or AMPA) and diethylenetriamine penta(methylene phosphonic acid), or DTPMP), showed markedly different results at the highest antiscalant concentration tested (85 mg/L and 100 mg/L, respectively, *i.e.* 17 mg/L and 20 mg/L in the hypothetical RO influent). Particle size measures taken directly after precipitation in the presence of 85 mg/L AMPA and 100 mg/L DTPMP showed a bimodal particle size distribution. Both peaks represented smaller sizes than peaks obtained for samples without antiscalant. Particle size measures taken 3 days after precipitation showed a significant increase in particle size, and results also varied as a function of pH. Photos taken with a light microscope (10x and 25x) confirm the results obtained from the laser granulometer. Particle counter results showed the majority of particles are within a particle size range of 2 – 50 micrometers; the number of larger particles (50 – 300 micrometers) increased with time for samples with 85 mg/L AMPA.

Light microscope photos also showed differences in particle shape and size with different antiscalants, as well as for solutions with and without antiscalant present. These photos, along with flux data manipulation, were used to provide explanations for the type of membrane fouling during dead end filtration.

Dead end filtration (MWCO = 0.1 micrometers, Millipore nitrocellulose membranes) was performed to compare the fouling potential of the precipitated solutions as a function of water composition, antiscalant type, and antiscalant concentration. A simplified concentrate, containing sodium chloride (NaCl), sodium bicarbonate ($NaHCO_3$), and calcium chloride ($CaCl_2*2H_2O$) was first tested. The water composition was then changed by separately adding magnesium chloride ($MgCl_2*6H_2O$) and sodium sulfate (Na_2SO_4). Finally, the full water data set was tested. Results showed differences between antiscalants and water compositions. AMPA and DTPMP showed flux decreases of approximately 30% over a period of 7 minutes for the simplified concentrate, while the same two antiscalants showed only a 19% flux decrease when magnesium was added to the simplified concentrate. For the four antiscalants tested, there was no consistent relationship between antiscalant concentration and membrane flux. For all antiscalants tested, while a greater antiscalant concentration resulted in less calcium precipitated, a greater antiscalant concentration did not result in lower permeate flux decline. Subsequent tests on solutions containing only antiscalant and deionized water indicated membrane fouling occurred through antiscalant adsorption onto the membrane surface.

Future experiments will focus on tests with real water samples to evaluate the effect of natural organic matter.

[1] *Chemical Pretreatment for RO and NF.* Hydranautics, Technical Application Bulletin No. 111, Rev. B, 2003.

637a Climate Change Legislation and Litigation Update

Mary Ellen Ternes, McAfee & Taft, 10th Floor, Two Leadership Square, Oklahoma City, OK 73102-7103

The 2008 Lieberman-Warner Climate Security Act S. 2191/S. 3036 failed in the 110th Congressional Session. But, climate change legislation is not far away. This discussion will review the elements of the proposed 2008 CSA, the criticisms, the politics, and any developments that may point to possible resolution of these conflicts. This discussion will also address intersections regarding energy and greenhouse gas emissions in other law, transitioning into the following speaker who will address the U. S. Clean Air Act and how that statute addresses greenhouse gas emissions. Finally, this discussion will address any new developments in litigation involving allegations based upon climate change.

637b Climate Change Under the Clean Air Act

Robert B. McKinstry Jr., Environmental Practice Group, Energy and Project Finance Practice Group, Ballard Spahr Andrews & Ingersoll, LLP, 1735 Market Street, 51st Floor, Philadelphia, PA 19103

In light of the Supreme Court's decision in Massachusetts v. Environmental Protection Agency that greenhouse gas emissions can be regulated under the Clean Air Act, it appears likely that the United States Environmental Protection Agency ("EPA")will be compelled to conclude that it must regulate these emissions under most sections of that Act. EPA has announced that it will soon issue a rule requiring that major sources report their GHG emissions and has represented that it will issue an Advance Notice of Proposed Rule-Making requesting comment on how to undertake regulation of GHG emissions under the Clean Air Act. The majority of states are already taking action individually or regionally to develop and implement multi-faceted GHG emissions reduction plans and a variety of regulatory and market based programs for GHG emissions control. If EPA develops a set of regulations specific to GHG emissions, the Clean Air Act, without amendment, can provide a flexible platform for coordinating state climate plans and regulatory programs under that state implementation plan mechanism with national programs such as cap and trade and emissions limitations for certain sources of emissions. Litigation may drive this process. The petitioners in Massachusetts v. EPA have filed a petition for a writ of mandamus requesting that the D.C. Court of Appeals require that EPA begin rulemaking. The question of whether GHG emissions must be regulated pollutants has been raised in a number of appeals from permits and other approvals for coal fired power plants.

637c Greenhouse Gas Cap and Trade Programs: Moving Forward

Michael E. Heintz, Porter Wright Morris & Arthur LLP, 41 South High Street, Columbus, OH 43215

The failed 2008 Lieberman-Warner Climate Security Act S. 2191/S. 3036 had adopted a Greenhouse Gas Cap and Trade System much like the European Union's (EU) Emission Trading Scheme (ETS). The EU ETS has been functioning for over three years as it enters the first phase of Kyoto applicability. As such, there is significant evidence as to the impacts a carbon trading system will have on large-scale economies, like that of the United States. This presentation will provide an overview of the EU ETS, including program specifics and applicability, and will then discuss the implications of a large scale carbon cap and trade system. The presentation will look to the results of the first period of carbon trading in Europe and draw analogies to a potential program in the United States. What impacts does a carbon trading system have on large economies? What can we learn from the three years of European data on carbon trading? And what can the United States expect from a nationwide carbon trading system based on accumulated data? This presentation will discuss these issues and attempt to provide answers to these questions which will be resurrected in the next Congressional session.

CALCULATION OF GREENHOUSE GAS EMISSIONS

Mike Bradford and Lan Cheah, Jacobs Consultancy Inc., Houston, TX, USA
Bill Keesom, Jacobs Consultancy Inc., Chicago, IL, USA

Introduction

There are multiple protocols that define how to calculate and report greenhouse gas (GHG) emissions. The purpose of the calculation and/or reporting must be defined and the protocol(s) followed must match that purpose. In some cases multiple purposes may apply, and the calculation/reporting may need to be modified for each applicable protocol.

The calculation of emissions—and the reporting of emissions in a way that provides information to management—is the first step in the reduction of GHG emissions.

This paper summarizes the major international and U.S. protocols that govern the calculation and reporting of GHG emissions. Calculation procedures are provided for the following units: fuel firing and energy purchases, including electricity, cogeneration, and mobile sources, as well as other miscellaneous sources.

Protocols for Calculating and Reporting GHG Emissions

Different protocols set different standards of accuracy, sometimes referred to as "tiers." The utilized calculation approach should match the most stringent protocol tier that applies. Depending on the use of the information, the calculations and report may need to be certified by an accredited 3rd party. ISO-14064 Part 3 requires that if a public statement is to be made concerning the data, a 3rd party should certify the emissions estimate.

Program protocols, guidelines and instructions that may apply include:

- DOE/EIA, Form EIA-1605, Voluntary Reporting of Greenhouse Gas, Instructions
- ISO14064 Part 1: Specification with Guidance at the Organizational Level for Quantification and Reporting Greenhouse Gas Emissions and Removals
- ISO14064-2: Specification with Guidance at the Project Level for Quantification, Monitoring and Reporting of Greenhouse Gas Emission reductions or Removal Enhancements
- IPCC 2006, IPCC Guidelines for National Greenhouse Gas Inventories, Prepared by the National Greenhouse Gas Inventories Programme.

- GHG Protocol for Project Accounting—World Resources Institute (WRI) and World Business Council for Sustainable Development (WBCSD)
- GHG Protocol Corporate Accounting and Reporting Standards, 2004
- California Climate Action Registry General Reporting Protocol, 2008
- The Climate Registry General Reporting Protocol, May 2008
- Compendium of Greenhouse Gas Emissions, Estimation Methodologies for the Oil and Gas Industry (API Compendium), 2004
- Petroleum Industry Guidelines for Reporting Greenhouse Gas Emissions (IPIECA), December 2003
- OpenEco Calculation Methodology
- U.S. Environmental Protection Agency (U.S. EPA). 2007a. Emissions & Generation Resource Integrated Database 2006 Version 2.1 (eGRID2006 Version 2.1). April.

The protocols divide the reporting into multiple categories. One of the major ways to separate the emissions is by direct, indirect, and other indirect, which are described below:

- Direct emissions—emissions that are generated, owned or controlled by the organization.
- Indirect emissions—these are emissions that are not generated by the organization, but are generated because of the organization and for the organization. The standard indirect emissions are energy purchases: electricity purchases, steam purchases, purchases from cogeneration, and heating & air conditioning purchases.
- Other indirect emissions—these are indirect emissions from sources other than energy purchases. Some protocols require reporting specific "other indirects," but most do not.

List of GHG Compounds and Global Warming Potential

Table 1 lists the compounds that are internationally recognized as GHGs. Also listed is the "global warming potential" (GWP) of each compound. The GWP reflects the relative impact each compound has on climate change. When the calculations are complete, each compound is multiplied by its GWP and added together to give an "equivalent CO_2 emission." For simplicity, ranges are shown in the table for classes of compounds. Many of the protocols list the specific GWP for specific compounds

Table 1. List of GHG Compounds

Greenhouse Gas	Global Warming Potential
CO_2	1
CH_4	21
N_2O	310
HFCs	140 to 9800
PFCs	4800 to 9200
SF_6	23,900

GHG Calculation Procedures

This section provides calculation information on the following major sources of GHG emissions:

- Fuel firing
- Energy purchases
- Cogeneration
- Mobil sources
- Miscellaneous other sources

Fuel Firing

The recommended approach to calculating GHG emissions from fuel firing is to measure the fuel consumption and the composition of the fuel. Measuring the composition of the fuel is especially important when firing offgases since the composition tends to be variable. Given the fuel firing rate and fuel composition, the CO_2 emissions can be directly calculated.

If the fuel composition is not known, standard factors can be used. DOE/EIA, the Climate Registry, and the API Compendium all provide standard factors. Methane and N_2O emissions cannot be directly calculated, so standard factors are always used.

Energy Purchases

Energy purchases include:

- Electricity
- Steam
- Heating and air conditioning

There are multiple approaches to accounting for GHG emissions from the purchase of electricity:

- Use specific emission factors from the electricity provider
- Use eGrid power pool factors
- Use factors for your state or area provided by the DOE/EIA or the API Compendium

For steam purchases, the emissions should be based on the actual fuel fired to generate the steam.

For purchases of heating and air conditioning, emission factors should be obtained from the supplier of the energy. If specific factors are not known, generic factors can be used. Some of the protocols provide generic factors.

Cogeneration—Standard Approach

For electricity and steam from cogeneration facilities, the emissions must be divided between power production and steam production, and also need to be allocated to the various steam levels. The API Compendium lists four different approaches to allocating the emissions. One approach tends to give most of the savings to the steam, whereas another gives the savings to the electricity. The remaining two approaches attempt to split the savings based on the relative efficiency of power generation and steam production. All of the approaches allocate the emissions based on enthalpy, with one using both enthalpy and entropy as the basis.

Mobile Sources

The estimating of mobile sources is complex, requiring information on some of the following: the types of mobile sources, fuel types, vehicle fleet age, maintenance procedures, operating conditions, operating frequency, emissions controls, and fuel consumption. The ideal is to have actual fuel use data and composition, although generic factors are available.

Miscellaneous Other Sources

Calculation procedures for many sources are available in the protocols. The API Compendium provides procedures for many of the refining units and some of the chemical processes. The Climate Registry provides procedures for many processes.

Summary

There are multiple protocols that define how to calculate and report GHG emissions. The purpose of the calculation and/or reporting must be defined and the protocol(s) followed that match that purpose. In some cases, multiple purposes may apply and the calculation/reporting may need to be modified for each matching protocol.

Different protocols set different standards of accuracy, sometimes referred to as "tiers." The calculation approach used should match the most stringent protocol tier that applies. Depending on the use of the information, the calculations and report may need to be certified by an accredited 3rd party.

The first step in the reduction of GHG emissions is the calculation of emissions and reporting the emissions in a way that provides information to management.

N_2O production by fungal denitrification in a semiarid soil

Lin Yang[1], **Vallejo Antonio**[2] and **Yulin Deng**[1]

(1)School of Life Science and Technology, Beijing Institute of Technology, 5 South Zhongguancun Street, Haidian District, Beijing, China

(2)ETSI Agrónomos, Universidad Politécnica de Madrid, Ciudad Universitaria, Madrid, Spain [2]

ETSI Agrónomos, Universidad Politécnica de Madrid, 28040 Madrid, Spain

1 Introduction

The increasing nitrous oxide (N_2O) concentration in the atmosphere causes concern due to its contribution to global warming.[1] Soils, especially agricultural soils, contribute approximately half of the world's anthropogenic N_2O emissions and currently this source of N_2O represents 2.4% of the European release of anthropogenic derived greenhouse gasses (GHG).[2,3] Furthermore, N_2O is also directly or indirectly involved in destruction of stratospheric ozone. [4] Modern agriculture must strive to mitigate N_2O emissions by cultivated soils.

The researches have shown that N_2O is released during the microbial nitrification and denitrification process, in principle.[5] Soil heterogeneity permits the coexistence of aerobic and anaerobic zones which allow organisms in the same soil aggregate to function simultaneously, and nitrification and denitrification can take place at the same time.[6,7]

Bacteria and fungi are the two most important microbes for emitting N_2O from agrarian soil. And both of them have the genetic potential to use organic and inorganic sources of N.[8,9] Although the basic mechanism of N_2O formation in soils is well known, there is not a better knowledge of the contribution of these microbes above.

Bacteria has been regarded as the unique microbe to denitrification for a long time, [10,11] but fungi were found to exhibit denitrifying activities in recently.[12-14] Actually, bacteria denitrify rapidly and completely, however it need a strict condition of dissolved oxygen (DO). In contrast, fungi can simultaneously perform denitrification under microanaerobic or aerobic conditions while bacteria.

In this laboratory study, we want to know the influence of different microbes to produce N_2O. Meanwhile we also identify the N_2O production pathways by use of cycloheximide and streptomycin, inhibitors of fungal and bacterial.

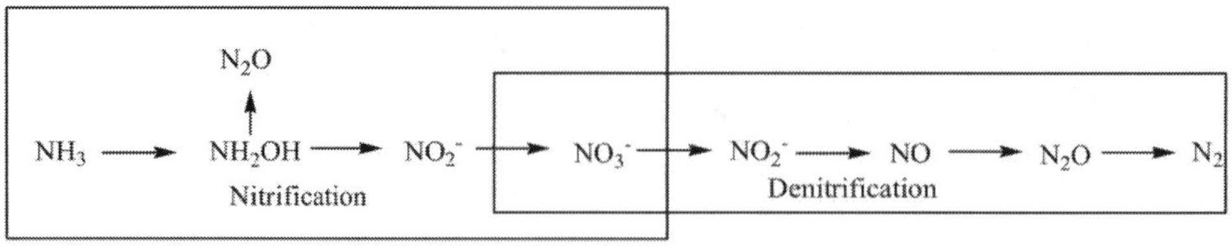

Fig. 1 The mechanism of denitrification and nitrification

2 Materials and methods

2.1 Soil and compost

Soil: surface (0-20 cm) clay loam soil was collected, arable but without plants, in April 2007 from "El Encín" field station, near Alcalá de Henares (Madrid, Spain) (latitude 40° 32'N, longitude 3° 17'W), in the middle of the Henares river basin. The soil was gently broken down by hand after transport to the laboratory, meanwhile it air-dried at room temperature during three days. Then it was sifted through a 2.5 mm sieve, and stored in plastic bag until use. Compost: Urban waste (Madrid, Spain). Inhibitor: streptomycin (Sigma-Aldrich, purity: 95%); cycloheximide (Sigma-Aldrich, purity: 94%).

2.2 Soil sampling and analysis

Soil NO_3^- and NH_4^+ content from each column (at the end of the experiment) and from the additional container (the day following to start the experiment) were determined by extracting 15 g of fresh soil with 50 ml 1M KCl solution; NO_3^- and NH_4^+ were analyzed by colorimetric methods.[15] Soluble organic C was also extracted and analyzed in samples obtained at the end of the experiment as described by Mulvaney et al.[16]

Gravimetric moisture contents for the columns were derived from the relationship between wet weight of the soil column and the dry weight of the soil column. Water filled pore space (WFPS) was calculated by dividing the volumetric water content by total soil porosity. Total soil porosity was calculated according to the relationship: soil porosity = (1- soil bulk density/2.65), assuming a particle density of 2.65 g cm^{-3}. Bulk densities were calculated from the volume of soil in the cores.

3 Laboratory experiments

3.1 Experiment 1

Weighed 2g of soil in vial (20ml). Prepared stock solution: the concentration of streptomycin was 3 mg/ml, solubility of cycloheximide was 1.5 mg/ml so the solution prepared should have dissolved it. Application of the different treatments will be carried out within the vial and these ones will be capped and left for few hours. The headspace will be sample by means of a syringe and this sample transferred to a smaller vial before being analyzed by gas chromatography.

Note: the tables reflect the order of addition of reagents and time of contact, for example: in table 3 soil was weighted and the biocides were added. These were left in contact for 1 hour and then the N solution was added. The headspace was flushed with He (after capping the vial) and the whole system was left in contact overnight. (Each sample repeated 3 times).

The protocol established for the following experiments is described. This was based on the results obtained from the preliminary experiments. Deal with the effect of the application of biocides to soil (bactericide/fungicide) on N_2O production.

3.1.1 The effect of the application of biocides to soil

Table 1. The application of biocides to soil

Sample	Soil (g)	Biocide	Atmosphere	Acetylene
1	2/20ml vial	6.7 ml water	He	No
2	2/20ml vial	C/2ml+4.7ml water	He	No
3	2/20ml vial	S/2ml+4.7ml water	He	No
4	2/20ml vial	C/2ml+S/2ml+2.7ml water	He	No
5	2/20ml vial	6.7ml water	He	1.8ml
6	2/20 ml vial	C/2ml+4.7ml water	He	1.8ml
7	2/20 ml vial	S/2ml+4.7ml water	He	1.8ml
8	2/20 ml vial	C/2ml+S/2ml+2.7ml water	He	1.8ml

C: Cycloheximide, S: streptomycin

Table 2. The resulting concentrations of sample 1-8

Sample	Area (N_2O)	ppm (N_2O)
1	8850.33	2.21
2	11695.33	3.34
3	17602.00	5.71
4	27525.67	9.68
5	103871.33	40.21
6	211922.00	83.43
7	169614.00	66.51

8	258109.00	101.91

The figure 2 shows the effect of both biocides on the fluxes, with and without acetylene. Because of little N and C in the soil and in both cases it is not clear there is inhibition when adding the biocides as the fluxes increased as compared with the blank (no added biocide). The addition of acetylene produced an increase on the fluxes in all the treatments. This could be evidence of the production of N_2 as acetylene blocks the last of denitrification. This could suggest that some of the biocides were being used by the microorganisms as a source of carbon and nitrogen or that materials leaking from the dead cells were used by the microorganisms.[17] The results must be looked at carefully as there is evidence of acetylene being used as a carbon source.[18]

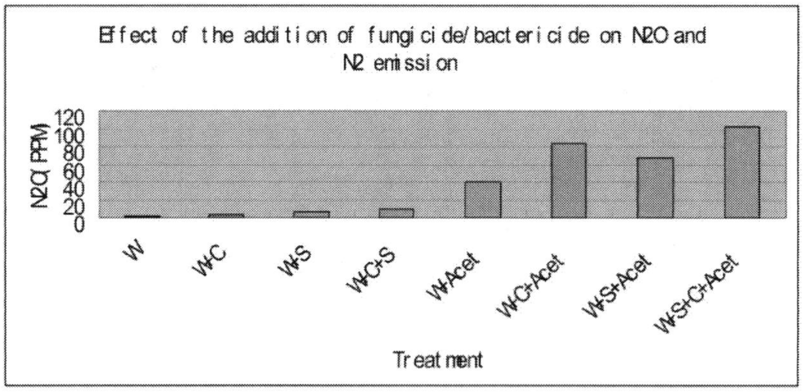

Fig. 2 The effect of biocides on the N_2O emission

3.1.2 The effect of the application of biocides to soil with glucose

Table 3. The application of biocides to soil with glucose

Sample	Soil (g)	Biocide	Atmosphere	Acetylene

9	2/20 ml vial	6ml water+10mg G	He	No
10	2/20 ml vial	C/2ml+4ml water+10mg G	He	No
11	2/20 ml vial	S/2ml+4ml water+10mg G	He	No
12	2/20 ml vial	C/2ml+S/2ml+2ml water+10mg G	He	No
13	2/20 ml vial	3ml water+10mg G	He	1.8ml
14	2/20 ml vial	C/1ml +2ml water+10mg G	He	1.8ml
15	2/20 ml vial	S/1ml +2ml water+10mg G	He	1.8ml
16	2/20 ml vial	C/1ml+S/1ml+1ml water+10mg G	He	1.8ml

G: Glucose

Table 4. The resulting concentrations of sample 9-16

Sample	Area (N_2O)	ppm (N_2O)
9	9794	4.45
10	5857.333	2.09
11	23462	12.65
12	7215.333	2.90
13	62591.33	36.13
14	76447.33	44.44
15	104198	80.54
16	105328	81.56

Figure 3 shows inhibition of the fluxes when using cycloheximide in the glucose treatment. The bottle with streptomycin produced an increase in the fluxes whereas the mixture of both biocides did not produce a significant effect. It showed that the streptomycin didn't work in this condition and it was used as nitrogen source by some fungal.

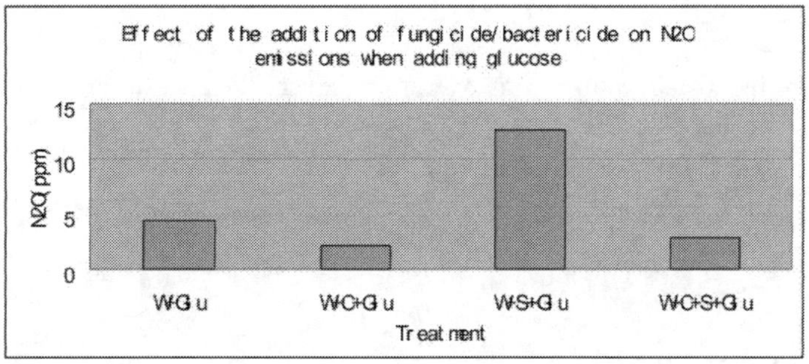

Fig. 3 The effect of biocides on the N_2O emission with glucose

Figure 4 shows mush larger fluxes when adding acetylene compared to no acetylene addition, and an increase in the fluxes was observed with the addition of streptomycin. Because streptomycin have abundant nitrogen and it can be easily translated to N_2O by microorganisms. And this result also confirms that fungal and bacteria out-of-run mixing with acetylene and He.

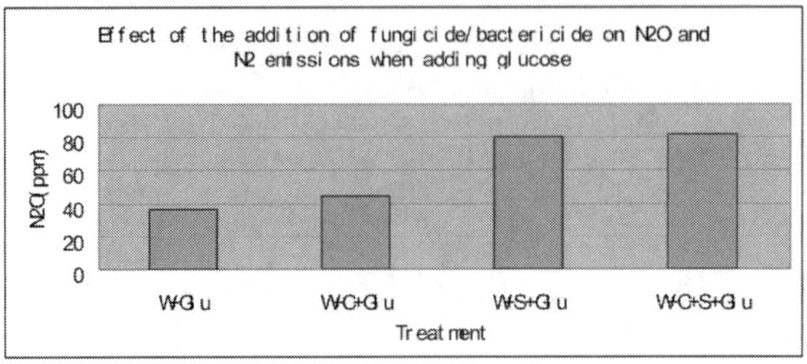

Fig. 4 The effect of biocides on the N_2O emission with acetylene and glucose

3.2 Experiment 2

Each experiment was conducted using 12 vitric jars (6.3 cm height, 5.4 cm diameter) with gas-tight lids fitted with a gas sampling port. And all of the sample bottles must be exhausted air by pump before collected gas simples.

70 g dry soil was weighed into each jar and additional distilled water with $(NH_4)_2SO_4$ 0.43g (200 kg N /ha) to achieve WFPS 60%. The concentrations of streptomycin (3.0mg/g soil) or cycloheximide (1.5mg/g soil) were tested in both soils to determine their optimal concentrations for inhibition. And all of bottles were divide into four groups:

(1) 210 mg (3.0mg/g soil) streptomycin perflask

(2) 105 mg (1.5mg/g soil) cycloheximide perflask

(3) 210mg streptomycin (3.0mg/g soil) and 105 mg streptomycin (1.5 mg/g soil) cycloheximide perflask

(4) N: Control

(Each treatment was replicated 3 times for gas analyses)

All treatments were incubated in the air uncovered for several hours (3h) after solution application, After that they were tightly sealed at room temperature. Gas samples were removed from the flasks on 0 min and 60 min. And soil WFPS has been maintained on a weight basis for 4 days.

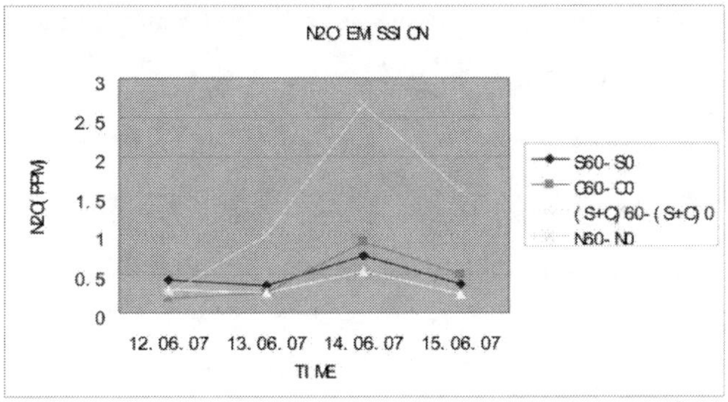

Fig. 5 The effect of time to N_2O emission

Figure 5 indicates that both of bacteria and fungal are work in air atmosphere, furthermore the max emission of N_2O in the 3rd day.

4 Conclusion

Results of this study suggest that the occurrence of fungal denitrification is of ecological significance as N_2O is the dominant gaseous product in this semiarid soil. As fungi have the ability to perform denitrification and O_2 respiration simultaneously in a range of O_2-stress conditions, the potential exists for fungi to produce N_2O in a wider range of soil aeration conditions than bacteria. Fungi are widely distributed in soils and water, hence the potential exists for fungi to make a significant contribution to the global N_2O budget.

References

[1] Hansen J, Greenhouse gas growth rates [J], Proc. Natl. Acad. Sci., 2004, 101(46):16109-16114.

[2] Church J A and Gregory J M, The Scientifie Basis [R], Intergovernmental Panel on Climate Change (IPCC), Cambridge University, UK. 2001.

[3] Avrahami S, Effect of soil ammonium concentration on N_2O release and on the community structure of ammonia oxidizers and denitrifiers [J], Appl. Environ.Microbiol, 2002, 68 (11):5685-5692.

[4] Dyominov and Zadorozhny, Contribution of natural and anthropogenic factors to the long-term changes of the Earth's ozone layer at the end of the XX century [J], Izv. Atmos. Ocean. Phys., 2005, 41:43-55.

[5] Renault P and P Stengel, Modelling oxygen diffusion in aggregated soils: I. Anaerobiosis inside the aggregates [J], Soil Sci. Soc. Am., 1994, 58:1017-1023.

[6] Stevens R and R J Laughlin, Cattle slurry affects nitrous oxide and dinitogen emissions from fertilizer nitrate [J], Soil Sci. Soc. Am., 2001, 65:1307-1314.

[7] Kuenen J G and Robertson L A, Combined nitrification denitrification process [J], FEMS Microbiol Rev., 1994, 15(2):109-117.

[8] Merrick M J and Edwards R A, Nitrogen control in bacteria [J], Microbiol. Rev., 1995, 59: 604-622.

[9] Marzluf Ga, Genetic regulation of nitrogen metabolism in the fungi [J]. Microbiol Mol Biol Rev, 1997, 61:17-32.

[10] Takaya N. and Shoun H, Nitric oxide reduction, the last. step of the fungal denitrification by Fusarium oxysporum is obligatorily mediated by cytochrome P450nor [J], Mol. Gen. Genet., 2000, 263:342-348.

[11] Anderson J and K H Domsch, Quantification of bacterialand fungal contributions to soil respiration [J], Arch. Mikrobiol., 1973, 93:113-127.

[12] Johnson C, Measuring bacterial and fungal substrate-induced respiration in dry soils [J], Soil Biol. Biochem., 1996, 28:427-432.

[13] Laughlin R J, Determining nitrogen-15 in ammonium by producing nitrous oxide [J], Soil Sci. Soc. Am., 1997, 61:462-465.

[14] Ruzicka S, D Edgerton, M Norman, and T Hill, The utility of ergosterol as a bioindicator of fungi in temperate soils [J], Soil Biol. Biochem, 2000, 32:989-1005.

[15] Harwood J E and Huysen D J, Automated analysis of ammonia in water [J], Water Res., 1970, 4:695-704.

[16] Mulvaney, Improved diffusion methods for determination of inorganic nitrogen in soil extracts and water [J], Biol. Fertil. Soils, 1997, 24:413-420.

[17] Badalucco L, Activity and degradation of streptomycin and cycloheximide in soil [J], Biol Fertil Soils, 1994, 18:334-340.

[18] Dendooven L, Carbon and nitrogen dynamics in a grassland soil with varying pH: effect of pH on the denitrification potential and dynamics of the reduction enzymes [J], Soil Biology and Biochemistry, 1998, 30:359-367.

718b Application of Fe-Pd Nanoparticles for Chloro-Organic Degradation

Abhijit Patil, Chemical and Materials Engg, University of Kentucky, 177 Anderson Hall, Lexington, KY 40506 and D. Bhattacharyya, Chemical and Materials Engineering, University of Kentucky, 177 F. Paul Anderson Tower, Lexington, KY 40506.

Over the years, toxic chlorinated organics that are present in soil and groundwater, have attracted much attention due to their adverse effect on human and environment. Various methods have been developed to destroy these organochlorines into environmentally benign compounds. Nanoparticles have been applied in various branches of catalysis for the high selectivity and reaction rates. Applicaton of multimetallic nanoparticle system as catalysts for this purpose offer a great advantage over other methods in terms of reaction rates and reaction conditions. Extensive studies have been done in area of destruction of these toxic compounds using Zero Valent Iron (ZVI). Permeable reactive barriers (PRBs) containing granular ZVI have been shown to be effective for the remediation of chlorinated organic solvents in contaminated aquifer. However, recent studies show that multimetallic nanoparticles, alternated reaction mechanism increase the catalytic activity and hence dechorination efficiency.

Objective of this research includes analysis of the background of dechlorination and nanoparticles in catalysis, surface functionalization of the inert membrane surface which acts as polymer matrix for Fe nanoparticles, prepare Fe nanoparticles in this polymer matrix, add Pd to this iron nanoparticle to get Fe-Pd bimetallic nanosized particles, characterize modified and unmodified membrane and the bimetallic nanoparticle for their concentration. Reaction meachnisms and kinetics of the dechlorination by these matrix embedded nanoparticles will be studied and also focus will be on detailed modeling of the chloro-organic degradation.

This work is funded by National Institute of Environmental Health Sciences(NIEHS-SBRP).

718c Catalytic Wet Peroxide Oxidation of Phenol : H2O2 Gradual Addition Study

Asghar Molaei Dehkordi, Amir Ali Ebrahimi, and Seyedehmansoureh Shahalami. Chemical Engineering, Sharif University of Technology, Tehran, Iran

The catalytic wet peroxide oxidation (CWPO) of phenol was studied in a batch system using a commercial activated carbon supplied by Jacobi Co. (Sweden, type of AquaSorb 101). The influences of various operating parameters including the initial pH of solution, temperature, catalyst loading, and the addition modes of hydrogen peroxide on the fractional conversion of phenol were investigated. The stability of hydrogen peroxide and consequently its effect on phenol conversion is dependent upon the pH of solution. To investigate this effect it was decided to measure the concentration of hydrogen peroxide and phenol vs. time at different pH of solution (2-10). The least consumption of H_2O_2 was observed at pH values of 2 - 4. Moreover, H_2O_2 has been decomposed rapidly at pH of around 10. The low consumption rate of H_2O_2 will result in the high phenol conversion; therefore, the high phenol conversion was obtained at pH around 4. Increasing the temperature over the range of 70-80 °C does not have profound effects on the conversion of phenol; therefore, the appropriate temperature could be 70 °C. It could be seen in all the experiments that the phenol concentration increased after a great initial reduction. This was happened during the beginning of the reaction (10 – 30 min) and proposed a reversibility mechanism in phenol degradation process. This reversibility was significantly higher in non catalytic reaction (figure 1). Two different modes of addition of H_2O_2, i.e. rapid and gradual addition with different flow rates were examined. The same amounts of H_2O_2 (77.65 mmol/L) were added to the reaction mixture with two different modes of operation to investigate the effect of H_2O_2 addition modes on the fractional conversion of phenol. The addition rate of H_2O_2 was adjusted at 0.3235, 0.3883 and 0.647 mmol/min. As can be seen from figure 1, the final conversion of phenol in the case of rapid addition of H_2O_2 is higher than that obtained with gradual addition of H_2O_2 which is contrary to what we expected. Furthermore, in the case of gradual addition of oxidant, high fractional conversions of phenol followed by the progressive decrease can be observed. This behavior is same as what we observed in the case of rapid addition in last experiments and may be attributed to the reversible mechanism of phenol degradation because of the lack of OH•. There are two reasons which cause the lack of OH•, (1) low catalyst loading, (2) low concentration of hydrogen peroxide. The first reason could be confirmed by considering the results of the catalyst loading variations which shows the reversibility was higher in the non catalytic reaction. The second explanation could be confirmed by considering the effect of increasing the addition rate of the oxidant on the phenol conversion which decreases the reversibility due to the increase in the OH• concentration. But there is still a problem why the phenol conversion is lower than that of rapid addition mode of the oxidant. To study whether such a trend could be obtained for other catalyst loading or not, a number of CWPO experiments with other catalyst loading were carried out while other operating conditions were kept unchanged. Figure 3 shows the variations of the phenol conversion with time as a function of catalyst loading and modes of H_2O_2 addition on the CWPO of phenol. As can be noticed from this figure, the catalyst loading plays an important role in the CWPO process. According to these experimental results for the catalyst loadings larger than 2 g/L, the aforementioned-progressive decrease in the fractional conversion of phenol could not be observed. Moreover, the fractional conversions of phenol for the catalyst loadings larger than 2 g/L were much higher than those of lower catalyst loadings. This behavior can be explained by two explanations, (1) an increase in the catalyst surface area available to the reactants which causes an increase in OH• concentration and (2) an increase in the catalyst surface area which increases the adsorption capacity and consequently hides the reversibility effect.

Therefore, it can be concluded that there is a minimum value of catalyst loading which above the performance of mode of gradual addition is much better than that of rapid addition mode under identical conditions particularly the total hydrogen peroxide consumed.

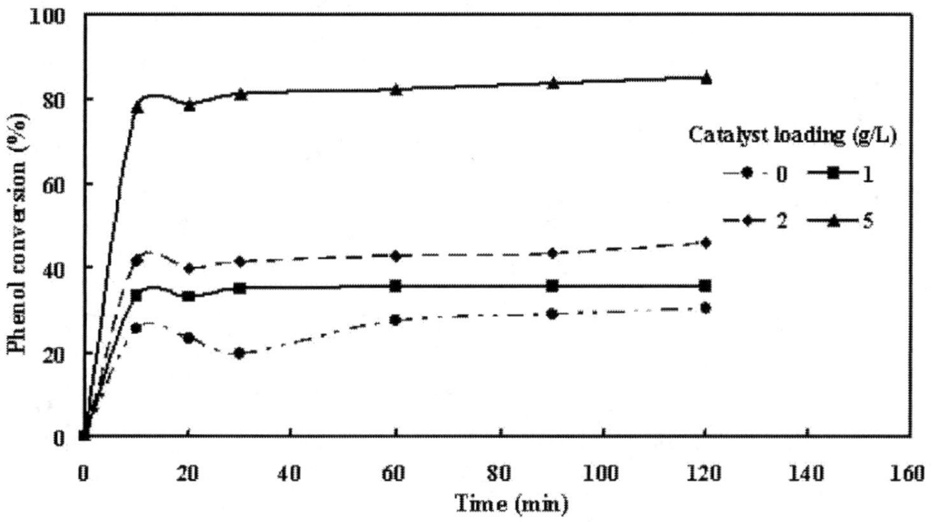

Figure 1. Effect of catalyst loading on the phenol conversion.

$T = 70$ °C, $C_{HP} = 77.65$ mmol/L, pH = 5.5, $C_{ph} = 500$ mg/L.

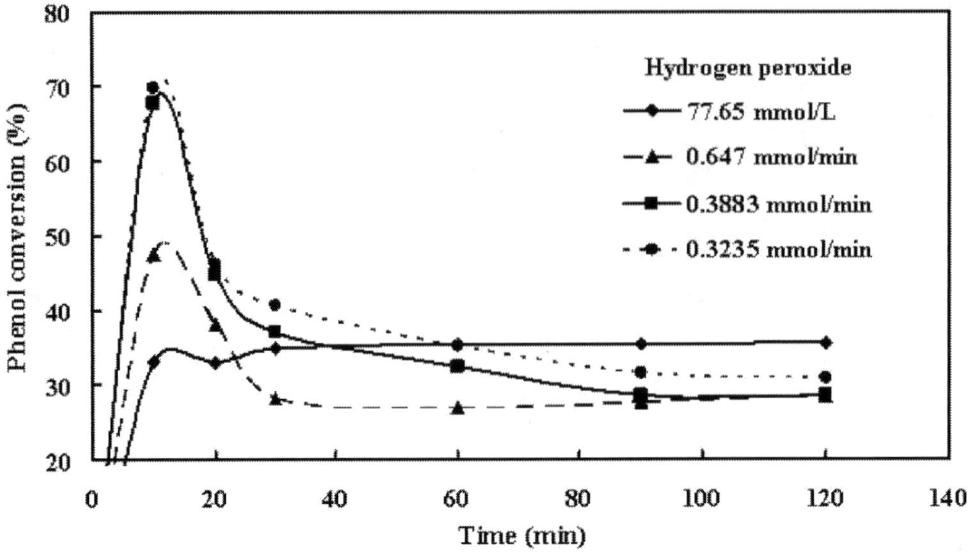

Figure 2. Effect of addition modes of hydrogen peroxide on the phenol conversion.

C_{ph} = 500 mg/L, catalyst loading = 1 g/L, pH = 5.5, T = 70 °C.

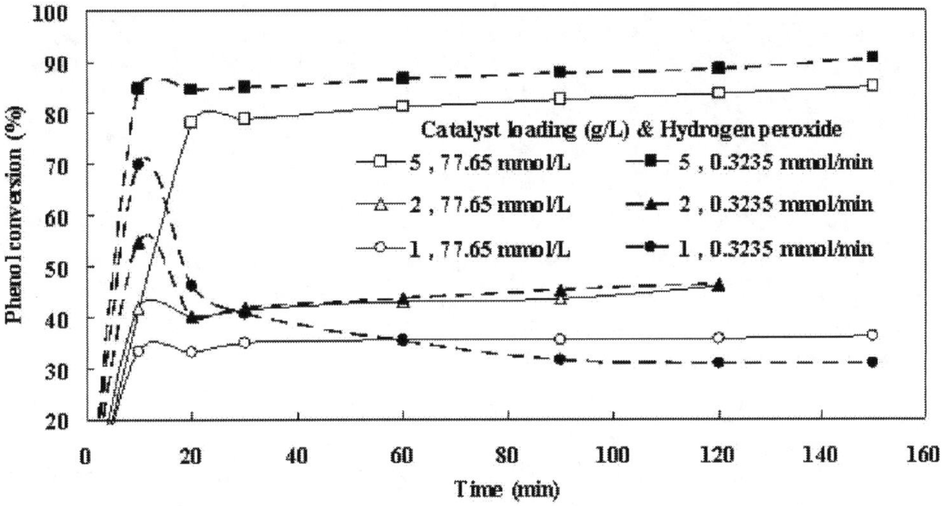

Figure 3. Effect of catalyst loading and addition modes of H_2O_2 on the phenol conversion.

C_{ph} = 500 mg/L, pH = 5.5, T = 70 °C.

Keywords:

Phenol oxidation; CWPO; Activated carbon; Wastewater treatment; Phenol

718d Catalytic Activities of Transition Metal Oxides for Plasma PM Removal

Shin Yamamoto, Shuiliang Yao, Satoshi Kodama, Chieko Mine, and Yuichi Fujioka. Chemical Research Group, Research Institute of Innovative Technology for the Earth (RITE), 9-2 Kizugawadai, Kizugawa-shi, Kyoto, 619-0292, Japan

Although diesel engines have a higher combustion efficiency and a lower CO_2 emission than gasoline engines, the harmful particulate matter (PM) emitted by the diesel engines is required to be removed. Plasma technologies have a potentiality as an effective method for PM removal. We have developed a dielectric barrier discharge (DBD) reactor driven by a pulse power supply to remove PM from diesel engines. It was found that PM was trapped within the DBD reactor due to electrostatic precipitation effects. In order to improve the PM removal ability of the DBD reactor, catalysts that can effectively promote the oxidation of PM trapped in the DBD reactor are desirable.

The efficiencies of PM oxidation catalyzed by ten different catalysts (seven transition metal oxides, two noble metals and aluminum oxide) have been measured using a thermo-gravimetrical analyzer under an atmosphere containing O_3 and NO_2 that can be generated in plasma discharges. The PM oxidation efficiency by MnO_2 has been found to exceed that of the noble metal catalysts even at a temperature less than 300 °C. The PM oxidative activities by transition metal oxides changed due to the presence of active oxygen species of O_3 and NO_2.

The efficiencies of PM oxidation catalyzed by seven transition metal oxides under a plasma discharge condition have been evaluated using a batch type of DBD reactor. The DBD reactor mainly consists of dielectric alumina plates and aluminum plates. The powders of each catalyst were deposited and fixed on the surfaces of alumina plates. PM collected from the exhaust gases of a diesel engine was then fixed on the surface of alumina plates with deposited catalyst powders. Experiments were carried out at 200 ° C.

The efficiency of PM oxidation using the DBD reactor with deposited Fe_2O_3 is approximately 35% higher than that without any catalysts. Fe_2O_3 deposited in the DBD reactor has the highest catalytic activity for the PM removal under the plasma discharge condition. The PM oxidative activities of the transition metal oxides under the plasma discharge condition are different from that observed by the thermo-gravimetrical analysis without plasma discharges. This difference is mainly due to photocatalytic effects by ultraviolet-light radiation generated within plasma discharges. The experimental results and discussion will be presented in detail.

This work was supported by the New Energy and Industrial Technology Development Organization, Japan.

Recovering Magnetic Powder from Sludge Discharged during Magnetic Separation of Oily Wastewater at Oil Production Sites

Tadashi Sano (1), Hisashi Isogami (2), Akira Mochizuki (2) and Norihide Saho (1)
(1) Mechanical Engineering Research Laboratory, Hitachi, Ltd., Japan
(2) Hitachi Plant Technologies, Ltd., Japan

1. Introduction

When bitumen is extracted from oil sand, a large amount of wastewater is discharged because a large amount of hot water is added to the oil sand to reduce its viscosity. Also, a large amount of wastewater is produced along with crude oil at various conventional oil production sites because water infusion is required in the oil production process.

That wastewater includes silt, clay (SS: suspended solids), and oil emulsion. However, the wastewater is being directly discarded without recovering oil and clarifying the water[1-2]. Therefore, we developed a system that separates SS, oil, and clarified water from wastewater to reuse the water and to reduce the impact on the environment.

2. Water treatment system with magnetic separation

Our system uses a technique for coagulation and magnetic separation[3-5] (Fig. 1). This technique is enhanced from coagulation sedimentation used in water treatment plants. First, a flocculant and magnetic powder are mixed into wastewater and agitated. Moments later, flocs of impurities and magnetic powder are formed. The flocs are then removed from the wastewater by magnets, and clarified water and sludge are produced.

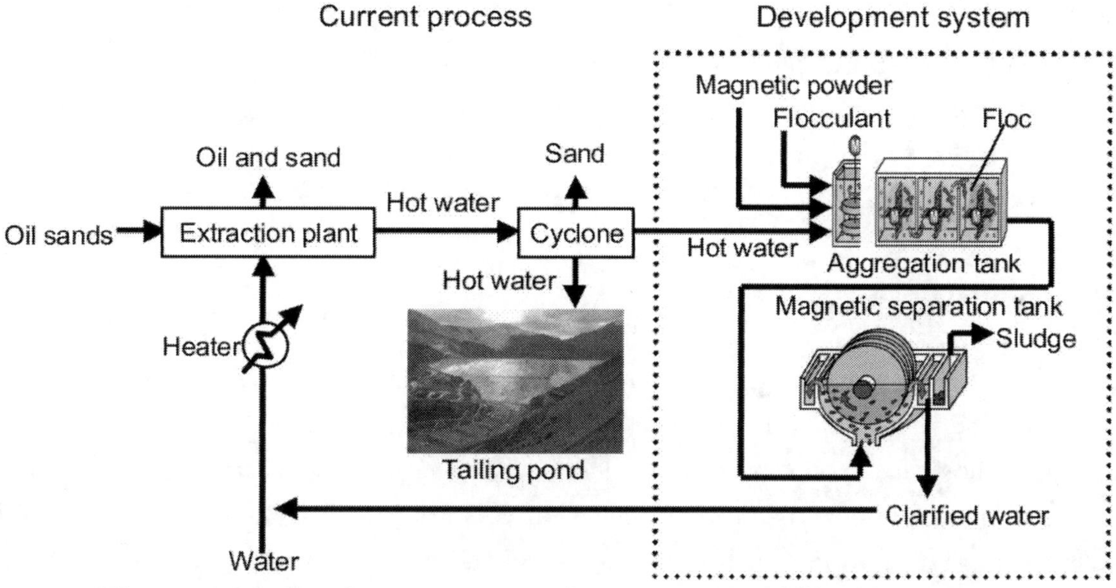

Figure 1 Application to wastewater treatment system for mining

The amount of magnetic powder used in this system increases in proportion to the amount of wastewater being treated. Because there is a large amount of target wastewater, a large amount of magnetic powder is drained with the wastewater when the magnetic powder is not recoverable. Therefore, recovering the magnetic powder from sludge is important.

3. Magnetic powder recovery system

In this study, we investigated the effectiveness of hydrothermal treatment on decomposing sludge that contains magnetic powder. When sludge is heated, flocculant is expected to lose its function of coagulation due to the chemical reaction, and magnetic powder can be recovered. We designed the system as follows.

The magnetic powder recovery system is shown in Fig. 2. First, sludge discharged from the magnetic separation equipment is discharged into a stainless steel tube by a high-pressure pump. Next, the sludge is preheated by a heat exchanger, and a chemical reaction with sludge is caused in a heated reactor. Then, the sludge passes through the heat exchanger and a back-pressure regulating valve to maintain the saturated vapor pressure. Finally, a magnetic separator mounted on a permanent magnet recovers the decomposed magnetic powder.

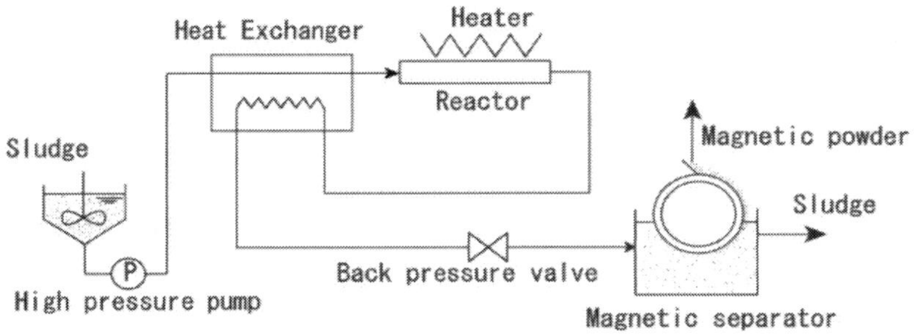

Figure 2 Magnetic powder recovery system

When the treatment temperature increases, not only the flocculant but also the magnetic powder is known to change in composition due to a chemical reaction. Therefore, we examined the decomposition temperature of the flocculant and the phase-change temperature of the magnetic powder and optimized the magnetic powder recycling system.

4. Experimental results and discussion

We examined the effect that temperature has on chemical change with flocculant and magnetic powder using a batch method of hydrothermal treatment. The flocculant used in the experiment is poly ferric chloride and the magnetic powder is magnetite.

Fig. 3 shows the hydrous rate of the sludge with the hydrothermal treatment temperature. Fig. 4 shows a picture of the treated sludge after stirring and standing for 30 min. The experimental results show that the hydrous rate sharply decreases at 200°C or more (Fig. 3). Also, the transparency of supernatant liquid sharply decreases at the same temperatures (Fig. 4). Because flocculant generally has a high water holding capacity, we conclude that the capability of the flocculant clearly deteriorated at these high temperatures.

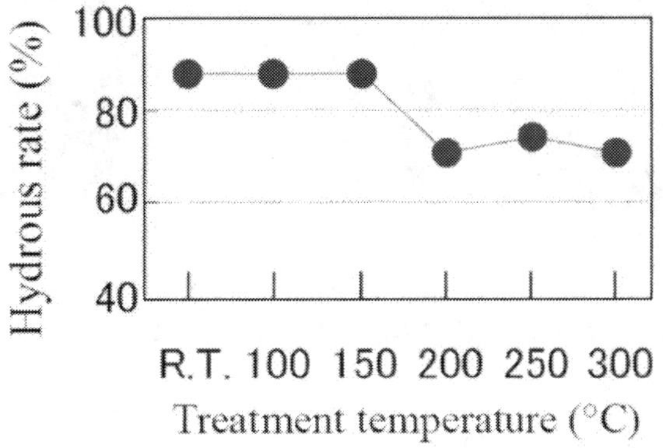

Figure 3 Hydrous rate of sludge with hydrothermal treatment temperature

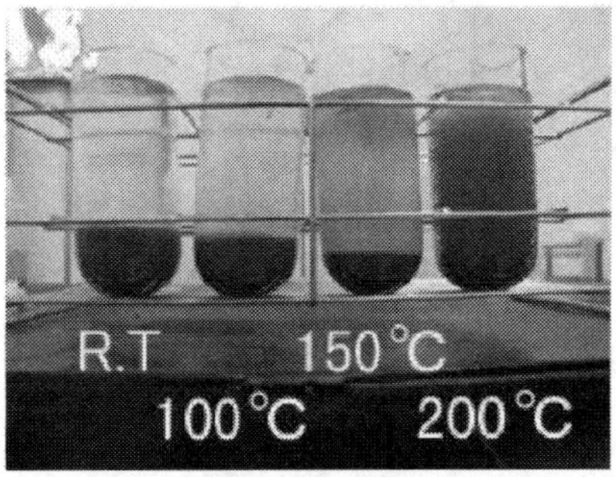

Figure 4 Treated sludge at various temperatures

Fig. 5 shows the saturation magnetization of the magnetic powder measured by a vibrating sample magnetometer with the hydrothermal treatment temperature. SEM pictures of the hydrothermally treated magnetic powder are shown in Fig. 6 (a)(raw material), Fig. 6 (b)(200°C), and Fig. 6(c)(300°C). Fig. 7 shows the XRD spectra of the hydrothermally treated magnetic powder.

The magnetic powder cannot help us detect a significant change in the saturation magnetization. But, the SEM picture of the magnetic powder heated at 300°C shows that the edges of the particles are more round than the raw material or the magnetic powder heated at 200°C. We observed the spectrum of hematite only in the XRD spectrum of the magnetic powder heated at 300°C (top of Fig.7). Because a minimal amount of hematite is produced at 300°C, we conclude that most of the magnetite changes into hematite when the treatment time is increased. In other words, the more the number of recycling increases, the more hematite is produced. Therefore we decided that the best temperature for a hydrothermal treatment of sludge is 200°C.

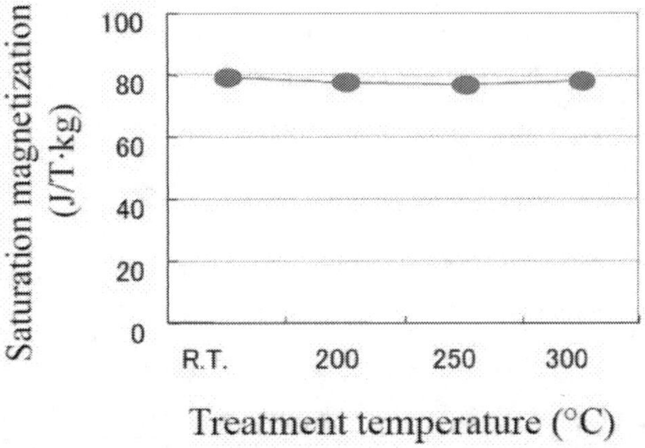

Figure 5 Saturated magnetic moment of magnetic powder with hydrothermal treatment temperature

(a) Raw material (b) 200°C (C) 300°C

Figure 6 SEM pictures of hydrothermally treated magnetic powder

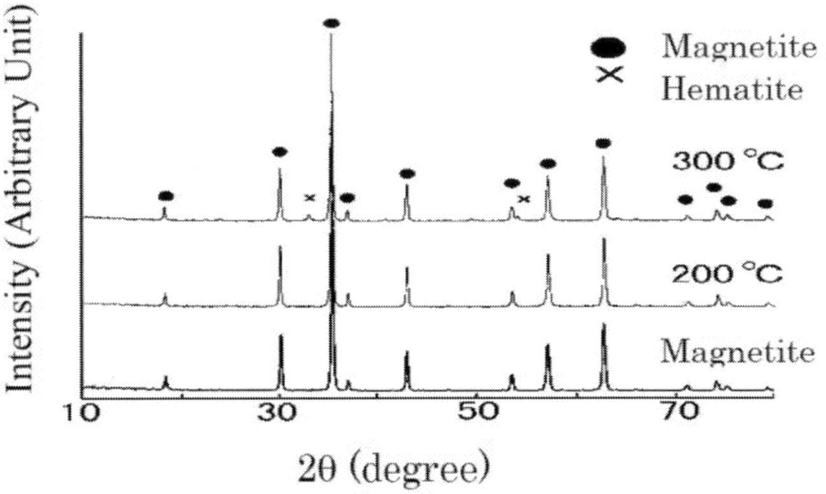

Figure 7 XRD spectra of hydrothermally treated magnetic powder

4. Conclusion

The major conclusions that can be derived from the present study are summarized as follows:

1) We designed a magnetic powder recovery system for sludge discharged during magnetic separation from oily water at oil production sites.
2) We found that the ideal hydrothermal treatment temperature for a magnetic powder recovering system is 200°C.

REFERENCES

1) J. Long: Energy & Fuels 19 (2005) 1440.
2) Janet Pelley: Environ. Sci. Technol. A-Pages, 41-14 (2007) 4838.
3) M.Morita: The Japan Society of Mechanical Engineers B, 70-696 (2004) 2071. (in Japanese).
4) M.Morita: The Japan Society of Mechanical Engineers B, 71-706 (2005) 1611. (in Japanese).
5) H.Isogami: The Japan Society of Mechanical Engineers B, 64-628 (1998) 4212. (in Japanese).

718f Mechanism of the Diesel PM Removal by Dielectric Barrier Discharges

Satoshi Kodama, Shuiliang Yao, Shin Yamamoto, Chieko Mine, and Yuichi Fujioka. Chemical Research Group, Research Institute of Innovative Technology for the Earth, 9-2 Kizugawadai, Kizugawa-shi, Kyoto 619-0292, Japan

It is desirable to change vehicles powered by gasoline to those by diesel to reduce the emission of the greenhouse gas CO_2, as the energy efficiency of vehicles powered by diesel is better than those by gasoline. Unfortunately, the emission of harmful particulate matter (PM) and nitrogen oxides (NO_x) from diesel vehicles is a problem; innovative after-treatment systems for removals of diesel PM and NO_x are required. Diesel particulate filters (DPF) have been developed for PM removal from diesel vehicles and it can remove PM efficiently. However, the pressure loss due to PM deposition in the DPF and additional fuel required for DPF regeneration create a fuel penalty. To resolve these problems, we have developed a plasma PM removal system aiming to reduce PM emission with a low fuel penalty. The plasma discharges are generated using a dielectric barrier discharge (DBD) reactor to activate O_2, H_2O and NO in diesel exhaust gases to O, OH, O_3, and NO_2. PM (the main component is carbon) reacts with these activated oxygen species to yield CO and CO_2. In this study, the mechanism of the diesel PM removal by the plasma discharges was investigated.

The PM samples were collected from the exhaust gases of a diesel engine. A lab-scale DBD reactor was used to generate plasma discharge. This DBD reactor consists of two pieces of flat alumina plates, two pieces of aluminum flat plates and spacers. Several parallel grooves are engraved on one side of each alumina plate. Each aluminum plate is placed on the flat side of the alumina plates. Each two alumina plates of grooves on their surfaces are assembled face to face, and between which the spacers are inserted to form discharge spaces. The aluminum plates are connected to the output side and earth terminal of a high-voltage pulse power supply. A mixture gas of PM, H_2O, N_2 and O_2 was introduced to the discharge spaces of the DBD reactor. PM removal ratios were calculated from concentrations of PM oxidation products (CO and CO_2) in the gas from the DBD reactor using gas chromatographs. After plasma discharges, PM samples were collected and analyzed using XPS to find the changes of carbon bonding ratios. The oxidation of PM (before and after plasma discharges) with O_3 was also measured by using a thermo gravimetric analyzer.

It was found that the ratio of sp2 bonding carbon in the PM sample decreased while the ratio of sp3 bonding carbon increased after plasma discharges. This finding suggested that the active oxygen species generated by plasma discharges reacted with the sp2 bonding carbon to form sp3 bonding carbon. On the other hand, the ratio of sp3 bonding carbon decreased while the ratio of sp2 bonding carbon increased after the reaction with O_3. This fact implied that PM is further oxidized by O_3 from the sp3 bonding carbon to CO and CO_2. The detail mechanism of plasma PM removal and the further development of this technology will be given.

This work was supported by the New Energy and Industrial Technology Development Organization, Japan.

Human Urine Utilization: A Waste Management Strategy

1.0 Introduction

Waste management is the collection, transportation, processing, recycling or disposal of waste materials. The term usually relates to materials produced by human activity, and is generally undertaken to reduce their effect on health; the environment or aesthetics.Human urine can be well harnessed through proper waste management method. Waste management is also carried out on human urine to reduce the materials' effect on the environment and to recover resources from them. Human urine is liquid waste product of the body secreted by the kidneys by a process of filtration from blood and excreted through the urethra. This waste is eventually expelled from the body in a process known as urination. Most commonly the excretion of urine serves for flushing waste molecules collected from the blood by the kidneys, and for the homeostasis of the body fluids. This liquid waste product of the body is a potential source of fertilizer which is more environmental-friendly compared with chemical fertilizer if properly harnessed.

The use of human urine is particular interest in the field of organic farming where fast-acting fertilizers are in demand. A new project was demonstrated which showed the potentiality for developing urine-separation toilets, and that from agricultural point of view-separated urine is comparable to liquid manure. This is a unique way of managing human urine.

1.01 Urine

Urine used directly or after storage is of high quality and low cost, alternative to N-rich mineral fertilizer in plant production. The nutrients in urine are in ionic form and there plant – availability compares well with chemical fertilizer (Johansson et al., 2001; kirchmann and Pettersson, 1995; Kvarmo, 1998; Richert Stintzing et al., 2001). Urine is best utilized as direct fertilizer for N-demanding crops and leafy vegetables. If crop and region – specific recommendations are available for the use of N fertilizers (urea, ammonium or nitrate), a good starting point for how to use urine is to translate the recommendations to urine. The translation is simplified if the N concentration of the urine is known. If not then as a rule of thumb, a concentration of 3-7 grams of N per litre of urine can be expected (vinneras, 2002; Jonsson & vinneras, 2004). Urine also contains large amounts of P and K, but due to its large content of N, its P/N and K/N ratios are lower than in many mineral fertilizers used for vegetable production.

The yield achieved when fertilizing with urine varies depending on many factors. One important aspect is the soil condition. The effect of urine, just as that of chemical fertilizers, is probably somewhat lower on a soil with a low content of organic substances than on a soil with a high organic content.

2.0 Broad Objectives

Testing the efficacy of the utilization of human urine as a fertilizer- a waste management strategy.

3.0Methods

This study was designed to determine the utilization of human urine in the context of waste management. The study was experimental in design and was conducted under the green house condition. Lemongrass was grown with three treatment media- inorganic fertilizer, urine, compost and ordinary soil as control. Each treatment was carried out by mixing 12.5g of organic manure with 5kg of soil, 0.25g of NPK with 5kg of soil, while 125mls of urine was mixed with 5kg of soil. The growth parameters which include number of leaves, number of stolons and height of the plant were monitored and determined for 18 weeks and compost made from city refuse. Lemongrass grown on ordinary soil was used as control. Each treatment was carried out by mixing 12.5g of organic manure (5tonnes/ha) with 5kg of soil. Furthermore, 0.25g of NPK (100kg/ha) was mixed with 5kg of soil while 125mls of urine was mixed with 5kg of soil which was applied at the rate of 4600l/ha.

The experiment was conducted in two sets and each treatment was carried out in triplicates. In the first set, urine was applied once while in the second set, the urine was repeatedly applied once every week for a period of 9-weeks at ratio 3:1 (water: urine). Meanwhile the whole of the experiment took 18-weeks after which it was terminated.

500ml of water was used for irrigation 3 – times in a week as this is the volume of water 5kg of the soil sample could take to avoid drainage. After the soil preparation, 17cm of lemongrass slips all of equal heights were planted at 5cm deep into the soil. The following parameters were taken weekly,

- The number of leaves - by counting
- The Height - using metric rule
- Number of stolon - By counting.

After 18 weeks, the leaves were harvested by cutting the plant at 15cm above ground level. One kilogram of fresh leaves from each treatment were milled and extracted for essential oil through hydro-distillation method. While some were oven dried at 65°C for 24hrs for laboratory analysis.

The soil samples were air-dried, sieved and packed for Nitrogen, Phosphorus and Potassium analysis both soil and plant samples were analyzed for NPK using standard laboratory method at International Institute for Tropical Agriculture Ibadan (IITA).

4.0 Results

The results showed that the mean number of leaves grown by the lemongrass treated with urine, inorganic fertilizer, compost and the control soil were 31.0, 30.7, 17.8, 14.6 respectively ($P<0.05$). The mean number of stolons of the lemongrass treated with urine, inorganic fertilizer, compost and the control soil were 4.9, 4.1, 2.7, 2.1 respectively ($P<0.05$). The mean height of the lemongrass in centimeter treated with urine, inorganic fertilizer, compost and the control were 71.6, 80.9, 93.3 and 63.9 respectively ($P<0.05$). The analysis showed that the growth of lemon grass were slightly higher with urine than with conventional fertilizer.

The total Nitrogen, Phosphorus and Potassium were determined from samples of lemongrass, including soil samples from each treatment. The results were analyzed using ANOVA and descriptive statistics. Table 1 to table 3 showed the significance difference within the group, the number of leaves, number of stolon and height of lemon grass treated with NPK, Urine, Organic fertilizer and the Control respectively.

Table 1

One-way Analysis of variance for first set of Lemon grass

Number of leaves:

Number of leaves	**Mean**	**Standard Deviation**	**F. Value**	**P.Value**
N.P.K	30.70	19.103		
Urine	31.00	22.059	6.634	0.000
Organic Fertilizer	17.81	12.689		
Control	14.63	13.290		
Total	23.54	18.554		

Table 2

Number of stolon

Number of stolon	Mean	Standard Deviation	F. Value	P.value
N.P.K	4.0741	2.38466		
Urine	4.9259	3.80208	7.051	0.000
Organic fertilizer	2.6667	1.66410		
Control	2.0741	1.73041		
Total	3.4352	2.75231		

Table 3

Height(cm)

Height	Mean	Standard Deviation	F value	P.value
N.P.K	80.9259	23.81003		
Urine	71.6667	13.76171	10.948	0.000
Organic Fertilizer	93.3333	16.20779		
Control	638889	23.76865		
Total	77.4537	22.50416		

Table 4

Table 4 to Table 6 show a multiple comparison between the groups.

Multiple Comparisons

Number of Leaves

Urine	**Treatment**	**Mean Difference**	**Standard Error**	**Sig.**
	N.P.K	.296	4.693	.950
	Organic fertilizer	13.185	4.693	.006
	Control	16.370	4.693	.001

Table 5

Number of Stolons

Urine	Treatment	Mean Difference	Standard Error	Sig.
	N.P.K	.85185	.69263	.222
	Organic Fertilizer	2.25926	.69263	.001
	Control	2.85185	.69263	.000

Table 6

Plant Height(cm)

Urine	Treatment	Mean Difference	Standard Error	Sig.
	N.P.K	-9.25926	5.41593	.090
	Organic Fertilizer	-21.66667	5.41593	.000
	Control.	7.77778	5.41593	.154

The statistical analysis of the first set showed an increase of 52.8% in number of leaves, 10.9% in height and 57.9% in number of stolon of lemongrass treated with urine compare with control ($p<0.05$), while there was an increase of 42.6% in number of leave, 45% in number of stolon of lemongrass treated with urine compare with organic fertilizer at ($p<0.05$).

Table 7

Analysis of Variance of soil Samples.

Nitrogen.

Treatment	**Mean**	**Standard Deviation**	**F value**	**P value**
N.P.K	.11400	.002646		
Urine	.12633	.011372	4.462	.040
Organic Fertilizer	.11033	.005686		
Control	.10600	.006083		
Total	.11417	.009998		

Table 8

Phosphorus.

Treatments	**Mean**	**Standard Deviation**	**F. Value**	**P. Value**
N.P.K	39.4667	17.59324		
Urine	33.3567	14.07438	1.694	.245
Organic Fertilizer	25.7433	1.66962		
Control	18.9600	7.40452		
Total	29.3817	12.96330		

Table 9

Potassium

Treatments	**Mean**	**Standard Deviation**	**F Value**	**P Value**
N.P.K	.2333	.03786		
Urine	.1933	.02517	2.367	.147
Organic fertilizer	.2233	.02309		
Control	.2667	.04509		
Total	.2292	.03988		

Table 10

Multiple Comparisons for Nitrogen

Nitrogen	Treatment	Mean Difference	Standard Error	Sig.
Urine	N.P.K	.012333	.005855	.068
	Organic Fertilizer	.016000	.005855	.026
	Control	.020333	.005855	.008

The mean difference is significant at P 0.05 level

There was no significant difference in Physical and Chemical characteristics of the soil samples except in Nitrogen content ($P < 0.05$). Nitrogen in urine treated soil is higher than that of the control and organic fertilizer.

Table 11

Analysis of Variance of Plant samples

Nitrogen.

Treatment	**Mean**	**Standard Deviation**	**F Value**	**Sig.**
N.P.K	2.06933	.041525		
Urine	2.47700	.222032	2.618	.146
Organic Fertilizer	2.11250	.284964		
Control	1.66000	.499361		
Total	2.03670	.402878		

Table 12

Phosphorus

Treatment	**Mean**	**Standard Deviation**	**F Value**	**Sig.**
N.P.K	.17900	.004000		
Urine	.17050	.006364	.313	**.816**
Organic Fertilizer	.18700	.007071		
Control	.16233	.051394		
Total	.17390	.026354		

Table 13

Potassium.

Treatment	Mean	Standard Deviation	F Value	Sig.
N.P.K	.31000	.057236		
Urine	.40250	.006364	3.426	.093
Organic Fertilizer	.46150	.062933		
Control	.37067	.056889		
Total	.37700	.071643		

5.0 Conclusion

In conclusion, human urine could use as a fertilizer for lemongrass through proper waste management method and the quality of fertilizer it produces is better than the conventional ones. There is significant positive effect of urine on lemongrass if the urine is applied once for each harvesting period. The non-significant different in quality and quantity of lemongrass treated with NPK and urine could be attributed to the ionic form of nutrient in both fertilizers. Meanwhile, chemical fertilizer has environmental hazard and health risks. The use of urine will reduce the cost of fertilizer and instead of its being indiscriminately disposed which has negative impact on environmental safety; it could be harnessed for agricultural purpose thus serving a dual purpose of ecosanitaton and soil fertilization. This in itself is a waste management technique.

718h Biodegradation of Refinery Wastewater Using *O. Intermedium*

Aditi Banerjee, Centre for Environment, Indian Institute of Technology Guwahati, Guwahati, India and **Aloke Kumar Ghoshal**, Chemical Engineering, Indian Institute of Technology Guwahati, Guwahati, India.

The objective of this work is to investigate an efficient method of aerobic biodegradation for wastewater generated in oil refinery.

Oil spillage and contaminated effluent are inevitable consequences of oil exploration, production and refining. As a result, the problem of pollution during production and transportation of refinery product had been a major issue during the last few decades. Organic compounds such as alkanes, cycloalkanes, aromatic hydrocarbons, or more complicated chemicals such as asphaltene are the main components of refinery effluents. Among these compounds, mono-aromatic hydrocarbons such as benzene, toluene and phenol are found in huge amounts [4]. These aromatic compound are highly toxic in nature and cause damage to the aquatic biota. Even for human lives, these compounds can cause liver and kidney damage, cardiac toxicity including weak pulse, cardiac depression and reduced blood pressure [1].

The microbiological degradation is a potential biological treatment technique which is based on the ability of microorganism (generally bacteria) to convert organic pollutant into water, carbon dioxide and biomass under aerobic or anaerobic condition. It appears to be the most environment friendly method of removal of oil pollutant, as no undesirable by-products or secondary emissions, such as chemical scrubbing or thermal waste gas treatment, are generated [2, 3]. Aerobic treatment is easier to implement on site than anaerobic process due to its less number of constraints on design parameters.

Thus, our present work is focused on, (a) to isolate and identify a microorganism which is capable of degrading refinery wastewater more efficiently than other reported microorganism [2], (b) to find whether the degradation process produces environment friendly byproducts, (c) to verify whether there are any kind of toxicity present in those byproducts and if yes, whether that can be controlled or manipulated by changing the degradation pathway or by any other means.

In our recent laboratory scale studies, we have isolated *O. intermedium* from the activated sludge of petroleum refinery. It has been observed that this organism is capable of degrading waste from the refineries and it has the ability to grow on and utilize some toxic compounds like phenol as sole carbon and energy source. The optimum temperature and pH required for maximum degradation are 37 ^{o}C and 7.0 respectively at 120 rpm. The microbe is capable of degrading high concentration of phenol (up to 2000mg/L) under aerobic condition. Microbial growth kinetic models shows the best fit for Haldane model with R^2=0.9978.

References

1. A., Nuhoglu, B., Yalcin, , Modeling of phenol removal in a batch reactor. Process Biochemistry) 40, 1233-1239, (2005)

2. N. Khoury, W. Dott, P. Kampfer, Anaerobic degradation of phenol in batch and continuous cultures by a denitrifying bacterial consortium. Appl. Microbiol. Biotechnol. 37, 524-528, (1992)

3. Schaefer, M., Juliane, F., The influence of earthworms and organic additives on the biodegradation of oil contaminated soil, Applied soil ecology, 36, 53-62, (2007)

4. Thangaraj, K., Kapley, A., Purohit, H.J., Characterization of diverse Acinetobacter isolates for utilization of multiple aromatic compounds, Bioresource Tech., 99, 2488–2494, (2008)

718i Decomposition of 2-Chlorophenol In Supercritical Water Oxidation Using Anticorrosive Continuous Reactor

Jae-Hyuk Lee, Tran Tan Viet, and **Chang-Ha Lee**. Department of Chemical and Biomolecular Engineering, Yonsei.university, 262 Sungsanno Seodaemun-Gu, Seoul, 120-749, South Korea

The 2-chlorophenol (2-CP) was decomposed in a continuous anti-corrosive supercritical water oxidation (SCWO) system. The effect of the corrosion of zirconium 702 on the decomposition efficiency was studied at the various conditions of feed concentration and reaction time. The surface chemical analysis of the corroded zirconium exposed to the SCWO was performed by the AES-SAM. According to AES depth profile, the oxygen penetration depth to zirconium was not proportional to the exposure time. It might be stemmed that the formation of zirconium oxide layer on the surface delayed the corrosion. However, the increase in feed concentration accelerated the corrosion of zirconium. The corrosion of zirconium at low feed concentration led to the improvement of decomposition efficiency due to the catalytic work of formed zirconium oxides while that at high feed concentration deteriorated the decomposition efficiency owing to large consumption of oxidant in corrosion.

718j Modeling and Design of Fluidized Bed Reactors for Biological Denitrification of Reverse Osmosis Brine Concentrates

Ilknur Ersever, Huan-Hao Tsai, **Varadarajan Ravindran**, and Massoud Pirbazari. Civil and Environmental Engineering, University of Southern California, Los Angeles, CA 90089-2531

The treatment of brine concentrates from reverse osmosis (RO) processes used for water recycling applications poses a major challenge. This research addresses the use of several bio-physicochemical processes for removing ammonia, nitrate and sulfate from RO brine rejects, and essentially consists of four different stages. The first stage involves nitrification of ammonia using a bioactive fluidized-bed reactor (FBR) process under aerobic conditions. The second stage deals with denitrification of effluent from the first FBR in a second FBR operated under anaerobic conditions. The third stage achieves simultaneous reduction of sulfate in the FBR performing denitrification. The final stage addresses the gas-phase biofiltration of hydrogen sulfide formed in the third stage. Among various phases of the treatment process, this work focuses the modeling and design of the denitrification and simultaneous sulfate reduction stages of the treatment with emphasis on process upscaling and design.

A mathematical model was developed for predicting the dynamic behavior of the FBR denitrification process in two stages along with simultaneous sulfate reduction under anaerobic conditions. The model involved the application of sorption and biodegradation phenomenon in an FBR system addressing liquid film and biofilm transport besides diffusion into the sorbent granular activated carbon (GAC) media particles. The modeling approach was also applied to non-adsorbing bioactive media particles such as sand. The model was intended to simulate or predict the dynamics and performances of FBR systems for denitrification and sulfate reduction under a variety of process and operating conditions. Sensitivity studies were undertaken to evaluate the effects of biokinetic, sorption and transport parameters on the FBR dynamics. More importantly, dimensional analysis and similitude techniques were employed to develop a process upscaling scenario. This process upscaling techniqueh was based on dimensionless groups associated with reaction kinetics, biofilm and particle diffusion transport as well as liquid film transport.

Laboratory-scale FBAR investigations were conducted to achieve maximum biological nitrification efficiencies. A series of batch studies were performed to determine the effects of several variables on the efficiency of the denitrification process including temperature, pH, total dissolved solids (TDS) and carbon-to-nitrogen (C:N) ratio. A series of chemostat tests were conducted to determine the Monod biokinetic parameters for the nitrification as well as the denitrification processes. It was found that nitrate was significantly more favorable as the main substrate for the denitrifying culture than nitrite, and that insufficient carbon source caused instability of the system due to inhibitory effect of nitrate and/or nitrite accumulation. The FBR denitrification experiments (first stage) were conducted under various conditions including different GAC quantities, hydraulic retention times and nitrate concentrations. These investigations showed that nitrate removal efficiencies as high as 96% to 100% could be achieved. Similar FBAR experiments conducted with sand showed that (i) completion of denitrification process took much longer time than with GAC with substantial nitrite accumulation, and (ii) significantly smaller biomass concentrations were obtained with sand as compared to GAC. It is important to note that the second FBAR achieved complete biological sulfate reduction to hydrogen sulfide, reducing the sulfate concentration almost to zero levels (third stage). The biofiltration process (fourth stage) effectively removed the hydrogen sulfide from the gas stream.

Model simulation studies demonstrated good agreement between the experimental data and model predictions. Model sensitivity analyses indicated that growth yield and maximum substrate utilization rate coefficients had the utmost influence on the process. The approach for upscaling the process from laboratory-scale to pilot-scale or full-scale is illustrated in this discussion.

Key words: biological denitrification, sulfate reduction, fluidized bed reactor modeling, process upscaling

718k Diversity and In Situ Quantification of Tributyltin-Degrading/resistant Bacteria Under Different Cultivating Conditions

Young Haeng Lee[1], Soohyun Park[1], Yongsu Choi[1], and Hee-Deung Park[2]. (1) Center for Environment Technology Research, Korea Institute of Science and Technology, 39-1 Howelgok-Dong, Sungbuk-Gu, Seoul, 130-650, South Korea, (2) R&D Institute, Kolon Engineering & Construction, 199-5 Chundae-Ri, Pogok-Myun, Yongin-si, 449-815, South Korea

All organotin compounds are toxic, but the effect varies according to the number and the type of organic moiety present, as butyl-group bearing organotin is more toxic to fungi and bacteria. Tributyltin (TBT) is an organometallic compound used as antifouling agent in marine paints.

Based on the literature review, a few researchers have reported evidence for the presence of TBT-resistant bacteria in natural seawater and marine sediment.

The objective of this study was to assess the diversity and community composition of TBT-resistant bacteria under different conditions (TBT, monobutyltin (MBT) known as microbial degradation product of TBT, and dextrin/peptone mixture as an external carbon source).

Resistant to TBT was examined in bacterial populations enriched using TBT-contaminated marine sediment (Ulsan Bay, South Korea) sample as inoculum, at the initial TBT concentration of 10 and 100 ppm which was much higher level than the observed TBT level in the contaminated sediment.

Phylogenetic trees based on the 16S rDNA sequence revealed that Marinobacter sp. and alpha proteobacterium sp. were identified for 100 ppm TBT-amended culture samples regardless of the presence or absence of carbon source and 100 ppm MBT. Those bacteria to date have never been reported from the TBT-contaminated marine sediments. Quantitative fluorescence in situ hybridization (FISH) analysis confirmed that the Alphaproteobacteria were the most dominant group in all 100 ppm TBT-amended cultures. For 10 ppm TBT-amended cultures, along with Marinobacter sp. and alpha proteobacterium sp., Halomonas sp. were identified in the absence of carbon source, and in the presence of carbon source, Marinobacterium sp., Desulfovibrio sp., Thiomicrospira sp., Soehngenia sp., Acholeplasma sp. were additionally identified. However, in the presence of 10 ppm MBT, phylogenetic analysis revealed that only Marinobacter sp. and alpha proteobacterium sp. were identified for all 10 ppm TBT-amended cultures. Similar to the results obtained in 100 ppm TBT-amended cultures, quantitative FISH analysis resulted that the Alphaproteobacteria were the most dominant group in all 10 ppm TBT-amended cultures. Both quantitative FISH and restriction fragment length polymorphism (RFLP) analyses resulted that the community composition of TBT-resistant bacteria changed significantly when the external carbon source and/or MBT were amended.

718I Characterization of Multilayer Ceramic Cake Filtration System: Collection Efficiency, the Porosity, and the Pressure Drop

Hern Kim, Department of Environmental Engineering and Biotechnology, Myongji University, San 38-2 Nam-dong Cheoin-gu, Yongin, Kyonggi-Do, 449-728, South Korea and In-Seob Lee, Mirae NET Co., 360-4 Yatop-dong Bundang, Seongnam, Kyonggi-do, 463-828, South Korea.

A new filtration system has been developed, which consists of reactive and/or selective functional multilayer ceramic cakes to remove fine dusts and toxic gases from a variety of air pollution sources. This study presents a method to predict the collection efficiency, cake thickness, cake porosity, and pressure drop across filter cakes during the particulate filtration of gases. The model predicts the cake collection efficiency, which was found to be strongly dependent on the ratio of the mean particle size to the mean pore size of the filter medium. The average cake porosity decreases with increasing cake thickness and the pressure drop increases when the mean particle diameter decreases. The cake height, the pressure profile and the local specific filtration resistance of the system are examined by experiment and theory. It was found that there was excellent agreement between the calculated and measured local pressures. It is possible to design a combination of ceramic filter cakes with the so-called a multilayer model.

718m Analysis and Retrofit Design of Wastewater Treatment Facilities Using Process Simulation Tools

Victor Papavasileiou[1], Charles Siletti[2], and Demetri Petrides[2]. (1) Intelligen Europe, Rijnsburgerweg 119, Leiden, 2334, Netherlands, (2) INTELLIGEN, INC., 700 Walton Ave., Mount Laurel, NJ 08054

Process simulators have been used in the petroleum and chemical industries for over four decades to facilitate the design of new processes and optimize the performance of existing ones. Similar benefits can be derived from the use of such tools in the environmental arena, particularly in the field of physical and biological treatment of municipal and industrial wastewater. Specifically, process simulators can be used to evaluate and improve options for: (1) more efficient removal of nutrients (e.g., organic nitrogen and phosphorous) that cause eutrofication, (2) estimation and control of volatile organic compound (VOC) emissions from open tanks, and (3) more efficient removal and control of hazardous compounds. The potential benefits will be illustrated with cases studies involving both municipal and industrial wastewater facilities.

Web Page: www.intelligen.com/literature

Chemical Speciation of Heavy Metals in Ground and Surface Waters

Hundal H. S., Kuldip- Singh and Dhanwinder- Singh
Department of Soils, Punjab Agricultural University,
Ludhiana, Punjab-140004, India

Metal speciation in natural waters is of increasing interest and importance because of toxicity, bioavailability, and environmental mobility. Biogeochemical behavior and potential risk in general are strongly dependent on the chemical species of metals. The present study aims to determine the distribution of chemical species of dissolved heavy metals in shallow and deep aquifers and surface waters from Buddha nala stream and Sutlej River of Punjab, north-west India. Attempt was also made to identify different possible minerals that may control the solubility of elements in these aqueous solutions. The study was performed at village Walipur in Ludhiana ($30^0 55^/$N, $75^0 54^/$E) district of Punjab, north-west India, where Buddha nala surface water stream loaded with heavy metals from municipal waste and Industrial effluents contaminate underground water. A sutlej River also flowed a couple of Km from the village site. Ground water samples were collected at four different occasions from the same sites of shallow aquifer (45 feet deep) and deep aquifer (200 feet deep) by the already installed hand pump and tube well adjacent to Buddha nala stream. Before sampling the water was drawn for half an hour to empty the hand pump and tube well pipes in order to collect the fresh water from the shallow and deep aquifers, respectively. Surface water samples on four different occasions were also taken from Buddha nala stream and Sutlej River. Polyethylene bottles of 50 ml volumes were used for collection of water samples. These bottles were repeatedly washed with the water to be sampled. Bottles were fully filled with water and closed avoiding air bubbles. Field measurements of redox potential (Eh) were also made at the time of water sampling. The redox meter checked with Zobell's solution several times during the investigation period. Water samples were taken from each location and brought back to Laboratory at Punjab Agricultural University for further analyses. The samples were put in a cooled ice box during transportation and stored in refrigerator until analysis (at 5 ^{0}C). The pH of water samples was determined using Elico model L 110 glass electrode in combination with calomel as reference electrode on pH meter. The electrical conductivity of water samples was measured by Elico model CM-84 Conductivity Bridge. The carbonates and bicarbonates concentrations were measured by titration a known volume of water against standard sulfuric acid using phenolphthalein and methyl red as indicators respectively. The chloride concentration in water samples were measured by titrating with a known volume against standard N/40 silver nitrate solution using potassium chromate as indicator. All samples were filtered and acidified prior to analysis. The acidified water samples were analyzed for Cadmium, Nickel, Lead, Zinc, copper, Iron, Manganese and sulfur on Inductively Coupled Argon Plasma of Thermo Electron corporation iCAP 6000 series. Mean of Metals and dominant anions composition of waters determined after four occasions for different sites are listed in Table 1. The higher concentration of Cd, Cu,Fe, Mn, Ni, and Pb in waste water of Buddha nala stream must be due to disposal of Industrial effluents. The ionic strength of waste water was also three times more than other water sources. Comparatively high concentration of Zinc (16.10 uM) was measured in water of shallow aquifer than water of deep aquifer and surface water of Buddha nala stream and Sutlej River. The geochemical speciation model Visual MINTEQA2 Version 3.11 was used to calculate the equilibrium concentrations of dissolved species for four solutions, representing the range of water compositions encountered in the shallow and deep aquifers as well as in surface waters of Buddha nala

Table1. Metals and dominant anions composition of waters collected from different sites used in speciation calculations

Element	Shallow aquifer Hand pump	Deep aquifer Tube well	Surface water Buddha nala	Surface water Sutlej River
Cu uM	0.10	0.06	2.14	0.11
Zn uM	16.10	0.79	12.80	0.53
Cd uM	0.03	0.02	0.07	0.01
Pb uM	0.04	0.07	0.55	0.14
Ni uM	0.18	0.13	5.25	0.27
Fe uM	9.39	0.01	250	7.51
Mn uM	0.61	0.20	4.06	0.34
Cl^-	45.71	22.86	170	28.57
CO_3^{2-}	110.20	58.10	190	25.00
SO_4^{2-}	880	340	650	460
Ionic strength (u) ML^{-1}	0.016	0.010	0.034	0.010
pH	7.50	7.90	7.30	8.0
Pe (mV)	47	75	-253	226

stream and sutlej Rivers, respectively. The speciation of trace elements in different sources of waters are summarizes in Table 2. In all water compositions, except waste water of Buddha nala Zn, Cd, Ni, and Mn were predominantly present (93-97%) as divalent cation forms. About 95 % of the total dissolved iron in shallow and deep aquifers was in ferrous (Fe^{2+}) form due the occurrence of reduced status in underground waters. In Sutlej River, 90 % of the total dissolved iron was computed to be in ferric (Fe^{3+}) form due to oxidized (+226 mV) water status. The free activity of Pb^{2+} ion were 52, 33, 39, and 34% of the total dissolved lead in waters of hand pump, tube well and surface waters of Buddha nala and Sutlej River streams, respectively. Significant contribution of $Me(OH)^+$ (29%) and $Me(CO)_3$ aq (27%) complexes were computed for all the metals in Buddha nala water stream and ultimately these complexes eventually contributes to the higher concentration of these elements in this aqueous system. Appreciable fraction of $Me(OH)^+$ were also found for Cu and Pb in shallow and deep aquifers and Sutlej River waters. The $Me(CO_3)$ aq species were also found for Cu and Pb in waters of shallow and deep aquifers abstracted by hand pump and tube well, respectively.

Analyzed heavy metals data were also used in Visual MINTEQA2 to determine geochemical signatures of solute sources in the aquitards of shallow and deep aquifers, and bed sediments of Sutlej River and Buddha Nala stream. In addition to bed sediments, discharges from industrial manufacturing activities into Buddha Nala stream also cause increases in heavy-metal concentrations in Waste water. Selected minerals with saturation indices in the -2 to +2 range are presented in Table 3. The results indicate that various minerals containing of heavy metals have the potential to precipitate or dissolve in surface and ground waters. A positive saturation index indicates that the mineral is super saturated and might be precipitate in solid phase form. A negative saturation indicates that mineral is under saturated and if is present in the aquifers materials or in the bed sediments is dissolving to maintained its equilibrated concentration in the aqueous medium. Among the copper minerals,

Table 2. Percent species distributions of copper, zinc, cadmium, lead, Nickel, Iron and Manganese in shallow, deep aquifers and in surface waters of Buddha Nala stream and Sutlej river in Punjab, North West India

Species	Hand Pump water							Tube Well Water						
	Cu	Zn	Cd	Pb	Ni	Fe	Mn	Cu	Zn	Cd	Pb	Ni	Fe	Mn
Me^{2+}	45	93	96	52	96	95	97	25	87	97	33	97	95	98
$Me(OH)^+$	31	2	-	28	-	1	1	49	5	-	50	1	2	-
$Me(OH)_2$	2	1	-	-	-	-	-	7	5	-	1	-	-	-
$Me(OH)_3$	-	-	-	-	-	-	-	-	-	-	-	-	-	-
$Me\ (SO_4)_{aq}$	2	3	4	4	3	4	3	-	2	2	1	2	2	1
$Me\ (CO_3)_{aq}$	20	-	-	13	-	-	-	18	1	-	14	-	-	1
$Me\ HCO_3^+$	-	-	-	2	1	-	-	-	-	-	1	-	-	-
Species	Buddha nala stream water							Satlej River Water						
Me^{2+}	39	39	39	39	39	39	39	28	86	95	34	96	90*	97
$Me(OH)^+$	29	29	29	29	29	29	29	54	5	-	54	1	-	-
$Me(OH)_2$	3	3	3	3	3	3	3	7	5	-	1	-	-	-
$Me(OH)_3$	-	-	-	-	-	-	-	-	-	-	-	-	4	-
$Me(OH)_4^-$	-	-	-	-	-	-	-	-	-	-	-	-	6	-
$Me\ (SO_4)_{aq}$	1	1	1	1	1	1	1	1	3	4	3	3	-	3
$Me\ (CO_3)_{aq}$	27	27	27	27	27	27	27	9	-	-	7	-		-
$Me\ HCO_3^+$	-	-	-	-	-	-	-	-	-	-	-	-		-

*Iron is present in ferric form

Table 3 Results from modeling with MINTEQ. All minerals with saturation indices (SI) from –2 to +2 are shown shallow, deep aquifers and Buddha Nala and Sutlej Rivers waters.

Mineral	Handpump	Tubewell	waste	Sutlej	Mineral	Handpump	Tubewell	waste	Sutlej
$Cu(OH)_2$	-1.85	-1.45	-0.47	-1.17	Malachite $Cu_2(OH)_2CO_3$	-1.56	-1.40		-1.19
$CuCO_3$	-	-	-1.50	-	Atacamite $Cu_2(OH)_3Cl$	-	-	-1.21	-
Langite $(Cu)_4(OH)_7HSO_4$	-	-	-1.28	-	Antlerite $Cu_3SO_4(OH)_4$	-	-	-1.40	-
Tenorite amp	-1.05	-0.65	0.33	-0.37	Azurite $Cu_3(OH)_2(CO_3)_2$	-	-	0.22	-
Tenorite (c)	-0.20	0.23	1.18	0.48	Brochantite	-	-	0.99	-
Siderite $FeCO_3$	-1.59	-	0.05		$Ni(OH)_2$ (c)	-	-	-1.18	
$Fe(OH)_2$ (c)	-	-	-1.60	-	Rhodochrosite $MnCO_3$	-	-	-1.32	
Smithsonite $ZnCO_3$	-1.05	-	-0.94	-	Hydrocerrusite$(Pb)_3(OH)_2(CO_3)_2$	-	-	0.07	
$Zn(OH)_2$	-1.78	-	-1.75	-1.72	Cerrusite $PbCO_3$	-1.57	-1.35	-0.30	-1.36
$NiCO_3$	-	-	-1.01	-	$Pb(OH)_2$	-1.01	-0.14	0.17	0.19

copper hydroxide [Cu(OH)2], amorphous and crystalline tenorite [CuO] were identified as the solid phase in aquifer materials of shallow and deep waters and in the bed sediments of Buddha Nala and Sutlej River waters. Malachite [Cu2(OH)2CO3] mineral was found in materials of shallow, deep water aquifers and in bed sediments of Sutlej Rivers (Table 3). Their could be the possibility of the existence of copper carbonate [CuCO3] and Langite [(Cu)4(OH)7HSO4], Atacamite [Cu2(OH)3Cl], Antlerite [Cu3SO4 (OH)4] and Brochantite [CuSO4·3Cu(OH)2] in wastewater of Buddha Nala. The occurrence of these different copper minerals in aquifer materials for ground water and bed sediments could eventually be the primary factors for controlling the concentration of copper in water sampled from different sites. Among zinc minerals, Smithsonite [ZnCO3] was found to be present in shallow aquifer materials and in the bed sediments of Buddha Nala waste water (Table 3). Zinc hydroxide [Zn(OH)2] was recorded in aquifer material of shallow water, bed sediments of Buddha Nala and Sutlej river waters. Nickel carbonate [NiCO3] and Nickel hydroxide [Ni(OH)2] mineral were identified in bed sediments of Buddha Nala waste water only. Similarly the Rhodochrosite [MnCO3] mineral for manganese and Hydrocerrusite [(Pb)3(OH)2(CO3)2] mineral for lead solid phases were identified in bed sediments materials of Buddha Nala waste water site. The most common minerals of lead Cerrusite [PbCO3] and lead hydroxide [Pb(OH)2] were found in aquifers material of shallow, deep waters as well as in the sediments of Buddha Nala waste and Sutlej Rivers waters. In case of iron, Siderite [FeCO3] mineral was identified in the shallow aquifer materials and in the sediments under waste water of Buddha Nala (Table 3). The reduced form of mineral for iron as ferrous hydroxide [Fe(OH)2 (c)] was found in the bed sediments of Buddha Nala waste water. The identification of this mineral illustrates that bed sediments under waste water of Buddha Nala were highly reduced conditions. The redox potential recorded for waste water at the time of sampling also showed low value of redox status (-253 mV).

Absorption of CO_2 on $CaSiO_3$

***Minghua Wang*[a], *Choong-Gon Lee*[b]*, *and Chong-Kul Ryu*[c]**

[a] *College of Material and Metallurgy, Northeastern University, Shenyang 110006 China*
[b] *Chemical Engineering, Hanbat National University, Daejeon 305-719, Korea*
E-mail : leecg@hanbat.ac.kr
[c] *Korea Electric Power Research Institute, Daejeon 305-380, Korea*

Abstract. $CaSiO_3$ was prepared through solid chemical reaction of $CaCO_3$ and SiO_2 at 800°C and investigated as a CO_2 sorbent via X-ray diffractometry (XRD) and thermal gravimetric analysis (TGA) techniques. The results revealed that $CaSiO_3$ commenced to absorb CO_2 at 400°C and ended at 800°C with about 28.72% CO_2 sorption efficiency while using 15% CO_2, 85% N_2 as flow gases. About ten cycles of CO_2 sorption-desorption were also exhibited. $CaSiO_3$ can be regenerated by calcination at 800°C within 10 hours, furthermore, the cheap $CaSiO_3$ can also be used to capture CO_2 eternally.

Keywords: Carbon dioxide capture; Calcium silicate; Sorbent; Thermal gravimetric analysis.

1. Introduction

It is well-known that the rising emission of green-house gases are exacerbating the global environment, bringing on drought, desertification, hot weather, and sea level increase. In order to maintain a compatible living surrounding, mankind have to control CO_2 emission into aerosphere, which is the principal ingredient of green-house gases. Flue gases occupy ca. 30% among CO_2 emission sources. Collecting CO_2 from flue gases is essential for the reduction of CO_2 emission.

There are various methods to collect CO_2 such as absorption with a solvent (Mandal et al., 2006), membrane separation (Kanksha et al., 2007) and using zeolite (Wirawan et al., 2006). These techniques are costly and consume much energy, even pollute environment again. The improved technique is chemical absorption of CO_2 with solid sorbent. Those researches are attracting much attention in recent years. Yamaguchi et al. investigated separation performance of CO_2 absorbing material based on lithium silicate, 33 wt% of a CO_2 was absorbed on the material (Yamaguchi et al., 2007). Researches on lithium silicates are numerous (Pfeiffer et al., 1998; Bretado et al., 2005; Cruz et al., 2006). Lee et al. studied alkali metal-based solid sorbents with different support such as activated carbon (AC), TiO_2, Al_2O_3, MgO, SiO_2 and various zeolites (Lee et al., 2006). Their CO_2 capture capacities are lower than 12 wt% and cycle properties of CO_2 sorption-desorption are poor owing to formation of intermediates though the operating temperature was lower than 400°C. Martavaltzi et al. prepared CaO sorbent by using two kinds of precursors, $Ca(OH)_2$ and $Ca(CH_3COO)_2$, confirming the former precursor provided a higher tortuosity to yielded CaO and that CaO produced by $Ca(CH_3COO)_2$ exhibited higher CO_2 uptake up to 50% by weight. They also found that using $Ca_{12}Al_{14}O_{33}$ as framework in CaO sorbent enhanced stability of CO_2 sorption-desorption process, and the amount of CO_2 fixed on CaO-$Ca_{12}Al_{14}O_{33}$ derived from $Ca(CH_3COO)_2$ was higher than 6 mol/kg sorbent even after 45 cycles of sorption-desorption (M artavaltzi et al., 2008). Shtepenko et al. investigated the effect of CO_2 on β-dicalcium silicate and Portland cement, revealing amount of CO_2 which reacted with dicalcium and Portland cement was 48 and 37%, respectively, in a cycled process at 40°C, 3 bar for 60 min (Shtepenko et al., 2006).

Herein, we explored sorption-desorption property of CO_2 on homemade $CaSiO_3$. $CaSiO_3$ is relatively cheap, thus it can be employed for CO_2 everlasting deposition. Moreover, the CO_2 sorption temperature is lower than that of CaO. Researches concerning $CaSiO_3$ have been rare so far.

2. Experimental

$CaCO_3$ (analysis purity) and SiO_2 (analysis purity) were weighed according to chemical stoichiometric ratio (molar ratio, 1:1), and then ball-milled with an instrument, SPEX 8000 Mixer/mill. The homogeneous admixture was pressed in a die via a pressing instrument into a disc shape with a diameter of 1 cm at 50 Mpa. The discs were calcined in a chamber furnace at 800°C for 10 h. The chemical reaction of CO_2 absorption/desorption should be as follows;

$$CaCO_3+SiO_2 = CaSiO_3+CO_2 \qquad (1)$$

The forward process is CO_2 desorption, and the reverse process belongs to CO_2 fixed process. Thus the reaction is reversible according to the thermodynamics. After that, the calcined specimen was ball-milled again. The powder was measured through XRD together with TGA (METTLER TOLEDO, Model TGA 851). A mixture of 15 vol % CO_2 with 85 vol % N_2 was purged during the TGA process. The ratio of CO_2 is the same as that of flue gas. XRD was measured with a Siemens D500 diffractometer employing Cu/Kα radiation.

3. Results and discussions

3.1 Characterizations of specimens

Comparing curves 1 and 2 in Fig.1, new peaks denoting $CaSiO_3$ occurred in curve 3, whereas most peaks from curve 1 and 2 vanished, indicating $CaSiO_3$ was synthesized via chemical reaction of SiO_2 and $CaCO_3$. Since $CaSiO_3$ is amorphous, its XRD reflection peaks are not distinct (Shtepenko et al., 2006).

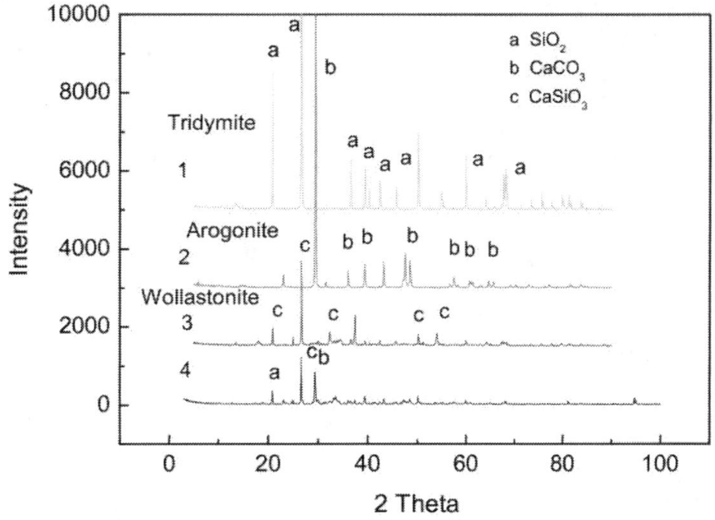

Fig.1 XRD patterns of calcium silicates at various states. 1 SiO_2 raw material; 2 $CaCO_3$ raw material; 3 synthesized $CaSiO_3$ at 800 °C; 4 $CaSiO_3$ after CO_2 absorption at 400-700°C.

Advantage of employing $CaCO_3$ other than CaO as feedstock lies in that the former can release CO_2 during the reaction, therefore, the product, $CaSiO_3$, is incompact. Sintering of product must be prevented, which provide larger specific surface area and are beneficial for latter reaction.

3.2 Effect of temperature on CO_2 capture capacity

It can be seen from Fig.2 that the mass gain of $CaSiO_3$ due to CO_2 absorption increased with increasing temperature. $CaSiO_3$ absorbs CO_2 sharply when temperature reached to 400°C. The highest CO_2 absorption ratio attained to be 28.72%, however, CO_2 absorption ratio decreased to 18.21% when measuring temperature reached 800°C and a valley appeared in mass gain curve. The temperature of 800°C was so high that newly produced $CaCO_3$ via CO_2 absorption reaction began to partially decompose, therefore, mass gain decreased until temperature descended. $CaSiO_3$ absorbed CO_2, mass gain ascended slightly again. The absorption process proceeded intensively that highest mass gain can be achieved within 45 minutes.

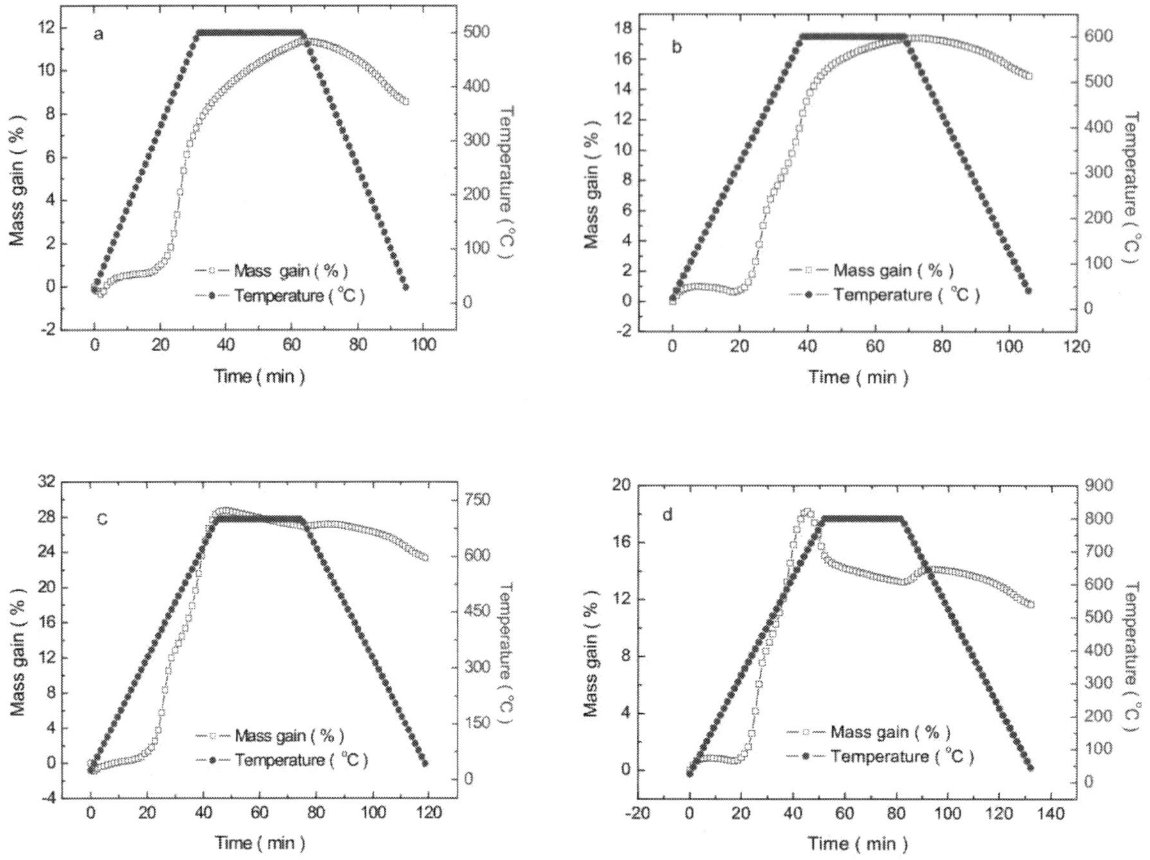

a Temperature range of 20-500°C; b 20-600 °C; c 20-700 °C; d 20-800 °C

Fig.2 Mass gain of $CaSiO_3$ during CO_2 absorption process

The highest CO_2 absorption ratio is 28.72%, obtained at 700°C, accordingly, 700°C is the optimal temperature for CO_2 absorption as $CaSiO_3$ is sorbent.

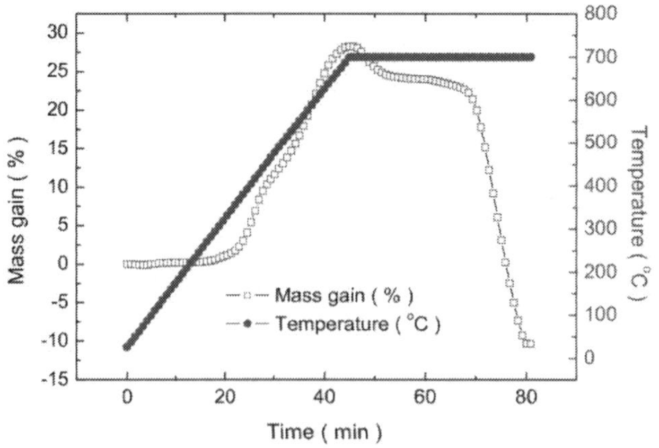

Fig.3 CO_2 sorption-desorption cycle

Experiments in Fig.2 were performed at CO_2/N_2 atmosphere all along, hence, mass dropped slowly because the temperature was decreased after mass gain reached the highest point. In order to perform desorption process, atmosphere alteration is essential. In Fig.3, CO_2/N_2 were changed into pure N_2 shortly after the highest mass gain was obtained, and then, mass dropped sharply to be even lower than the original value, releasing the adverse process of reaction (Mandal et al., 2006), namely CO_2-desorption process, could also proceed. The specimen derived from the CO_2-desorption process was characterized via XRD (curve 4, Fig.1). The peak in curve 4 at about 30 degree was attributed to $CaCO_3$, the feedstock, proving that the whole reaction is reversible. Hence, the adverse process can be achieved by atmosphere alteration at 700°C.

3.3 Cycle property of CO_2 sorption-desorption

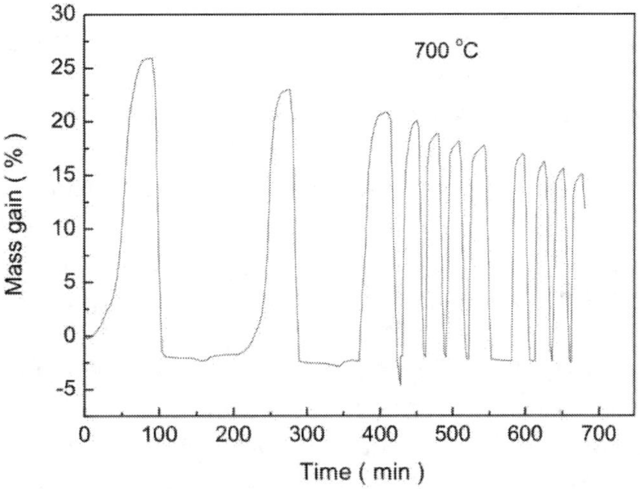

Fig.4 Cycle property of CO_2 sorption-desorption on $CaSiO_3$ at 700°C.

Capacity of CO_2 sorption/desorption descended drastically from 26.09%, 23.04% to 15.28% with the increasing cycle time (Fig.4). The decay rate is about 3%. Fig.4 showed certain cycle capacity though it is much inferior to that of CaO (Martavaltzi et al., 2008). The figure also indicated that CO_2-desorption reaction required long time, which decreased the whole cycle efficiency. We observed that the specimen after cycle experiment became more rigid than before. It was sintered additionally, which resulted in decrease of reaction specific area, consequently, low reaction activity.

On the other hand, the spent sorbent can be regenerated via sintering at 800°C for 10 hours to ensure the reaction of $CaCO_3$ with SiO_2 completely like that have been done during $CaSiO_3$ preparation process. About 95% of CO_2 sorption-desorption efficiency can be resumed by the above process. Regardless of the low cycle capacity, $CaSiO_3$ is a principal ingredient as wollastonite that is widely existing in the earth with cheap, thus it can be employed to deposit CO_2 everlasting since CO_2 utilization ratio is so low even less than 1% of emission quantity (Yang, 2005), which is also superior to Li_4SiO_4 (Yamaguchi et al., 2007).

4. Conclusions

In the present work, $CaSiO_3$ was investigated as a CO_2 sorbent via XRD and TGA techniques, releasing that $CaSiO_3$ commenced to absorb CO_2 at 400°C and ended at 800°C wi th about 28.72% CO_2 sorption efficiency while using 15% CO_2, 85% N_2 as flow gases. The ten cycles of CO_2 sorption-desorption were also exhibited well. $CaSiO_3$ can be regenerated by the calcination at 800°C, furthermore, the cheap $CaSiO_3$ can also be applied to deposit CO_2 et ernally.

References

Bretado M.E. et al., A new synthesis route to Li_4SiO_4 as CO_2 catalytic/sorbent, *Catalysis Today*, 2005, 107-108: 863-867

Cruz D. et al., Kinetic analysis of the thermal stability of lithium silicates Li_4SiO_4 and Li_2SiO_3, *Journal of Solid Chemistry*, 2006, 179: 909-916

Kanksha A., Pant A.A., Srivastava V.K., Carbon dioxide absorption into monoethanolamine in a continuous film contactor, *Chemical Engineering Journal*, 2007, 133:229-237

Lee S.C., Choi B.Y. et al., CO_2 absorption and regeneration of alkali metal-based solid sorbents, *Catalysis Today*, 2006, 111: 385-390

Mandal B.P., Bandyopadhyay S.S., Absorption of carbon dioxide into aqueous blends of 2-amino-2-methyl-1-propanal and monoethanolamine, *Chemical Engineering Science*, 2006, 61: 5440-5447

Martavaltzi C.S., Lemonidou A.A., Development of new CaO basedsorbent materials for CO_2 removal at high temperature, *Microporous and Mesoporous Materials*, 2008, 110: 119-127

Pfeiffer H. et al., Synthesis of lithium silicates, *Journal of Nuclear Materials*,1998, 257: 309-317

Shtepenko O., Hills C. et al., The effect of carbon dioxide on β-dicalcium silicate and Portland cement, *Chemical Engineering Journal*, 2006, 118: 107-118

Wirawan S.K., Creaser D., CO_2 adsorption on silicate-1 and cation exchanged ZSM-5 zeolites using a step change response method, *Microsporous and Mesoporous Materials*, 2006, 91: 196-205

Yamaguchi Y., Nittsuma T. et al., Lithium silicate based membranes for high temperature CO_2 separation, *Journal of Membrane Science*, 2007, 294: 16-21

Yang M., Experimental study on CO_2 elimination of tail gases from power generation via membrane and chemical sorption techniques [D], *Master thesis of Zhejiang university*, 2005.1

718p Modeling of a Closed-Loop Heat Pump Dryer for Cabin Solid Waste on Long-Duration Space Missions

J.M.R. Apollo Arquiza[1], Jean Hunter[1], Robert Morrow[2], and Ross Remiker[2]. (1) Biological and Environmental Engineering, Cornell University, Riley-Robb Hall, Ithaca, NY 14853, (2) Orbital Technologies Corporation, 1212 Fourier Drive, Madison, WI 53717

During long-term space missions, astronauts generate wet trash, including food residues, moist hygiene wipes and wet paper towels. This waste gives rise to two problems: first, the generation of odors and potential health hazards by microbial growth; second, the diversion of water from the available recovery loops. We have designed and built a prototype of an energy-efficient closed air-loop heat-pump dryer to remove water from wet material. The equipment is designed to stop microbial activity by both pasteurization and desiccation, and to recover humidity from the closed air loop in a gravity-independent Porous Membrane Condensing Heat Exchanger. A computational model of the system consisting of the dryer, condenser, and heat recovery modules has been developed. This is based on the conservation equations for energy and moisture applied to the air and solid phases and its formulation includes the unique waste characteristic of having both dry and wet solids. It also incorporates the heat and mass transfer coefficients for the system that have been measured. The experimentally determined moisture sorption equilibrium relationship for the wet material is also used. The resulting system of differential equations is solved by the finite-volume method as implemented by the commercial software COMSOL Multiphysics. The model is used to predict the system performance and energy cost per unit of recovered water under a range of operating conditions including air temperature, air flowrate, and initial moisture content of the trash.

718q Dynamic Optimization Approaches for the Determination of Bottom-up and Top-down Control In Lakes and Reservoirs

Vanina Estrada, Chemical Engineering, Planta Piloto de Ingenieria Quimica - Universidad Nacional del Sur, Camino La Camino La Carrindanga Km 7, Bahia Blanca, 8000, Argentina, Elisa Parodi, Biology, Biochemistry and Pharmacy, Instituto Argentino de Oceanografía (IADO) – Universidad Nacional del Sur, CONICET, Camino La Carrindanga Km 7, Bahia Blanca, 8000, Argentina, and Maria Soledad Diaz, Chemical Engineering, Planta Piloto de Ingenieria Quimica (PLAPIQUI), Universidad Nacional del Sur - CONICET, Camino La Carrindanga Km 7, Bahia Blanca, 8000, Argentina.

Biogeochemical processes that take place in water bodies can be represented through a set of complex nonlinear partial differential algebraic equations resulting from dynamic mass balances for phytoplankton and nutrients. Once calibrated, the model can be integrated into a dynamic optimization framework for the determination of optimal control policies to prevent eutrophication of the water body.

In this work, we develop a dynamic optimization model for a water body to determine optimal concentrations of zooplankton required to decrease phytoplankton concentration in the lake, associated to fish removal, as well as optimal limiting nutrient inflow profiles to the lake, with the associated flowrate of the inflows that has to be derived to a nearby wetland for remediation.

The model is based on horizontally averaged concentrations and its main parameters have been previously estimated (Estrada et al., 2007) based on collected data sets corresponding to an entire year (Parodi et al., 2004). Dynamic mass balances in phytoplankton (in the form of diatoms, green algae and cyanobacteria), dissolved oxygen and nutrients, which include nitrate, ammonium, organic nitrogen, phosphate and organic phosphorus, have been formulated. Biogeochemical processes include growth, respiration and death processes for phytoplankton, as well as mineralization, nitrification and uptake and release of nutrients. Algebraic equations represent profiles for temperature, solar radiation and river inflows, in addition to the calculation of most factors that affect rate equations. The resulting partial differential algebraic equations (PDE) model is transformed into an ordinary differential equations system by spatially discretizing the PDE into sets of ordinary differential-algebraic equations (DAE). The objective function is the minimization of the offset between phytoplankton concentration and the desired concentration in the water body along the first year, subject to the previously described DAE system and upper bounds on phytoplankton and nutrients concentrations, to remain below eutrophication levels.

The dynamic optimization problem is solved through a simultaneous approach by transforming it into a large-scale nonlinear programming (NLP) problem discretizing state and control variables applying collocation over finite elements (Raghunathan et al., 2004). The discretized NLP model has more than 17,000 variables and it has been solved with a reduced successive quadratic programming algorithm (Kameswaran and Biegler, 2006). The study has been performed on Paso de las Piedras Lake, which supplies drinking water for more than 400,000 inhabitants in Bahia Blanca (Argentina).

References

Estrada V., E. Parodi, S. Diaz, Ecological Studies and Dynamic Parameter Estimation for Eutrophication Models, ECCE-6, Denmark, 16-21 September 2007.

Kameswaran, S., L.T. Biegler (2006), Simultaneous dynamic optimization strategies: Recent advances and challenges, Comp. & Chem.Eng., 30, 1560–1575

Parodi, E.., Estrada, V., Trobbiani, N., Argañaraz Bonini, G., 2004, Análisis del estado trófico del Embalse Paso de las Piedras. Ecología en tiempos de Cambio. 178.

Raghunathan A., S. Diaz, L.T. Biegler, 2004, An MPEC Formulation for Dynamic Optimization of Distillation Operations, Comp. & Chem.Eng., 28, 2037.

718r Fractional Uptake Efficiency of Ozone In Human Subjects

Tim Brenza and James Ultman. Chemical Engineering, The Pennsylvania State University, University Park, PA 16802

Ozone (O_3) is an urban outdoor pollutant primarily generated by a photochemical reaction of atmospheric oxygen with oxides of nitrogen and unburned hydrocarbons from automobile emissions. The inhalation of O_3 is associated with health effects that include decrements in pulmonary function, pulmonary inflammation, and alterations in lung development. The purpose of the current research was to determine the influence of smoking on the retention of O_3 in the adult human lung.

Thirty smokers and thirty nonsmokers each participated in two research sessions. During an O_3 exposure session, subjects exercised on a cycle ergometer for one hour while breathing room air containing 0.3 ppm O_3. By making periodic adjustments in the ergometer workload, the experimenter targeted a minute ventilation (MV) of 15 L/min/m^2 of body surface. During a control session, carried out at least one week prior to the O_3 exposure session, the subject followed the same procedure while breathing room air not containing O_3. During both sessions, the subject breathed through an oral mask fitted with sensors that monitored respiratory flow and O_3 concentration throughout the one-hour session.

The fractional uptake efficiency (UE), defined as the ratio of retained amount of O_3 to inhaled amount of O_3, was determined for each breath and averaged at five minute increments. UE decreased for both smokers and nonsmokers during the one-hour exposure, and the smoking subjects had a somewhat higher UE than the nonsmoking subjects. An ANCOVA of UE using time as the covariate and smoking status as the fixed variable revealed that smoking status was indeed a significant predictor of uptake efficiency ($p=0.02$).

Respiratory flow data was processed to obtain MV, frequency (FR), and tidal volume (VT). Minute-by-minute values of the MV exhibited minimal variation within either the control or O_3 exposure sessions, and the session-averaged MV values were not significantly different between the control and exposure sessions or between the smokers and nonsmokers. Relative to the control session, there was an increase in FR and a decrease in TV associated with O_3 exposure for both smokers and nonsmokers. Smokers breathed at a slightly higher VT and lower FR than the nonsmokers. We hypothesized it was these variations in breathing pattern that were responsible for the observed variations in UE.

In previous work, the UE values observed in health nonsmoking subjects during a one-hour O_3 exposure were simulated with a single-path respiratory transport model. Simulations were performed for individual subjects by incorporating a square wave respiratory flow pattern with VT and FR varying in accordance with their observed minute-by-minute values. By adjusting the rate constant for O_3 reaction in the mucous layer (k_r) on a subject-by-subject basis, these simulations closely matched the measured UE for all subjects.

To test our hypothesis, the single-path model was employed to separately simulate the current data grouped into four subpopulations: male nonsmokers, female nonsmokers, male smokers, and female smokers. The squared error between the observed and simulated UE was minimized by adjustment of k_r for each subpopulation. The final values of k_r were 1.06E6 s^{-1} for female nonsmokers, 2.25E6 s^{-1} for male nonsmokers, 2.55E6 s^{-1} for female smokers, and 2.60E6 s^{-1} for male smokers. Thus, by taking breathing pattern alone into account, it was possible to simulate differences in the observed UE between the smokers and nonsmokers.

718s Consideration of Toxicological Aspects in the Development of Lubricants Based on Synthetic Esters

Bibiane Erlenkaemper, Esm-CS, EvonikDegussa GmbH, Rodenbacher Chaussee 4, D-63457 Hanau, Germany, Wolfgang Dott, Institute for Hygiene and Environmental Medicine, RWTH Aachen, Pauwelsstr. 30, D-52074 Aachen, Germany, Marcel A. Liauw, Itmc, RWTH Aachen University, Worringerweg 1, Aachen, 52074, Germany, Sven Eichholz, RWTH Aachen, Institute of Technical and Macromolecular Chemistry, Worringerweg 1, Aachen, 52074, Germany, and Adolf Eisentraeger, Federal Environment Agency Germany, Woerlitzer Platz 1, D-06844 Dessau, Germany.

Up to now the development of new lubricants mainly aimed at an optimal technical performance. The consideration of toxicological characteristics only became relevant when the product was brought to market.

This study contributes to the equal implementation of toxicological and performance characteristics in the early stages of development of lubricants. In terms of sustainable development the application of environmentally acceptable lubricants in different technical applications is promoted. To enhance their technical performance and avoid the use of additives, bio-based lubricants are often chemically modified. These processes do not only influence their technical characteristics, but also their ecotoxicological behaviour. By the combined application of chemical analyses and standardised bioassays relevant compounds causing toxic effects were identified and an optimization of the synthesis was achieved. The applied analytical methods include the determination of acid number, saponification number, iodine number, peroxide number, epoxide number and viscosity, as well as gas chromatography, 1H-NMR, 13C-NMR, ATR-IR and reaction calorimetry. For the determination of ecotoxicological effects, aqueous extracts of the lubricants were prepared and applied to bioassays with algae and daphnids.

One example for the integrated approach is the detection of highly acidic compounds leached from the solid acid catalyst. These toxicologically detrimental substances were removed by steam treatment in the 20L technical setup.

The authors thank the DFG for financial support (SFB 442).

753a Biofuel Generation from Cheese Whey In a Two –Stage Anaerobic Process

Georgia Antonopoulou, Katerina Stamatelatou, and **Gerasimos Lyberatos**. Department of Chemical Engineering, University of Patras, 1 Karatheodori St., Patras, 26500, Greece

Global energy and environmental concerns due to the increasing use of fossil - derived fuels, lead many scientists to the search of for alternative energy sources. Among others, much attention has been paid to the hydrogen gas, and its potential use as a fuel for transport purposes and electricity generation. Hydrogen is a clean and environmentally friendly fuel, which produces water instead of greenhouse gases, when combusted. It possesses a high-energy yield (122 kJ/g) due to its light weight. Hydrogen can be produced by renewable raw materials, such as biomass, through biological or thermochemical processes. Anaerobic fermentative hydrogen production, has received special attention during the last decade. It is well founded that carbohydrates are the main source of hydrogen during fermentative processes and therefore wastes/wastewater or agricultural residues rich in carbohydrates can be considered as potential sources of hydrogen.

From the wastewater treatment point of view, fermentative hydrogen production mainly includes sugars transformation into volatile fatty acids, without much effect on the organic content of the wastewater. Thus, a complementary stage would be necessary for chemical oxygen demand elimination. It is well known that the formation of volatile fatty acids during the acidogenesis of the organic matter is actually the precursor to methanogenesis. Therefore, the hydrogen production process could be efficiently coupled with a subsequent anaerobic digestion step with the conversion of the remaining organic content to biogas (mainly methane and carbon dioxide), which may also be used as a fuel for the production of electricity.

As it has already been reported, wastewaters rich in carbohydrates can be used as raw materials for bio hydrogen production. Cheese whey is the lactose-rich watery by-product of cheese manufacturing and could be an excellent candidate for fermentation processes. Cheese whey represents about 85–95% of the milk volume and contains nutrients, such as lactose (4.5-5% w/v), soluble proteins (0.6-0.8% w/v), lipids (0.4-0.5% w/v) and mineral salts (8-10% of dried extract) (Sizo, 1996). Because of its high organic content, cheese whey disposal constitutes a serious environmental problem with lactose being mainly responsible for its high COD values.

The present study focuses on hydrogen and subsequent methane generation from cheese whey in a two-stage anaerobic process. The fermentative hydrogen production process was conducted in a mesophilic continuous stirred tank type bioreactor with an active volume of 3L, which was operated at hydraulic retention time (HRT) of 24 h and was started up and fed with cheese whey. For start-up, the reactor was filled up with cheese whey and operated anaerobically at a batch mode for 24 h in order to activate the indigenous microflora contained in the waste and was subsequently switched to the continuous mode at the an HRT of 24 h.

It has been proved that continuous fermentative hydrogen production is possible and stable using the indigenous microflora contained in cheese whey. The percentage of hydrogen in the gas phase was 29.3 ± 1.6 % while the hydrogen production rate reached 7.53 LH2 per day. The efficiency of glucose consumption was 68.6% since the concentration of carbohydrates in the mixed liquor was 11.30 ± 0.5 g / L. Based on the above observations, the yield of hydrogen produced per mole of glucose consumed was approximately 0.9 ± 0.1 mol H2 / mol glucose consumed and, consequently, that of hydrogen produced per L of cheese whey was 2.49 L H2/ L cheese whey, respectively. This corresponds to an energy yield of 24.85 kJ/ L cheese whey (assuming that the energy yield from hydrogen is 122000 kJ/ kg).The obtained yield of hydrogen production is lower than the theoretical yield of 2 or 4 mol H2 / mol glucose consumed, which corresponds to the maximum theoritical production of butyric or acetic acid, from glucose degradation

The effluent from the hydrogenogenic reactor was subsequently fed to a Periodic Anaerobic Buffled Reactor (PABR), an innovative reactor, initially developed by Skiadas and Lyberatos (1998) and Skiadas et al 2000. The anaerobic digester was operated at mesophilic conditions at hydraulic retention times of 20, 10 and 4.4 d. The biogas and methane production rates were higher at the HRT of 4.4 d and were equal to 105.9 L biogas /d and 75.6 L CH4 /d respectively. The percentage of methane in the gas phase lied between 71.4 and 74.9 %, while the culture pH was approximately stable at all HRTs. The COD removal efficiency was more than 94 % at all HRTs, implying that the performance of the anaerobic digester was not kinetically limited.

Consequently, this work demonstrated that continuous fermentative hydrogen production is possible and stable using the indigenous microflora contained in cheese whey. In addition, it was proven that biohydrogen production can be very efficiently coupled with a subsequent step of methane production and that cheese whey could be an ideal substrate for a combined gaseous biofuels generation.

Literature cited

Gonzalez Siso, M. I (1996). The biotechnological utilization of cheese whey: a review. Biores. Technol. 57, 1-11

Skiadas, I.V.; Lyberatos, G. (1998) The periodic anaerobic baffled reactor. Wat. Sci. Technol. 38, 401-408

753b Biocrude Production from Switchgrass Using Subcritical Water

Sandeep Kumar and Ram B. Gupta. Chemical Engineering, Auburn University, Auburn, AL 36849-5127

Subcritical water is considered as an environmentally attractive solvent for organic matters and it can be used to convert biomass to liquid form (biocrude), which is a mixture of oxygenated hydrocarbons with a wide molecular weight distribution. The biocrude produced at the biomass production site can be pumped through pipelines to a centralized bio-refining facility for the production of variety of chemicals and renewable fuel.

Studies have been conducted on the liquefaction of switchgrass in sub-critical water in a flow through reactor. Effect of heterogeneous catalyst K2CO3, and temperature on the liquefaction of switchgrass is studied. The finely grounded switchgrass (40-60 mesh size) is packed in a tubular reactor, and preheated compressed water / K2CO3-water solution is passed through the packed bed at constant pressure. Reactor temperature is increased using electrical furnace. Studies have been conducted for the temperature range 230-260 °C. Liquid product is analyzed for total organic carbon (TOC) content and yields phenols, monomers, degradation products of glucose and organic acids.

Web Page: www.eng.auburn.edu/~gupta

Noncatalytic Reformation of Sucrose in Supercritical Water

Jason Picou, Jonathan Wenzel, Michael Stever, Jared Bouquet, and Sunggyu Lee,
Missouri University of Science and Technology, Rolla, MO

Abstract

The effect of space time and temperature on the non-catalytic reformation of sucrose in supercritical water was studied in a specially designed 0.4-L Haynes Alloy 230 tubular reactor. Experiments were performed non-catalytically in a continuous mode of operation at a constant pressure of 24.05±0.04 MPa and at temperatures varying from 600°C to 760°C and space times between 55 to 245 seconds. Sucrose is a renewable, biological energy resource; the reformation of which would enable on-site and on-demand hydrogen production. Increasing space time and temperature increases the gasification percentage of sucrose along with the production of hydrogen gas, with temperature having a larger effect than space time. Gasification percentages of 99% were achieved using two different combinations of temperature and space time. A maximum of 7.9 moles of hydrogen gas was produced per mole of sucrose fed, which is 23% of the theoretical maximum possible, occurred at the highest temperature studied.

Introduction

Sucrose, $C_{12}H_{22}O_{11}$, commonly known as table sugar, is a naturally occurring disaccharide of glucose and fructose. It is produced in plants, and is commercially harvested from sugar cane and sugar beets due to its high concentration in these plants.[1] In the current energy market, sucrose is used to produce ethanol through fermentation. This ethanol must be further processed to remove the water and other contaminants in order to make a renewable liquid fuel. While the supercritical water reformation of other sugars such as glucose has been studied since the late 1970's, there is a dearth of literature on the non-catalytic supercritical water reformation of sucrose.[2-5] The purpose of these experiments are to investigate sucrose reformation to produce hydrogen, an emerging alternative fuel.

Apparatus & Chemicals

The water used in the reformer was deionized water, and the sucrose used was commercially available table sugar. The water and sucrose was mixed together to a predetermined ratio; for these experiments a ratio of 18.5 grams of water to one gram of sucrose was used throughout. This corresponds to a molarity of 0.16 mole of sucrose per liter of water, or a ratio of about 170 moles of water for every mole of sucrose.

The supercritical water system consists of a liquid feed system, preheat, reactor, reactor heaters, sample collection system, and data acquisition and control system, of which a schematic process flow diagram is illustrated in Figure 1 The supercritical water reactor has a volume of 383 mL, and is constructed of Haynes Alloy 230, which is an alloy of mostly nickel, chromium, tungsten, and molybdenum, among with other elements.[6] This allows the reactor to operate at temperatures up to 800 °C at a pressure of 36.2 MPa. Analysis of the gaseous effluent was performed using a HP 5890 Series A gas chromatograph with a thermal conductivity detector. The gas chromatograph is calibrated to detect hydrogen, nitrogen, carbon monoxide, methane, carbon dioxide, acetylene, ethylene and ethane. The total carbon content of the liquid effluent was analyzed with a Dohrmann DC-190 total organic carbon analyzer.

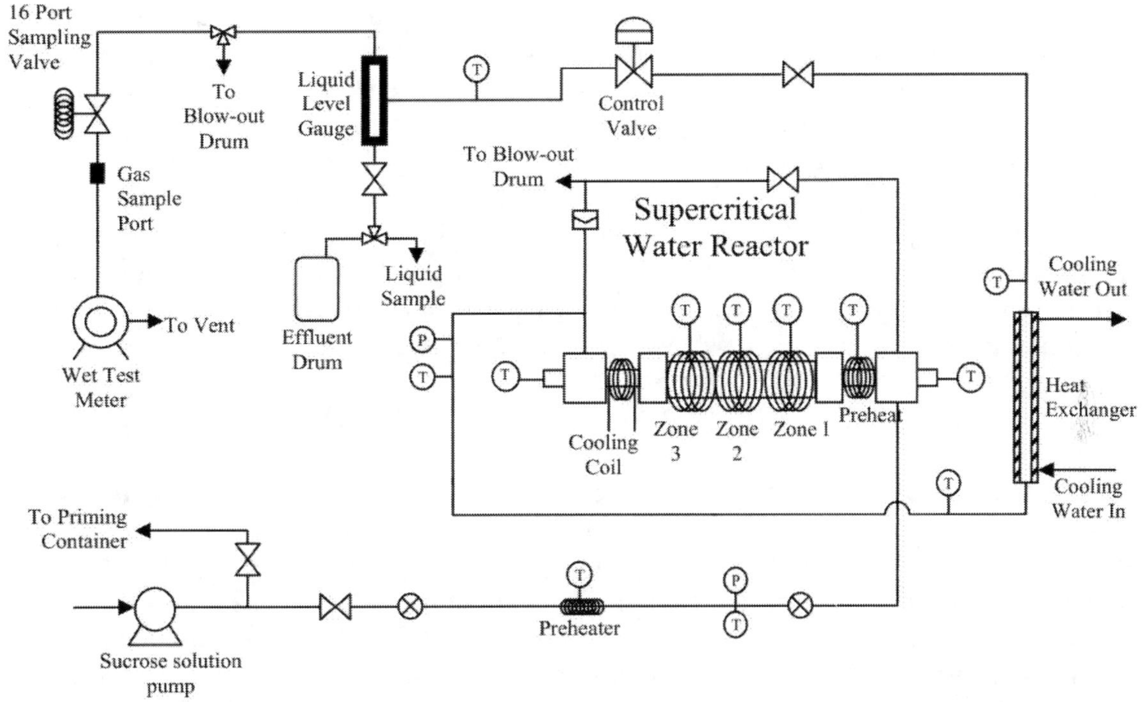

Figure 1. A schematic of the supercritical water reaction system at Missouri University of Science and Technology.

Chemical Reactions

When sucrose-water mixtures are heated, a variety of chemical reactions can take place. At temperatures near the atmospheric melting point of sucrose (185°C), sucrose splits into glucose and fructose, the two monosaccharides of this disaccharide sugar.[7] Also around this temperature, pyrolysis begins to occurs, whereby these sugars are further split into a variety of smaller fractions, creating gaseous hydrocarbons. It will be assumed throughout this paper that any gaseous hydrocarbons are the result of the endothermic pyrolysis reaction, although there may be other contributing reactions, which will be discussed later in this section. At higher temperatures, polymeric substances are formed via radical polymerization.[8] Repeated pyrolysis leads to hydrogen deficient fragments that, together with polymerization, function as coke precursors.[9]

Reformation, an endothermic reaction in which a hydrocarbon or sugar reacts with water to produce hydrogen gas and carbon monoxide, is an important industrial reaction and a promising avenue for hydrogen production. Currently, the widest use of reformation is in the reformation of natural gas to produce hydrogen for the chemical industry.[10] For present industrial applications, the reforming reaction is a catalytic reactions which is carried out at temperatures of 700 to 1100°C and pressures from 345 to 2400 kPa using a nickel catalyst.[11] The natural gas must be cleaned of sulfur and chlorine before being reformed to avoid catalyst poisoning.[12] The experiments conducted for this paper were all carried out non-catalytically, in that no catalyst was placed inside the reactor. It has been demonstrated that metallic reactor walls can potentially function as a catalyst in the reformation of various hydrocarbons in supercritical water, but these effects will have to be analyzed in future work.[3,13-15] In this paper, the production of carbon monoxide and hydrogen is indicative of the reformation reaction.

In addition to these reactions, the forward water gas shift (WGS) reaction can also occur. The water gas shift reaction is the reaction of carbon monoxide and water to form carbon dioxide and hydrogen gas. This would be a highly desirable reaction, since additional hydrogen is produced. The forward reaction is thermodynamically favored at temperatures of 815°C or below.[16] Most carbon dioxide present in the effluent gas is assumed to be the result of the water gas shift reaction. Total reformation, combined with all carbon monoxide going through the water gas shift reaction, could produce 35 moles of hydrogen gas per mole of sucrose fed, which is shown below.

$$C_{12}H_{22}O_{11} + 24\,H_2O \rightarrow 35\,H_2 + 12\,CO_2 + 5\tfrac{1}{2}O_2 \tag{1}$$

The theoretical maximum amount of hydrogen that could be produced is therefore 35 moles of hydrogen gas per mole of sucrose fed. The theoretical maximum also assumes that the oxygen contained in the sucrose would not react with either the carbon monoxide or hydrogen, which is very unlikely given the structure of disaccharide as well as the nature of supercritical water oxidation.[17,18] Various other reactions, like methanation or the Boudouard reaction, could also be possible. The methanation reaction involves one mole of carbon monoxide and three moles of hydrogen gas producing one mole of methane and water. This would be counterproductive since the goal is to produce hydrogen gas. Previous studies involving the supercritical water gas shift reaction saw no methanation at conditions similar to those encountered during in this paper.[19] The Boudouard reaction is a reversible reaction that involves two moles of carbon monoxide reacting to produce one each of carbon dioxide and elemental carbon. The high temperatures used in this paper should limit the Boudouard reaction; the equilibrium constant favors carbon monoxide production at temperatures above 680°C.[16,20-23] Another reaction conceivable is direct dehydration of disaccharide, thus generating carbon, or lighter hydrocarbons, and water. The large excess of water used in these experiments would favor reformation or water gas shift over other reactions. The discussion was limited to the aforementioned reactions for simplicity and because they effectively and accurately describe all of the products observed.[24]

Experimental

The temperature of the reactor as well as the space time of the solution in the reactor was varied to investigate the effects upon the effluent gas composition as well as sucrose gasification. Experiments were performed non-catalytically in a continuous mode of operation at a constant pressure of 24.05±0.04 MPa and at a constant water-to-sucrose ratio of 18.5 grams of water per gram of sucrose. The temperatures were varied from 600°C to 760°C at a constant space time of 95±6 seconds. The space time was varied from 55 to 245 seconds at a constant temperature of 683±2°C. The space time was varied by changing the inlet solution flow rate and calculated using the Peng-Robinson equation of state based upon the reactor temperature, pressure and the properties of pure water. Table 1 outlines the reactant solution flow rate, temperature, pressure, and space time. Run numbers 1, 2, 3 and 5 have a varying temperature with a constant space time, while the others have a constant temperature with a varying space time. Run numbers 3 and 5 are duplicates.

Table 1. Experimental matrix for sucrose reformation in supercritical water. Water-to-sucrose ratio is constant at 18.5-to-1 by mass and pressure at 24.05±0.04 MPa. The space time and temperature are varied.

Run #	Temp (°C)	Pressure (MPa)	Solution Flow (g/min)	Space Time (sec)
1	595	24.00	14.9	112
3	688	24.14	15.3	94
5	685	23.94	15.6	92
2	758	24.12	15.4	84
4	681	24.12	6.0	245
6	677	23.96	26.5	55

Results and Discussion

The results of the experiments outlined in Table 1 are shown in Table 2, which lists the liquid and gaseous components of the reactor effluent. More detailed analysis of the gas effluent is done later in this paper. If the carbon present in the inlet sucrose does not leave as gas it either comes out in the liquid effluent or stays behind as solid pyrolysis remnants, so any carbon not accounted for in the liquid or gas remains in the reactor as solid.

Table 2. Liquid and gas effluent compositions and the percent of carbon accounted for in liquid and gas for the experiments conducted.

Run #	Liquid total organic carbon (ppm)	% of inlet carbon in liquid	Gas flow rate (L/min)							% of carbon accounted for in liquid and gas
			H_2	CO	CH_4	CO_2	C_2H_4	C_2H_6	Total	
1	3804	11.1%	0.06	0.22	0.08	0.10	0.010	0.020	0.48	81.1%
3	949	2.8%	0.23	0.08	0.15	0.30	0.005	0.041	0.80	95.4%
5	709	2.1%	0.28	0.05	0.15	0.30	0.007	0.033	0.81	86.9%
2	46	0.1%	0.44	0.03	0.19	0.42	N.A.	0.017	1.10	99.9%
4	84	0.2%	0.13	0.01	0.07	0.15	N.A.	0.017	0.38	99.5%
6	1056	3.1%	0.40	0.17	0.24	0.46	0.012	0.055	1.34	90.1%

As seen in Table 2, temperature has an effect on the percentage of carbon that exits the reactor as liquid. As temperature increases, that percentage decreases, going from 11% at 595°C to 0.1% at 758°C. Space time also has an effect, but less so than temperature. The largest percentage of carbon in the liquid occurs at the shortest space time, run number 6, while the longest space time, run number 4, has very little carbon in the liquid. Any carbon not accounted for in the liquid or gas phase must remain in the reactor as solid. On occasion this solid was removed from the reactor, but at this time has not been analyzed.

Effect of space time

Run numbers 3 through 6 investigate the effect that space time has on the supercritical water reformation of sucrose. The net effect of the variation in space time on the gas composition and sucrose gasification is illustrated in Figure 2. The sucrose gasification percentage, the left axis of Figure 2, is the molar flow rate of carbon that was present in all gases divided by the molar flow rate of carbon in the inlet sucrose. The species gas yield, the right axis of Figure 2, is shown in dimensionless numbers, and is calculated from the molar effluent flow rate of each gas species divided by the inlet molar flow rate of sucrose, to account for variations in inlet flow rate.

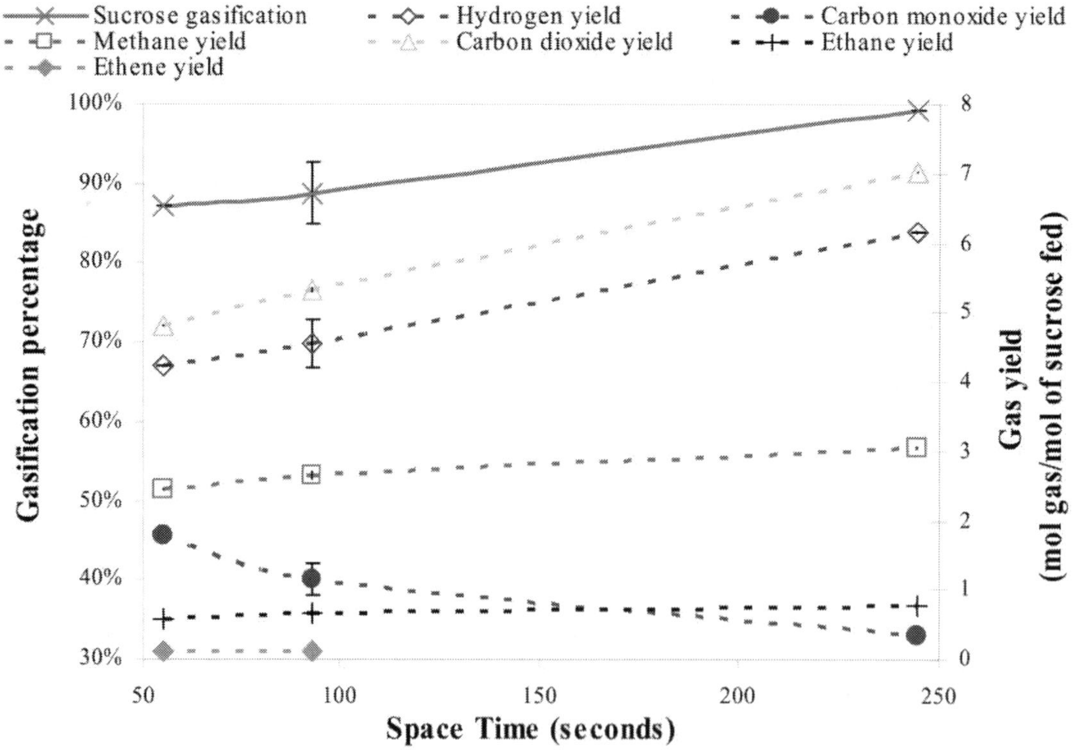

Figure 2. Dimensionless gas yield and sucrose gasification percentage as a function of space time. Temperature constant at 683±2°C, pressure 24.04±0.05 MPa, with an 18.5-to-1 water-to-sucrose ratio by mass.

Because there were duplicate runs performed at a space time of 93±1 seconds, the results were averaged and error bars are used for the standard error. The sucrose gasification percentage increases with increasing space time in the reactor, going from 87% at the shortest space time to 99% for the longest (245 seconds), an increase of 14%. Referring to Table 2, the percent of carbon accounted for in the liquid and gas for increasing space time increased, reaching 99.5% at a space time of 245 seconds, meaning less solid coke is produced with longer space times.

The hydrogen and carbon dioxide yield increases with space time, with hydrogen increasing 46% from the smallest to larges space time, while carbon dioxide also increases 46% over the same interval. The carbon monoxide yield decreases with increasing space time, which could be attributable to the progress of the water gas shift reaction. This reaction would also partially contribute to the increasing hydrogen and carbon dioxide flow rates. The methane yield did not change as much as the other yields for increasing space time, increasing 24% over the interval. Ethane also increased, 32%, with increasing space time. Ethene gas was only present in trace amounts for the two shortest space times, and was not detectable for the longest. It could be that the pyrolysis reaction occurs fast enough that the increase in space time has little effect. Because hydrogen gas yield is increasing, it is assumed that methanation plays a very small, if any role, in the methane gas yield. Longer space times seem to favor reformation and the water gas shift reaction, as seen by the increasing hydrogen, carbon dioxide and sucrose gasification percentage, and the decreasing carbon monoxide yield. Pyrolysis is effected less by time in the reactor, based on the smaller increase in the methane and ethane yield.

As stated in the chemical reactions section, the theoretical maximum production of hydrogen gas would be 35 moles per mole of sucrose fed. As seen in Figure 2, the most that was produced was 6.1 moles of hydrogen per mole of sucrose, which represents 18% of the theoretical maximum. Much of the missing hydrogen is in the form of the gaseous hydrocarbons, methane, ethane and ethene. Also, any carbon monoxide in the effluent gas, being capable of producing hydrogen through the WGS reaction, should be considered. If these gaseous products were to be totally reformed to hydrogen, including the water gas shift reaction, the dimensionless hydrogen yield would increase to 29.3 moles of hydrogen gas per mole of sucrose, which accounts for 84% of the theoretical maximum. The remaining hydrogen is either in the liquid effluent or is oxidized by the oxygen contained in the sucrose to form water. A small amount may remain in the reactor as a component of the coke formed during pyrolysis.

Effect of temperature

Temperature is the other variable of interest. Run numbers 1, 2, 3 and 5 investigate the effect of increasing temperature while the pressure was held constant at 24.05±0.05 MPa and at a space time of 95±6 seconds, the results of which are presented in Figure 3.

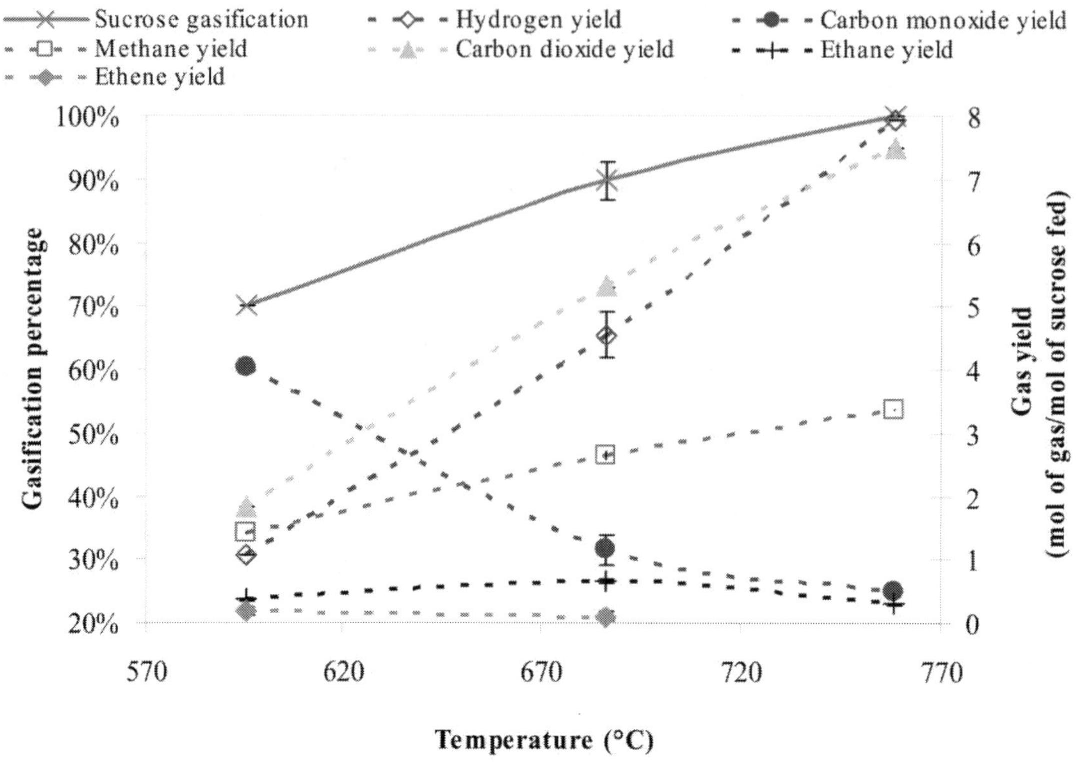

Figure 3. Dimensionless gas yield and sucrose gasification percentage as a function of temperature. Space time constant at 95±6 seconds, pressure 24.05±0.05 MPa, with an 18.5-to-1 water-to- sucrose ratio by mass.

Again, because run numbers 3 and 5 are duplicates, the results are averaged and error bars for the 686±1°C point representing the standard error are used. The sucrose gasification percentage increases with increasing temperature, going from 70% at 595°C to 99% at 758°C, a 43% increase. As stated above, the percent of carbon accounted for in the liquid and gas effluent increases as temperature increases, meaning less solid formation inside the reactor. Dimensionless hydrogen gas yield increased from 1.1 to 7.9, a 646% increase with increasing temperature. Carbon dioxide saw a 307% increase, while methane increased 136% with increasing temperature over the range studied. Carbon monoxide decreased 87% with increasing temperature, having a dimensionless yield that decreased from 4.0 to 0.51 with increasing temperature, perhaps due to greater activity of the water gas shift reaction. Other gaseous species did not exhibit a straight line increase like the previous examples. Ethane increases 82% before decreasing 54% for the three increasing temperatures. There is no ethene detectable at 758°C, and it decreases 50% from 595°C to 686°C.

The amount of hydrogen gas produced at the highest temperature, 7.9 moles per mole of sucrose fed, is again far below the theoretical maximum. This is 23% of the theoretical maximum hydrogen production. The increase in sucrose gasification, along with hydrogen and carbon dioxide yield and the decrease in carbon monoxide, indicates that both reformation and the water gas shift reaction are temperature dependant, increasing with increasing temperature. Pyrolysis also increases with increasing temperature, but less than the reformation products, as indicated by the smaller increases in methane and ethane yields compared to hydrogen or carbon dioxide. Pyrolysis does not seem to be as strongly temperature dependant as the reformation or water gas shift reaction. This finding is also in agreement with that with supercritical water reformation of jet fuel.[9,24]

Over the range of temperatures and space times that were analyzed in these experiments, temperature had a significantly larger effect on the sucrose gasification percentage, hydrogen, carbon monoxide and carbon dioxide flow rates than space time. During the space time increase from 55 to 255 seconds, gasification increased 14% while hydrogen and carbon dioxide yields increased 46%. When the temperature was increased from 600 to 760°C, gasification increased 43%, hydrogen yield 646% and carbon dioxide 307%.

Conclusion

The non-catalytic reformation of sucrose in supercritical water was studied in a 0.4-L Haynes Alloy 230 tubular flow reactor. Three space times, 55, 92 and 255 seconds, were evaluated at 683±2°C, a pressure of 24.05±0.04 MPa, with an 18.5-to-1 water-to-sucrose ratio by mass. In general, as the space time increased the sucrose gasification percentage increased, as did the effluent hydrogen and carbon dioxide yield. Carbon monoxide yield decreases with increasing space time, while methane and ethane increased, but increased less than the hydrogen or carbon dioxide yield. Longer space times seem to favor reformation and water gas shift more than pyrolysis. A maximum of 6.1 moles of hydrogen gas per mole of sucrose fed (18% of the theoretical value), along with a gasification percentage of 99%, was achieved at a space time of 255 seconds. Experiments were performed at temperatures of 595, 686, and 758°C at the same pressure and water-to-sucrose ratio as the previous experiments. The sucrose gasification percentage, as well as the hydrogen and carbon dioxide yield, all increase with increasing temperature, which may be due to the reformation reaction becoming more active at higher temperatures. The increases in hydrogen and carbon dioxide yields, along with the modest increase in methane and ethane yields, indicates that the reformation reaction is more dependant on temperature than the pyrolysis reaction. A maximum of 7.9 moles of hydrogen gas per mole of sucrose fed (23% of the theoretical maximum), along with a gasification percentage of 99%, was achieved at 758°C. Increasing temperature and space time increased the percentage of carbon accounted for in the liquid and gas, meaning less solid was produced and remained in the reactor at these conditions. Increasing temperature, over the range studied, had more effect than increasing space time on the gasification percentage and the hydrogen, carbon monoxide and carbon dioxide yields.

References

(1) Vollhardt, K.P.C.; Schore, N.E. Organic Chemistry 2nd Ed. W.H. Freeman and Co. New York, 1994; pp 967.

(2) Modell, M.; Reid, R.C.; Amin, S.I. Gasification Process. U.S. Patent 4,113,446, 1978.

(3) Yu, D.; Aihara, M.; Antal, J.M. Hydrogen Production by Steam Reforming Glucose in Supercritical Water. *Energy Fuels* **1993**, *7*, 574-577.
(4) Hao, X.H.; Guo, L.J.; Mao, X.; Zhang, X.M.; Chen, X.J. Hydrogen production from glucose used as a model compound of biomass gasified in supercritical water. *Int. J. Hydr. Energy* **2003**, *28*, 55-64.
(5) Cortright, R.D.; Davda, R.R.; Dumesic, J.A. Hydrogen from catalytic reforming of biomass-derived hydrocarbons in liquid water. *Nature* **2002**, *418*, 964-967
(6) *Haynes® 230 Alloy*, Haynes High-Temperature Alloys International, 2004 pp. 2.
(7) Simkovic, I.; Surina, I.; Vrican, M. Primary reactions of sucrose thermal degradation. *J. Anal. Appl. Pyrolysis* **2003**, *70*, 493-504.
(8) Claude, J.; Ubbink, J. Thermal degradation of carbohydrate polymers in amorphous states: A physical study including colorimetry. *Food Chemistry* **2006**, *96*, 402-410.
(9) Lee, S., Lanterman, H., Wenzel, J., Edwards, N., Adams, A., Wootton, J., Garcia, A. Noncatalytic direct reformation of JP-8 fuel in supercritical water. *Preprints-American Chemical Society, Division of Petroleum Chemistry* **2006**, 51, 521-523
(10) The National Hydrogen Association, Frequently Asked Questions. http://www.hydrogenassociation.org/general/faqs.asp#howmuchproduced (accessed Dec 2007).
(11) Beurden, V. P. *On the Catalytic Aspects of Steam-Methane Reforming, A Literature Survey*; Report ECN-I--04-033; Energy Research Center of the Netherlands: Petten, The Netherlands, 2004.
(12) Hydrogen Production-Steam Methane Reforming. In *Hydrogen Fact Sheet*, New York State Energy Research and Development Authority: Albany NY, 2004.
(13) Arita, T.; Nakahara, K.; Nagami, K.; Kajimoto, O. Hydrogen generation from ethanol in supercritical water without catalyst. *Tetrahedron Letters* **2003**, *44*, 1083-1086.
(14) Lachance, R.P. A Fundamental Study of Model Fuel Conversion Reactions in Sub and Supercritical Water. Ph.D. Massachusetts Institute of Technology, Cambridge, MA, June 2005.
(15) Yang, H.H.; Eckert, C.A. Homogeneous Catalysis in the Oxidation of p-Chlorophenol in Supercritical Water. *Ind. Eng. Chem. Res.* **1988**, *27*, 2009-2014.
(16) Lee, S. *Alternative Fuels*, Taylor & Francis, Philadelphia, **1996**; pp.140-151.
(17) Lilac, W. D. Controlled Depolymerization of Polypropylene via Selective Partial Oxidation in a Supercritical Water Medium. Ph.D. Dissertation, University of Missouri-Columbia, Columbia, MO, Dec 1999.
(18) Hong, G. T.; Spritzer, M. H. Supercritical Water Partial Oxidation. Presentation to the Department of Energy Hydrogen Program Annual Review, Berkeley, CA, May 19-22, 2003.
(19) Picou, J.W.; Wenzel, J.; Lanterman, H.B.; Lee, S. Kinetics of Noncatalytic Water Gas Shift Reaction in a Supercritical water Medium. Presented at the AIChE Spring National Meeting, New Orleans, LA, April 2008.
(20) McIntosh, S.; Gorte, R.J. Direct Hydrocarbon Solid Oxide Fuel Cells. *Chem. Rev.* **2004**, *104*, 4845-4865.
(21) Murray, E. P.; Tsai, T.; Barnett, S. A. Adirect-methane fuel cell with a ceria-based anode *Nature* **1999**, *400*, 649.

(22) Garcia-Serna, J.; Garcia-Merino, E.; Cocero, M.J. Gasification of charcoal using supercritical CO_2 at high pressure. *J. of Supercritical Fluids* **2007**, *43*, 228-235.
(23) Mondal, K.; Lorethova, H.; Hippo, E.; Wiltowski, T.; Lalvani, S.B. Reduction of iron oxide in carbon monoxide atmosphere-reaction controlled kinetics. *Fuel Processing Technology* **2004**, *86*, 33-47.
(24) Picou, J. W. Autothermal Non-Catalytic Reformation of Jet Fuel in a Supercritical Water Medium. M.S. Thesis, Missouri University of Science and Technology, Rolla, MO, June 2008.

SYNTHESIS OF BIODEGRADABLE LUBRICANT FROM JATROPHA OIL WITH HIGH CONTENT OF FREE FATTY ACIDS

Gunam Resul, M. F. M.[1], Mohd. Ghazi, T. I.[1*], Muhammad Syam, A.[1], and Idris, A.[1]

[1]Department of Chemical and Environmental Engineering, Faculty of Engineering, Universiti Putra Malaysia 43400, Serdang, Selangor Darul Ehsan, Malaysia.
[*]Corresponding author: Phone: +603-8946 4427, Fax: +603-8656 7120
Email: tinia@eng.upm.edu.my

ABSTRACT

Plant oil extracted from a non edible plant has the potential application as biodegradable lubricant. Such plant as toxic contained *Jatropha curcas* inhibits human consumption and thus making it suitable for such application since it does not compete with the food production. However, high content of free fatty acids (FFA) in the jatropha oil (triglycerides) is found to have significant effect in the production of bio-lubricant. The formation of fatty acids occurs when the triglycerides are hydrolysed with moisture due to poor storage and handling. This paper discusses the techniques performed to produce high yield of bio-lubricant from jatropha oil containing high content of FFA up to 30%. It is reported that more than 80% yield of jatropha bio-lubricant was successfully achieved. Lubricity characteristics such as the pour point, viscosity and thermal oxidative stability for the produced jatropha bio-lubricant are found to be comparable with the other vegetable oil based lubricants.

Keywords: jatropha curcas, free fatty acids, trimethylolpropane, biodegradable lubricant.

Introduction

In recent years, increasing attention has again been paid to natural triglycerides, fatty acids derived from them, and their potential applications (Vizintin et al., 2000; Arnsek and Vizintin, 2000). This raw material is cheap and renewable. Moreover, natural fats and products derived from them are generally environmentally friendly. They have the attractions of being natural, renewable, non-toxic, non polluting and cheaper than synthetics. In general, synthetic lubricants cost four to eight times higher than the mineral oil-based lubricants

(Bartz 1993). Vegetable oil-based lubricants are 50% cheaper than synthetic oils but are easily degradable and have poor lubricity. Nevertheless, they are plentiful in supply and can be used as biodegradable lubricant base stock. Lubricants based on vegetable oils still comprise a narrow segment; however, they are finding their way into such applications as chainsaw bar lubricants, drilling muds and oils, straight metalworking fluids, food industry lubricants, open gear oils, biodegradable grease, hydraulic fluids, marine oils and outboard engine lubricants, oils for water and underground pumps, rail flange lubricants, shock absorber lubricants, tractor oils, agricultural equipment lubricants, elevator oils, mould release oils, two stroke engine lubricants and other. Hydraulic fluids, which are consumed at 5 million metric tons per year rate in the US (Padavich and Honary, 1995), have the highest need for biodegradable lubricants. Although vegetable oils have reputation of a low-cost candidate for the biodegradable replacement of mineral oils due to their high inherent biodegradability (Battersby et al., 1992; Randles and Wright, 1992), unsatisfactory performance curbs their more widespread utilization as lubricants.

However, glycerol, a component of the triglyceride molecule is readily destructible at high temperatures. This disadvantageous property stems from the presence of hydrogen atoms in position b relative to the hydroxyl group in the glycerol molecule. This structural feature is conducive to the partial defragmentation of the molecule and the formation of unsaturated compounds (Bu€unemann et al., 2000). The compounds formed undergo polymerisation, increasing the liquids viscosity and resulting in the formation of precipitate particles. The problem can be solved by replacing glycerol with another polyhydric alcohol which does not contain b-hydrogen atoms, i.e., neopentyl glycol (NPG), trimethylol propane (TMP) or pentaerythritol. Although such alcohols also decompose at high temperatures, their thermal decomposition has a radical character and proceeds slowly. Synthetic esters produced from natural acids cannot be used at extremely high temperatures, but they are very suitable in less extreme applications such as two-stroke engine oils, chain bar oils, cutting oils, concrete mould release agents and cosmetics ingredients.

The properties of esters also depend on the structure of the constituent fatty acids and alcohols, i.e., on the length of their aliphatic chain and the number and relative position of unsaturated bonds. Saturated acids are highly resistant to oxidation and high temperature, but their esters pour point is high due to the linear structure of the acids. The high pour point of saturated fatty acids makes it possible to separate the latter from unsaturated acids by physical methods, e.g., through freezing out. Polyunsaturated fatty acids, especially those which contain conjugated bonds, are the most susceptible to oxidation and thermal degradation. As a rule, these acids, particularly their cis-isomers, are characterized by very low pour points down to -30°C.

The objective of this research was to explore the possibilities of producing esters of TMP. Jatropha curcas were used as the sources of fatty acids. The esters evaluated in this study were synthesized through transesterification.

JME preparation

Four hundred and fifty grams (450g) of the Jatropha triglyceride and 0.5 g of NaOH dissolved in 35 g of methanol were placed in a 1000 ml glass bath reactor equipment with a magnetic stirrer and capped with a reflux condenser. The reactor temperature was maintained between 65 and 70 C, slightly below the boiling point of methanol. The reaction was carried out for 60 minute. The reaction mixture was cooled down to ambient temperature and the resulting two layers were separated. The upper layer contained a solution of methyl esters.

Chromatographic analysis

Analysis of JME was performed on a HP 5890II gas chromatograph fitted with a capillary inlet system and flame ionization detector. The capillary column, HP INNOWAX, was 30 m long, with 0.32 mm i.d. and 0.25 lm film thickness. A split injection system 10:1 was used with nitrogen carrier gas. The column temperature was held at 80 C throughout the entire 6 min run. Temperature programming was initiated at 80 C at the rate of 10 C min^{1} to reach a temperature of 250 C. The inlet temperature was 250 C and the detector temperature was 280 C. The oil reference standard, AOCS for Low Erucic Rapeseed Oil from Sigma—Aldrich, Steinheim, Germany was used as an external standard for quantitative analysis. For one sample, the GC analysis was performed in duplicate. Relative standard deviations of duplicate measurements of fatty acids composition was below 5%.

Physicochemical properties determination

The physical properties of the prepared esters were determined according to the following standard test methods, i.e., acid number to ASTM D 974, viscosity to ASTM D 445, viscosity index to ASTM D 2270 and pour point to ASTM D 97.

Kinematic viscosities of the fluids were measured according D 445 (ASTM, 1991b). Cannon-Fenske capillary viscometers and the constant temperature bath were from Cannon Instruments, State College, PA. Temperature of 40+-0.2°C was maintained throughout the experiments. The viscometers were calibrated for 10–50 cSt viscosity range using calibration fluids provided by the manufacturer.

Pour points were measured strictly according D 97 (ASTM, 1991a), which determines pour points by placing a test jar containing 50 ml of the sample into a metal cylinder submerged into cooling media. Sample temperature was measured in 3°C increments at the top of the sample until it stopped pouring, whereas the temperature of cooling media was kept constant below the sample temperature. When the sample temperature reached the specified range (e.g. three of the ranges are +9 to −6, −6 to −24 and −24 to −42°C), the temperature of cooling media was also reduced to the specified value (−18, −33 and −51°C, respectively). The pour point is defined as temperature where sample still pours. Statistically, the test has shown satisfactory accuracy; as described in a precision statement (ASTM, 1991a,b), the difference between two test results from independent laboratories exceeding 6°C in only one case of 20 with repeatability of 2.87°C at 95% confidence.

Results and discussion

TMP is a well-known neopentylpolyol-like compound used in lubrication fields [10]. The transesterification with TMP took place in accordance with the following reaction (Fig. 1). Sodium Methoxide at a concentration of 1.0%(wt) was used as catalyst in the reaction mixture. The conditions of the reaction are the following:

- branched methyl esters/TMP ratio: 3.9:1
- temperature: 110–120 °C
- reaction time: 3 h.

The reaction was performed under vacuum (10 mbar) up to 120 °C. The methanol was continuously recovered in a condenser. The catalysts were removed by filtration treatment. Conversion yields in triesters of TMP is higher than 80%(wt) and the remaining branched methyl esters are less than 15%. This new product was compared with some other TMP esters (Table 1). This lubricant base coming from TMP has very interesting physical and characteristics with a Viscosity Index higher than 180, and noteworthy properties of thermal stability and cold behaviours often better than the commercial products.

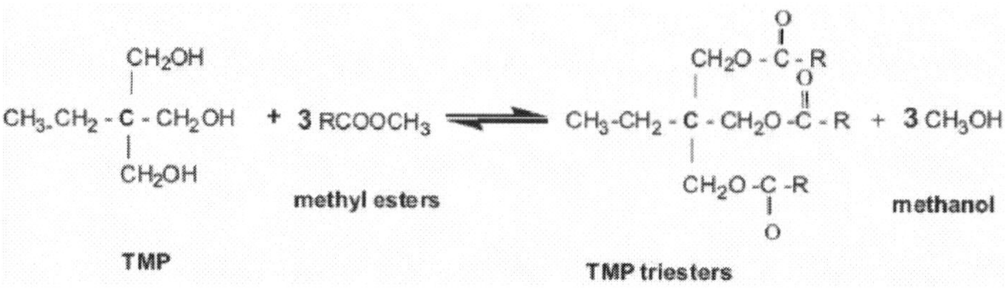

Fig. 1: Reaction of transesterification of TMP with methyl esters.

Table 1: Physical characteristics of the different lubricant bases

Fluid		Standard Requirement by ISO VG46	Mineral Oil	Synthetic*		Bio-based Synthetic	
				Olefin Oligomers	Propane Alkylated Aromatics	Oil Palm	Jatropha
Kinematic Viscosity (cSt)	40°C	>41.4	568	399	109	52.4	43.9
	100°C	>4.1	27.8	40	12.4	10.2	8.71
Viscosity Index (VI)		>90	62	147	105	186	180
Pour point (°C)		-10	-4	-36	-36	-5	-6

The GC analysis of the JME is given in Table 2. Jatropha oil, whose major constituent is oleic acid, contains a relatively small amount of linoleic acid (7.2%), and about 14% saturated fatty acids. In comparison to Jatropha oil, rapeseed oil has a lower saturated acid content (8.3%), which is advantageous, and a higher polyunsaturated fatty acids content (28%). The presence of large amounts of polyunsaturated acids results in lowering the resistance to oxidative-thermal treatment. Lard-based FAMEs have the lowest polyunsaturated acids content, but the saturate content is as high as 50%. The transesterification of JME with TMP, being actually an opposite reaction, proceeds at a much lower rate. Sodium methoxide, which is sparingly soluble and can be easily separated from the reaction products by filtration, was used as the catalyst. Generally, TMP esters of fatty acids have higher viscosity probably due to the fact that there are three acid groups in the structure of TMP esters and only two such groups in the molecules of NPG esters. Both the synthesized esters and native triglycerides studied have a high viscosity index, around 150 and higher. Moreover, the viscosity index is slightly higher for the synthesized esters than that of native triglycerides. This is a highly desirable property in view of their use as lubricants. The viscosity of liquid with high viscosity indices changes very little with temperature. High viscosity index is characteristic of compounds which have long, unbranched hydrocarbon chains, e.g. fatty acids of most natural triglycerides.

Generally, the synthesized esters show moderate thermal-oxidative stability. This is a characteristic feature of most synthetic oils produced from natural triglycerides. In similar thermal-oxidative conditions, the viscosity and acid number of hydrocarbon oils increased only slightly. However, in many technological applications high stability is not required. Oils synthesized from natural products are generally approved of as environmentally friendly functional fluids and therefore they can be used widely.

Initially, kinematic viscosities at 40°C and pour points of the fluids were determined, results shown in Table 1. It appears that low temperature properties of vegetable oils are much more inferior to those of synthetic basestocks or even mineral oil.

Chemical modification is necessary to improve these performance limitations with the focus on optimal extent of structural alteration for improved low temperature performance. Sufficient experimental evidence is not yet available, however, it can be suggested that such modification pathways as alkylation, alkoxylation, interesterification and oligomerization might appreciably improve overall lubricant properties of vegetable oils.

Conclusion

This new synthesis of branched fatty acids methyl esters from Jatropha oils has been largely optimised and good conversion yields have been achieved. After transesterification with TMP this product behaves as an excellent lubricant. This could be a new way of development of Jatropha oil, which are renewable feedstock. Other fields of applications than lubrication can be considered, in particular detergency where the soaps of branched fatty acids have noteworthy foaming, antistatic and emulsifying properties. The cosmetic industry can be also considered. The oxidative stability and low temperature performance of raw Jatropha oils is poor. Thus,

References

[1] K.D. Haase, G. Taylor, P.A. Smith, Seiffen-öle-Fette-Wachse 7 (1988) 231

[2] F.J.B. Weghorst, Fette-Seifen, Anstrichmittel 67 (1965) 447.

[3] G.D.A. Hata, Bull. Chem. Soc. Jpn 40 (1967) 3754.

[4] A.C.L. Su, J. Organomet. Chem. 36 (1972) 177.

[5] S. Muthukumaru, G.L. Pillai, M. Ravindranathan, J. Mol. Catal. 84 (1993) 77

[6] A. Laufenberg, A. Behr, Patents WO 91/11427 (24/01/1990), WO 91/11428 (24/01/1990).

[7] R. Stern, A. Kotoko, G. Hillion, Y. Chauvin, US Patents Nos. 5,434,282 (18/07/1995) & 5,476,956 (19/12/1995).

[8] G. Hillion, H. Olivier, K. Siepen, R. Stern, US Patent No. 6,153,772 (28/112000).

[9] I. Bach, G. Hillion, H. Olivier, French Patent 99/16 508 (24/12/1999).

753e Wastewater Treatment Systems: The Next Biorefinery

Jaclyn Hall[1], W. Todd French[1], Rafael Hernandez[1], Andro H. Mondala[1], Earl Alley[1], William E. Holmes[2], Ashli Brown[1], and Mark G. White[1]. (1) Chemical Engineering, Mississippi State University, Box 9595, Mississippi State, MS 39762, (2) Mississippi State Chemical Laboratory, Mississippi State University, Box 9595, Mississippi State University, MS 39762

Oleaginous yeasts are a type of microorganism that can produce up to 70 percent of their body weight in oil. The use of this type of microorganism as a source of oils for biofuel production is limited by the availability of cost effective carbon, water and other nutrient sources. A growth medium that has not been previously considered for cultivating oleaginous microorganisms is municipal and industrial wastewaters. This study is designed to determine the concentration of indigenous microorganisms that effect the growth and oil accumulation of oleaginous microorganisms when cultivated on wastewater. By using wastewater, the oleaginous microorganisms could produce an abundant amount of oil while treating this water to Environmental Protection Agency (EPA) required treatment levels. The cell mass yield, oil percentage, and oil yield are measured using well-established techniques such as plate counting, Bligh& Dyer, and transesterification. In addition, water quality is measured by determining the biochemical oxygen demand (BOD), the chemical oxygen demand (COD), and the amount of nitrates in the wastewater after the yeast growth stabilizes. These quantities will determine the efficacy of the oleaginous microorganisms assist in removing wastewater contaminants. Preliminary results show that the microorganisms can be cultivated on wastewater, accumulating approximately 10% of their weight in oil. Wastewater treatment facilities may be operated with oleaginous yeasts as the main consumers of BOD. This modification could potentially generate billions of gallons of oil for producing biofuels.

AUTHOR INDEX

AUTHOR INDEX

AUTHOR INDEX

AUTHOR INDEX

AUTHOR INDEX

AUTHOR INDEX

AUTHOR INDEX